Springer Series in Optical Sciences   Volume 64

Editor: A.E. Siegman

Springer-Verlag Berlin Heidelberg GmbH

Springer Series in Optical Sciences

Editorial Board: A. L. Schawlow A. E. Siegman T. Tamir

Managing Editor: H. K. V. Lotsch

V.G. Dmitriev   G.G. Gurzadyan
D.N. Nikogosyan

Handbook of Nonlinear Optical Crystals

Second, Revised and Updated Edition

With 39 Figures

 Springer

Professor VALENTIN G. DMITRIEV, Ph.D.
R&D Institute "Polyus", Vvedenskogo St. 3
Moscow, 117342, Russia

GAGIK G. GURZADYAN, Ph.D.
Yerevan State University
Yerevan, Armenia

Professor DAVID N. NIKOGOSYAN, Ph.D.
Institute of Spectroscopy
Russian Academy of Sciences
Troitzk, Moscow Region, 142092, Russia

ISSN 0342-4111

Library of Congress Cataloging-in-Publication Data. Gurzadian, G.G. (Gagik Grigor 'evich), 1957- [Nelineino-opticheskie kristally. English] Handbook of nonlinear optical crystals/V.G. Dmitriev, G.G. Gurzadyan, D.N. Nikogosyan. p. cm. – (Springer series in optical sciences; v. 64) Gurzadian's name appears first on earlier edition. Includes bibliographical references and index.

DOI 10.1007/978-3-540-68392-6
1. Laser
materials – Handbooks, manuals, etc. 2. Optical materials – Handbooks, manuals, etc. 3. Crystals – Handbooks, manuals, etc. 4. Nonlinear optics – Handbooks, manuals, etc. I. Dmitriev, V.G. (Valentin Georgievich) II. Nikogosyan, D.N., 1946- . III. Title. IV. Series. QC374.G8713 1997 621.36'6–dc20 97-23159

Cover design: Design & Production GmbH, Heidelberg
Typesetting: Scientific Publishing Services (P) Ltd, Madras
SPIN: 10480587   54/3144/SPS – 5 4 3 2 1 0 – Printed on acid-free paper
www.springer.com/mycopy

To our Parents

Preface

Four years ago when we had finished our work on the first edition of our Handbook we didn't even suppose that three years later it would become necessary to greatly revise and update the material. It happened because of the following developments.

1. The invention and tremendous development of modern nonlinear optical crystals such as BBO, LBO, KTP, $ZnGeP_2$, etc.
2. Rapid progress in laser techniques (femtosecond CPM laser, Ti: sapphire laser, diode-pumped solid-state lasers, etc.).
3. The appearence of numerous organic crystals which can be synthesized with predictable properties.
4. Progress in the theory of nonlinear frequency conversion utilizing biaxial crystals, femtosecond pulses, etc.
5. Accumulation of new data on the properties of nonlinear optical crystals.

In accordance with the above, we have made many changes in the text. The first chapter was revised by D.N. Nikogosyan, the second one by V.G. Dmitriev and D.N. Nikogosyan, and the fourth one by G.G. Gurzadyan. The third chapter, containing the main reference material on 77 nonlinear optical crystals was completely rewritten and updated by D.N. Nikogosyan. The Appendix containing the list of most commonly used laser wavelengths was compiled by D.N. Nikogosyan. We would appreciate any valuable comments and recommendations that will allow us to further improve the Handbook.

We would like to thank H.K.V. Lotsch for fruitful and long-lasting cooperation.

<table>
<tr><td>Moscow, Yerevan, Troitzk</td><td>V. G. Dmitriev</td></tr>
<tr><td>Russia, Armenia</td><td>G. G. Gurzadyan</td></tr>
<tr><td>December 1995</td><td>D. N. Nikogosyan</td></tr>
</table>

Preface to the First Edition

Since the invention of the first laser 30 years ago, the frequency conversion of laser radiation in nonlinear optical crystals has become an important technique widely used in quantum electronics and laser physics for solving various scientific and engineering problems. The fundamental physics of three-wave light interactions in nonlinear optical crystals is now well understood. This has enabled the production of various harmonic generators, sum- and difference-frequency generators, and optical parametric oscillators based on nonlinear optical crystals that are now commercially available. At the same time, scientists continue an active search for novel, highly efficient nonlinear optical materials.

Therefore, in our opinion, there is a great need for a handbook of nonlinear optical crystals, intended for specialists and practitioners with an engineering background. This book contains a complete description of the properties and applications of all nonliner optical crystals of practical importance reported in the literature up to the beginning of 1990. In addition, it contains the most important equations for calculating the main parameters (such as phase-matching direction, effective nonlinearity, and conversion efficiency) of nonlinear frequency converters.

Dolgoprudnyi, Yerevan, Troitzk
USSR
October 1990

V. G. Dmitriev
G. G. Gurzadyan
D. N. Nikogosyan

Contents

List of Abbreviations

a	Aperture
c	Cut
cont	Continuum
cr	Critical
cw	Continuous wave
DF	Difference frequency
DFG	Difference-frequency generation
dif	Diffraction
DROPO	Doubly-resonant optical parametric oscillation
ds, dis	Dispersive spreading
e	Extraordinary
eff	Effective
exp	Experimental
f	Fast
fcg	Free-carrier generation
FIHG	Fifth-harmonic generation
FOHG	Fourth-harmonic generation
ICSFG	Intracavity sum-frequency generation
ICSHG	Intracavity second-harmonic generation
int	Internal
IR	Infrared
L	Linear
NCPM	Non-critical phase matching
NL	Nonlinear
no pm	No phase matching
o	Ordinary
OPO	Optical parametric oscillation
OR	Optical rectification
p	Pulse
PM, pm	Phase matching
PL	Parametric luminescence
pr	Photorefraction
qs	Quasistatic
s	Slow
SF	Sum frequency

SFG	Sum-frequency generation
SFM	Sum-frequency mixing
SH	Second harmonic
SHG	Second-harmonic generation
SIHG	Sixth-harmonic generation
SROPO	Singly-resonant optical parametric oscillation
SRS	Stimulated Raman scattering
theor	Theoretical
THG	Third-harmonic generation
thr	Threshold
tsa	Thermal self-action
TWOPO	Traveling-wave optical parametric oscillation
unc	Unconverted
UV	Ultraviolet

1 Introduction

In 1960, *Maiman* (USA) created the first source of coherent optical radiation, namely, a ruby laser emitting in the red spectral region ($\lambda = 0.6943\,\mu$m) [1.1]. Several years later a great family of lasers was already in existence. The following types were known:

1) solid-state lasers, e.g., $Nd:CaWO_4$ laser emitting at 1.065 μm [1.2], neodymium glass laser ($\lambda = 1.06\,\mu$m) [1.3], Nd:YAG laser ($\lambda = 1.064\,\mu$m) [1.4]
2) gas lasers, e.g., He-Ne laser ($\lambda = 0.6328,\ 1.1523,\ 3.3913\,\mu$m) [1.5], argon ion laser ($\lambda = 0.4880, 0.5145\,\mu$m) [1.6], CO_2 laser ($\lambda = 9.6,\ 10.6\,\mu$m) [1.7];
3) dye lasers [1.8,9]
4) semiconductor lasers [1.10–12];

and so on. The wavelengths of the above mentioned lasers were either fixed or tunable over a small range. It was a matter of practical importance to widen the range of wavelengths generated by laser sources.

The propagation of electromagnetic waves through nonlinear media gives rise to vibrations at harmonics of the fundamental frequency, at sum and difference frequencies, and so on. In the optical frequency range, the same effect is observed when light waves propagate through weakly nonlinear optical dielectrics. When one or two sufficiently powerful beams of laser radiation pass through these dielectrics, the radiation frequency may be transformed to the second, third, and higher harmonics and to combination (sum and difference) frequencies. In this way, the range of wavelengths generated by a certain laser source can be considerably increased. For instance, the second harmonic of the ruby laser radiation lies in the UV region ($\lambda = 0.34715\,\mu$m), whereas the second harmonic of the neodymium glass laser radiation lies in the green spectral range ($\lambda = 0.53\,\mu$m).

As early as in 1961, *Franken* et al. [1.13] observed a radiation at the doubled frequency when a ruby laser light was directed into a quartz crystal. However, because of phase mismatch of the waves at the fundamental and doubled frequencies upon propagation in a quartz crystal, the efficiency of conversion to the second harmonic proved to be very low, less than 10^{-12}.

In 1962, *Giordmaine* [1.14] and *Maker* et al. [1.15] simultaneously proposed an ingenious method of matching the phase velocities of the waves at the fundamental and doubled frequencies. Their technique used the difference between the refractive indices of the waves with different polarizations in an optically anisotropic (uniaxial or biaxial) nonlinear crystal (phase-matching method), and with it the efficiency of conversion of laser radiation to the second harmonic was enhanced to several ten percent.

At the beginning of the 1960s, parallel to the research on second-harmonic generation, first experiments were carried out on the generation of optical radiation at combination frequencies, namely: sum-frequency generation of radiation from two lasers [1.16], sum-frequency generation of radiation from a laser and a noncoherent source [1.17], and difference-frequency generation [1.18,19]. We should specially mention optical parametric oscillation, which is a nonlinear effect that allows one to obtain continuously tunable coherent optical radiation [1.20].

The ferroelectrics ADP and KDP used in electro-optic and elasto-optic devices were the first crystals applied for nonlinear frequency conversion (nonlinear optical crystals) [1.21]. They were grown by conventional techniques. However, some special nonlinear optical problems called for crystals with improved properties (better transparency, higher nonlinearity, lower hygroscopicity, etc.). The resulting intensive scientific search for new materials has led to the synthesis of a number of nonlinear crystals of high optical quality: $LiNbO_3$ in 1964 [1.22], $BaNaNb_5O_{15}$ in 1967 [1.23], proustite in 1967 [1.24], $LiIO_3$ in 1969 [1.25], KTP in 1976 [1.26], and others. The first reviews comparing the properties of various nonlinear optical crystals have been published [1.27,28].

Very recently two new nonlinear crystals from the borate family, of excellent quality, were invented by *Chen* et al.: BaB_2O_4 (BBO) in 1985 [1.29] and LiB_3O_5 (LBO) in 1989 [1.30].

2 Optics of Nonlinear Crystals

This chapter introduces the main concepts of the physics of nonlinear optical processes: three-wave interactions, phase matching and phase-matching angle, role of phase mismatch for the interaction of quasi-plane waves, group-velocity mismatch and interaction of ultrashort light pulses, optics of uniaxial and biaxial crystals, crystal symmetry and effective nonlinearity, "walk-off" angle, phase-matching bandwidths (angular, temperature, spectral), thermal effects, and so on. It presents the main material required for calculating of phase-matching angles and for an assessment (as a rule, in approximation of quasi-plane light waves) of frequency conversion efficiency in the case of generation of optical harmonics and combination (sum and difference) frequencies, and optical parametric oscillation in nonlinear optical crystals. For convenience, the so-called "effective lengths" are introduced for the corresponding processes: by comparing the nonlinear crystal's length with the effective length of the corresponding process, we may conclude whether this process must be taken into account for the calculation of the conversion efficiency or not. The chapter contains many tables with the equations for calculating phase-matching and "walk-off" angles, bandwidths, effective nonlinearity and conversion efficiency.

2.1 Three- and Four-Wave (Three- and Four-Frequency) Interactions in Nonlinear Media

Conversion of a light-wave frequency (multiplication, division, mixing) is possible in *nonlinear optical crystals* for which the refraction index n is a function of the electric field strength vector E of the light wave

$$n(E) = n_0 + n_1 E + n_2 E^2 + \dots , \tag{2.1}$$

where n_0 is the refractive index in the absence of the electric field (this quantity is used in conventional "linear" optics), and n_1, n_2, and so on are the coefficients of the series expansion of $n(E)$.

In nonlinear optics a vector of *dielectric polarization* P (dipole moment of unit volume of the matter) is introduced. It is related to the field E by the *matter equation* [2.1–4]

$$P(E) = \kappa(E)E = \kappa_0 E + \chi^{(2)} E^2 + \chi^{(3)} E^3 + \ldots, \tag{2.2}$$

where κ is the *linear dielectric susceptibility* (denoted as κ_0 in the absence of the electric field), and $\chi^{(2)}$, $\chi^{(3)}$, and so on are the *nonlinear dielectric susceptibility coefficients* (square, cubic, and so on, respectively). The following equations hold true:

$$\kappa_0 = \frac{1}{4\pi}(\varepsilon_0 - 1) = \frac{1}{4\pi}(n_0^2 - 1) \; ;$$

$$\chi^{(2)} \cong \frac{1}{2\pi} n_0 n_1 \; ; \quad \chi^{(3)} \cong \frac{1}{2\pi} n_0 n_2 \; ; \tag{2.3}$$

where ε_0 is the dielectric constant in the absence of the electric field. In the general case of anisotropic crystals, the quantities ε_0, n, κ, and χ are the tensors of the corresponding ranks [2.4].

The square nonlinearity takes place ($\chi^{(2)} \neq 0$) only in acentric crystals, i.e., in crystals without symmetry center; in crystals with symmetry center and as well as in isotropic matter $\chi^{(2)} \equiv 0$. On the contrary the cubic nonlinearity exists in all crystalline and isotropic materials.

Propagation of two monochromatic waves with frequencies ω_1 and ω_2 in crystals with square nonlinearity gives rise to new light waves with combination frequencies $\omega_{3,4} = \omega_2 \pm \omega_1$; the sign *plus* corresponds to *sum frequency*, the sign *minus* - to *difference frequency* (three-wave or three-frequency interaction). Sum-frequency generation (SFG) is frequently used for conversion of long-wave radiation, for instance, infrared (IR) radiation, to short-wave radiation, namely, ultraviolet (UV) or visible light. Difference-frequency generation (DFG) is used for conversion of short-wave radiation to long-wave radiation.

At $\omega_1 = \omega_2$ we obtain two special cases of conversion, namely, *second-harmonic generation* (SHG) as a special case of SFG, $\omega_3 = 2\omega_1$, and *optical rectification* (OR) as a special case of (DFG), $\omega_4 = 0$.

The effect of *parametric luminescence* (PL), or *optical parametric oscillation* (OPO), is the opposite process to SFG and involves the appearance of two light waves with the frequencies $\omega_{1,2}$ in the field of the intense light wave with frequency $\omega_3 = \omega_1 + \omega_2$.

Generation of more complex combination frequencies is possible with successive SFG and/or SHG processes. For example, the *third-harmonic generation* (THG) can be realized by using the following SFG process:

$$\omega_3 = 3\omega_1 = \omega_1 + 2\omega_1 \; ; \tag{2.4}$$

the *fourth-harmonic generation* (FOHG, $\omega_4 = 4\omega_1$) can be realized as SHG process of frequency $2\omega_1$. In a similar manner, the *fifth-* and *sixth-harmonic generations* (FIHG and SIHG) can be realized:

$$\omega_5 = 5\omega_1 = \omega_1 + 4\omega_1 \tag{2.5}$$

or

$$\omega_5 = 5\omega_1 = 2\omega_1 + 3\omega_1 \; ; \tag{2.6}$$

$$\omega_6 = 6\omega_1 = \omega_1 + 5\omega_1 \tag{2.7}$$

or

$$\omega_6 = 6\omega_1 = 2\omega_1 + 4\omega_1 \ . \tag{2.8}$$

Propagation of two light waves with frequencies $\omega_{1,2}$ in substance with cubic nonlinearity gives rise to new light waves with combination frequencies $2\omega_1 \pm \omega_2$ and $\omega_1 \pm 2\omega_2$ (four-wave or four-frequency interaction). The special cases with $\omega_1 = \omega_2$ are the *direct* THG process, $\omega_3 = 3\omega_1$, and the process of *self-action*, $\omega_4 = 2\omega_1 - \omega_1 = \omega_1$, or the generation of the same frequency ω_1.

Because of the relatively seldom usage of frequency conversion in cubic substances (as a rule, $\chi^{(3)} E \ll \chi^{(2)}$), in this chapter we shall consider only the three-wave interactions occuring in the crystals with square nonlinearity ($\chi^{(2)} \not\equiv 0$).

2.2 Phase-Matching Conditions

Under usual conditions all optical media are weakly nonlinear, i.e., the inequalities $\chi^{(3)} E^2 \ll \chi^{(2)} E \ll \kappa_0$ are valid. Noticeable nonlinear effects can be observed only when light propagates through fairly long crystals and the so-called *phase-matching conditions* are fulfilled:

$$k_3 = k_2 + k_1 \tag{2.9}$$

or

$$k_4 = k_2 - k_1 \tag{2.10}$$

where k_i are the wave vectors corresponding to the waves with frequencies ω_i $(i = 1, 2, 3, 4)$:

$$\mid k_i \mid = k_i = \frac{\omega_i n(\omega_i)}{c} = \frac{\omega_i}{v(\omega_i)} = \frac{2\pi n_i}{\lambda_i} = 2\pi n_i \nu_i \ , \tag{2.11}$$

where the quantities v_i, $n_i = n(\omega_i)$, λ_i and ν_i are the phase velocity, refractive index, wavelength, and wave number at the frequency ω_i, respectively.

The relative location of the wave vectors under phase matching can be either *collinear* (scalar phase matching) or *noncollinear* (vector phase matching) (Fig. 2.1)

Under scalar (collinear) phase matching we have for SFG

$$k_3 = k_2 + k_1, \ \text{or} \ \omega_3 n_3 = \omega_2 n_2 + \omega_1 n_1 \tag{2.12}$$

and for SHG ($\omega_1 = \omega_2$; $\omega_3 = 2\omega_1$) :

$$k_3 = 2k_1 \ \text{or} \ n_3 = n_1 \ . \tag{2.13}$$

The physical sense of phase-matching conditions (2.9,10) is the space resonance of the propagating waves, namely, between the wave of nonlinear *dielectric*

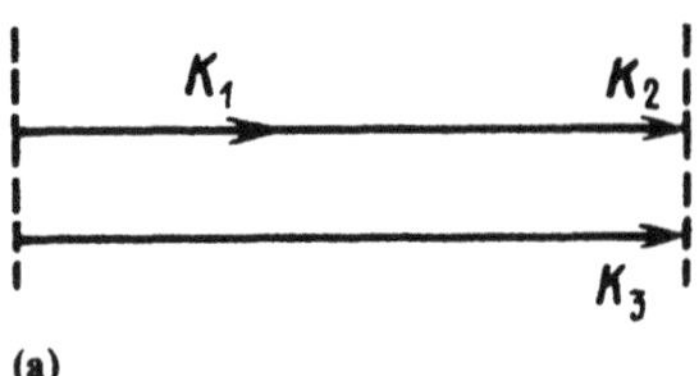

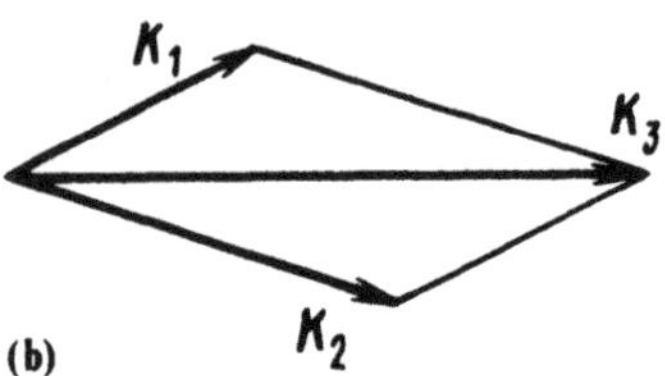

Fig. 2.1. Collinear or scalar (**a**) and noncollinear or vector (**b**) phase matching for three-wave interactions

polarization at the frequency ω_3 for SFG (or ω_4 for DFG) and produced by her *light wave* at the same frequency ω_3 (or ω_4, respectively). Note that in the optical transparency region in isotropic crystals (and also in anisotropic crystals for identically polarized waves), the equality (2.13) for SHG is never fulfilled because of normal dispersion ($n_1 < n_3$). The use of anomalous dispersion is almost impossible since energy absorption is very high. The phase-matching conditions are fulfilled only in anisotropic crystals under interaction of differently polarized waves.

Combination of nonzero square nonlinearity of an optically transparent crystal with phase matching is the necessary and sufficient condition for an effective three-wave interaction.

2.3 Optics of Uniaxial Crystals

In *uniaxial crystals* a special direction exists called the *optic axis* (Z axis). The plane containing the Z axis and the wave vector $\mathbf{k}$ of the light wave is termed the *principal plane*. The light beam whose polarization (i.e., the direction of the vector $\mathbf{E}$ oscillations) is normal to the principal plane is called an *ordinary beam* or an o-*beam* (Fig. 2.2). The beam polarized in the principal plane is known as an *extraordinary beam* or e-*beam* (Fig. 2.3). The refractive index of the o-beam does not depend on the propagation direction, whereas for the e-beam it does. Thus, the refractive index in anisotropic crystals generally depends both on light polarization and propagating direction.

The difference between the refractive indices of the ordinary and extraordinary beams is known as *birefringence* Δn. The value of Δn is equal to zero along the optic axis Z and reaches a maximum in the direction normal to this axis. The refractive indices of the ordinary and extraordinary beams in the plane normal to the Z axis are termed the *principal values* of the *refractive*

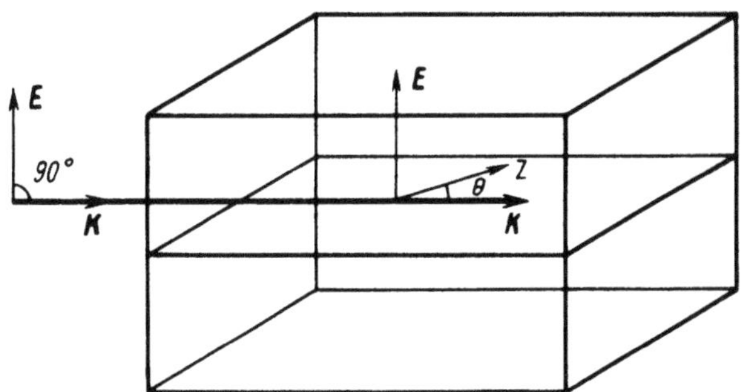

Fig. 2.2. Principal plane of the crystal (kZ) and ordinary beam

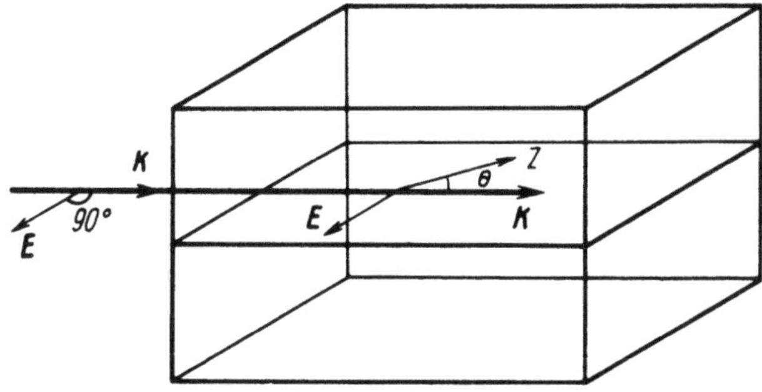

Fig. 2.3. Principal plane of the crystal (kZ) and extraordinary beam

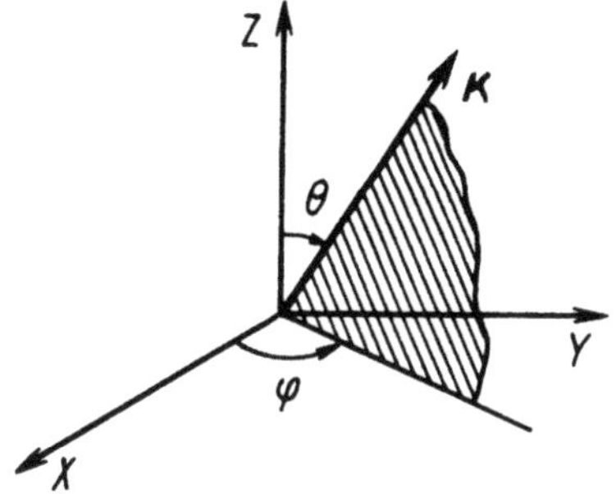

Fig. 2.4. Polar coordinate system for description of refraction properties of uniaxial crystal (k is the light propagation direction, Z is the optic axis, θ and ϕ are the coordinate angles)

index and are denoted by n_o and n_e, respectively; the value n_o should not be confused with the refractive index value n_0 in the absence of electric field in (2.1). The refractive index of the extraordinary wave is, in general, a function of the polar angle θ between the Z axis and the vector k (Fig. 2.4). It is determined by the equation (index e in this case is written as a superscript):

$$n^e(\theta) = n_o \sqrt{\frac{1 + \tan^2 \theta}{1 + (n_o/n_e)^2 \tan^2 \theta}} \tag{2.14}$$

The following equations are evident:

$$n^{\mathrm{o}}(\theta) \equiv n_{\mathrm{o}} \; , \tag{2.15}$$

$$n^{\mathrm{e}}(\theta = 0^\circ) = n_{\mathrm{o}} \; , \tag{2.16}$$

$$n^{\mathrm{e}}(\theta = 90^\circ) = n_{\mathrm{e}} \; , \tag{2.17}$$

$$\Delta n(\theta = 0^\circ) = 0 \; , \tag{2.18}$$

$$\Delta n(\theta = 90^\circ) = n_{\mathrm{e}} - n_{\mathrm{o}} \; , \tag{2.19}$$

$$\Delta n(\theta) = n^{\mathrm{e}}(\theta) - n_{\mathrm{o}} \; . \tag{2.20}$$

If $n_{\mathrm{o}} > n_{\mathrm{e}}$, the crystal is *negative*; if $n_{\mathrm{o}} < n_{\mathrm{e}}$, it is *positive*. The quantity n^{e} does not depend on the azimuthal angle ϕ (the angle between the projection of k onto the XY plane and the X axis – see Fig. 2.4). The dependence of the refractive index on light propagation direction inside the uniaxial crystal (index surface) is a combination of a sphere with radius n_{o} (for an ordinary beam) and

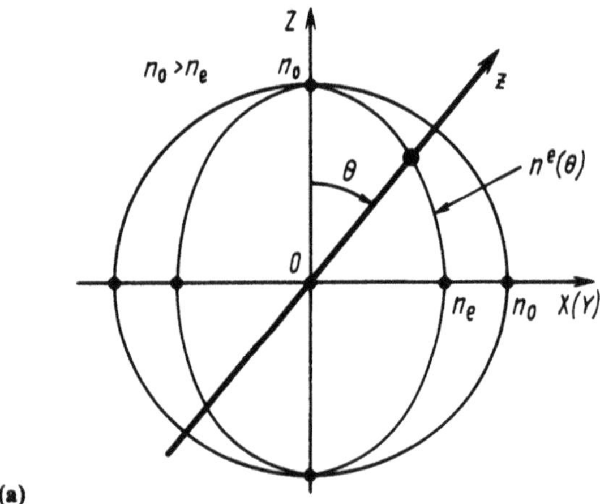

(a)

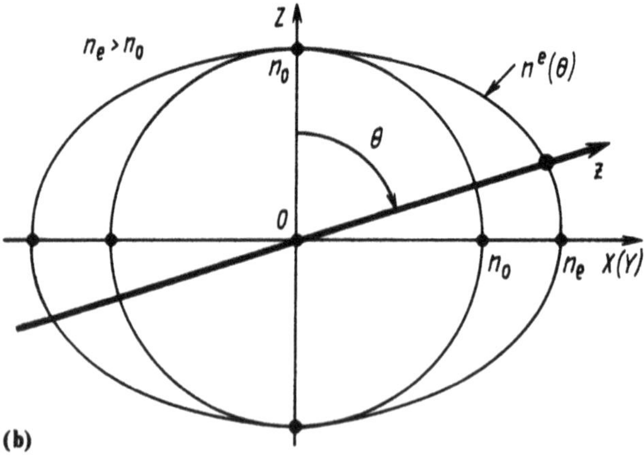

(b)

Fig. 2.5. Dependence of refractive index on light propagation direction and polarization (index surface) in negative (a) and positive (b) uniaxial crystals

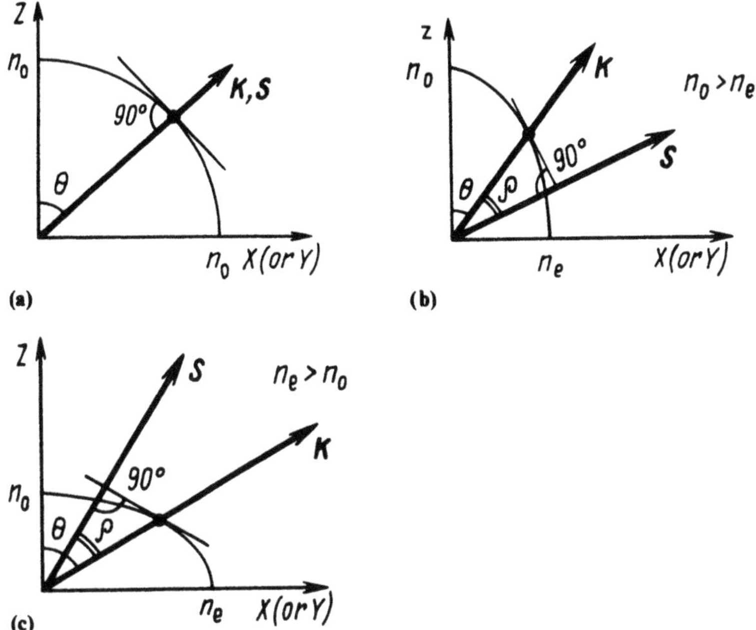

Fig. 2.6. Disposition of the wave (k) and beam (s) vectors in an isotropic medium (**a**) and anisotropic negative (**b**) and positive (**c**) uniaxial crystals (ρ is the birefringence angle)

an ellipsoid of rotation with semiaxes n_o and n_e (for an extraordinary beam, the axis of the ellipsoid of rotation is the Z axis). In the Z axis direction the sphere and ellipsoid are in contact with each other. In a negative crystal the ellipsoid is inscribed in the sphere (Fig. 2.5a), whereas in a positive crystal the sphere is inscribed in the ellipsoid (Fig. 2.5b).

When a plane light wave propagates in a uniaxial crystal, the direction of propagation of the wave phase (vector k) generally does not coincide with that of the wave energy (vector s). The direction of s can be defined as the normal to the tangent drawn at the point of intersection of vector k with the $n(\theta)$ curve. For an ordinary wave the $n(\theta)$ dependence is a sphere with radius n_o. Therefore, the normal to the tangent coincides with the wave vector k. For an extraordinary wave the normal to the tangent (with the exception of the cases $\theta = 0$ and $\theta = 90°$) does not coincide with the wave vector k but is rotated from it by the *birefringence or "walk-off" angle* (Fig. 2.6):

$$\rho(\theta) = \pm \arctan[(n_o/n_e)^2 \tan \theta] \mp \theta \;, \tag{2.21}$$

where the upper signs refer to a negative crystal and the lower signs to a positive one.

The correlation between ρ and θ may serve as the basis of a simple way to orient uniaxial single crystals [2.5]. Let a laser beam with an arbitrary linear polarization fall normal to the input face of a crystal of thickness L. After

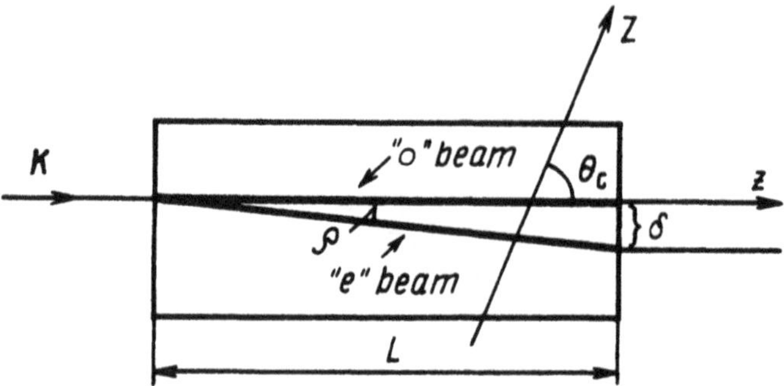

Fig. 2.7. Determination of the cut angle θ_c for the uniaxial crystal

passing through the crystal, the beam is divided onto two orthogonally polarized beams that, at the output face of the crystal, are separated by (Fig. 2.7)

$$\delta = L \tan \rho \qquad (2.22)$$

The crystal *cut angle* θ_c, which is the angle between the optic axis Z and the normal to the crystal surface, corresponds to one of two values

$$\theta_c = \text{arc } \tan\left(\frac{\mid n_o^2 - n_e^2 \mid L}{2\delta n_o^2} \pm \left|\frac{(n_o^2 - n_e^2)^2 L^2}{4\delta^2 n_o^4} - \frac{n_o^2}{n_e^2}\right|^{1/2}\right) . \qquad (2.23)$$

A more rigorous consideration of crystal optics of anisotropic media is given in [2.6].

2.4 Types of Phase Matching in Uniaxial Crystals

To fulfill the phase-matching condition in three-frequency interaction, differently polarized waves should be used. Let us consider the case of SFG. If the mixing waves have the same polarization, the radiation at sum frequency (SF) will be polarized in the perpendicular direction; in this case type I phase matching is realized. In negative crystals,

$$k_{o1} + k_{o2} = k_3^e \qquad (2.24)$$

(this is called "ooe" *phase matching* or "ooe" *interaction* or type $I^{(-)}$ *phase matching*). In positive crystals,

$$k_1^e(\theta) + k_2^e(\theta) = k_{o3} \qquad (2.25)$$

("eeo" *phase matching* or "eeo" *interaction* or type- $I^{(+)}$ *phase matching*). Here and below for SFG the first symbol in the expressions ooe, eeo, eoe, and so on, refers to the wave with the lower frequency, the third symbol to the wave with the higher frequency. Note that the *wave number* of the ordinary wave k_o should not be confused with linear dielectric susceptibility coefficient κ_0; see above in (2.2–2.3).

If the mixing waves are of orthogonal polarizations, type II phase matching takes place and the SF wave corresponds to an extraordinary wave in negative crystals:

$$k_{o1} + k_2^e(\theta) = k_3^e(\theta) \tag{2.26}$$

("oee" *phase matching* or "oee" *interaction* or type II$^{(-)}$ *phase-matching*) or

$$k_1^e(\theta) + k_{o2} = k_3^e(\theta) \tag{2.27}$$

("eoe" *phase matching* or "eoe" *interaction* or type II$^{(-)}$ *phase-matching*); and to an ordinary wave in positive crystals:

$$k_{o1} + k_2^e(\theta) = k_{o3} \tag{2.28}$$

("oeo" *phase matching* or "oeo" *interaction* or type II$^{(+)}$ *phase-matching*), or

$$k_1^e(\theta) + k_{o2} = k_{o3} \tag{2.29}$$

("eoo" *phase matching* or "eoo" *interaction* or type II$^{(+)}$ *phase-matching*).

All the above refers also to parametric luminescence (optical parametric oscillation). Here the wave with the higher frequency ω_3 is the *pump wave*; the other two waves - namely, *idler* ω_1 and *signal* ω_2 - are the waves of parametric luminescence (oscillation).

To use the equations of this section for DFG, the indices of n in the equations should be interchanged: $2 \rightarrow 4$, $3 \rightarrow 2$ (or $1 \rightarrow 4$, $3 \rightarrow 2$, $2 \rightarrow 1$).

Note that in the general case the *noncollinear* or *vector* phase matching takes place (Fig. 2.1b). In practice, however, *collinear* or *scalar* phase matching, which is the special case, is widely used (Fig. 2.1a).

Figure 2.8 illustrates how we can find the direction of collinear phase matching for the type I$^{(-)}$ of SHG ($\omega_3 = 2\omega_1$) in uniaxial negative crystals. For the ooe interaction,

$$n_{o1}(\omega_1) = n_3^e(2\omega_1, \theta_{pm}^{(1)}) \tag{2.30}$$

or

$$2k_{o1}(\omega_1) = k_3^e(2\omega_1, \theta_{pm}^{(1)}) \ . \tag{2.31}$$

Therefore, the *phase-matching direction* Oz (z is the propagation direction, it should not be confused with the optical axis Z!) for this case is formed when the circle of the ordinary refractive index at frequency ω_1 crosses the ellipse of the extraordinary refractive index at frequency $2\omega_1$ (Fig. 2.8a), or when the circle $2k_{o1}$ intersects the ellipse $k_3^e(\theta)$ (Fig. 2.8b).

Type I$^{(-)}$ vector phase matching with phase-matching angle $\theta_{pm}^{(2)}$ (Fig. 2.9) can be realized only within the region of angles $\theta_{pm}^{(1)} \le \theta_{pm}^{(2)} \le \pi - \theta_{pm}^{(1)}$, i.e., in the region of specific "anomalous" dispersion, because the inequality $n_3^e(2\omega_1) \le n_{o1}(\omega_1)$ is valid for these angles.

Figure 2.10 demonstrates the positions of scalar (angle $\theta_{pm}^{(3)}$) and vector (angle $\theta_{pm}^{(4)}$) phase matching of type II for SHG in a negative uniaxial crystal. The phase-matching direction in the former case is determined by intersection

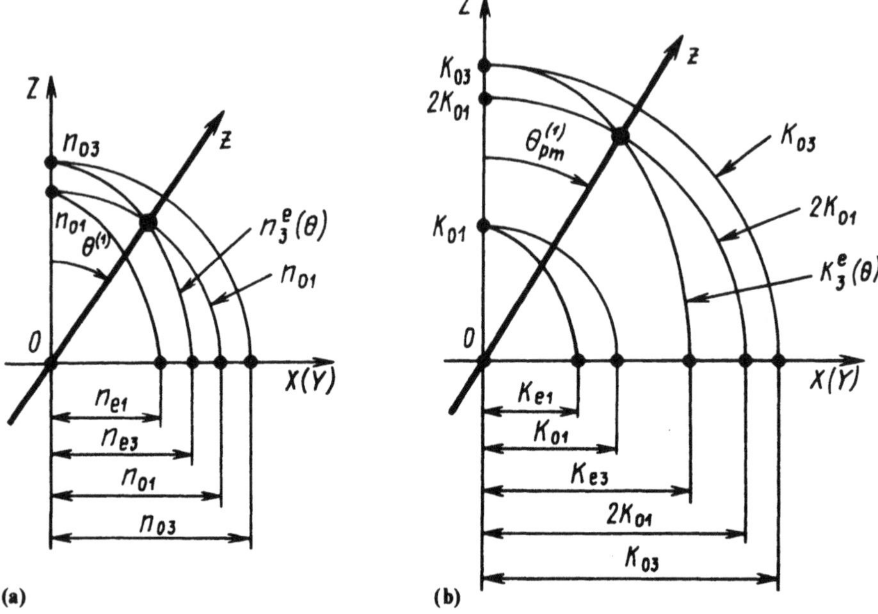

Fig. 2.8. Scalar (collinear) phase matching of type I ("ooe") in a uniaxial negative crystal in coordinates of refractive indices (**a**) and wave vectors (**b**) in the first quadrant of the XZ (YZ) plane

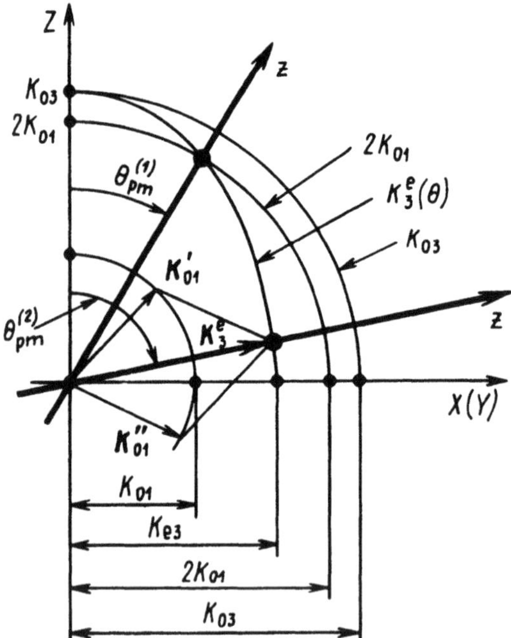

Fig. 2.9. Scalar (collinear) and vector (noncollinear) phase matching of type I ("ooe") in a uniaxial negative crystal

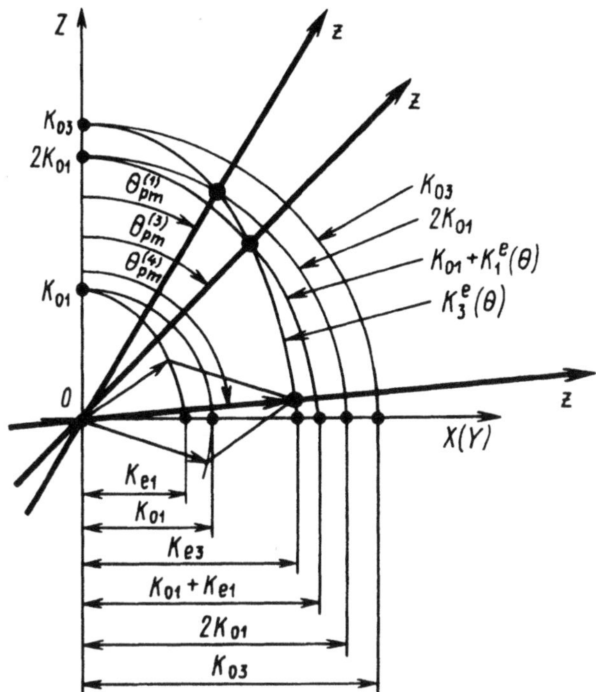

Fig. 2.10. Scalar (collinear) phase matching of type I ("ooe") and type II ("eoe"), and vector (noncollinear) phase matching of type II ("eoe") in a uniaxial negative crystal

of the ellipse $k_3^e(\theta)$ with the quasi-ellipse $k_{o1} + k_1^e(\theta)$. Type II vector phase matching is possible in the region $\theta_{pm}^{(3)} < \theta_{pm}^{(4)} < \pi - \theta_{pm}^{(3)}$.

If collinear phase matching is realized at $\theta_{pm} = 90°$, vector phase matching of the same type is absent. Besides, if $\theta_{pm}^{(1)} = 90°$, no phase matching of type II is realized.

2.5 Calculation of Phase-Matching Angles in Uniaxial Crystals

The dispersion properties of uniaxial nonlinear crystals are determined only by the polar angle θ. Therefore, to find the phase-matching direction in uniaxial crystals, it is sufficient to determine the corresponding phase-matching angle θ_{pm} for a certain three-wave interaction. Table 2.1 lists the precise analytical expressions for θ_{pm}^{ooe}, θ_{pm}^{oeo}, and θ_{pm}^{eoo}, and approximate expressions for θ_{pm}^{eoe}, θ_{pm}^{oee}, and θ_{pm}^{eeo} (the determination accuracy $\sim 0.1°–0.2°$) [2.7].

Phase matching is realized independently of the azimuthal angle ϕ, i.e., on the surface of the cone with an apex angle $2\theta_{pm}$. At the same time the efficiency of the nonlinear conversion process is determined by both θ_{pm} and ϕ.

Table 2.1. Equations for calculating phase-matching angles in uniaxial crystals

Negative uniaxial crystals	Positive uniaxial crystals
$\tan^2 \theta_{\text{pm}}^{\text{ooe}} = (1 - U)/(W - 1)$	$\tan^2 \theta_{\text{pm}}^{\text{eeo}} \cong (1 - U)/(U - S)$
$\tan^2 \theta_{\text{pm}}^{\text{eoe}} \cong (1 - U)/(W - R)$	$\tan^2 \theta_{\text{pm}}^{\text{oeo}} = (1 - V)/(V - Y)$
$\tan^2 \theta_{\text{pm}}^{\text{oee}} \cong (1 - U)/(W - Q)$	$\tan^2 \theta_{\text{pm}}^{\text{eoo}} = (1 - T)/(T - Z)$

Notations:
$U = (A + B)^2/C^2$; $W = (A + B)^2/F^2$; $R = (A + B)^2/(D + B)^2$;
$Q = (A + B)^2/(A + E)^2$; $S = (A + B)^2/(D + E)^2$; $V = B^2/(C - A)^2$;
$Y = B^2/E^2$; $T = A^2/(C - B)^2$; $Z = A^2/D^2$;
$A = n_{\text{o}1}/\lambda_1$; $B = n_{\text{o}2}/\lambda_2$; $C = n_{\text{o}3}/\lambda_3$;
$D = n_{\text{e}1}/\lambda_1$; $E = n_{\text{e}2}/\lambda_2$; $F = n_{\text{e}3}/\lambda_3$.

The expressions presented in Table 2.1 can be generalized to the noncollinear phase matching. In this case, for example, the phase-matching angle $\theta_{\text{pm}}^{\text{ooe}}$ is determined from the above presented equation using the new coefficients U and W:

$$U = (A^2 + B^2 + 2AB\cos\gamma)/C^2, \quad W = (A^2 + B^2 + 2AB\cos\gamma)/F^2$$

where γ is the angle between wave vectors k_1 and k_2.

2.6 Reflection and Refraction of Light Waves at the Surfaces of Uniaxial Crystals

Reflection and refraction of light waves at the vacuum-dielectric interface must be taken into account. Therefore, we shall give the equations for the refraction angles and for the reflection coefficients for different incidence angles and polarizations of the light wave incident on the plane surface of an uniaxial nonlinear crystal. In all cases the reflection angles are equal to the incident angles.

1) The E vector is perpendicular to the principal plane, α is the incidence angle, ψ_{o} is the refraction angle (Fig. 2.11a); ψ_{o} can be found from

$$\sin\alpha = n_{\text{o}} \sin\psi_{\text{o}}; \tag{2.32}$$

the reflection coefficient is

$$R^{\text{o}} = \frac{\sin^2(\alpha - \psi_{\text{o}})}{\sin^2(\alpha + \psi_{\text{o}})} . \tag{2.33}$$

For normal incidence ($\alpha = \psi_{\text{o}} = 0$)

$$R^{\text{o}}_{\alpha=0} = \frac{(n_{\text{o}} - 1)^2}{(n_{\text{o}} + 1)^2} . \tag{2.34}$$

2) The E vector is in the main plane, α is the incident angle, ψ^{e} is the refraction angle (Fig. 2.11b,c); ψ^{e} can be found from

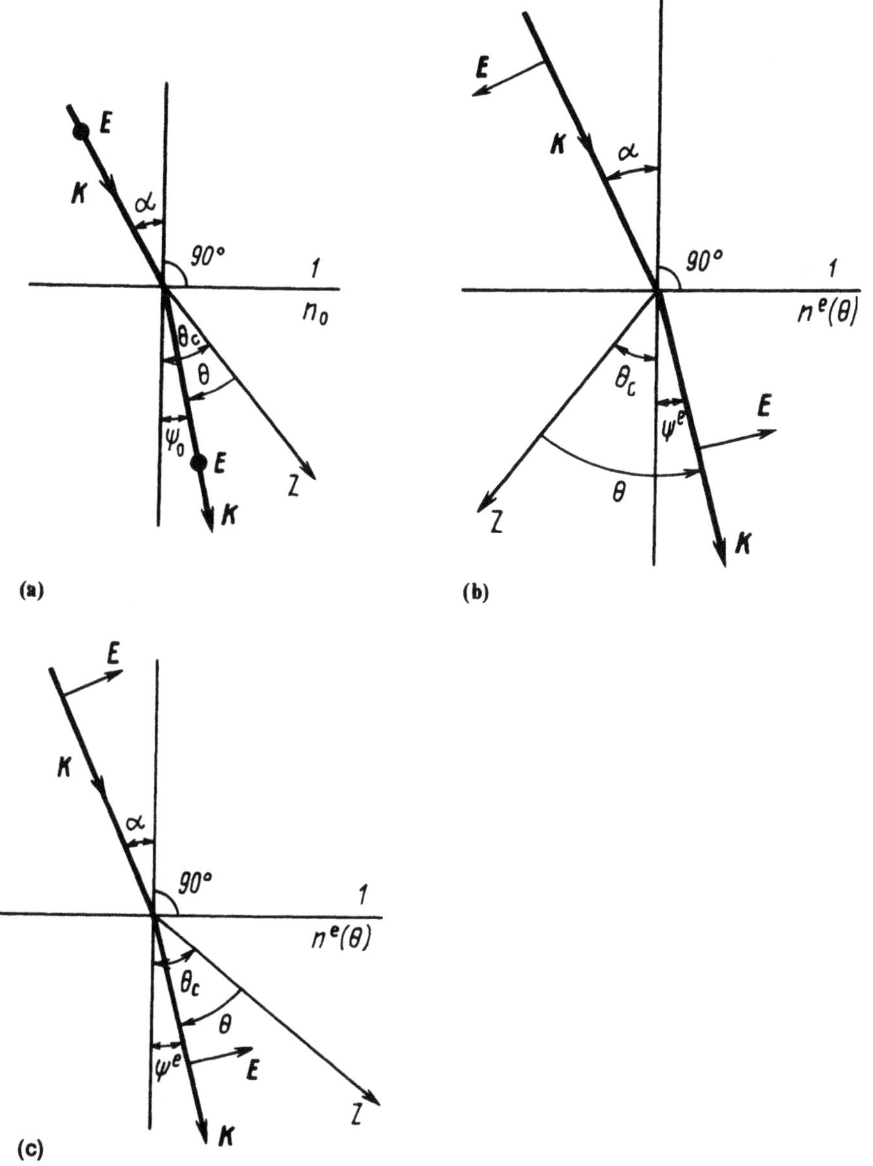

Fig. 2.11. Reflection and refraction of waves on the surface of an uniaxial crystal for incident waves with ordinary (**a**) and extraordinary (**b,c**) polarizations

$$\frac{\sin \alpha}{\sin \psi^{\mathrm{e}}} = n^{\mathrm{e}}(\theta) = \frac{n_0[1 + \tan^2(\theta_{\mathrm{c}} \pm \psi^{\mathrm{e}})]^{1/2}}{[1 + (n_0/n_{\mathrm{e}})^2 \tan^2(\theta_{\mathrm{c}} \pm \psi^{\mathrm{e}})]^{1/2}} \,, \tag{2.35}$$

where θ_{c} is the cut angle of the crystal, and θ is the angle between the optical axis Z and vector $\boldsymbol{k}$ in the crystal. If the vector $\boldsymbol{k}$ and optic axis Z lie on different sides of the normal to the crystal surface (Fig. 2.11b), the *plus* sign is used in (2.23). When the vector k and optic axis Z are on the same side of the normal to the crystal surface (Fig. 2.11c), the *minus* sign is used. The reflection coefficient is

$$R^e = \frac{\tan^2(\alpha - \psi^e)}{\tan^2(\alpha + \psi^e)} \ . \tag{2.36}$$

For normal incidence ($\alpha = \psi^e = 0$)

$$R^e_{\alpha=0} = \frac{[n^e(\theta) - 1]^2}{[n^e(\theta) + 1]^2} \ , \tag{2.37}$$

where

$$n^e(\theta) = n^e(\theta_c) = n_o \left(\frac{1 + \tan^2 \theta_c}{1 + (n_o/n_e)^2 \tan^2 \theta_c} \right)^{1/2} \ . \tag{2.38}$$

Note that the reflection coefficients on the input and output surfaces of the uniaxial crystal are identical, as with an isotropic dielectric.

The equations given here can be used for calculating the external rotation crystal angles necessary for angular tuning of the phase-matching conditions and for evaluating the reflection losses.

2.7 Optics of Biaxial Crystals

For the *biaxial* crystals the dependence of the refractive index on light propagation direction and it's polarization (index surface) corresponds to a much more complex function than for the uniaxial crystals. The resulting surface has a *bilayer* structure with four points of interlayer contact through which two optic axes pass [2.6]. Similar to the case of a uniaxial crystal, the propagation direction of plane light wave is defined by two angles: polar θ and azimuthal ϕ.

Note that the use of terms ordinary (o) and extraordinary (e) waves for the general case of light propagation inside a biaxial crystal is senseless. We shall use in consideration below the terms *slow* (symbol s) and *fast* (symbol f) waves ($n_s > n_f$, and $v_s < v_f$, respectively). The use of old terminology (o and e waves) has some meaning only in the principal planes of a biaxial crystal.

For simplicity we confine ourselves to the case of light propagation in the *principal planes XY, YZ* and *XZ*. In these planes the dependences of the refractive index on the propagation direction of two waves with orthogonal polarizations represent a combination of an ellipse and a circle (Fig. 2.12a,b). We shall relate *dielectric* (X, Y, Z) and *crystallographic* (a,b,c) axes in a biaxial crystal in such a way that the optic axes, whose directions are given by the intersection points of the ellipse and circle, will always lie in the principal plane XZ.

Consider one of two possible cases: $n_X < n_Y < n_Z$ (Fig. 2.12a), where n_X, n_Y, and n_Z are the principal values of the refractive indices. The angle V_Z formed by one of the optic axes with the axis Z can be found from the expression

$$\sin V_Z = \frac{n_Z(n_Y^2 - n_X^2)^{1/2}}{n_Y(n_Z^2 - n_X^2)^{1/2}} \cdot \tag{2.39}$$

The angle between optical axes in the plane XZ is equal to $2V_Z$. In the plane XY the refractive index of the wave polarized normally to this plane is constant and equals n_Z, and that of the wave polarized in this plane changes from n_Y to n_X with ϕ varying from 0° to 90°. Hence, a biaxial crystal with $n_X < n_Y < n_Z$ in the plane XY is similar to a negative uniaxial crystal with $n_o = n_Z$ and

$$n^e(\phi) = n_Y \frac{(1 + \tan^2 \phi)^{1/2}}{[1 + (n_Y/n_X)^2 \tan^2 \phi]^{1/2}} \cdot \tag{2.40}$$

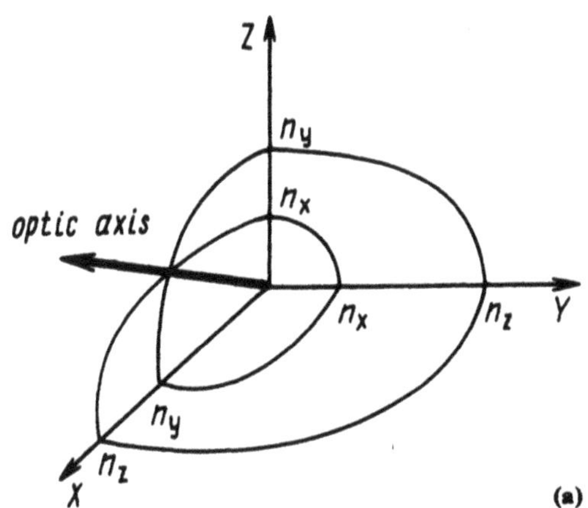

(a)

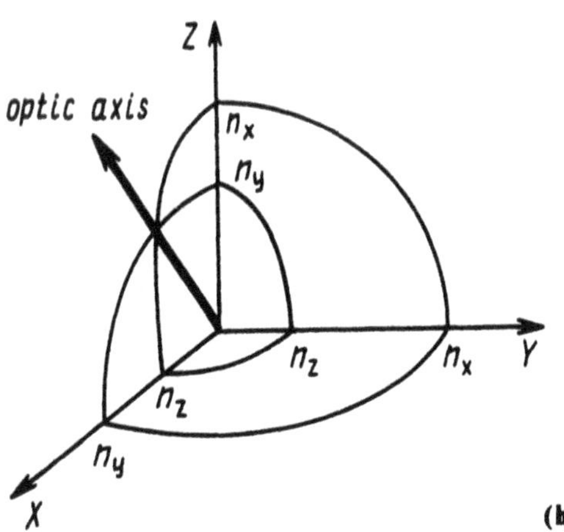

(b)

Fig. 2.12. Dependence of refractive index on light propagation direction and polarization (index surface) in biaxial crystals under the following relations between principal values of refractive indices: a) $n_X < n_Y < n_Z$; b) $n_X > n_Y > n_Z$.

In the plane YZ the refractive index of the wave polarized normally to this plane is constant and equals n_X, whereas for the wave polarized in this plane the refractive index changes from n_Y to n_Z with θ varying from $0°$ to $90°$. Hence, a biaxial crystal with $n_X < n_Y < n_Z$ in the plane YZ is similar to a positive uniaxial crystal with $n_{\mathrm{o}} = n_X$ and

$$n^{\mathrm{e}}(\theta) = n_Y \frac{(1 + \tan^2 \theta)^{1/2}}{[1 + (n_Y/n_Z)^2 \tan^2 \theta]^{1/2}} \; . \tag{2.41}$$

We can also see that in the plane XZ at $\theta < V_Z$ a biaxial crystal with $n_X < n_Y < n_Z$ is similar to a negative uniaxial crystal and, at $\theta > V_Z$, to a positive uniaxial crystal.

A biaxial crystal with $n_X > n_Y > n_Z$ can be considered in a similar way (Fig. 2.12b). Here the angle V_Z between the optic axis and the axis Z is expressed as

$$\cos V_Z = \frac{n_X (n_Y^2 - n_Z^2)^{1/2}}{n_Y (n_X^2 - n_Z^2)^{1/2}} \; . \tag{2.42}$$

The biaxial crystal is said to be optically positive if the bisectrix of the acute angle between optic axes coincides with $n_{\max}$, and optically negative if the bisectrix coincides with $n_{\min}$.

To estimate the "walk-off" angle in biaxial crystal it is possible, as a first approximation, to use the formulae for uniaxial crystals, (2.21); the accurate calculation of "walk-off" angle along the phase-matching direction in a nonlinear biaxial crystal is given, for example, in [2.8]. The above-presented formulae (2.32–38) for uniaxial crystals can also be used for the calculation of reflection and refraction of light waves at the surfaces of a biaxial crystal, especially for light propagation in principal planes; however, the accurate expressions for biaxial crystals are much more complicated.

2.8 Types of Phase Matching in Biaxial Crystals

It can be shown, that in biaxial crystals only three general types of phase matching take place: ss-f, sf-f, and fs-f (third index corresponds to higher frequency ω_3); the ss-f case we shall mark as type I phase matching, and sf-f or fs-f cases - as type II phase matching. Note that in [2.9, 10] not only types I and II phase matching are discussed, but types I, II and III phase matching are introduced (sf-f type is type II, and fs-f type is type III), but this designation in our opinion is not very successful because it leads to non-uniform classification of phase-matching types. Such classification has some meaning only in the principal planes in accordance with the "sign" of a biaxial crystal (*negative* or *positive*) in these planes.

Hobden [2.11] considered the 14 possible cases of phase matching in the biaxial crystals, *Stepanov* et al. [2.10] generalized thus consideration for SFG and DFG and found 30 cases of collinear phase matching; in [2.9] the complete classification and calculation of direction loci in the 72 possible classes of collinear phase matching in uniaxial and biaxial nonlinear crystals is given.

In the case of SHG ($\omega_3 = 2\omega_1$) in all the principal planes of a biaxial crystal only two types of collinear phase matching take place, namely:

$$n_{\mathrm{s}}(\omega_1) = n_{\mathrm{s}1} = n_{\mathrm{f}}(\omega_3) = n_{\mathrm{f}3} \tag{2.43}$$

(ss-f or type I *of phase matching*);

$$n_{\mathrm{s}1} + n_{\mathrm{f}1} = 2n_{\mathrm{f}3} \tag{2.44}$$

(sf-f or type II *of phase matching*).
The difference between these types of phase matching for different principal planes lies in their "signs" (*plus* or *minus* phase matching) and in their accordance to ooe, oee, eeo or eoo types of phase matching in the case of a uniaxial crystal. For the case $n_X < n_Y < n_Z$ in the plane XY we have *minus*-types of phase matching (types I$^{(-)}$ and II$^{(-)}$), in plane YZ – *plus*-types (types I$^{(+)}$ and II$^{(+)}$), in plane XZ with $\theta < V_Z$ – *minus*-types, and with $\theta < V_Z$ – *plus*-types of phase matching. The same wave (slow or fast) may be an o-wave as well as an e-wave in dependence on position in coordinate space. The similar consideration of phase matching types can be done for the case $n_X > n_Y > n_Z$.

Similar to uniaxial crystals, the existence of one kind of phase matching or another depends on the relation between the principal values of the refraction index (i.e., on birefringence). For example, in the case $n_X < n_Y < n_Z$ the type I$^{(-)}$ phase matching in the plane XY takes place by fulfilling the following inequality: $n_Z(\omega_1) < n_Y(\omega_3)$; for more details see [2.9–11].

2.9 Calculation of Phase-Matching Angles in Biaxial Crystals

Table 2.2 gives the equations for calculating phase-matching angles θ_{pm} or ϕ_{pm} upon collinear propagation of interacting waves in the principle planes of a biaxial crystal [2.12]. Note that some equations are approximate.

For accurate calculation of phase-matching angles, i.e., of dependence $\theta_{\mathrm{pm}}(\phi_{\mathrm{pm}})$, a general approach given by *Hobden* [2.11], *Stepanov* et al. [2.10], and *Kashke* and *Koch* [2.13] can be used. Considering the generalized Fresnel equation we can calculate the phase velocities of slow (s) and fast (f) waves for arbitrary direction with angles θ, ϕ [2.13]:

$$v_j^{\mathrm{s,f}} = \frac{P_j}{2} \pm \left(\frac{P_j^2}{4}\right)^{1/2} - Q_j \,, \tag{2.45}$$

where j = 1, 2, 3 (remember, that $\omega_3 = \omega_1 + \omega_2$ for a three-frequency interaction),

Table 2.2. Equations for calculating phase-matching angles in biaxial crystals upon light propagation in the principal planes

$(a) n_X < n_Y < n_Z$

Principal plane	Type of interaction	Equations	Notations
1	2	3	4
	ooe	$\tan^2\phi = \dfrac{1-U}{W-1}$	$U = \left(\dfrac{A+B}{C}\right)^2;\ W = \left(\dfrac{A+B}{F}\right)^2;\ A = \dfrac{n_{Z1}}{\lambda_1},\ B = \dfrac{n_{Z2}}{\lambda_2},\ C = \dfrac{n_{Y3}}{\lambda_3},\ F = \dfrac{n_{X3}}{\lambda_3}$
XY	eoe	$\tan^2\phi \cong \dfrac{1-U}{W-R}$	$U = \left(\dfrac{A+B}{C}\right)^2;\ W = \left(\dfrac{A+B}{F}\right)^2;\ R = \left(\dfrac{A+B}{D+B}\right)^2;\ A = \dfrac{n_{Y1}}{\lambda_1},\ B = \dfrac{n_{Z2}}{\lambda_2},\ C = \dfrac{n_{Y3}}{\lambda_3},\ D = \dfrac{n_{X1}}{\lambda_1},\ F = \dfrac{n_{X3}}{\lambda_3}$
	oee	$\tan^2\phi \cong \dfrac{1-U}{W-Q}$	$U = \left(\dfrac{A+B}{C}\right)^2;\ W = \left(\dfrac{A+B}{F}\right)^2;\ Q = \left(\dfrac{A+B}{A+E}\right)^2;\ A = \dfrac{n_{Z1}}{\lambda_1},\ B = \dfrac{n_{Y2}}{\lambda_2},\ C = \dfrac{n_{Y3}}{\lambda_3},\ E = \dfrac{n_{X2}}{\lambda_2},\ F = \dfrac{n_{X3}}{\lambda_3}$
	eeo	$\tan^2\theta \cong \dfrac{1-U}{U-S}$	$U = \left(\dfrac{A+B}{C}\right)^2;\ S = \left(\dfrac{A+B}{D+E}\right)^2;\ A = \dfrac{n_{Y1}}{\lambda_1},\ B = \dfrac{n_{Y2}}{\lambda_2},\ C = \dfrac{n_{X3}}{\lambda_3},\ D = \dfrac{n_{Z1}}{\lambda_1},\ E = \dfrac{n_{Z2}}{\lambda_2}$
YZ	oeo	$\tan^2\theta = \dfrac{1-V}{V-Y}$	$V = \left(\dfrac{B}{C-A}\right)^2;\ Y = \left(\dfrac{B}{E}\right)^2;\ A = \dfrac{n_{X1}}{\lambda_1},\ B = \dfrac{n_{Y2}}{\lambda_2},\ C = \dfrac{n_{X3}}{\lambda_3},\ E = \dfrac{n_{Z2}}{\lambda_2}$
	eoo	$\tan^2\theta = \dfrac{1-T}{T-Z}$	$T = \left(\dfrac{A}{C-B}\right)^2;\ Z = \left(\dfrac{A}{D}\right)^2;\ A = \dfrac{n_{Y1}}{\lambda_1},\ B = \dfrac{n_{X2}}{\lambda_2},\ C = \dfrac{n_{X3}}{\lambda_3},\ D = \dfrac{n_{Z1}}{\lambda_1}$
XZ	ooe	$\tan^2\theta = \dfrac{1-U}{W-1}$	$U = \left(\dfrac{A+B}{C}\right)^2;\ W = \left(\dfrac{A+B}{F}\right)^2;\ A = \dfrac{n_{Y1}}{\lambda_1},\ B = \dfrac{n_{Y2}}{\lambda_2},\ C = \dfrac{n_{X3}}{\lambda_3},\ F = \dfrac{n_{Z3}}{\lambda_3}$
$\theta < V_Z$	eoe	$\tan^2\theta \cong \dfrac{1-U}{W-R}$	$U = \left(\dfrac{A+B}{C}\right)^2;\ W = \left(\dfrac{A+B}{F}\right);\ R = \left(\dfrac{A+B}{D+B}\right)^2;\ A = \dfrac{n_{X1}}{\lambda_1},\ B = \dfrac{n_{Y2}}{\lambda_2},\ C = \dfrac{n_{X3}}{\lambda_3},\ D = \dfrac{n_{Z1}}{\lambda_1},\ F = \dfrac{n_{Z3}}{\lambda_3}$
	oee	$\tan^2\theta \cong \dfrac{1-U}{W-Q}$	$U = \left(\dfrac{A+B}{C}\right)^2;\ W = \left(\dfrac{A+B}{F}\right)^2;\ Q = \left(\dfrac{A+B}{A+E}\right)^2;\ A = \dfrac{n_{Y1}}{\lambda_1},\ B = \dfrac{n_{X2}}{\lambda_2},\ C = \dfrac{n_{X3}}{\lambda_3},\ E = \dfrac{n_{Z2}}{\lambda_2},\ F = \dfrac{n_{Z3}}{\lambda_3}$
XZ	eeo	$\tan^2\theta \cong \dfrac{1-U}{U-S}$	$U = \left(\dfrac{A+B}{C}\right)^2;\ S = \left(\dfrac{A+B}{D+E}\right)^2;\ A = \dfrac{n_{X1}}{\lambda_1},\ B = \dfrac{n_{X2}}{\lambda_2},\ C = \dfrac{n_{Y3}}{\lambda_3},\ D = \dfrac{n_{Z1}}{\lambda_1},\ E = \dfrac{n_{Z2}}{\lambda_2}$
$\theta > V_Z$	oeo	$\tan^2\theta = \dfrac{1-V}{V-Y}$	$V = \left(\dfrac{B}{C-A}\right)^2;\ Y = \left(\dfrac{B}{E}\right)^2;\ A = \dfrac{n_{Y1}}{\lambda_1},\ B = \dfrac{n_{X2}}{\lambda_2},\ C = \dfrac{n_{Y3}}{\lambda_3},\ E = \dfrac{n_{Z2}}{\lambda_2}$
	eoo	$\tan^2\theta = \dfrac{1-T}{T-Z}$	$T = \left(\dfrac{A}{C-B}\right)^2;\ Z = \left(\dfrac{A}{D}\right)^2;\ A = \dfrac{n_{X1}}{\lambda_1},\ B = \dfrac{n_{Y2}}{\lambda_2},\ C = \dfrac{n_{Y3}}{\lambda_3},\ D = \dfrac{n_{Z1}}{\lambda_1}$

(b) $n_X > n_Y > n_Z$

	eeo	$\tan^2\phi \cong \dfrac{1-U}{U-S}$	$U = \left(\dfrac{A+B}{C}\right)^2$; $S = \left(\dfrac{A+B}{D+E}\right)^2$; $A = \dfrac{n_{Y1}}{\lambda_1}$; $B = \dfrac{n_{Y2}}{\lambda_2}$; $C = \dfrac{n_{Z3}}{\lambda_3}$; $D = \dfrac{n_{X1}}{\lambda_1}$; $E = \dfrac{n_{X2}}{\lambda_2}$
XY	oeo	$\tan^2\phi = \dfrac{1-V}{V-Y}$	$V = \left(\dfrac{B}{C-A}\right)^2$; $Y = \left(\dfrac{B}{E}\right)^2$; $A = \dfrac{n_{Z1}}{\lambda_1}$; $B = \dfrac{n_{Y2}}{\lambda_2}$; $C = \dfrac{n_{Z3}}{\lambda_3}$; $E = \dfrac{n_{X2}}{\lambda_2}$
	eoo	$\tan^2\phi = \dfrac{1-T}{T-Z}$	$T = \left(\dfrac{A}{C-B}\right)^2$; $Z = \left(\dfrac{A}{D}\right)^2$; $A = \dfrac{n_{Y1}}{\lambda_1}$; $B = \dfrac{n_{Z2}}{\lambda_2}$; $C = \dfrac{n_{Z3}}{\lambda_3}$; $D = \dfrac{n_{X1}}{\lambda_1}$
	ooe	$\tan^2\theta = \dfrac{1-U}{W-1}$	$U = \left(\dfrac{A+B}{C}\right)^2$; $W = \left(\dfrac{A+B}{F}\right)^2$; $A = \dfrac{n_{X1}}{\lambda_1}$; $B = \dfrac{n_{X2}}{\lambda_2}$; $C = \dfrac{n_{Y3}}{\lambda_3}$; $F = \dfrac{n_{Z3}}{\lambda_3}$
YZ	eoe	$\tan^2\theta \cong \dfrac{1-U}{W-R}$	$U = \left(\dfrac{A+B}{C}\right)^2$; $W = \left(\dfrac{A+B}{F}\right)^2$; $R = \left(\dfrac{A+B}{D+B}\right)^2$; $A = \dfrac{n_{Y1}}{\lambda_1}$; $B = \dfrac{n_{X2}}{\lambda_2}$; $C = \dfrac{n_{Y3}}{\lambda_3}$; $D = \dfrac{n_{Z1}}{\lambda_1}$; $F = \dfrac{n_{Z3}}{\lambda_3}$
	oee	$\tan^2\theta \cong \dfrac{1-U}{W-Q}$	$U = \left(\dfrac{A+B}{C}\right)^2$; $W = \left(\dfrac{A+B}{F}\right)^2$; $Q = \left(\dfrac{A+B}{A+E}\right)^2$; $A = \dfrac{n_{X1}}{\lambda_1}$; $B = \dfrac{n_{Y2}}{\lambda_2}$; $C = \dfrac{n_{Y3}}{\lambda_3}$; $E = \dfrac{n_{Z2}}{\lambda_2}$; $F = \dfrac{n_{Z3}}{\lambda_3}$
XZ	eeo	$\tan^2\theta \cong \dfrac{1-U}{U-S}$	$U = \left(\dfrac{A+B}{C}\right)^2$; $S = \left(\dfrac{A+B}{D+E}\right)^2$; $A = \dfrac{n_{X1}}{\lambda_1}$; $B = \dfrac{n_{X2}}{\lambda_2}$; $C = \dfrac{n_{Y3}}{\lambda_3}$; $D = \dfrac{n_{Z1}}{\lambda_1}$; $E = \dfrac{n_{Z2}}{\lambda_2}$
$\theta < V_Z$	oeo	$\tan^2\theta = \dfrac{1-V}{V-Y}$	$V = \left(\dfrac{B}{C-A}\right)^2$; $Y = \left(\dfrac{B}{E}\right)^2$; $A = \dfrac{n_{Y1}}{\lambda_1}$; $B = \dfrac{n_{X2}}{\lambda_2}$; $C = \dfrac{n_{Y3}}{\lambda_3}$; $E = \dfrac{n_{Z2}}{\lambda_2}$
	eoo	$\tan^2\theta = \dfrac{1-T}{T-Z}$	$T = \left(\dfrac{A}{C-B}\right)^2$; $Z = \left(\dfrac{A}{D}\right)^2$; $A = \dfrac{n_{X1}}{\lambda_1}$; $B = \dfrac{n_{Y2}}{\lambda_2}$; $C = \dfrac{n_{Y3}}{\lambda_3}$; $D = \dfrac{n_{Z1}}{\lambda_1}$
XZ	ooe	$\tan^2\theta = \dfrac{1-U}{W-1}$	$U = \left(\dfrac{A+B}{C}\right)^2$; $W = \left(\dfrac{A+B}{F}\right)^2$; $A = \dfrac{n_{Y1}}{\lambda_1}$; $B = \dfrac{n_{Y2}}{\lambda_2}$; $C = \dfrac{n_{X3}}{\lambda_3}$; $F = \dfrac{n_{Z3}}{\lambda_3}$
$\theta > V_Z$	eoe	$\tan^2\theta \cong \dfrac{1-U}{W-R}$	$U = \left(\dfrac{A+B}{C}\right)^2$; $W = \left(\dfrac{A+B}{F}\right)^2$; $R = \left(\dfrac{A+B}{D+B}\right)^2$; $A = \dfrac{n_{X1}}{\lambda_1}$; $B = \dfrac{n_{Y2}}{\lambda_2}$; $C = \dfrac{n_{X3}}{\lambda_3}$; $D = \dfrac{n_{Z1}}{\lambda_1}$; $F = \dfrac{n_{Z3}}{\lambda_3}$
	oee	$\tan^2\theta \cong \dfrac{1-U}{W-Q}$	$U = \left(\dfrac{A+B}{C}\right)^2$; $W = \left(\dfrac{A+B}{F}\right)^2$; $Q = \left(\dfrac{A+B}{A+E}\right)^2$; $A = \dfrac{n_{Y1}}{\lambda_1}$; $B = \dfrac{n_{X2}}{\lambda_2}$; $C = \dfrac{n_{X3}}{\lambda_3}$; $E = \dfrac{n_{Z2}}{\lambda_2}$; $F = \dfrac{n_{Z3}}{\lambda_3}$

$$P_j = s_X^2 v_{jY}^2 + s_X^2 v_{jZ}^2 + s_Y^2 v_{jX}^2 + s_Y^2 v_{jZ}^2 + s_Z^2 v_{jX}^2 + s_Z^2 v_{jY}^2 \ , \tag{2.46}$$

$$Q_j = s_X^2 v_{jY}^2 v_{jZ}^2 + s_Y^2 v_{jX}^2 v_{jZ}^2 + s_Z^2 v_{jX}^2 v_{jY}^2 \ . \tag{2.47}$$

$s_{X,Y,Z}$ are the projections of unit wave vector $\mathbf{k}/k$ on axes X, Y, Z:

$$s_X = \cos\theta\cos\phi, \quad s_Y = \sin\theta\sin\phi, \quad s_Z = \cos\theta. \tag{2.48}$$

In (2.45) the sign "*plus*" is for fast wave, "*minus*" is for slow wave; for collinear phase matching:

$$\frac{1}{v_3^{\mathrm{f}}} = \frac{1}{v_1^{\mathrm{s,f}}}\frac{\omega_1}{\omega_3} + \frac{1}{v_2^{\mathrm{s,f}}}\frac{\omega_2}{\omega_3} \ . \tag{2.49}$$

After substitution of (2.45) into (2.49), the required dependence $\theta_{\mathrm{pm}} = \theta_{\mathrm{pm}}(\phi_{\mathrm{pm}})$ can be obtained (in the general case, only numerically); for such calculation it is necessary to know only the principal values n_{jX}, n_{jY}, n_{jZ}.

Note that the cases sf-f and fs-f are essentially different if $\omega_1 \neq \omega_2$.

More simple analytical approximate equations for collinear phase-matching directions, i.e., for dependence $\theta_{\mathrm{pm}}(\phi_{\mathrm{pm}})$ in biaxial crystals in the case of SHG ($\omega_3 = 2\omega_1$) were found in [2.14] with accuracy less than 8%. For example, for type I phase matching in a *positive* biaxial crystal the following expression is valid [2.14]:

$$\sin^2\theta_{\mathrm{pm}} = \frac{1}{K}\left[\frac{1}{n_{3Y}^2} - \frac{1}{n_{1X}^2} + \left(\frac{1}{n_{1X}^2} - \frac{1}{n_{1Y}^2} + \frac{1}{n_{3X}^2} - \frac{1}{n_{3Y}^2}\right)\sin^2\phi_{\mathrm{pm}}\right] , \tag{2.50}$$

where

$$K = \frac{1}{n_{1Z}^2} - \frac{1}{n_{1X}^2} + \left(\frac{1}{n_{1X}^2} - \frac{1}{n_{1Y}^2}\right)\sin^2\phi_{\mathrm{pm}} \ . \tag{2.51}$$

In order to obtain the corresponding expression for type I phase matching in a *negative* biaxial crystal it is necessary to interchange the indices 1 and 3 in (2.50–51).

For type II phase matching in a positive crystal the following approximation is used [2.14]:

$$\left[\frac{\sin^2\phi}{n_X^2} + \frac{\cos^2\phi}{n_Y^2}\right]^{-1/2} \approx n_Y\left[1 - (n_Y - n_X)\frac{\sin^2\phi}{n_X}\right] \tag{2.52}$$

and after this we have:

$$\sin^2\theta_{\mathrm{pm}} = K^{-1}\{[2n_{3Y} - n_{1Y} + (2n_{3X} - n_{1X} - 2n_{3Y}$$
$$+ n_{1Y})\sin^2\phi_{\mathrm{pm}}]^{-2} - n_{1X}^{-2} + (n_{1X}^{-2} - n_{1Y}^{-2})\sin^2\phi_{\mathrm{pm}}\} \ . \tag{2.53}$$

For type II phase matching in a **negative** crystal one can obtain:

$$\left\{\left[\frac{1}{n_{1Z}^2}-\frac{\cos^2\phi_{\rm pm}}{n_{1X}^2}-\frac{\sin^2\phi_{\rm pm}}{n_{1Y}^2}\right]+\frac{\cos^2\phi_{\rm pm}}{n_{1X}^2}\right.$$

$$\left.+\frac{\sin^2\phi_{\rm pm}}{n_{1Y}^2}\right\}^{-1/2}+\left[\frac{\sin^2\phi_{\rm pm}}{n_{1X}^2}+\frac{\cos^2\phi_{\rm pm}}{n_{1Y}^2}\right]^{-1/2}$$

$$=2\left\{\left[\frac{1}{n_{3Z}^2}-\frac{\cos^2\phi_{\rm pm}}{n_{3X}^2}-\frac{\sin^2\phi_{\rm pm}}{n_{3Y}^2}\right]\sin^2\theta_{\rm pm}\right.$$

$$\left.+\frac{\cos^2\phi_{\rm pm}}{n_{3X}^2}+\frac{\sin^2\phi_{\rm pm}}{n_{3Y}^2}\right\}^{-1/2} \tag{2.54}$$

2.10 Crystal Symmetry and Effective Nonlinearity: Uniaxial Crystals

For anisotropic media the dielectric susceptibility coefficients κ_0 and $\chi^{(2)}$ in (2.2), are in general case the tensors of the second and third ranks, respectively. Below we shall consider the uniaxial crystals. In dielectric reference frame X, Y, Z, where Z is the optic axis, the tensors κ_0 and ε_0 are diagonal. The following components:

$$\varepsilon_{0XX}=\varepsilon_{0YY}=n_{\rm o}^2\ ;$$
$$\varepsilon_{0ZZ}=n_{\rm e}^2\ ; \tag{2.55}$$

are nonzero components of the linear dielectric polarization tensor ε_0. In practice the tensor d_{ijk} is used instead of tensor χ_{ijk}, the two tensors being interrelated by the equation

$$\chi_{ijk}=2d_{ijk}\ . \tag{2.56}$$

Unlike tensor ε_0, tensors χ and d can be given only in a three dimensional representation. Usually a "plane" representation of tensor d_{ijk} in the form d_{il} is used, where i $=$ 1 corresponds to (X), i $=$ 2 to (Y), i $=$ 3 to (Z), and l takes the following values:

$$XX \quad YY \quad ZZ \quad YZ=ZY \quad XZ=ZX \quad XY=YX$$
$$l=\ 1 \quad\ \ 2 \quad\ \ 3 \quad\quad 4 \quad\quad\quad\ 5 \quad\quad\quad\ 6 \quad\ . \tag{2.57}$$

The expression (2.2) can be rewritten in a reduced form (with respect to the components):

$$P_i=\kappa_{0ik}E_k+2d_{il}E_l^2+\ldots, \tag{2.58}$$

where E_l^2 is the six-dimensional vector of the field products (summation over the repeating indices is carried out). For SFG in matrix form we have:

$$\left[\begin{pmatrix} P_X \\ P_Y \\ P_Z \end{pmatrix}\right] = \begin{bmatrix} d_{11} & d_{12} & d_{13} & d_{14} & d_{15} & d_{16} \\ d_{21} & d_{22} & d_{23} & d_{24} & d_{25} & d_{26} \\ d_{31} & d_{32} & d_{33} & d_{34} & d_{35} & d_{36} \end{bmatrix} \cdot \begin{bmatrix} E_X^2 \\ E_Y^2 \\ E_Z^2 \\ 2E_Y E_Z \\ 2E_X E_Z \\ 2E_X E_Y \end{bmatrix} \cdot \tag{2.59}$$

The total number of the components of the square nonlinearity tensor d_{il} is 18. In centrosymmetrical crystals (where the center is a symmetry element) all the components of the square nonlinearity tensor d are equal to zero. The noncentrosymmetrical crystals comprising 21 crystallographic classes out of 32 usually have one or more symmetry elements (axes or planes of different orders), which considerably decrease the number of independent components of the tensor d_{il}.

Kleinman [2.15] has established additional *symmetry conditions* for the case of no dispersion of electron nonlinear polarizability. When the Kleinman symmetry conditions are valid (in the great majority of practical cases), the number of independent components of the tensor d_{il} decreases from 18 to 10, because

$$\begin{aligned} d_{21} &= d_{16}; \quad d_{24} = d_{32}; \quad d_{31} = d_{15}; \\ d_{13} &= d_{35}; \quad d_{12} = d_{26}; \quad d_{32} = d_{24}; \\ d_{14} &= d_{36} = d_{25} \ . \end{aligned} \tag{2.60}$$

Since any linearly polarized wave in a uniaxial crystal can be represented as a superposition of two waves with "ordinary" and "extraordinary" polarizations, we provide the components of a unit polarization vector p given in polar coordinates θ and ϕ along the dielectric axes X, Y, Z, where Z is the optic axis and $| p | = 1$:

$$\begin{aligned} p_{oX} &= -\sin\phi, & p_X^e &= \cos\theta\cos\phi \\ p_{oY} &= \cos\phi, & p_Y^e &= \cos\theta\sin\phi \\ p_{oZ} &= 0, & p_Z^e &= -\sin\theta \ . \end{aligned} \tag{2.61}$$

The equations for calculating the conversion efficiency use the *effective nonlinearity* d_{eff}, which comprises all the summation operations along the polarization directions of the interacting waves:

$$d_{\mathrm{eff}} = p_1 d p_3 p_2 = p_2 d p_3 p_1 = p_3 d p_1 p_2 \tag{2.62}$$

The quantity d_{eff} represents a scalar product of the first vector in (2.62) and a tensor-vector product of the $d\,pp$ type, which is also a vector. Depending on the type of interaction (ooe, oee, and so on), the vector components p_i are calculated by (2.61), and the product (2.62) is found by the known rules of vector

Table 2.3. Expressions for d_{eff} in uniaxial crystals of different point groups when Kleinman symmetry relations are valid

Point group	Type of interaction	
	ooe, oeo, eoo	eeo, eoe, oee
$\bar{4}2m(D_{2d})$	$d_{36}\sin\theta\sin 2\phi$	$d_{36}\sin 2\theta\cos 2\phi$
$3m(C_{3v})$	$d_{31}\sin\theta - d_{22}\cos\theta\sin 3\phi$	$d_{22}\cos^2\theta\cos 3\phi$
$4(C_4)$ $4mm(C_{4v})$ $6(C_6)$ $6mm(C_{6v})$	$d_{31}\sin\theta$	0
$\bar{4}(S_4)$	$(d_{36}\sin 2\phi + d_{31}\cos 2\phi)\sin\theta$	$(d_{36}\cos 2\phi - d_{31}\sin 2\phi)\sin 2\theta$
$3(C_3)$	$(d_{11}\cos 3\phi - d_{22}\sin 3\phi)\cos\theta + d_{31}\sin\theta$	$(d_{11}\sin 3\phi + d_{22}\cos 3\phi)\cos^2\theta$
$32(D_3)$	$d_{11}\cos\theta\cos 3\phi$	$d_{11}\cos^2\theta\sin 3\phi$
$\bar{6}(C_{3h})$	$(d_{11}\cos 3\phi - d_{22}\sin 3\phi)\cos\theta$	$(d_{11}\sin 3\phi + d_{22}\cos 3\phi)\cos^2\theta$
$\bar{6}m2(D_{3h})$	$d_{22}\cos\theta\sin 3\phi$	$d_{22}\cos^2\theta\cos 3\phi$
$422(D_4)$	0	0
$622(D_6)$	0	0

algebra. Table 2.3 illustrates the values of d_{eff} determined in this way for nonlinear uniaxial crystals of 13 point groups [2.3, 16].

The inclusion of the birefringence or "walk-off" angle (Fig. 2.6) leads to the change of the expressions for the *nonlinear coupling coefficients* (see below) and for the *effective nonlinearity*. Although the angle θ is defined as the angle between axis Z and light propagation direction z, the unit polarization vectors $\boldsymbol{p}_i$ are perpendicular to the direction of propagation of the wave energy s. Therefore, it is necessary to correct the expressions for components of the unit polarization vector $\boldsymbol{p}$ given by (2.61). The sign of *birefringence angle* ρ in these formulas will depend on "walk-off" direction, i.e., in the case of a uniaxial crystal, on the sign of the crystal (Fig. 2.6): for a negative crystal the angle ρ must be added to θ, for a positive crystal it must be subtracted from θ. So, in formulas (2.61) the angle θ must be changed for $(\theta + \rho)$ for the negative crystal and for $(\theta - \rho)$ for the positive one. Remember that the value ρ is the function of the angle θ for the uniaxial crystal (2.21). The dispersion of ρ should be also taken into account. Therefore, in (2.61), instead of θ, we have substitute $\theta \pm \rho(\omega, 2\omega)$, and the corresponding changes should be done also in expressions for d_{eff} (Table 2.3).

2.11 Crystal Symmetry and Effective Nonlinearity: Biaxial Crystals

For a biaxial crystal the nonzero components of the tensor ε_0 in dielectric reference frame X, Y, Z are equal to

$$\varepsilon_{OXX} = n_X^2; \quad \varepsilon_{OYY} = n_Y^2; \quad \varepsilon_{OZZ} = n_Z^2 \ . \tag{2.63}$$

The optical indicatrix in biaxial crystals is a three-axes ellipsoid with three different semi-axes n_X, n_Y, n_Z:

$$\frac{X^2}{n_X^2} + \frac{Y^2}{n_Y^2} + \frac{Z^2}{n_Z^2} = 1 \tag{2.64}$$

To find the polarization vector directions for *slow* (s) and *fast* (f) light waves it is necessary to define the directions of axes of an elliptical cross-section of this ellipsoid normal to light propagation direction [2.17, 18]. The equations for components of unit polarization vector p (so-called *direction cosines*) can be given in form:

$$
\begin{aligned}
p_X^s &= \cos\theta\cos\phi\cos\delta - \sin\phi\sin\delta \ , \\
p_Y^s &= \cos\theta\sin\phi\cos\delta + \cos\phi\sin\delta \ , \\
p_Z^s &= -\sin\theta\cos\delta \ , \\
p_X^f &= -\cos\theta\cos\phi\sin\delta - \sin\phi\cos\delta \ , \\
p_Y^f &= -\cos\theta\sin\phi\sin\delta + \cos\phi\cos\delta \ , \\
p_Z^f &= -\sin\theta\sin\delta \ ,
\end{aligned}
\tag{2.65}
$$

where an angle δ can be found from equation:

$$\cot\delta = \frac{\cot^2 V_Z \sin^2\theta + \sin^2\phi - \cos^2\theta\cos^2\phi}{\cos\theta\sin 2\phi} \ . \tag{2.66}$$

For $n_X < n_Y < n_Z$ the angle δ is defined in the range $0 < \delta < \pi/2$, for $n_X > n_Y > n_Z$ – in the range $-\pi/2 < \delta < 0$. Note that the angle δ is introduced only for the sake of convenience [2.17–23]. Three angles: θ, ϕ, and δ determine the polarization vector directions of slow and fast waves in biaxial crystals.

Let us consider the derivation of the formula for d_{eff} for the case of mm2 point group which is the most widespread class of biaxial crystals. In the crystallographic orthogonal coordinate system (a,b,c) the nonzero components of the square polarizability tensor d_{ijK} for such crystals in the general case have the form [2.24]:

$$
\begin{aligned}
d_{caa} &= d_{31}; \quad d_{cbb} = d_{32}; \quad d_{ccc} = d_{33}; \\
d_{aac} &= d_{15}; \quad d_{bbc} = d_{24} \ .
\end{aligned}
\tag{2.67}
$$

If the Kleinman symmetry relations [2.15] are valid, all the subscripts in the d_{ijk} tensor can be permutated so that $d_{15} = d_{31}$ and $d_{24} = d_{32}$.

In Table 2.4 the components of d_{ijk} in the dielectric coordinate system (X, Y, Z) corresponding to different assignments between the dielectric (X, Y, Z) and crystallographic (a,b,c) reference frames are given.

After substitution of components d_{ijk} from Table 2.4 and components of vector p (2.65) in (2.62), which has the following full form:

Table 2.4. Components of d_{ijk} tensor in the dielectric coordinate system for different assignments between the dielectric and crystallographic reference frames

N	Assignment	$d_{caa} = d_{31}$	$d_{cbb} = d_{32}$	$d_{ccc} = d_{33}$	$d_{aac} = d_{15}$	$d_{bbc} = d_{24}$
1	$X, Y, Z \to a, b, c$	d_{ZXX}	d_{ZYY}	d_{ZZZ}	d_{XXZ}	d_{YYZ}
2	$X, Y, Z \to b, a, c$	d_{ZYY}	d_{ZXX}	d_{ZZZ}	d_{YYZ}	d_{XXZ}
3	$X, Y, Z \to a, c, b$	d_{YXX}	d_{YZZ}	d_{YYY}	d_{XXY}	d_{ZZY}
4	$X, Y, Z \to b, c, a$	d_{YZZ}	d_{YXX}	d_{YYY}	d_{ZZY}	d_{XXY}
5	$X, Y, Z \to c, b, a$	d_{XZZ}	d_{XYY}	d_{XXX}	d_{ZZX}	d_{YYX}
6	$X, Y, Z \to c, a, b$	d_{XYY}	d_{XZZ}	d_{XXX}	d_{YYX}	d_{ZZX}

$$d_{\text{eff}} = \sum_i p_i \sum_j \sum_k d_{ijk}\, p_i\, p_k, \tag{2.68}$$

we shall receive the values for d_{eff} for a biaxial crystal of the mm2 point group under different assignments between (X, Y, Z) and (a, b, c) reference frames, see Table 2.5 [2.22], where

$$A = \sin\theta; \quad B = \cos\theta; \quad C = \sin\phi;$$
$$D = \cos\phi; \quad E = \sin\delta; \quad F = \cos\delta . \tag{2.69}$$

The correct expressions for $d_{\text{eff}}^{\text{ss-f}}$ and $d_{\text{eff}}^{\text{sf-f}}$ for the particular case of the coincidence between two coordinate systems (X, Y, Z) and (a, b, c) were first obtained by *Lavrovskaya* et al. [2.21]. Note that corresponding expressions published earlier in [2.17–19] are incorrect. In [2.22] the general expressions for a biaxial crystal of the mm2 point group in the case of 6 different assignements between coordinate systems were found. The expressions for the special case $X, Y, Z \Rightarrow a, b, c$ are given below in the open form:

$$\begin{aligned}
d_{\text{eff}}^{\text{ss-f}} =\;& (d_{32} - d_{31}) \cos\theta \sin\theta \sin 2\phi \cos\delta \sin^2\delta \\
&+ (d_{15} - d_{24}) \cos\theta \sin\theta \sin 2\phi \cos\delta \cos 2\delta \\
&+ (d_{32} \cos^2\phi + d_{31} \sin^2\phi) \sin\theta \sin^3\delta \\
&+ (d_{32} \sin^2\phi + d_{31} \cos^2\phi) \cos^2\theta \sin\theta \cos^2\delta \sin\delta \\
&- 2(d_{24} \cos^2\phi + d_{15} \sin^2\phi) \sin\theta \cos^2\delta \sin\delta \\
&+ 2(d_{24} \sin^2\phi + d_{15} \cos^2\phi) \cos^2\theta \sin\theta \cos^2\delta \sin\delta \\
&+ d_{33} \sin^3\theta \cos^2\delta \sin\delta ,
\end{aligned} \tag{2.70}$$

$$\begin{aligned}
d_{\text{eff}}^{\text{sf-f}} =\;& (d_{32} - d_{31}) \cos\theta \sin\theta \cos\phi \sin\phi \sin\delta \cos 2\delta \\
&+ (d_{24} - d_{15}) \cos\theta \sin\theta \cos\phi \sin\phi \sin\delta (4\cos^2\delta - 1) \\
&- (d_{31} \cos^2\phi + d_{32} \sin^2\phi) \cos^2\theta \sin\theta \cos\delta \sin^2\delta \\
&+ (d_{31} \sin^2\phi + d_{32} \cos^2\phi) \sin\theta \cos\delta \sin^2\delta \\
&- 2(d_{15} \cos^2\phi + d_{24} \sin^2\phi) \cos^2\theta \sin\theta \cos\delta \sin^2\delta \\
&- (d_{15} \sin^2\phi + d_{24} \cos^2\phi) \sin\theta \cos\delta \cos 2\delta \\
&- d_{33} \sin^3\theta \cos\delta \sin^2\delta .
\end{aligned} \tag{2.71}$$

Table 2.5. The effective nonlinearity of mm2 point group biaxial crystal for the different assignments between the dielectric and crystallographic coordinate systems

Assignment	$d_{\text{eff}}^{\text{ss-f}}$ (Type I)	$d_{\text{eff}}^{\text{sf-f}}$ (Type II)
$X, Y, Z \to a, b, c$	$2d_{15}AH(BDH - CE)(BDE + CH)$ $+2d_{24}AH(BCE - DH)(BCH + DE)$ $+d_{31}AE(BDH - CE)^2$ $+d_{32}AE(BCH + DE)^2$ $+d_{33}A^3H^2E$	$-d_{15}[AH(BDE + CH)^2$ $+AE(BDH - CE)(BDE + CH)]$ $-d_{24}[AH(BCE - DH)^2$ $+AE(BCE - DH)(BCH + DE)]$ $-d_{31}AE(BDH - CE)(BDE + CH)$ $-d_{32}AE(BCE - DH)(BCH + DE)$ $-d_{33}A^3E^2H$
$X, Y, Z \to b, a, c$	$2d_{15}AH(BCE - DH)(BCH + DE)$ $+2d_{24}AH(BDH - CE)(BDE + CH)$ $+d_{31}AE(BCH + DE)^2$ $+d_{32}AE(BDH - CE)^2$ $+d_{33}A^3H^2E$	$-d_{15}[AH(BCE - DH)^2$ $+AE(BCE - DH)(BCH + DE)]$ $-d_{24}[AH(BDH - CE)^2$ $+AE(BDH - CE)(BDE + CH)]$ $-d_{31}AE(BCE - DH)(BCH + DE)$ $-d_{32}AE(BDH - CE)(BDE + CH)$ $-d_{33}A^3E^2H$
$X, Y, Z \to a, c, b$	$2d_{15}AH(BDH - CE)(BDE + CH)$ $-2d_{24}A^2EH(BCH + DE)$ $-d_{31}(BCE - DH)(BDH - CE)^2$ $-d_{32}A^2H^2(BCE - DH)$ $-d_{33}(BCE - DH)(BCH + DE)^2$	$d_{15}[(BCH + DE)(BDE + CH)^2$ $+(BCE - DH)(BDH - CE)(BDE + CH)]$ $+d_{24}[A^2EH(BCE - DH)$ $+A^2E^2(BCH + DE)]$ $+d_{31}(BCE - DH)(BDH - CE)(BDE + CH)$ $+d_{32}A^2EH(BCE - DH)$ $+d_{33}(BCE - DH)^2(BCH + DE)$
$X, Y, Z \to b, c, a$	$-2d_{15}A^2EH(BCH + DE)$ $+2d_{24}AH(BDH - CE)(BDE + CH)$ $-d_{31}A^2H^2(BCE - DH)$ $-d_{32}(BCE - DH)(BDH - CE)^2$ $-d_{33}(BCE - DH)(BCH + DE)^2$	$d_{15}[A^2EH(BCE - DH)$ $+A^2E^2(BCH + DE)]$ $+d_{24}[(BCH + DE)(BDE + CH)^2$ $+(BCE - DH)(BDH - CE)(BDE + CH)]$ $+d_{31}A^2EH(BCE - DH)$ $+d_{32}(BCE - DH)(BDH - CE)(BDE + CH)$ $+d_{33}(BCE - DH)^2(BCH + DE)$
$X, Y, Z \to c, b, a$	$-2d_{15}A^2EH(BDH - CE)$ $-2d_{24}(BCE - DH)$ $\times(BDH - CE)(BCH + DE)$ $-d_{31}A^2H^2(BDE + CH)$ $-d_{32}(BCH + DE)^2(BDE + CH)$ $-d_{33}(BDH - CE)^2(BDE + CH)$	$d_{15}[A^2E^2(BDH - CE) + A^2EH(BDE + CH)]$ $+d_{24}[(BCE - DH)^2(BDH - CE)$ $+(BCE - DH)(BCH + DE)(BDE + CH)]$ $+d_{31}A^2EH(BDE + CH)$ $+d_{32}(BCE - DH)(BCH + DE)(BDE + CH)$ $+d_{33}(BDH - CE)(BDE + CH)^2$
$X, Y, Z \to c, a, b$	$-2d_{15}(BCE - DH)$ $\times(BDH - CE)(BCH + DE)$ $-2d_{24}A^2EH(BDH - CE)$ $-d_{31}(BCH + DE)^2(BDE + CH)$ $-d_{32}A^2H^2(BDE + CH)$ $-d_{33}(BDH - CE)^2(BDE + CH)$	$d_{15}[(BCE - DH)^2(BDH - CE)$ $+(BCE - DH)(BCH + DE)(BDE + CH)]$ $+d_{24}[A^2E^2(BDH - CE) + A^2EH(BDE + CH)$ $+d_{31}(BCE - DH)(BCH + DE)(BDE + CH)$ $d_{32}A^2EH(BDE + CH)$ $d_{33}(BDH - CE)(BDE + CH)$

As it was mentioned above the existence of both the nonzero d_{eff} values and of phase-matching direction (θ_{pm}, ϕ_{pm}), is the *necessary* and *sufficient* condition for an effective three-wave interaction. It should be emphasized that when varying θ, ϕ, δ together with d_{eff} some other parameters of three-wave interaction such as angular, thermal (temperature), and spectral bandwidths, anisotropy ("walk-off") angle, etc., are also changed. Therefore the maximum value of d_{eff} in the general case does not correspond to the maximum efficiency of interaction.

From the practical point of view the calculation of d_{eff} in the particular case of light propagation in the principal planes of a biaxial crystal (XY, YZ, ZX; in the ZX plane two different cases: $\theta < V_Z$ and $\theta > V_Z$ should be distinguished) is of significant interest. The corresponding expressions can be deduced from Table 2.5 using values of the angles θ, ϕ, δ and coefficients A, B, C, D, E, H for light propagation in principal planes from Table 2.6.

It should be noted that when calculating the principal plane values of the angle δ it is necessary to evaluate correctly the arising indeterminate form of (2.66), each time taking into account the definition range of the angle δ.

The "sign" (negative or positive) of the principal plane determinates the assignment between "s,f" and "o,e" indices. For instance, for the case with $n_X < n_Y < n_Z$ in the ZX plane, an ordinary wave corresponds to a slow wave at $\theta < V_Z$ and to a fast wave at $\theta > V_Z$ (Fig. 2.12a); for the case $n_X > n_Y > n_Z$ the situation is opposite (Fig. 2.12b). Tables 2.7 and 2.8 list the possible types of interactions and Tables 2.9 and 2.10 contain the calculated expressions for d_{eff} for the cases of light propagation in principal planes. To use these tables (remember they correspond to the biaxial crystals of the mm2 point group!) it is necessary first to determine the assignment between the coordinate systems (X, Y, Z) and (a, b, c). Then using the data of Table 2.5 for the given assignment, the general expressions for d_{eff} and for ss-f or sf-f interactions could

Table 2.6. Meaning of the angles and coefficients for the formulae from Table 2.5 in the case of light propagation in the principal planes of mm2 point group biaxial crystal

Angles and coefficients	Principal plane			
	XY	YZ	XZ	
			$\theta < V_Z$	$\theta > V_Z$
θ	$\pi/2$	θ	θ	θ
A	1	$\sin\theta$	$\sin\theta$	$\sin\theta$
B	0	$\cos\theta$	$\cos\theta$	$\cos\theta$
ϕ	ϕ	$\pi/2$	0	0
C	$\sin\phi$	1	0	0
D	$\cos\phi$	0	1	1
		$n_X < n_Y < n_Z$		
δ	0	0	$\pi/2$	0
E	0	0	1	0
H	1	1	0	1
		$n_X > n_Y > n_Z$		
δ	$-\pi/2$	$-\pi/2$	0	$-\pi/2$
E	-1	-1	0	-1
H	0	0	1	0

Table 2.7. The possible types of phase matching in the principal planes of the mm2 point group biaxial crystal for the case $n_X < n_Y < n_Z$

| Assignment | Principal plane | | | |
| | XY | YZ | XZ | |
			$\theta < V_Z$	$\theta > V_Z$
$X,Y,Z \to a,b,c$	II$^{(-)}$	II$^{(+)}$	I$^{(-)}$	II$^{(+)}$
or $\to b,a,c$	oe-e,eo-e	oe-o, eo-o	oo-e	oe-o, eo-o
$X,Y,Z \to a,c,b$	I$^{(-)}$	II$^{(+)}$	II$^{(-)}$	I$^{(+)}$
or $\to b,c,a$	oo-e	oe-o,eo-o	oe-e,eo-e	ee-o
$X,Y,Z \to c,b,a$	I$^{(-)}$	I$^{(+)}$	I$^{(-)}$	II$^{(+)}$
or $\to c,a,b$	oo-e	ee-o	oo-e	oe-o,eo-o

Table 2.8. The possible types of phase matching in the principal planes of the mm2 point group biaxial crystal for the case $n_X > n_Y > n_Z$

| Assignment | Principal plane | | | |
| | XY | YZ | XZ | |
			$\theta < V_Z$	$\theta > V_Z$
$X,Y,Z \to a,b,c$	I$^{(+)}$	I$^{(-)}$	II$^{(+)}$	I$^{(-)}$
or $\to b,a,c$	ee-o	oo-e	oe-o,eo-o	oo-e
$X,Y,Z \to a,c,b$	II$^{(+)}$	I$^{(-)}$	I$^{(+)}$	II$^{(-)}$
or $\to b,c,a$	oe-o,eo-o	oo-e	ee-o	oe-e,eo-e
$X,Y,Z \to c,b,a$	II$^{(+)}$	II$^{(-)}$	II$^{(+)}$	I$^{(-)}$
or $\to c,a,b$	oe-o,eo-o	oe-e,eo-e	oe-o,eo-o	oo-e

be determined. For the concretization of these expressions it is necessary to substitute the coefficients A, B, etc., using (2.69). Note that the angles θ and ϕ determine the direction of three-wave phase-matched collinear interaction of light waves inside the biaxial crystal whereas the angle δ is deduced from (2.66) using the given values θ, ϕ angles and taking into account the definition range of δ. In the case of light propagation in the principal planes, Tables 2.7–2.10 should be employed. First using the data of Tables 2.7,8 for the given assignment between the coordinate systems and relation between the principal values of the refraction index, the possible types of phase matching are determined, then from Tables 2.9,10 the formulae for d_{eff} can be found.

The above-discussed method of calculation of d_{eff} values for mm2 point group crystals can be applied to the nonlinear biaxial crystals of other point groups. The calculations performed in the case of the biaxial crystals of the 222 point group show that upon the validity of Kleinman symmetry relations the single nonzero component d_{XYZ} exists for all possible assignments between two reference frames (Table 2.11).

Table 2.9. The d_{eff} expressions for the principal planes of the mm2 point group biaxial crystal in the case $n_X < n_Y < n_Z$

Assignment	Plane	$d_{\text{eff}}^{\text{ss-f}}$ (Type I)	$d_{\text{eff}}^{\text{sf-f}}$ (Type II)
$X,Y,Z \rightarrow a,b,c$	XY	0	$d_{15}\sin^2\phi + d_{24}\cos^2\phi$
	YZ	0	$d_{15}\sin\theta$
	$XZ,\ \theta < V_Z$	$d_{32}\sin\theta$	0
	$XZ,\ \theta > V_Z$	0	$d_{24}\sin\theta$
$X,Y,Z \rightarrow b,a,c$	XY	0	$d_{24}\sin^2\phi + d_{15}\cos^2\phi$
	YZ	0	$d_{24}\sin\theta$
	$XZ,\ \theta < V_Z$	$d_{31}\sin\theta$	0
	$XZ,\ \theta > V_Z$	0	$d_{15}\sin\theta$
$X,Y,Z \rightarrow a,c,b$	XY	$d_{32}\cos\phi$	0
	YZ	0	$d_{15}\cos\theta$
	$XZ,\ \theta < V_Z$	0	$d_{24}\sin^2\theta + d_{15}\cos^2\theta$
	$XZ,\ \theta > V_Z$	$d_{32}\sin^2\theta + d_{31}\cos^2\theta$	0
$X,Y,Z \rightarrow b,c,a$	XY	$d_{31}\cos\phi$	0
	YZ	0	$d_{24}\cos\theta$
	$XZ,\ \theta < V_Z$	0	$d_{15}\sin^2\theta + d_{24}\cos^2\theta$
	$XZ,\ \theta > V_Z$	$d_{31}\sin^2\theta + d_{32}\cos^2\theta$	0
$X,Y,Z \rightarrow c,b,a$	XY	$d_{31}\sin\phi$	0
	YZ	$d_{31}\sin^2\theta + d_{32}\cos^2\theta$	0
	$XZ,\ \theta < V_Z$	$d_{32}\cos\theta$	0
	$XZ,\ \theta > V_Z$	0	$d_{24}\cos\theta$
$X,Y,Z \rightarrow c,a,b$	XY	$d_{32}\sin\phi$	0
	YZ	$d_{32}\sin^2\theta + d_{31}\cos^2\theta$	0
	$XZ,\ \theta < V_Z$	$d_{31}\cos\theta$	0
	$XZ,\ \theta > V_Z$	0	$d_{15}\cos\theta$

Concerning the biaxial crystals of 2 point group it should be mentioned that in [2.25] the expressions for effective nonlinearity in the dielectric reference frame (X, Y, Z) using nonlinear coefficients defined in crystallographic reference frame (a, b, c) were deduced for **MAP** crystal. In all other ensuing works (see, for instance, [2.26–28]) the determination of **d**-tensor coefficients was made directly in dielectric coordinate system (X, Y, Z). Table 2.12 presents the expressions for d_{eff} and possible types of phase matching for biaxial crystals of the 2 point group when Kleinman symmetry relations are valid and nonlinear coefficients are measured in dielectric reference frame.

The inclusion of birefringence (anisotropy) in the calculation of d_{eff} for light propagation into a biaxial crystal is complicated enough and we haven't done it here. It is possible, however, as a first approximation, (2.65), to substitute instead of θ, the values $(\theta \pm \rho)$, depending on the "sign" of the crystal. Usually we have $\rho \ll \theta_{\text{pm}}$; but the inclusion of angle ρ is necessary for completeness of the physical picture as well as for the increase of calculation accuracy.

In conclusion, note that the lack of adherence to uniform nomenclature and conventions in nonlinear crystal optics (first of all, for the biaxial crystals) has resulted in growing confusion in the literature. In [2.29] the standards were

Table 2.10. The d_{eff} expressions for the principal planes of the mm2 point group biaxial crystal in the case $n_X > n_Y > n_Z$

Assignment	Plane	$d_{\text{eff}}^{\text{ss-f}}$ (Type I)	$d_{\text{eff}}^{\text{sf-f}}$ (Type II)
$X,Y,Z \rightarrow a,b,c$	XY	$d_{31} \sin^2 \phi + d_{32} \cos^2 \phi$	0
	YZ	$d_{31} \sin \theta$	0
	$XZ, \theta < V_Z$	0	$d_{24} \sin \theta$
	$XZ, \theta > V_Z$	$d_{32} \sin \theta$	0
$X,Y,Z \rightarrow b,a,c$	XY	$d_{32} \sin^2 \phi + d_{31} \cos^2 \phi$	0
	YZ	$d_{32} \sin \theta$	0
	$XZ, \theta < V_Z$	0	$d_{15} \sin \theta$
	$XZ, \theta > V_Z$	$d_{31} \sin \theta$	0
$X,Y,Z \rightarrow a,c,b$	XY	0	$d_{24} \cos \phi$
	YZ	$d_{31} \cos \theta$	0
	$XZ, \theta < V_Z$	$d_{32} \sin^2 \theta + d_{31} \cos^2 \theta$	0
	$XZ, \theta > V_Z$	0	$d_{24} \sin^2 \theta + d_{15} \cos^2 \theta$
$X,Y,Z \rightarrow b,c,a$	XY	0	$d_{15} \cos \phi$
	YZ	$d_{32} \cos \theta$	0
	$XZ, \theta < V_Z$	$d_{31} \sin^2 \theta + d_{32} \cos^2 \theta$	0
	$XZ, \theta > V_Z$	0	$d_{15} \sin^2 \theta + d_{24} \cos^2 \theta$
$X,Y,Z \rightarrow c,b,a$	XY	0	$d_{15} \sin \phi$
	YZ	0	$d_{15} \sin^2 \theta + d_{24} \cos^2 \theta$
	$XZ, \theta < V_Z$	0	$d_{24} \cos \theta$
	$XZ, \theta > V_Z$	$d_{32} \cos \theta$	0
$X,Y,Z \rightarrow c,a,b$	XY	0	$d_{24} \sin \phi$
	YZ	0	$d_{24} \sin^2 \theta + d_{15} \cos^2 \theta$
	$XZ, \theta < V_Z$	0	$d_{15} \cos \theta$
	$XZ, \theta > V_Z$	$d_{31} \cos \theta$	0

Table 2.11. Expressions for d_{eff} and possible types of phase matching in the principal planes of the 222 point group biaxial crystal when Kleinman symmetry relations are valid

Plane	$n_X < n_Y < n_Z$	$n_X > n_Y > n_Z$
XY	$d_{14} \sin 2\phi$, type II$^{(-)}$	$-d_{14} \sin 2\phi$, type I$^{(+)}$
YZ	$d_{14} \sin 2\theta$, type I$^{(+)}$	$-d_{14} \sin 2\theta$, type II$^{(-)}$
$XZ, \theta < V_Z$	$-d_{14} \sin 2\theta$, type II$^{(-)}$	$d_{14} \sin 2\theta$, type I$^{(+)}$
$XZ, \theta > V_Z$	$-d_{14} \sin 2\theta$, type I$^{(+)}$	$d_{14} \sin 2\theta$, type II$^{(-)}$

proposed in order to eliminate any ambiguity in the definition of nonlinear tensor components, effective nonlinearity, "walk-off" angle, and so on.

2.12 Theory of Nonlinear Frequency-Conversion Efficiency

The initial equation for calculation of the nonlinear frequency-conversion efficiency is the wave equation derived from the Maxwell equations [2.1–4]

Table 2.12. Expressions for d_{eff} and possible types of phase matching in the principal planes of the biaxial crystal of 2 point group when Klienman symmetry relations are valid and nonlinear coefficients are defined in dielectric reference frame

Plane	$n_X < n_Y < n_Z$	$n_X > n_Y > n_Z$
XY	$d_{25} \sin 2\phi$, type II$^{(-)}$	$d_{25} \sin 2\phi$, type I$^{(+)}$
	$d_{23} \cos \phi$, type I$^{(-)}$	$d_{23} \cos \phi$, type II$^{(+)}$
YZ	$d_{21} \cos \theta$, type II$^{(+)}$	$d_{21} \cos \theta$, type I$^{(-)}$
	$d_{25} \sin 2\theta$, type I$^{(+)}$	$d_{25} \sin 2\theta$, type II$^{(-)}$
XZ, $\theta < V_z$	$d_{21} \cos^2 \theta + d_{23} \sin^2 \phi$	$d_{21} \cos^2 \theta + d_{23} \sin^2 \phi$
	$+ d_{25} \sin 2\theta$, type II$^{(-)}$	$+ d_{25} \sin 2\theta$, type I$^{(+)}$
XZ, $\theta > V_z$	$d_{21} \cos^2 \theta + d_{23} \sin^2 \phi$	$d_{21} \cos^2 \theta + d_{23} \sin^2 \phi$
	$+ d_{25} \sin 2\theta$, type I$^{(+)}$	$+ d_{25} \sin 2\theta$, type II$^{(-)}$

$$\text{curl curl } E(r,t) + \frac{\varepsilon_0}{c^2}\frac{\partial^2 E(r,t)}{\partial t^2} = -\frac{4\pi}{c^2}\frac{\partial^2 P_{\text{NL}}(r,t)}{\partial t^2} \tag{2.72}$$

in combination with (2.2) for nonlinear polarization (in the approximation of square nonlinearity)

$$P_{\text{NL}}(r,t) = \chi^{(2)} E^2(r,t) \tag{2.73}$$

and with initial and boundary conditions for the electric field $E(r,t)$. In (2.72, 73) r is the radius vector, t is time, and c is light velocity.

Let us present the field E as a superposition of three interacting waves

$$E(r,t) = \frac{1}{2}\sum_{n=1}^{3}(p_n A_n(r,t)\exp[j(\omega_n t - k_n \cdot r)] + \text{C.C.}), \tag{2.74}$$

where $A(r,t)$ are the complex wave amplitudes; ω_n and k_n are frequencies and wave vectors, respectively; and C.C. means "complex conjugate". Substituting of (2.74) into (2.72) with allowance for (2.73) and using the *method of slowly varying amplitudes* gives the following *truncated equations* for complex amplitudes [2.4]:

$$\hat{M}_1 A_1 = j\sigma_1 A_3 A_2^* \exp(j\Delta kz) , \tag{2.75}$$

$$\hat{M}_2 A_2 = j\sigma_2 A_3 A_1^* \exp(j\Delta kz) , \tag{2.76}$$

$$\hat{M}_3 A_3 = j\sigma_3 A_1 A_2 \exp(-j\Delta kz) , \tag{2.77}$$

where operator M_n has the form

$$\hat{M}_n = \frac{\partial}{\partial z} + \rho\frac{\partial}{\partial x} + \frac{j}{2k_n}\left(\frac{\partial^2}{\partial x^2} + \frac{\partial^2}{\partial y^2}\right) + u_n^{-1}\frac{\partial}{\partial t}$$

$$+ jg_n\frac{\partial^2}{\partial t^2} + \delta_n + Q_n(A). \tag{2.78}$$

The calculation is carried out in the Cartesian coordinates x, y, z, where z is the propagation direction (not to be confused with the dielectric axes X, Y, Z). In

(2.75–78) ρ_n are the *birefringence* (or "*walk-off*") angles (the "*walk-off*" of an extraordinary beam being assumed to be in the XZ plane), σ_n are the *nonlinear coupling coefficients*, u_n are the *group velocities*, g_n are the *dispersive spreading coefficients*, Δk is the total *wave mismatch*, δ_n are the *linear absorption coefficients*, and Q_n takes into account nonlinear (commonly two-photon) *absorption*. The following relations take place:

$$\sigma_{1,2} = 4\pi k_{1,2} \, n_{1,2}^{-2} \, \mathbf{p}_{1,2} \, \mathbf{d} \, \mathbf{p}_{2,1} \mathbf{p}_3 \ , \tag{2.79}$$

$$\sigma_3 = 2\pi k_3 n_3^{-2} \, \mathbf{p}_3 \, \mathbf{d} \mathbf{p}_1 \, \mathbf{p}_2 \ , \tag{2.80}$$

$$\delta_n = k_n (2n_n^2)^{-1} \mathbf{p}_n [\mathrm{Im}\{\varepsilon_0(\omega_n)\}] \mathbf{p}_n \ , \quad n = 1, 2, 3 \tag{2.81}$$

$$g_n = \frac{1}{2}\left(\frac{\partial^2 k}{\partial \omega^2}\right)_{\omega=\omega_n} , \tag{2.82}$$

$$u_n = \left(\frac{\partial \omega}{\partial k}\right)_{\omega=\omega_n} = c\left[\frac{\partial \omega}{\partial(n\omega)}\right]_{\omega=\omega_n}$$

$$= c\left[n_n + \omega_n\left(\frac{\partial n}{\partial \omega}\right)_{\omega=\omega_n}\right]^{-1} . \tag{2.83}$$

In (2.81) $\mathrm{Im}\{\varepsilon_0(\omega_n)\}$ is the imaginary part of the linear dielectric susceptibility tensor responsible for linear absorption of radiation. The sequence of writing vectors and tensors in (2.79–81) should not be violated. Finally

$$\Delta k = \Delta k_{\mathrm{L}} + \Delta k_{\mathrm{tsa}} + \Delta k_{\mathrm{pr}} + \Delta k_{\mathrm{fcg}} \ , \tag{2.84}$$

where Δk_{L} is the linear wave mismatch:

$$\Delta k_{\mathrm{L}} = \mid \Delta k_{\mathrm{L}} \mid_z = \mid \mathbf{k}_1 + \mathbf{k}_2 - \mathbf{k}_3 \mid_z \ ; \tag{2.85}$$

Δk_{tsa} is the mismatch due to *thermal self-actions* (tsa) in nonlinear crystal, Δk_{fcg} is the mismatch due to free-carrier generation in the conduction band because of a nonlinear absorption. Thermal mismatch appears in the thermal conductivity equation, which has the following form for a stationary (with respect to heat) process:

$$\Delta_r\{\Delta k_{\mathrm{tsa}}(r, z)\} = -cP_{\mathrm{cr}}^{-1} \sum_{n=1}^{3} \delta_n n_n \langle a_n^2(r, t)\rangle \ . \tag{2.86}$$

Here Δ_r is the Laplace operator with respect to the transverse coordinates $r = (x, y)$; P_{cr} is the critical power of self-focusing equal to

$$P_{\mathrm{cr}} = \gamma \lambda_1 [4L(\partial B/\partial T)_{T=T_0}]^{-1} \ , \tag{2.87}$$

where B is the so-called *dispersive birefringence*, for SHG $B = n_{\mathrm{o}1} - n_2^{\mathrm{e}}$; T is the temperature; T_0 is the temperature of the crystal at which the z-axis (the normal to the input crystal surface) coincides with the phase-matching direction; γ is the *thermal conductivity coefficient* of the crystal; and L is the length of the crystal. A mean square of real amplitudes $a_n = \mid A_n \mid$ is equal to

$$\langle a_n^2(r,t)\rangle = f \int_{-\infty}^{\infty} a_n^2(r,z,t)\,dt \;, \tag{2.88}$$

where f is the pulse-repetition frequency of laser radiation.

The appearance of *heat mismatch* is physically related to non-uniform (over the beam cross section) radiative heating of a nonlinear crystal. Thermal conductivity equation given by (2.86) can be solved with the corresponding boundary conditions and with truncated equations (2.75–77). Nonlinear absorption in the crystal ($Q_n \neq 0$) must be taken into account not only in the truncated equations but mainly in the value of δ_n in (2.86).

For SHG ($a_1 = a_2$), we have [2.4]

$$Q_1 = \beta_{12}a_3^2; \qquad Q_2 = 2\beta_{12}a_1^2 + \beta_{22}a_3^2 \;; \tag{2.89}$$

where β_{12} and β_{22} are the coefficients of mixed ($h\omega_1 + h\omega_2 > E_g$) and two-photon ($2h\omega_3 > E_g$) *nonlinear absorption* (E_g is the value of the forbidden energy band, i.e., the *band gap*). Note that for great nonlinear (usually two-photon) absorption at a maximum (sum) frequency ω_3, total (linear and nonlinear) absorption at frequencies $\omega_{1,2}$ and ω_3 are not equal. This may result in *asymmetry* and even in *hysteresis* of the temperature dependence of the resulting radiation power (near the temperature of phase matching).

Photorefraction (the photorefractive effect) arises in some nonlinear crystals (of lithium niobate type) and consists in a radiation-induced change of the refractive index. In the case of continuous irradiation of lithium niobate at a frequency ω_3 with a power density $S_3 \geq 200$ W cm^{-2}, $\Delta B = \Delta(n_o - n^e) \approx 10^{-3}$. For pulse irradiation of lithium niobate with $S_3 \leq 10^8$ W cm^{-2}, $\Delta B \approx \beta S_3^{-1/2}$, where $\beta = 6 \times 10^{-9}$cm W$^{-1/2}$. At small ΔB, the value of Δk_{pr} can be compensated for at the expense of Δk_L, i.e., by *phase mismatching* (this can always be realized in practice). Remember that the photorefraction may result in coloration of the crystal, increase of absorption, and thermal self-actions.

Nonlinear absorption is accompanied with electron transitions from the valency band to the conductivity band, i.e., *free-carrier generation* (Δk_{fcg}). The fcg-effect leads in turn to two phenomena: an additional absorption at all three frequencies (absorption on free carriers) and an additional wave mismatch. The wave mismatch Δk_{fcg} is proportional to the square of the power density (i.e., to the fourth power of the amplitude) of two-photon absorbed radiation, generally at a maximum frequency

$$\Delta k_{fcg} = -qa_3^4 \;, \tag{2.90}$$

where q is a coefficient depending on the nonlinear absorption parameters, lifetime of free carriers, and so on. The fcg-effect must be taken into account when crystals of lithium and barium sodium niobates are used; then the absorption on free carriers at all three frequencies may be neglected, but the mismatch Δk_{fcg} is left in the equations.

Now we shall reconsider the operator M_n (2.78). Its first term (the derivative with respect to z) describes changes of the amplitudes in the process of their propagation and interaction. The second term (the derivative with respect to x) describes the influence of *crystal anisotropy* (the " *walk-off*" of an extraordinary beam along the x axis). The third term, containing second derivatives with respect to transverse coordinates x and y, corresponds to the *diffraction effect* (the diffractive spreading of the beam). The fourth term (the derivative with respect to time) describes the effect of *temporary modulation* (the pulse mode), including the effect of *group-velocity mismatch* of the pulses. The term containing the second derivative with respect to time corresponds to the effect of the *dispersive spreading of pulses*. The terms δ_n and $Q_n(A)$ describe *linear* and *nonlinear absorption*, respectively.

The right-hand parts of eqs (2.75–77) describe *nonlinear interaction* of the waves.

An exact calculation of the efficiency of SHG, SFG, and DFG convertors according to (2.75–77) is very complex and generally requires the numerical calculation. Only in some simple cases do analytical solutions allow one to evaluate roughly the conversion efficiency. For proper evaluation of the efficiency, the parameters of the initial (convertible) radiations and of the crystal converter must be known, and an adequate calculation procedure must be chosen on the basis of the recommendations below.

Let us introduce the *effective lengths* of the interaction process:

1) *Aperture length L_a:*

$$L_a = d_0/\rho \tag{2.91}$$

where d_0 is the characteristic diameter of the beam and ρ is the anisotropy ("walk-off") angle.

2) *Quasi-static interaction length L_{qs}:*

$$L_{qs} = \tau/v \tag{2.92}$$

where τ is the radiation pulse duration and v is the *inverse group-velocity mismatch*. For SHG

$$v = u_1^{-1} - u_3^{-1} \tag{2.93}$$

where u_1 and u_3 are the group velocities at the corresponding wavelengths (2.83).

3) *Diffraction length L_{dif}:*

$$L_{dif} = kd_0^2 \tag{2.94}$$

4) *Dispersive spreading length L_{ds}:*

$$L_{ds} = \tau^2/g \, , \tag{2.95}$$

where g is the *dispersive spreading coefficient* (2.82). A *nonlinear interaction length L_{NL}* is also introduced:

$$L_{\mathrm{NL}} = \frac{1}{\sigma a_0} \, , \tag{2.96}$$

where σ is the nonlinear coupling coefficient (2.79,80) and a_0 is expressed by the equation

$$a_0 = [a_1^2(0) + a_2^2(0) + a_3^2(0)]^{1/2} \, , \tag{2.97}$$

where $a_n(0)$ are the wave amplitudes at the input surface of the crystal (at $z = 0$).

Whether or not a given effect must be taken into account in the mathematical description of nonlinear conversion is determined by a comparison of the *crystal length L* with the corresponding *effective length L_{eff}* from (2.91–96). If $L < L_{\mathrm{eff}}$, the corresponding effect can be neglected. For instance, when $L < L_a$, one may neglect the anisotropy effect and put the second term in operator $\hat{M}_n$ equal to zero; when $L < L_{\mathrm{dif}}$, the diffractive spreading of the beam can be neglected; and so on.

Note the role of the nonlinear interaction length L_{NL}. When the condition $L < L_{\mathrm{NL}}$ is fulfilled, the so-called *fixed-field approximation* is realized; for instance, for SFG it means that the SF field amplitude is

$$a_3(z) \ll a_{1,2}(0) \, , \tag{2.98}$$

and the nonlinear equations given by (2.75–77) are transformed into linear (with respect to the real field amplitudes $a = |A|$) equations. In particular, for the SF field amplitude we have

$$\hat{M}_3 = -\mathrm{j}\sigma_3 a_1(0)a_2(0) \exp[\mathrm{j}(-\Delta kz - \phi_1 - \phi_2)] \, , \tag{2.99}$$

where $\phi_{1,2}$ are the wave phases. When $L \geq L_{\mathrm{NL}}$, we must solve exact (nonlinear) equations.

When $L < L_{\mathrm{NL}}$ and all of $L_{\mathrm{eff}} = \infty$, we have *plane-wave fixed-field approximation*. With $L < L_{\mathrm{qs}} \ll L_{\mathrm{dis}}$ (the inequality $L_{\mathrm{qs}} \ll L_{\mathrm{dis}}$ is valid always) the *quasi-static approximation* takes place, as well as with $L < L_{\mathrm{qs}}, L < L_{\mathrm{a}}$ and $L < L_{\mathrm{dif}}$ we have *quasi-plane-wave approximation*. The very important difference between *plane-wave* and *quasi-plane-wave* approximations lies in the fact that in *quasi-plane-wave* case it is necessary to take into account the inhomogeneity of spatial (beam) and temporal (pulse) intensity distributions of interacting waves (by simple integrating with respect to the time for pulses or over the area for beams, see below). But in this case it's not necessary to take into account *"walk-off" angle, diffraction, group-velocity mismatch* and *dispersive spreading*.

Thus, before calculation one should

1) determine all the effective lengths L_{eff} of the process, compare them with the length L of a nonlinear crystal, and find out all of the effects that must be taken into account;

2) calculate the nonlinear interaction length L_{NL}, compare it with the crystal length L, and determine whether the fixed-field approximation is valid or exact nonlinear equations must be solved.

Here are some practical cases with corresponding recommendations.

Under continuous-wave laser irradiation, we may neglect the group-velocity mismatch ($L_{qs} = \infty$) and the dispersive spreading of pulses ($L_{ds} = \infty$). In the practically used crystals with $L \cong 1$ cm we may neglect the following: the diffraction and anisotropy for the beams of $d_0 \simeq 1$ cm in diameter; group velocity mismatch at $\tau \geq 10^{-9}$ s; dispersive spreading at $\tau \geq 10^{-12}$ s; and nonlinear absorption and the fcg-effect at $2\hbar\omega_3 < E_g$. The photorefractive effect may be neglected in calculations, because the value Δk_{pr} (but not an additional absorption due to photorefractive effect!) is easily compensated for by an additional turn of the crystal (in lithium niobate crystals the photorefractive effect disappears completely at $T_0 \geq 170°C$). Diffraction must be taken into account only for conversion of focused beams [2.4,30].

If the crystal length L is smaller than each effective length, the operator $\hat{M}_n$ in (2.78) has the form

$$\hat{M}_n = \delta_n + \frac{\mathrm{d}}{\mathrm{d}z} \; . \tag{2.100}$$

When in this case ($L < L_{\mathrm{eff}}$) the radiations being converted are temporally and spatially modulated (pulse duration τ, beam diameter d_0) and the modulation shape is nonuniform (for instance, Gaussian beams, Gaussian pulses), the following calculation procedure can be used within fairly good accuracy.

The beam (or pulse) envelope of the radiation being converted is approximated by a step-wise function (Fig. 2.13), the field amplitude inside of each step being constant. For each step – i.e., for each field amplitude value,– the conversion efficiency is calculated by the equations for plane waves. Then the results are summed (integrated) with respect to transverse coordinates (or time), and the power (or energy) of the beam (or pulse) of the resulting radiation is determined.

If the condition $P_0 < P_{cr}$ is fulfilled, where P_0 is an average (or continuous) power of the radiation being converted at the input surface of the crystal, then the effects of thermal self-actions may be neglected (and the thermal conductivity equation need not be solved). If an opposite inequality is valid, truncated equations must be solved together with the thermal conductivity equation; two variants are possible. In one there is no dispersion of the absorption coefficients ($\delta_1 = \delta_2 = \delta_3$) and therefore the thermal conductivity equation as a first approximation can be solved independently on the truncated equations. If in this case the heat contact of the crystal with the outer medium (thermostat) is ideal – i.e., there is no *temperature jump* at the crystal-thermostat interface – we have for temperature mismatch

$$\Delta k_{\mathrm{tsa}} = \frac{4\pi\gamma}{P_{cr}}[T(r) - T_0] \; , \tag{2.101}$$

where

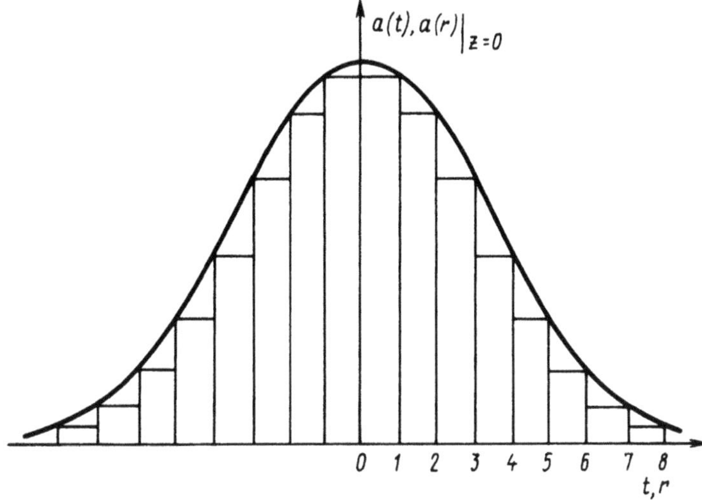

Fig. 2.13. Pulse (beam) approximation with a step-wise function for calculating the conversion efficiency in quasi-static (for a pulse) and diffraction-free (for a beam) approximations

$$T(r) = T(0) - \frac{\delta P_0}{2\pi\gamma}\left[\ln\left(2\frac{r^2}{w_0^2}\right) - \mathrm{Ei}\left(-2\frac{r^2}{w_0^2}\right) + C\right].\tag{2.102}$$

Here $T(0)$ is the temperature on the beam axis, w_0 is the characteristic radius of the convertible radiation beam, $C = 0.5772\ldots$ is the *Euler-Mascheroni constant*, and $\mathrm{Ei}(x) = \int_{-x}^{x}\frac{\exp(y)}{y}\,\mathrm{d}y$ is the *integral exponential function* [2.31]. For instance, for a LiNbO$_3$ crystal ($\delta \cong 0.01$ cm^{-1}, $\gamma = 2.6 \times 10^{-3}$ Wcm^{-1}K^{-1}) at $P_0 \simeq 10$W the temperature gradient between the crystal axis and beam boundary may be about 2K, which exceeds the temperature bandwidth (see below).

The temperature mismatch calculated from (2.101) is substituted into the truncated equations. They are solved for each value of the transverse co-ordinate r, and then the summation over the surface area is carried out to determine the power of the resulting radiation.

In the second variant, when $\delta_{1,2} \neq \delta_3$, the thermal conductivity equation cannot be solved independently of the truncated equations, and the solution can be found only by using numerical calculation. The situation is similar for a temperature jump at the crystal-thermostat interface.

Figure 2.14 illustrates typical dependences of the SHG conversion efficiency on the average input power of the fundamental laser radiation $P_1(0)$ is the widely used CDA and DCDA crystals with a typical absorption. A decrease of losses in nonlinear crystals is a cardinal way of eliminating heat self-action effects.

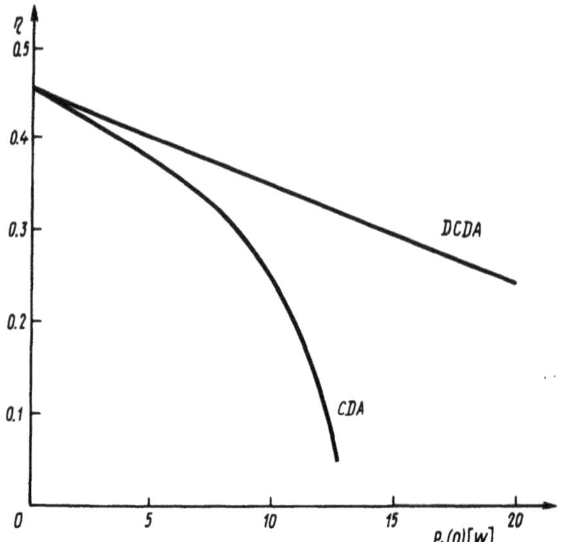

Fig. 2.14. SHG conversion efficiency versus the average power of fundamental laser radiation in 3 cm long CDA and DCDA crystals ($\lambda_1 = 1.06$ μm)

2.13 Wave Mismatch and Phase-Matching Bandwidth

In real frequency converters the situation is far from ideal: the convertible radiation is not a plane wave – i.e., it is divergent, pulse and nonmonochromatic – and the temperature of the crystal converter is unstable. Therefore, in practice we must calculate the following parameters of nonlinear frequency converters: *angular, spectral,* and *temperature bandwidths* corresponding to maximum permissible divergence, spectral width of the convertible radiation, and instability of temperature.

The value Δk is a function of crystal temperature T, frequencies of the interacting waves v_n, and deviation from the phase-matching angle $\delta\theta = \theta - \theta_{\mathrm{pm}}$. The dependence of Δk of these parameters in the first (linear) approximation is determined by first derivatives:

$$\Delta k(T, \delta\theta, v) \simeq \Delta k(0) + \frac{\partial(\Delta k)}{\partial T}\Delta T + \frac{\partial(\Delta k)}{\partial(\delta\theta)}\Delta\theta + \frac{\partial(\Delta k)}{\partial v}\Delta v , \tag{2.103}$$

where $\Delta k(0)$ is the mismatch for the exact phase matching (therefore $\Delta k(0) = 0$), and partial derivatives with respect to one argument are taken under the condition that the other two arguments are constant.

We show that the power of the resulting radiation in the *fixed-field approximation* is halved if wave mismatch is equal to

$$\Delta k = 0.886\frac{\pi}{L} . \tag{2.104}$$

This makes possible the evaluation of the angular ($\Delta\theta$), temperature (ΔT), and spectral (Δv) bandwidths:

Table 2.13. Equations for calculating the SHG internal angular bandwidth for the different types of interaction

Type of interaction	Internal angular bandwidth for SHG ($\omega_1 + \omega_1 = \omega_2$)
ooe	$\Delta\theta = \dfrac{0.443\lambda_1 \left[1 + (n_{o2}/n_{e2})^2 \tan^2\theta\right]}{L\tan\theta \left|1 - (n_{o2}/n_{e2})^2\right| n_2^e(\theta)}$
eoe, oee	$\Delta\theta = \dfrac{0.886}{L\tan\theta} \left| \dfrac{n_1^e(\theta)\left[1 - (n_{o1}/n_{e1})^2\right]}{\lambda_1\left[1 + (n_{o1}/n_{e1})^2 \tan^2\theta\right]} - \dfrac{n_2^e(\theta)\left[1 - (n_{o2}/n_{e2})^2\right]}{\lambda_2\left[1 + (n_{o2}/n_{e2})^2 \tan^2\theta\right]} \right|^{-1}$
eeo	$\Delta\theta = \dfrac{0.443\lambda_1 \left[1 + (n_{o1}/n_{e1})^2 \tan^2\theta\right]}{L\tan\theta \left[1 - (n_{o1}/n_{e1})^2\right] n_1^e(\theta)}$
eoo, oeo	$\Delta\theta = \dfrac{0.886\lambda_1 \left[1 + (n_{o1}/n_{e1})^2 \tan^2\theta\right]}{L\tan\theta \left[1 - (n_{o1}/n_{e1})^2\right] n_1^e(\theta)}$

Table 2.14. Equations for calculating the SFG internal angular bandwidth for the different types of interaction

Type of interaction	Internal angular bandwidth for SFG ($\omega_1 + \omega_2 = \omega_3$)
ooe	$\Delta\theta = \dfrac{0.886\lambda_3 \left[1 + (n_{o3}/n_{e3})^2 \tan^2\theta\right]}{L\tan\theta \left|1 - (n_{o3}/n_{e3})^2\right| n_3^e(\theta)}$
eoe	$\Delta\theta = \dfrac{0.886}{L\tan\theta} \left| \dfrac{n_1^e(\theta)\left[1 - (n_{o1}/n_{e1})^2\right]}{\lambda_1\left[1 + (n_{o1}/n_{e1})^2 \tan^2\theta\right]} - \dfrac{n_3^e(\theta)\left[1 - (n_{o3}/n_{e3})^2\right]}{\lambda_3\left[1 + (n_{o3}/n_{e3})^2 \tan^2\theta\right]} \right|^{-1}$
oee	$\Delta\theta = \dfrac{0.886}{L\tan\theta} \left| \dfrac{n_2^e(\theta)\left[1 - (n_{o2}/n_{e2})^2\right]}{\lambda_2\left[1 + (n_{o2}/n_{e2})^2 \tan^2\theta\right]} - \dfrac{n_3^e(\theta)\left[1 - (n_{o3}/n_{e3})^2\right]}{\lambda_3\left[1 + (n_{o3}/n_{e3})^2 \tan^2\theta\right]} \right|^{-1}$
eeo	$\Delta\theta = \dfrac{0.886}{L\tan\theta} \left\{ \dfrac{n_1^e(\theta)\left[1 - (n_{o1}/n_{e1})^2\right]}{\lambda_1\left[1 + (n_{o1}/n_{e1})^2 \tan^2\theta\right]} + \dfrac{n_2^e(\theta)\left[1 - (n_{o2}/n_{e2})^2\right]}{\lambda_2\left[1 + (n_{o2}/n_{e2})^2 \tan^2\theta\right]} \right\}^{-1}$
eoo	$\Delta\theta = \dfrac{0.886\lambda_1 \left[1 + (n_{o1}/n_{e1})^2 \tan^2\theta\right]}{L\tan\theta \left[1 - (n_{o1}/n_{e1})^2\right] n_1^e(\theta)}$
oeo	$\Delta\theta = \dfrac{0.886\lambda_2 \left[1 + (n_{o2}/n_{e2})^2 \tan^2\theta\right]}{L\tan\theta \left[1 - (n_{o2}/n_{e2})^2\right] n_2^e(\theta)}$

Table 2.15. Equations for calculating the SHG internal angular 90° phase-matching bandwidth for all types of interaction

Types of Interaction	Internal angular 90° phase-matching bandwidth for SHG ($\omega_1 + \omega_1 = \omega_2$)
ooe	$\Delta\theta = 2\left(\dfrac{0.443\lambda_1}{Ln_{e2}[1 - (n_{e2}/n_{o2})^2]}\right)^{1/2}$
eoe oee	$\Delta\theta = 2\left(\dfrac{0.886}{L}\left|\dfrac{n_{e1}}{\lambda_1}\left[1 - \left(\dfrac{n_{e1}}{n_{o1}}\right)^2\right] - \dfrac{n_{e2}}{\lambda_2}\left[1 - \left(\dfrac{n_{e2}}{n_{o2}}\right)^2\right]\right|^{-1}\right)^{1/2}$
eeo	$\Delta\theta = 2\left(\dfrac{0.443\lambda_1}{Ln_{e1}\left|1 - (n_{e1}/n_{o1})^2\right|}\right)^{1/2}$
eoo oeo	$\Delta\theta = 2\left(\dfrac{0.886\lambda_1}{Ln_{e1}\left|1 - (n_{e1}/n_{o1})^2\right|}\right)^{1/2}$

Table 2.16. Equations for calculating the SFG internal angular 90° phase-matching bandwidth for all types of interaction

Types of interaction	Internal angular 90° phase-matching bandwidth for SFG ($\omega_1 + \omega_2 = \omega_3$)
ooe	$\Delta\theta = 2\left(\dfrac{0.886\lambda_3}{Ln_{e3}\left[1 - (n_{e3}/n_{o3})^2\right]}\right)^{1/2}$
eoe	$\Delta\theta = 2\left(\dfrac{0.886}{L}\left|\dfrac{n_{e1}}{\lambda_1}\left[1 - \left(\dfrac{n_{e1}}{n_{o1}}\right)^2\right] - \dfrac{n_{e3}}{\lambda_3}\left[1 - \left(\dfrac{n_{e3}}{n_{o3}}\right)^2\right]\right|^{-1}\right)^{1/2}$
oee	$\Delta\theta = 2\left(\dfrac{0.886}{L}\left|\dfrac{n_{e2}}{\lambda_2}\left[1 - \left(\dfrac{n_{e2}}{n_{o2}}\right)^2\right] - \dfrac{n_{e3}}{\lambda_3}\left[1 - \left(\dfrac{n_{e3}}{n_{o3}}\right)^2\right]\right|^{-1}\right)^{1/2}$
eeo	$\Delta\theta = 2\left(\dfrac{0.886}{L}\left|\dfrac{n_{e1}}{\lambda_1}\left[1 - \left(\dfrac{n_{e1}}{n_{o1}}\right)^2\right] + \dfrac{n_{e2}}{\lambda_2}\left[1 - \left(\dfrac{n_{e2}}{n_{o2}}\right)^2\right]\right|^{-1}\right)^{1/2}$
eoo	$\Delta\theta = 2\left(\dfrac{0.886\lambda_1}{Ln_{e1}\left|1 - (n_{e1}/n_{o1})^2\right|}\right)^{1/2}$
oeo	$\Delta\theta = 2\left(\dfrac{0.886\lambda_2}{Ln_{e2}\left|1 - (n_{e2}/n_{o2})^2\right|}\right)^{1/2}$

Table 2.17. Equations for calculating the SHG temperature band-
with for the different types of interaction

Type of interaction	Temperature bandwidth for SHG $\omega_1 + \omega_1 = \omega_2$
ooe	$\Delta T = \dfrac{0.443\lambda_1}{L}\left\|\dfrac{\partial n_{o1}}{\partial T} - \dfrac{\partial n_2^e(\theta)}{\partial T}\right\|^{-1}$
eoe, oee	$\Delta T = \dfrac{0.886\lambda_1}{L}\left\|\dfrac{\partial n_1^e(\theta)}{\partial T} + \dfrac{\partial n_{o1}}{\partial T} - \dfrac{2\partial n_2^e(\theta)}{\partial T}\right\|^{-1}$
eeo	$\Delta T = \dfrac{0.443\lambda_1}{L}\left\|\dfrac{\partial n_1^e(\theta)}{\partial T} - \dfrac{\partial n_{o2}}{\partial T}\right\|^{-1}$
eoo, oeo	$\Delta T = \dfrac{0.886\lambda_1}{L}\left\|\dfrac{\partial n_1^e(\theta)}{\partial T} + \dfrac{\partial n_{o1}}{\partial T} - \dfrac{2\partial n_{o2}}{\partial T}\right\|^{-1}$

$$\Delta\theta = 1.772\frac{\pi}{L}\left[\frac{\partial(\Delta k)}{\partial(\delta\theta)}\right]^{-1}_{\theta=\theta_{pm}}, \tag{2.105}$$

$$\Delta T = 1.772\frac{\pi}{L}\left[\frac{\partial(\Delta k)}{\partial T}\right]^{-1}_{T=T_{pm}}, \tag{2.106}$$

$$\Delta\nu = 1.772\frac{\pi}{L}\left[\frac{\partial(\Delta k)}{\partial\nu}\right]^{-1}_{\nu=\nu_{pm}}. \tag{2.107}$$

The derivatives used in (2.105–107) depend on the dispersion of the refractive indices and on the type of phase matching. Note that the expressions (2.104–107) are valid, in the strict sense, only in the *fixed-field approximation*, but nevertheless they can be successfully used for quantitative assessments.

Table 2.18. Equations for calculating the SFG temperature bandwidth for the different types of interaction

Type of interaction	Temperature bandwidth for SFG $\omega_1 + \omega_2 = \omega_3$
ooe	$\Delta T = \dfrac{0.886}{L}\left\|\dfrac{1}{\lambda_1}\dfrac{\partial n_{o1}}{\partial T} + \dfrac{1}{\lambda_2}\dfrac{\partial n_{o2}}{\partial T} - \dfrac{1}{\lambda_3}\dfrac{\partial n_3^e(\theta)}{\partial T}\right\|^{-1}$
eoe	$\Delta T = \dfrac{0.886}{L}\left\|\dfrac{1}{\lambda_1}\dfrac{\partial n_1^e(\theta)}{\partial T} + \dfrac{1}{\lambda_2}\dfrac{\partial n_{o2}}{\partial T} - \dfrac{1}{\lambda_3}\dfrac{\partial n_3^e(\theta)}{\partial T}\right\|^{-1}$
oee	$\Delta T = \dfrac{0.886}{L}\left\|\dfrac{1}{\lambda_1}\dfrac{\partial n_{o1}}{\partial T} + \dfrac{1}{\lambda_2}\dfrac{\partial n_2^e(\theta)}{\partial T} - \dfrac{1}{\lambda_3}\dfrac{\partial n_3^e(\theta)}{\partial T}\right\|^{-1}$
eeo	$\Delta T = \dfrac{0.886}{L}\left\|\dfrac{1}{\lambda_1}\dfrac{\partial n_1^e(\theta)}{\partial T} + \dfrac{1}{\lambda_2}\dfrac{\partial n_2^e(\theta)}{\partial T} - \dfrac{1}{\lambda_3}\dfrac{\partial n_{o3}}{\partial T}\right\|^{-1}$
eoo	$\Delta T = \dfrac{0.886}{L}\left\|\dfrac{1}{\lambda_1}\dfrac{\partial n_1^e(\theta)}{\partial T} + \dfrac{1}{\lambda_2}\dfrac{\partial n_{o2}}{\partial T} - \dfrac{1}{\lambda_3}\dfrac{\partial n_{o3}}{\partial T}\right\|^{-1}$
oeo	$\Delta T = \dfrac{0.886}{L}\left\|\dfrac{1}{\lambda_1}\dfrac{\partial n_{o1}}{\partial T} + \dfrac{1}{\lambda_2}\dfrac{\partial n_2^e(\theta)}{\partial T} - \dfrac{1}{\lambda_3}\dfrac{\partial n_{o3}}{\partial T}\right\|^{-1}$

Table 2.19. Equations for calculating the SHG spectral bandwidth for the different types of interaction

Type of interaction	Spectral bandwidth for SHG $(\omega_1 + \omega_1 = \omega_2)$
ooe	$\Delta\nu_1 = \dfrac{0.443}{\lambda_1 L}\left\|\dfrac{\partial n_{o1}}{\partial\lambda_1} - \dfrac{\partial n_2^e(\theta)}{\partial\lambda_2}\right\|^{-1}$
eoe, oee	$\Delta\nu_1 = \dfrac{0.886}{\lambda_1 L}\left\|\dfrac{\partial n_{o1}}{\partial\lambda_1} + \dfrac{\partial n_1^e(\theta)}{\partial\lambda_1} - 2\dfrac{\partial n_2^e(\theta)}{\partial\lambda_2}\right\|^{-1}$
eeo	$\Delta\nu_1 = \dfrac{0.443}{\lambda_1 L}\left\|\dfrac{\partial n_1^e(\theta)}{\partial\lambda_1} - \dfrac{\partial n_{o2}}{\partial\lambda_2}\right\|^{-1}$
eoo, oeo	$\Delta\nu_1 = \dfrac{0.886}{\lambda_1 L}\left\|\dfrac{\partial n_{o1}}{\partial\lambda_1} + \dfrac{\partial n_1^e(\theta)}{\partial\lambda_1} - 2\dfrac{\partial n_{o2}}{\partial\lambda_2}\right\|^{-1}$

Tables 2.13,14 contain the equations for calculating the internal (inside the crystal) angular bandwidth ($\Delta\theta$) for SHG and SFG. The equations used for SFG can also be applied to DFG if polarization designations for the interacting waves are made in the order of increasing frequency.

For $\theta_{pm} = 90°$ (90° *phase matching*) the first derivative $\partial(\Delta k)/\partial(\delta\theta)$ becomes equal to zero and the corresponding second derivative becomes important. Hence, the 90° phase-matching internal angular bandwidth is

$$\Delta\theta\big|_{\theta_{pm}=90°} \simeq 2\left[0.886\frac{\pi}{L}\left(\frac{\partial^2(\Delta k)}{\partial(\delta\theta)^2}\right)^{-1}\right]^{1/2} . \tag{2.108}$$

Table 2.20. Equations for calculating the SFG spectral bandwidth when the lower-frequency interacting wave has a wide-band spectrum

Type of interaction	Spectral bandwidth for SFG $(\omega_1 + \omega_2 = \omega_3)$ λ_1: wide-band spectrum; λ_2: fixed wavelength
ooe	$\Delta\nu_1 = \dfrac{0.886}{L}\left\|n_{o1} - n_3^e(\theta) - \lambda_1\dfrac{\partial n_{o1}}{\partial\lambda_1} + \lambda_3\dfrac{\partial n_3^e(\theta)}{\partial\lambda_3}\right\|^{-1}$
eoe	$\Delta\nu_1 = \dfrac{0.886}{L}\left\|n_1^e(\theta) - n_3^e(\theta) - \lambda_1\dfrac{\partial n_1^e(\theta)}{\partial\lambda_1} + \lambda_3\dfrac{\partial n_3^e(\theta)}{\partial\lambda_3}\right\|^{-1}$
oee	$\Delta\nu_1 = \dfrac{0.886}{L}\left\|n_{o1} - n_3^e(\theta) - \lambda_1\dfrac{\partial n_{o1}}{\partial\lambda_1} + \lambda_3\dfrac{\partial n_3^e(\theta)}{\partial\lambda_3}\right\|^{-1}$
eeo	$\Delta\nu_1 = \dfrac{0.886}{L}\left\|n_1^e(\theta) - n_{o3} - \lambda_1\dfrac{\partial n_1^e(\theta)}{\partial\lambda_1} + \lambda_3\dfrac{\partial n_{o3}}{\partial\lambda_3}\right\|^{-1}$
eoo	$\Delta\nu_1 = \dfrac{0.886}{L}\left\|n_1^e(\theta) - n_{o3} - \lambda_1\dfrac{\partial n_1^e(\theta)}{\partial\lambda_1} + \lambda_3\dfrac{\partial n_{o3}}{\partial\lambda_3}\right\|^{-1}$
oeo	$\Delta\nu_1 = \dfrac{0.886}{L}\left\|n_{o1} - n_{o3} - \lambda_1\dfrac{\partial n_{o1}}{\partial\lambda_1} + \lambda_3\dfrac{\partial n_{o3}}{\partial\lambda_3}\right\|^{-1}$

Table 2.21. Equations for calculating the SFG spectral bandwidth when the higher-frequency interacting wave has a wide-band spectrum

Type of interaction	Spectral bandwidth for SFG ($\omega_1 + \omega_2 = \omega_3$) λ_1: fixed wavelength; λ_2: wide-band spectrum
ooe	$\Delta v_2 = \dfrac{0.886}{L}\left\lvert n_{o2} - n_3^e(\theta) - \lambda_2\dfrac{\partial n_{o2}}{\partial\lambda_2} + \lambda_3\dfrac{\partial n_3^e(\theta)}{\partial\lambda_3}\right\rvert^{-1}$
eoe	$\Delta v_2 = \dfrac{0.886}{L}\left\lvert n_{o2} - n_3^e(\theta) - \lambda_2\dfrac{\partial n_{o2}}{\partial\lambda_2} + \lambda_3\dfrac{\partial n_3^e(\theta)}{\partial\lambda_3}\right\rvert^{-1}$
oee	$\Delta v_2 = \dfrac{0.886}{L}\left\lvert n_2^e(\theta) - n_3^e(\theta) - \lambda_2\dfrac{\partial n_2^e(\theta)}{\partial\lambda_2} + \lambda_3\dfrac{\partial n_3^e(\theta)}{\partial\lambda_3}\right\rvert^{-1}$
eeo	$\Delta v_2 = \dfrac{0.886}{L}\left\lvert n_2^e(\theta) - n_{o3} - \lambda_2\dfrac{\partial n_2^e(\theta)}{\partial\lambda_2} + \lambda_3\dfrac{\partial n_{o3}}{\partial\lambda_3}\right\rvert^{-1}$
eoo	$\Delta v_2 = \dfrac{0.886}{L}\left\lvert n_{o2} - n_{o3} - \lambda_2\dfrac{\partial n_{o2}}{\partial\lambda_2} + \lambda_3\dfrac{\partial n_{o3}}{\partial\lambda_3}\right\rvert^{-1}$
oeo	$\Delta v_2 = \dfrac{0.886}{L}\left\lvert n_2^e(\theta) - n_{o3} - \lambda_2\dfrac{\partial n_2^e(\theta)}{\partial\lambda_2} + \lambda_3\dfrac{\partial n_{o3}}{\partial\lambda_3}\right\rvert^{-1}$

For 90° phase matching the angular bandwidth of phase matching for SHG and SFG can be calculated by the equations given in Tables 2.15,16. Temperature and spectral bandwidths of phase matching are calculated by the

Table 2.22. Equations for calculating the angular tuning of phase matching in the case of SHG for the different types of interaction

Type of interaction	Angular tuning of phase matching for SHG ($\omega_1 + \omega_1 = \omega_2$)
ooe	$\dfrac{\partial v_1}{\partial\theta} = \dfrac{n_2^e(\theta)\left\lvert 1-(n_{o2}/n_{e2})^2\right\rvert\tan\theta}{\lambda_1^2\left[1+\left(\dfrac{n_{o2}}{n_{e2}}\right)^2\tan^2\theta\right]\left\lvert\dfrac{\partial n_{o1}}{\partial\lambda_1} - \dfrac{\partial n_2^e(\theta)}{\partial\lambda_2}\right\rvert}$
eoe, oee	$\dfrac{\partial v_1}{\partial\theta} = \dfrac{\left\lvert\dfrac{[1-(n_{o1}/n_{e1})^2]n_1^e(\theta)}{1+(n_{o1}/n_{e1})^2\tan^2\theta} - 2\dfrac{[1-(n_{o2}/n_{e2})^2]n_2^e(\theta)}{1+(n_{o2}/n_{e2})^2\tan^2\theta}\right\rvert\tan\theta}{\lambda_1^2\left\lvert\dfrac{\partial n_{o1}}{\partial\lambda_1} + \dfrac{\partial n_1^e(\theta)}{\partial\lambda_1} - \dfrac{2\partial n_2^e(\theta)}{\partial\lambda_2}\right\rvert}$
eeo	$\dfrac{\partial v_1}{\partial\theta} = \dfrac{n_1^e(\theta)\left[1-(n_{o1}/n_{e1})^2\right]\tan\theta}{\lambda_1^2\left[1+\left(\dfrac{n_{o1}}{n_{e1}}\right)^2\tan^2\theta\right]\left\lvert\dfrac{\partial n_1^e(\theta)}{\partial\lambda_1} - \dfrac{\partial n_{o2}}{\partial\lambda_2}\right\rvert}$
eoo, oeo	$\dfrac{\partial v_1}{\partial\theta} = \dfrac{n_1^e(\theta)\left[1-(n_{o1}/n_{e1})^2\right]\tan\theta}{\lambda_1^2\left[1+\left(\dfrac{n_{o1}}{n_{e1}}\right)^2\tan^2\theta\right]\left\lvert\dfrac{\partial n_{o1}}{\partial\lambda_1} + \dfrac{\partial n_1^e(\theta)}{\partial\lambda_1} - \dfrac{2\partial n_2^e(\theta)}{\partial\lambda_2}\right\rvert}$

Table 2.23. Equations for calculating the angular tuning of phase matching in the case of SFG when the lower-frequency interacting wave has a wide-band spectrum

Type of interaction	Angular tuning of phase matching for SFG ($\omega_1 + \omega_2 = \omega_3$) λ_1: wide-band spectrum; λ_2: fixed wavelength
ooe	$$\frac{\partial v_1}{\partial \theta} = \frac{n_3^{\mathrm{e}}(\theta)\left|1 - (n_{\mathrm{o}3}/n_{\mathrm{e}3})^2\right|\tan\theta}{\lambda_3\left[1 + \left(\frac{n_{\mathrm{o}3}}{n_{\mathrm{e}3}}\right)^2\tan^2\theta\right]\left|n_{\mathrm{o}1} - n_3^{\mathrm{e}}(\theta) - \lambda_1\frac{\partial n_{\mathrm{o}1}}{\partial\lambda_1} + \lambda_3\frac{\partial n_3^{\mathrm{e}}(\theta)}{\partial\lambda_3}\right|}$$
eoe	$$\frac{\partial v_1}{\partial \theta} = \frac{\left|\dfrac{n_1^{\mathrm{e}}(\theta)\left[1 - (n_{\mathrm{o}1}/n_{\mathrm{e}1})^2\right]}{\lambda_1\left[1 + (n_{\mathrm{o}1}/n_{\mathrm{e}1})^2\tan^2\theta\right]} - \dfrac{n_3^{\mathrm{e}}(\theta)\left[1 - (n_{\mathrm{o}3}/n_{\mathrm{e}3})^2\right]}{\lambda_3\left[1 + (n_{\mathrm{o}3}/n_{\mathrm{e}3})^2\tan^2\theta\right]}\right|\tan\theta}{\left|n_1^{\mathrm{e}}(\theta) - n_3^{\mathrm{e}}(\theta) - \lambda_1\dfrac{\partial n_1^{\mathrm{e}}(\theta)}{\partial\lambda_1} + \lambda_3\dfrac{\partial n_3^{\mathrm{e}}(\theta)}{\partial\lambda_3}\right|}$$
oee	$$\frac{\partial v_1}{\partial \theta} = \frac{\left|\dfrac{n_2^{\mathrm{e}}(\theta)\left[1 - (n_{\mathrm{o}2}/n_{\mathrm{e}2})^2\right]}{\lambda_2\left[1 + (n_{\mathrm{o}2}/n_{\mathrm{e}2})^2\tan^2\theta\right]} - \dfrac{n_3^{\mathrm{e}}(\theta)\left[1 - (n_{\mathrm{o}3}/n_{\mathrm{e}3})^2\right]}{\lambda_3\left[1 + (n_{\mathrm{o}3}/n_{\mathrm{e}3})^2\tan^2\theta\right]}\right|\tan\theta}{\left|n_{\mathrm{o}1} - n_3^{\mathrm{e}}(\theta) - \lambda_1\dfrac{\partial n_{\mathrm{o}1}}{\partial\lambda_1} + \lambda_3\dfrac{\partial n_3^{\mathrm{e}}(\theta)}{\partial\lambda_3}\right|}$$
eeo	$$\frac{\partial v_1}{\partial \theta} = \frac{\left\{\dfrac{n_1^{\mathrm{e}}(\theta)\left[1 - (n_{\mathrm{o}1}/n_{\mathrm{e}1})^2\right]}{\lambda_1\left[1 + (n_{\mathrm{o}1}/n_{\mathrm{e}1})^2\tan^2\theta\right]} + \dfrac{n_2^{\mathrm{e}}(\theta)\left[1 - (n_{\mathrm{o}2}/n_{\mathrm{e}2})^2\right]}{\lambda_2\left[1 + (n_{\mathrm{o}2}/n_{\mathrm{e}2})^2\tan^2\theta\right]}\right\}\tan\theta}{\left|n_1^{\mathrm{e}}(\theta) - n_{\mathrm{o}3} - \lambda_1\dfrac{\partial n_1^{\mathrm{e}}(\theta)}{\partial\lambda_1} + \lambda_3\dfrac{\partial n_{\mathrm{o}3}}{\partial\lambda_3}\right|}$$
eoo	$$\frac{\partial v_1}{\partial \theta} = \frac{n_1^{\mathrm{e}}(\theta)\left[1 - (n_{\mathrm{o}1}/n_{\mathrm{e}1})^2\right]\tan\theta}{\lambda_1\left[1 + \left(\frac{n_{\mathrm{o}1}}{n_{\mathrm{e}1}}\right)^2\tan^2\theta\right]\left|n_1^{\mathrm{e}}(\theta) - n_{\mathrm{o}3} - \lambda_1\frac{\partial n_1^{\mathrm{e}}(\theta)}{\partial\lambda_1} + \lambda_3\frac{\partial n_{\mathrm{o}3}}{\partial\lambda_3}\right|}$$
oeo	$$\frac{\partial v_1}{\partial \theta} = \frac{n_2^{\mathrm{e}}(\theta)\left[1 - (n_{\mathrm{o}2}/n_{\mathrm{e}2})^2\right]\tan\theta}{\lambda_2\left[1 + \left(\frac{n_{\mathrm{o}2}}{n_{\mathrm{e}2}}\right)^2\tan^2\theta\right]\left|n_{\mathrm{o}1} - n_{\mathrm{o}3} - \lambda_1\frac{\partial n_{\mathrm{o}1}}{\partial\lambda_1} + \lambda_3\frac{\partial n_{\mathrm{o}3}}{\partial\lambda_3}\right|}$$

equations presented in Tables 2.17–21. In Tables 2.22–27 the equations are given for calculation of the derivatives $\partial v/\partial\theta$ and $\partial v/\partial T$ describing *angular* and *temperature tuning*. These derivatives characterize a change of the convertible radiation frequency v with variations in the angle of temperature, respectively. Tables 2.20,21,23,24,26,27 contain the equations for SFG when one of the interacting waves has a wide-band spectrum.

Table 2.24. Equations for calculating the angular tuning of phase matching in the case of SFG when the higher-frequency interacting wave has a wide-band spectrum

Type of interaction	Angular tuning of phase matching for SFG ($\omega_1 + \omega_2 = \omega_3$) λ_1 : fixed wavelength; λ_2: wide-band spectrum
ooe	$$\frac{\partial v_2}{\partial \theta} = \frac{n_3^e(\theta)\left|1 - (n_{o3}/n_{e3})^2\right|\tan\theta}{\lambda_3\left(1+\dfrac{n_{o3}^2}{n_{e3}^2}\tan^2\theta\right)\left|n_{o2} - n_3^e(\theta) - \lambda_2\dfrac{\partial n_{o2}}{\partial\lambda_2} + \lambda_3\dfrac{\partial n_3^e(\theta)}{\partial\lambda_3}\right|}$$
eoe	$$\frac{\partial v_2}{\partial \theta} = \frac{\left|\dfrac{n_1^e(\theta)(1-n_{o1}^2/n_{e1}^2)}{\lambda_1[1+(n_{o1}/n_{e1})^2\tan^2\theta]} - \dfrac{n_3^e(\theta)(1-n_{o3}^2/n_{e3}^2)}{\lambda_3[1+(n_{o3}/n_{e3})^2\tan^2\theta]}\right|\tan\theta}{\left|n_{o2}-n_3^e(\theta)-\lambda_2\dfrac{\partial n_{o2}}{\partial\lambda_2}+\lambda_3\dfrac{\partial n_3^e(\theta)}{\partial\lambda_3}\right|}$$
oee	$$\frac{\partial v_2}{\partial \theta} = \frac{\left|\dfrac{n_2^e(\theta)(1-n_{o2}^2/n_{e2}^2)}{\lambda_1[1+(n_{o2}/n_{e2})^2\tan^2\theta]} - \dfrac{n_3^e(\theta)(1-n_{o3}^2/n_{e3}^2)}{\lambda_3[1+(n_{o3}/n_{e3})^2\tan^2\theta]}\right|\tan\theta}{\left|n_2^e(\theta)-n_3^e(\theta)-\lambda_2\dfrac{\partial n_2^e(\theta)}{\partial\lambda_2}+\lambda_3\dfrac{\partial n_3^e(\theta)}{\partial\lambda_3}\right|}$$
eeo	$$\frac{\partial v_2}{\partial \theta} = \frac{\left|\dfrac{n_1^e(\theta)(1-n_{o1}^2/n_{e1}^2)}{\lambda_1[1+(n_{o1}/n_{e1})^2\tan^2\theta]} - \dfrac{n_2^e(\theta)(1-n_{o2}^2/n_{e2}^2)}{\lambda_2[1+(n_{o2}/n_{e2})^2\tan^2\theta]}\right|\tan\theta}{\left|n_2^e(\theta)-n_{o3}-\lambda_2\dfrac{\partial n_2^e(\theta)}{\partial\lambda_2}+\lambda_3\dfrac{\partial n_{o3}}{\partial\lambda_3}\right|}$$
eoo	$$\frac{\partial v_2}{\partial \theta} = \frac{n_1^e(\theta)(1-n_{o1}^2/n_{e1}^2)\tan\theta}{\lambda_1\left(1+\dfrac{n_{o1}^2}{n_{e1}^2}\tan^2\theta\right)\left|n_{o2}-n_{o3}-\lambda_2\dfrac{\partial n_{o2}}{\partial\lambda_2}+\lambda_3\dfrac{\partial n_{o3}}{\partial\lambda_3}\right|}$$
oeo	$$\frac{\partial v_2}{\partial \theta} = \frac{n_2^e(\theta)(1-n_{o2}^2/n_{e2}^2)\tan\theta}{\lambda_1\left(1+\dfrac{n_{o2}^2}{n_{e2}^2}\tan^2\theta\right)\left|n_2^e(\theta)-n_{o3}-\lambda_2\dfrac{\partial n_2^e(\theta)}{\partial\lambda_2}+\lambda_3\dfrac{\partial n_{o3}}{\partial\lambda_3}\right|}$$

Table 2.25. Equations for calculating the temperature tuning of phase matching in the case of SHG for the different types of interaction

Type of interaction	Temperature tuning of phase matching for SHG ($\omega_1 + \omega_1 = \omega_2$)
ooe	$$\frac{\partial v_1}{\partial T} = \frac{\left|\partial n_{o1}/\partial T - \partial n_2^e(\theta)/\partial T\right|}{\lambda_1^2\left|\partial n_{o1}/\partial\lambda_1 - \partial n_2^e(\theta)/\partial\lambda_2\right|}$$
eoe, oee	$$\frac{\partial v_1}{\partial T} = \frac{\left|\partial n_{e1}/\partial T + \partial n_{o1}/\partial T - 2\partial n_2^e(\theta)/\partial T\right|}{\lambda_1^2\left|\partial n_1^e(\theta)/\partial\lambda_1 + \partial n_{o1}/\partial\lambda_1 - 2\partial n_2^e(\theta)/\partial\lambda_2\right|}$$
eeo	$$\frac{\partial v_1}{\partial T} = \frac{\left|\partial n_{e1}/\partial T - \partial n_{o2}/\partial T\right|}{\lambda_1^2\left|\partial n_1^e(\theta)/\partial\lambda_1 - \partial n_{o2}/\partial\lambda_2\right|}$$
eoo, oeo	$$\frac{\partial v_1}{\partial T} = \frac{\left|\partial n_1^e(\theta)/\partial T + \partial n_{o1}/\partial T - 2\partial n_{o2}/\partial T\right|}{\lambda_1^2\left|\partial n_1^e(\theta)/\partial\lambda_1 + \partial n_{o1}/\partial\lambda_1 - 2\partial n_{o2}/\partial\lambda_2\right|}$$

Table 2.26. Equations for calculating the temperature tuning of phase matching in the case of SFG when the lower-frequency interacting wave has a wide-band spectrum.

Type of interaction	Temperature tuning of phase matching for SFG $(\omega_1 + \omega_2 = \omega_3)$; λ_1: wide-band spectrum; λ_2: fixed wavelength
ooe	$$\frac{\partial v_1}{\partial T} = \frac{\left| \dfrac{1}{\lambda_1}\dfrac{\partial n_{o1}}{\partial T} + \dfrac{1}{\lambda_2}\dfrac{\partial n_{o2}}{\partial T} - \dfrac{1}{\lambda_3}\dfrac{\partial n_3^e(\theta)}{\partial T} \right|}{\left| n_{o1} - n_3^e(\theta) - \lambda_1 \dfrac{\partial n_{o1}}{\partial \lambda_1} + \lambda_3 \dfrac{\partial n_3^e(\theta)}{\partial \lambda_3} \right|}$$
eoe	$$\frac{\partial v_1}{\partial T} = \frac{\left| \dfrac{1}{\lambda_1}\dfrac{\partial n_1^e(\theta)}{\partial T} + \dfrac{1}{\lambda_2}\dfrac{\partial n_{o2}}{\partial T} - \dfrac{1}{\lambda_3}\dfrac{\partial n_3^e(\theta)}{\partial T} \right|}{\left| n_1^e(\theta) - n_3^e(\theta) - \lambda_1 \dfrac{\partial n_1^e(\theta)}{\partial \lambda_1} + \lambda_3 \dfrac{\partial n_3^e(\theta)}{\partial \lambda_3} \right|}$$
oee	$$\frac{\partial v_1}{\partial T} = \frac{\left| \dfrac{1}{\lambda_1}\dfrac{\partial n_{o1}}{\partial T} + \dfrac{1}{\lambda_2}\dfrac{\partial n_2^e(\theta)}{\partial T} - \dfrac{1}{\lambda_3}\dfrac{\partial n_3^e(\theta)}{\partial T} \right|}{\left| n_{o1} - n_3^e(\theta) - \lambda_1 \dfrac{\partial n_1^e(\theta)}{\partial \lambda_1} + \lambda_3 \dfrac{\partial n_3^e(\theta)}{\partial \lambda_3} \right|}$$
eeo	$$\frac{\partial v_1}{\partial T} = \frac{\left| \dfrac{1}{\lambda_1}\dfrac{\partial n_1^e(\theta)}{\partial T} + \dfrac{1}{\lambda_2}\dfrac{\partial n_2^e(\theta)}{\partial T} - \dfrac{1}{\lambda_3}\dfrac{\partial n_{o3}}{\partial T} \right|}{\left| n_1^e(\theta) - n_{o3} - \lambda_1 \dfrac{\partial n_1^e(\theta)}{\partial \lambda_1} + \dfrac{1}{\lambda_3}\dfrac{\partial n_{o3}}{\partial \lambda_3} \right|}$$
eoo	$$\frac{\partial v_1}{\partial T} = \frac{\left| \dfrac{1}{\lambda_1}\dfrac{\partial n_1^e(\theta)}{\partial T} + \dfrac{1}{\lambda_2}\dfrac{\partial n_{o2}}{\partial T} - \dfrac{1}{\lambda_3}\dfrac{\partial n_{o3}}{\partial T} \right|}{\left| n_1^e(\theta) - n_{o3} - \lambda_1 \dfrac{\partial n_1^e(\theta)}{\partial \lambda_1} + \dfrac{1}{\lambda_3}\dfrac{\partial n_{o3}}{\partial \lambda_3} \right|}$$
oeo	$$\frac{\partial v_1}{\partial T} = \frac{\left| \dfrac{1}{\lambda_1}\dfrac{\partial n_{o1}}{\partial T} + \dfrac{1}{\lambda_2}\dfrac{\partial n_2^e(\theta)}{\partial T} - \dfrac{1}{\lambda_3}\dfrac{\partial n_{o3}}{\partial T} \right|}{\left| n_{o1} - n_{o3} - \lambda_1 \dfrac{\partial n_{o1}}{\partial \lambda_1} + \dfrac{1}{\lambda_3}\dfrac{\partial n_{o3}}{\partial \lambda_3} \right|}$$

2.14 Calculation of Nonlinear Frequency-Conversion Efficiency in Some Special Cases

An accurate calculation of the frequency-converter efficiency in the general case with allowance to the accompanying factors is very complex. The analytical solving can be derived only for some simple special cases, but they can be used for evaluating the limiting efficiency of nonlinear frequency converters. Below, some analytical equations are given for calculating the conversion efficiency.

Table 2.27. Equations for calculating the temperature tuning of phase matching in the case of SFG when the higher-frequency interacting wave has a wide-band spectrum

Type of interaction	Temperature tuning of phase matching for SFG $(\omega_1 + \omega_2 = \omega_3)$; λ_1: fixed wavelength; λ_2: wide-band spectrum
ooe	$$\frac{\partial v_2}{\partial T} = \frac{\left|\dfrac{1}{\lambda_1}\dfrac{\partial n_{o1}}{\partial T} + \dfrac{1}{\lambda_2}\dfrac{\partial n_{o2}}{\partial T} - \dfrac{1}{\lambda_3}\dfrac{\partial n_3^e(\theta)}{\partial T}\right|}{\left|n_{o2} - n_3^e(\theta) - \lambda_2\dfrac{\partial n_{o2}}{\partial \lambda_2} + \lambda_3\dfrac{\partial n_3^e(\theta)}{\partial \lambda_3}\right|}$$
eoe	$$\frac{\partial v_2}{\partial T} = \frac{\left|\dfrac{1}{\lambda_1}\dfrac{\partial n_1^e(\theta)}{\partial T} + \dfrac{1}{\lambda_2}\dfrac{\partial n_{o2}}{\partial T} - \dfrac{1}{\lambda_3}\dfrac{\partial n_3^e(\theta)}{\partial T}\right|}{\left|n_{o2} - n_3^e(\theta) - \lambda_2\dfrac{\partial n_{o2}}{\partial \lambda_2} + \lambda_3\dfrac{\partial n_3^e(\theta)}{\partial \lambda_3}\right|}$$
oee	$$\frac{\partial v_2}{\partial T} = \frac{\left|\dfrac{1}{\lambda_1}\dfrac{\partial n_{o1}}{\partial T} + \dfrac{1}{\lambda_2}\dfrac{\partial n_2^e(\theta)}{\partial T} - \dfrac{1}{\lambda_3}\dfrac{\partial n_3^e(\theta)}{\partial T}\right|}{\left|n_2^e(\theta) - n_3^e(\theta) - \lambda_2\dfrac{\partial n_2^e(\theta)}{\partial \lambda_2} + \lambda_3\dfrac{\partial n_3^e(\theta)}{\partial \lambda_3}\right|}$$
eeo	$$\frac{\partial v_2}{\partial T} = \frac{\left|\dfrac{1}{\lambda_1}\dfrac{\partial n_1^e(\theta)}{\partial T} + \dfrac{1}{\lambda_2}\dfrac{\partial n_2^e(\theta)}{\partial T} - \dfrac{1}{\lambda_3}\dfrac{\partial n_{o3}}{\partial T}\right|}{\left|n_2^e(\theta) - n_{o3} - \lambda_2\dfrac{\partial n_2^e(\theta)}{\partial \lambda_2} + \lambda_3\dfrac{\partial n_{o3}}{\partial \lambda_3}\right|}$$
eoo	$$\frac{\partial v_2}{\partial T} = \frac{\left|\dfrac{1}{\lambda_1}\dfrac{\partial n_1^e(\theta)}{\partial T} + \dfrac{1}{\lambda_2}\dfrac{\partial n_{o2}}{\partial T} - \dfrac{1}{\lambda_3}\dfrac{\partial n_{o3}}{\partial T}\right|}{\left|n_{o2} - n_{o3} - \lambda_2\dfrac{\partial n_{o2}}{\partial \lambda_2} + \lambda_3\dfrac{\partial n_{o3}}{\partial \lambda_3}\right|}$$
oeo	$$\frac{\partial v_2}{\partial T} = \frac{\left|\dfrac{1}{\lambda_1}\dfrac{\partial n_{o1}}{\partial T} + \dfrac{1}{\lambda_2}\dfrac{\partial n_2^e(\theta)}{\partial T} - \dfrac{1}{\lambda_3}\dfrac{\partial n_{o3}}{\partial T}\right|}{\left|n_2^e(\theta) - n_{o3} - \lambda_2\dfrac{\partial n_2^e(\theta)}{\partial \lambda_2} + \lambda_3\dfrac{\partial n_{o3}}{\partial \lambda_3}\right|}$$

2.14.1 Plane-Wave Fixed-Field Approximation

In this approximation we can neglect such restricting factors as diffraction, anisotropy, group-velocity mismatch, and dispersive spreading. In addition, we neglect heat effects, linear and nonlinear absorption (and hence the fcg-effect). In other words, in this approximation the following conditions must be fulfilled:

$$L < L_{NL} , \tag{2.109}$$
$$L < L_{eff} , \tag{2.110}$$
$$P_0 < P_{cr} , \tag{2.111}$$
$$2\hbar\omega_3 < E_g . \tag{2.112}$$

In particular, (2.109) means for SFG (it is the *fixed-field approximation*)

$$a_3(z) \ll a_{1,2}(0) \tag{2.113}$$

for DFG,

$$a_1(z) \ll a_{2,3}(0) \tag{2.114}$$

or

$$a_2(z) \ll a_{1,3}(0) \tag{2.115}$$

and for SHG,

$$a_3(z) \ll a_1(0) = a_2(0) \ . \tag{2.116}$$

The conditions (2.110–112) signify the *plane-wave approximation* and together with (2.109) the *plane-wave fixed-field approximation*. We assume also that the temporal and spatial distributions of beam and pulse are homogenous.

Tables 2.28,29 illustrate the equations for calculating the conversion efficiency for SHG, SFG and DFG with the use of (2.94–97) in the SI and CGS systems. Here d_{eff} is the *effective nonlinearity* in the phase-matching direction (see Tables 2.3,14); n_i are the refractive indices at wavelengths λ_i in the phase-matching direction with allowance for wave polarizations; $A = \pi w_0^2$ is the cross-sectional area of the laser beam with a radius w_0 (the areas of the beams of all interacting waves are assumed to be equal); and P_i are the powers of the corresponding waves with frequencies ω_i. For pulsed (quasi-static) irradiation, by P_i we mean the pulse powers

$$P_i = E_i \tau_i^{-1} \ , \tag{2.117}$$

where E_i are the pulse energies at frequencies ω_i, and τ_i are the corresponding pulse durations. When the powers of the mixing waves are almost equal, the conversion efficiency for SFG is determined by the equation

Table 2.28. Equations for calculating SHG, SFG, and DFG conversion efficiency in the plane-wave fixed-field approximation in the SI system

Nonlinear process	Conversion efficiency
SHG $\omega_1 + \omega_1 = \omega_2$ $\Delta k = 2k_1 - k_2$	$\dfrac{P_2}{P_1} = \dfrac{2\pi^2 d_{\text{eff}}^2 L^2 P_1}{\varepsilon_o c n_1^2 n_2 \lambda_2^2 A} \operatorname{sinc}^2(|\Delta k| L / 2)$
SFG $\omega_1 + \omega_2 = \omega_3$ $\Delta k = k_1 + k_2 - k_3$	$\dfrac{P_3}{P_1} = \dfrac{2^3 \pi^2 d_{\text{eff}}^2 L^2 P_2}{\varepsilon_o c n_1 n_2 n_3 \lambda_3^2 A} \operatorname{sinc}^2(|\Delta k| L / 2)$
DFG $\omega_1 = \omega_3 - \omega_2$ $\Delta k = k_1 + k_2 - k_3$	$\dfrac{P_1}{P_3} = \dfrac{2^3 \pi^2 d_{\text{eff}}^2 L^2 P_2}{\varepsilon_o c n_1 n_2 n_3 \lambda_1^2 A} \operatorname{sinc}^2(|\Delta k| L / 2)$

$[d_{\text{eff}}] = \text{m/V}; \ [P] = \text{W}; \ [L] = \text{m}; \ [\lambda] = \text{m}; \ [A] = \text{m}^2;$
$\varepsilon_o = 8.854 \times 10^{-12} \text{ As/Vm}; \ c = 3 \times 10^8 \text{m/s}$

Table 2.29. Equations for calculating SHG, SFG, and DFG conversion efficiency in the plane-wave fixed-field approximation in the CGS system

Nonlinear process	Quantum conversion efficiency
SHG $\omega_1 + \omega_1 = \omega_2$ $\Delta k = 2k_1 - k_2$	$\dfrac{P_2}{P_1} = \dfrac{2^7 \pi^5 d_{\mathrm{eff}}^2 L^2 P_1}{c n_1^2 n_2 \lambda_2^2 A}\,\mathrm{sinc}^2(|\Delta k|L/2)$
SFG $\omega_1 + \omega_2 = \omega_3$ $\Delta k = k_1 + k_2 - k_3$	$\dfrac{P_3}{P_1} = \dfrac{2^9 \pi^5 d_{\mathrm{eff}}^2 L^2 P_2}{c n_1 n_2 n_3 \lambda_3^2 A}\,\mathrm{sinc}^2(|\Delta k|L/2)$
DFG $\omega_1 = \omega_3 - \omega_2$ $\Delta k = k_1 + k_2 - k_3$	$\dfrac{P_1}{P_3} = \dfrac{2^9 \pi^5 d_{\mathrm{eff}}^2 L^2 P_2}{c n_1 n_2 n_3 \lambda_1^2 A}\,\mathrm{sinc}^2(|\Delta k|L/2)$ $[d_{\mathrm{eff}}] = \mathrm{cm/dyn}^{1/2}; [P] = \mathrm{erg/s}; [L] = \mathrm{cm}; [\lambda] = \mathrm{cm};$ $[A] = \mathrm{cm}^2; c = 3 \times 10^{10}\,\mathrm{cm/s}$

$$\eta = \frac{P_3}{\sqrt{P_1 P_2}}\,, \tag{2.118}$$

and for DFG by

$$\eta = \frac{P_1}{\sqrt{P_2 P_3}}\,, \tag{2.119}$$

It follows from Tables 2.28,29 that the conversion efficiency is proportional to the pump power density, to the square of the crystal length, to the square of the effective nonlinearity [i.e., to the *"quality parameter"* of a nonlinear crystal $d_{\mathrm{eff}}^2 (n_1 n_2 n_3)^{-1}$] and also the term

$$\mathrm{sinc}^2 1/2|\Delta k|L = \frac{\sin^2(1/2\|L)}{(1/2|\Delta k|L)^2} \tag{2.120}$$

characterizing the effect of the wave mismatch on the conversion efficiency. When $\Delta k = \pm 0.886\,\pi/L$, the conversion efficiency is halved, as in the derivation of (2.105–107).

The conversion-efficiency equation for SFG, after substituting $\omega_1 = \omega_2$, $\omega_3 = 2\omega_1$, and $P_1 = P_2$ into it, gives a conversion-efficiency value four times more than that obtained by the equation for SHG. This paradox results because SHG has only one wave at frequency ω_1, whereas, in SFG, at $\omega_1 = \omega_2$ two waves with the same power interact. The resulting radiation intensity for SFG is determined by the product of the powers of the convertible waves, giving the factor 4 in the SFG equation.

Table 2.30 gives the equations for the *quantum conversion efficiency* (i.e., for the ratio of the quantum numbers of the resulting and initial radiations) in the case of exact phase matching ($\Delta k = 0$) for SFG and DFG. These equations can be used for evaluating the length of a nonlinear crystal required to attain 100% quantum conversion efficiency.

Table 2.30. Equations for calculating the quantum conversion efficiency in the case of exact phase matching for SFG and DFG in the SI and CGS systems

Nonlinear process	Quantum conversion efficiency	
SFG	$\dfrac{P_3\lambda_3}{P_1\lambda_1} = \sin^2\left(2\pi d_{\mathrm{eff}}L\sqrt{\dfrac{2P_2}{\varepsilon_o cn_1 n_2 n_3 \lambda_1 \lambda_3 A}}\right)$	SI system
$\omega_1 + \omega_2 = \omega_3$	$\dfrac{P_3\lambda_3}{P_1\lambda_1} = \sin^2\left(16\pi^2 d_{\mathrm{eff}}L\sqrt{\dfrac{2\pi P_2}{cn_1 n_2 n_3 \lambda_1 \lambda_3 A}}\right)$	CGS system
DFG	$\dfrac{P_1\lambda_1}{P_3\lambda_3} = \sin^2\left(2\pi d_{\mathrm{eff}}L\sqrt{\dfrac{2P_2}{\varepsilon_o cn_1 n_2 n_3 \lambda_1 \lambda_3 A}}\right)$	SI system
$\omega_1 = \omega_3 - \omega_2$	$\dfrac{P_1\lambda_1}{P_3\lambda_3} = \sin^2\left(16\pi^2 d_{\mathrm{eff}}L\sqrt{\dfrac{2\pi P_2}{cn_1 n_2 n_3 \lambda_1 \lambda_3 A}}\right)$	CGS system

2.14.2 Fundamental Wave Depletion ("Nonlinear Regime")

At an arbitrary L/L_{NL} ratio we must take into account the power of exhaustion of the wave at the fundamental frequency (fundamental wave depletion). In some cases we can derive analytical equations for calculating the conversion efficiency in the *nonlinear regime*, using the *plane-wave approximation*.

The power-conversion efficiency for SHG with exact phase matching (zero mismatch) for the initial condition $a_3(0) = 0$, $a_{1,2}(0) = a_0$ is given by

$$\eta(L) = P_3(L)/P_1(0) = \tanh^2(\sigma_3 a_0 L) = \tanh^2(L/L_{\mathrm{NL}}) , \qquad (2.121)$$

where $\tanh x$ is the *hyperbolic tangent* (a tabulated special function) [2.31]. At nonzero mismatch in the same approximations we have

$$\eta(L) = \kappa \, \mathrm{sn}^2(u; \kappa) . \qquad (2.122)$$

Here $\mathrm{sn}(u; \kappa)$ is the *elliptic two-parametric sine* (a tabulated special *Jacoby function* [2.31]),

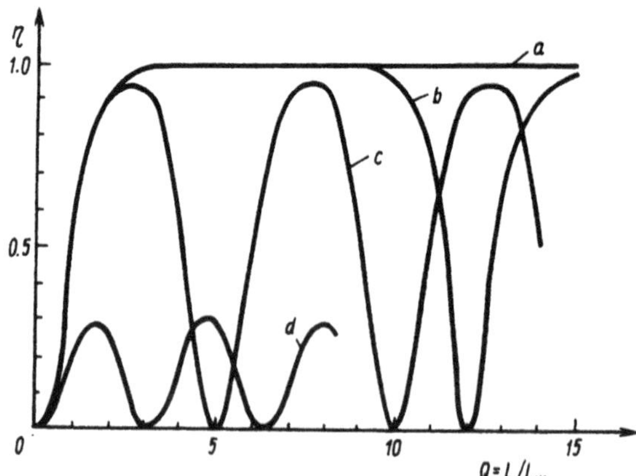

Fig. 2.15. SHG conversion efficiency versus L/L ratio for different κ^2 values: (a) $\kappa^2 = 1$, (b) $\kappa^2 = 1 - 10^{-4}$, (c) $\kappa^2 = 0.9$, (d) $\kappa^2 = 0.1$

$$u = \sigma_3 a_0 L \kappa^{-1/2} = \kappa^{-1/2} L / L_{\mathrm{NL}} \ , \tag{2.123}$$

$$\kappa = [(1 + \Delta_0^2/4)^{1/2} - \Delta_0/2]^2 \ , \tag{2.124}$$

$$\Delta_0 = \Delta k (2\sigma_3 a_0)^{-1} = 1/2 \Delta k L_{\mathrm{NL}} \ . \tag{2.125}$$

If $\Delta k = 0$, then $\Delta_0 = 0$, $\kappa = 1$ and the *elliptic sine* transforms into a *hyperbolic tangent* (Fig. 2.15a). If $\Delta_0 \gg 1$, then $\kappa \to 0$, and the *elliptic sine* transforms into an ordinary *circular sine* (Fig. 2.15d). The procedure for calculation of the SHG efficiency by (2.121–125) is as follows. First L_{NL} is calculated and then Δ_0 is evaluated, using (2.125). If $\Delta_0 \ll 1$, (2.121) or plot a in Fig. 2.15 can be used. If $\Delta_0 \gg 1$, the calculation is performed with the equation given in tables 2.28,29 or with equivalent equation

$$\eta(L) = (L/L_{\mathrm{NL}})^2 \ \mathrm{sinc}^2(1/2|\Delta k|L) \ . \tag{2.126}$$

To calculate L_{NL}, we must determine a_0 and $\sigma_3 [L_{\mathrm{NL}} = 1/(\sigma_3 a_0)]$ found from the known power $P_1(0)$ of the radiation at the fundamental frequency (fundamental radiation) by the equation

$$a_0 = \left[\frac{752 P_1(0)}{\pi w_0^2 n} \right]^{1/2} \ , \tag{2.126}$$

where the following dimensions of the quantities are used: $[a_0] = \mathrm{V\,cm^{-1}}$; $[w_0] = \mathrm{cm}$; and $[P_1(0)] = \mathrm{W}$. The parameter σ_3 is calculated in accordance with (2.59):

$$\sigma_3 = 8\pi^2 d_{\mathrm{eff}}/n\lambda_1 \ , \tag{2.127}$$

where the following dimensions of the quantities are used: $[d_{\mathrm{eff}}] = \mathrm{mV^{-1}}$, $[\sigma_3] = \mathrm{V^{-1}}$, $[\lambda_1] = \mathrm{m}$.

For exact phase matching where SHG is described by (2.121), 100% conversion to the second harmonic can be realized at $L \geq 1.5 L_{\mathrm{NL}}$ (Fig. 2.15a). For a KDP crystal, for instance, when the input power density of the fundamental radiation $P_1(0)/\pi w_0^2 = 5 \times 10^8 \ \mathrm{Wcm^{-2}}$, $\lambda_1 = 10^{-4} \mathrm{cm}$ (neodymium laser), and $n = 1.5$ we have $a_0 = 5 \times 10^5 \ \mathrm{Vcm^{-1}}$, $\sigma_3 \simeq 10^{-6} \mathrm{V^{-1}}$, and $L_{\mathrm{NL}} = 2 \ \mathrm{cm}$. Thus, 100% conversion to the second harmonic is possible when $L = 3 \ \mathrm{cm}$ in the absence of all limiting factors (except that the fundamental wave depletion is taken into account). Of course, to attain this limiting efficiency, both the fundamental radiation and the crystal converter must have ideal parameters.

Note that when $\kappa^2 \leq 0.3$ or, which is the same, $\Delta_0 \geq 0.6$, the elliptic sine in (2.122) may be replaced with a circular sine

$$\eta(L) \simeq \kappa \ \sin^2 u \ , \tag{2.129}$$

where κ and u are calculated by (2.123–125). The use of (2.129) instead of (2.122) gives a relative error no greater than 10%. This approximation is analogous to the so-called *fixed-intensity approximation* (Sect. 2.14.5).

2.14.3 SHG of a Divergent Fundamental Radiation Beam in the Fixed-Field Approximation

The equations of the previous sections are suitable for plane waves. In the geometrical approximation these equations can be used for calculating the SHG efficiency for the divergent beam at the fundamental frequency with an axisymmetric divergence angle

$$\phi_0 \gg \phi_{\text{diff}} = 0.61 \, \lambda_1 w_0^{-1} \, , \tag{2.130}$$

where ϕ_{diff} is the *diffractive-divergence angle*. In the approximation of geometrical optics the SHG processes in the partial beams are independent. Each partial beam propagates in its own direction (i.e., at its own angle θ to the axis Z inside the angle ϕ_0). Therefore, each value of θ may be related to the corresponding mismatch value [2.4]:

$$\Delta k(\theta) = \left. \frac{\partial(\Delta k)}{\partial \theta} \right|_{\theta=\theta_{\text{pm}}} (\theta - \theta_{\text{pm}}) = \gamma_1(\theta - \theta_{\text{pm}}) \, . \tag{2.131}$$

Here γ_1 is the *angular dispersive coefficient* of the first order; near 90° phase matching the angular dispersive coefficient of the second order should be taken into account (2.108). For SHG under type I phase matching ("ooe") the parameter γ_1 is calculated by

$$\gamma_1 = 2n_{\text{o}1}\omega_1[(n_{\text{o}1}^2 - n_{\text{e}3}^2)(n_{\text{o}3}^2 - n_{\text{e}3}^2)]^{1/2}(cn_{\text{o}3}n_{\text{e}3})^{-1} \, . \tag{2.132}$$

The integral SHG efficiency is found by integration of (2.126) over the angle within the divergence limit ϕ_0:

$$\begin{aligned}
\eta(L, \phi_0) &= P_3(L, \phi_0)/P_1(0) \\
&= (L/L_{\text{NL}})^2[\Omega^{-1}\text{Si}(2\Omega) - \Omega^{-2}\sin^2\Omega] \, . \tag{2.133}
\end{aligned}$$

Here $\Omega = \gamma_1\phi_0 L/4$ (the beam bisectrix is assumed to coincide with the phase-matching direction), and the function Si x is the *integral sine* (a tabulated special function [2.31]):

$$\text{Si}(x) = \int_0^x y^{-1}\sin y \, dy \, . \tag{2.134}$$

Here the limiting cases are:

1) $\Omega \ll 1$ (divergence, crystal length, and angular dispersive coefficient are small); then Si $(2\,\Omega) \simeq 2\,\Omega$, sinc $\Omega \simeq 1$, and

$$\eta(L, \Omega) \simeq (L/L_{\text{NL}})^2 \tag{2.135}$$

which coincides with the case of exact phase matching ($\Delta k = 0$).

2) $\Omega \gg 1$ (divergence, crystal length, and angular dispersive coefficient are large); then Si $(2\,\Omega) \simeq \pi/2$, sinc $\Omega \simeq 0$, and

$$\eta(L, \Omega) \simeq \frac{\pi}{2\,\Omega} \frac{L^2}{L_{\mathrm{NL}}^2} \; .$$ (2.136)

2.14.4 SHG of a Divergent Fundamental Radiation Beam in the Nonlinear Regime

The *nonlinear regime* in the plane-wave approximation for SHG is described by (2.122–125). Integration of (2.122) over the angle gives (here $\phi = \theta - \theta_{\mathrm{pm}}$)

$$\eta(L, \phi_0) = \phi_0^{-1} \int_{-\phi_0/2}^{+\phi_0/2} \kappa(\phi) \mathrm{sn}^2[u(\phi); \kappa(\phi)] \mathrm{d}\phi$$ (2.137)

In approximation (2.129), (2.137) acquires a form suitable for operative calculations:

$$\begin{aligned}
\eta(\Omega, Q) \simeq \frac{Q^2}{3\Omega} \Bigg\{ & Q - \frac{Q^2}{v} + \frac{1}{Q}\sin^2 Q + \frac{1}{2}\sin(2Q) \\
& - \frac{Q^2}{v^3}\left[\sin^2 v + \frac{v}{2}\sin(2v)\right] \\
& + (3 - 2Q^3)\left[\mathrm{Si}(2v) - \frac{1}{v}\sin^2 v - \mathrm{Si}(2Q) + \frac{1}{Q}\sin^2 Q\right] \Bigg\}
\end{aligned}$$ (2.138)

where

$$Q = L/L_{\mathrm{NL}}, \quad v = [Q^2 + (\Omega/2)^2]^{1/2} + \Omega/2 \; .$$ (2.139)

When $Q \ll 1$ or $\Omega/Q \gg 1$, (2.138) transforms into (2.133).

Note the following: using the approximation (2.129) valid for $\Delta_0 \geq 0.6$ with an accuracy of no less than 10% and integrating with respect to the angle $\phi = \theta - \theta_{\mathrm{pm}}$, we also include the region $[-0.6 \leq \Delta_0 \leq +0.6]$, in which (2.133) is incorrect. However, the contribution of this region to the error at $Q \leq 2$ is insignificant and the use of (2.138) gives an error of no more than 10%. The error increase to $25 - 30\%$ only in the region where large values of Q are combined with small values of Ω ($Q \geq 2$; $\Omega \leq 1, \ldots, 2$), i.e., where (2.138) is suitable only for estimates.

Figures 2.16,17 show the results of computation of (2.137) as curves of equal values of SHG efficiency in the coordinates Ω and Q. Figure 2.16 shows the results with respect to power for continuous-wave irradiation. Figure 2.17 shows the curves with respect to pulse energy for pulsed irradiation with a Gaussian shape of the pulse at the fundamental frequency. Since the parameters Ω and Q are proportional to the crystal length L, the plots in Figs. 2.16,17 remain unchanged when L varies (but the scale of the axis does change!). The efficiency values used in practice ($\eta \geq 0.2$) at $Q \approx 1$ are concentrated in the region of small Ω ($\Omega \leq 5, \ldots, 7$). From here the known recommendations follow (for the accepted approximations): to obtain a high

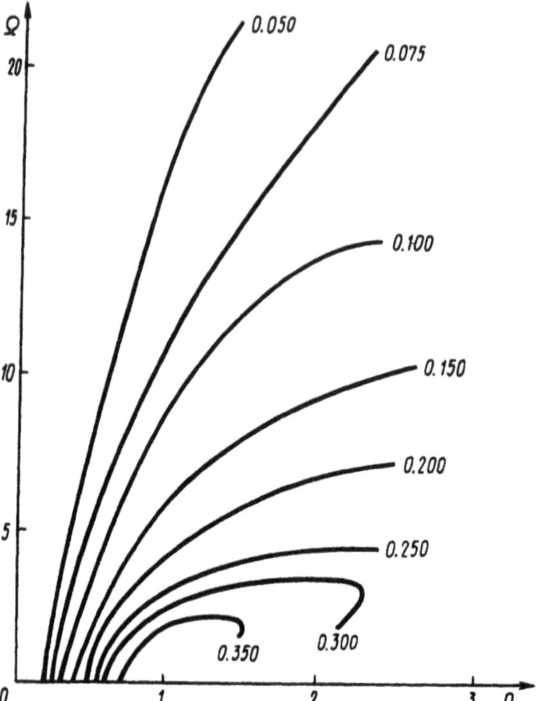

Fig. 2.16. Curves plotted for equal values of power-conversion efficiency in coordinates Ω, Q for SHG in the nonlinear regime under continuous-wave irradiation

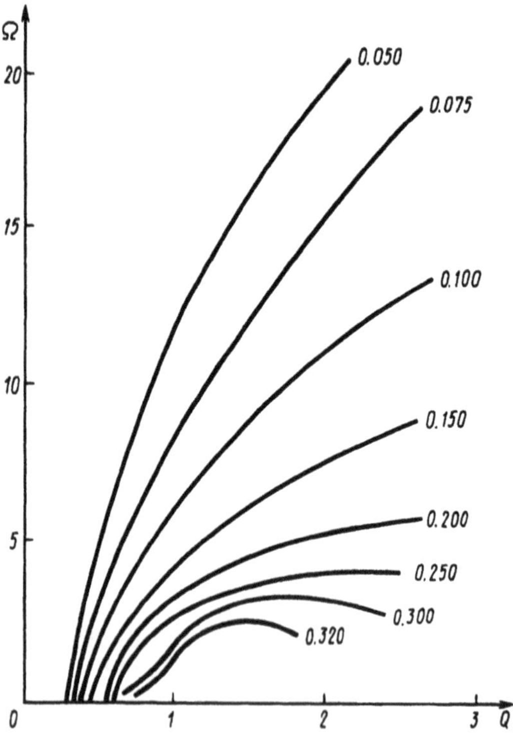

Fig. 2.17. Curves plotted for equal values of energy-conversion efficiency in coordinates Ω, Q for SHG in the nonlinear regime under pulse irradiation in approximation of Gaussian pulse shape of laser radiation at the fundamental frequency

SHG efficiency, one must use high densities of the fundamental radiation powers ($Q \geq 1$ up to the breakdown values) and low values of Ω ($\Omega < 5$). In other words, one must use crystals with small angular dispersive coefficients (crystals of CDA type) and decrease the divergence of the laser radiation as much as possible.

Note that in the last two subsections we neglect absorption (linear and nonlinear), diffraction, anisotropy, group-velocity mismatch, dispersive spreading of the pulse, thermal self-actions, photorefraction, and fcg-effect. The rigorous inclusion of these phenomena except in some of the simplest cases requires the numerical calculations.

2.14.5 Fixed-Intensity Approximation

The *fixed-field approximation* described above is widely used in nonlinear theory of frequency conversion for evaluation of efficiency. Evidently in this approximation the *complex amplitude* of fundamental radiation (of the "strong" wave) is constant; for SHG we have $a_1(z) = a_2(z) \equiv a(0)$, $\phi_1(z) = \phi_2(z) \equiv \phi(0)$. This approximation permits us to simplify the calculations, but here very important information loses about the nonlinear mechanism of interaction, i.e. about waves phases. In the *fixed-intensity approximation* only amplitude, but not phase, of the fundamental wave is assumed to be constant, therefore this approximation is more warranted [2.4]:

$$a_{1,2}(z) \equiv a(0); \quad \phi_{1,2}(z) \not\equiv \phi(0) \ . \tag{2.140}$$

Hence, in this approximation only *intensity* (i.e., square of the real amplitude) of a "strong" wave, but not it's *complex amplitude* is assumed to be constant. It is a next step for the derivation of an approximate analytical formula for more rigorous evaluation of conversion efficiency.

In the absence of losses, let us rewrite (2.75–77) for SHG in the *plane-wave approximation* in following way:

$$\frac{\mathrm{d}A_1}{\mathrm{d}z} = -\mathrm{j}\sigma_1 A_1^* A_3 \exp(-\mathrm{j}\Delta kz) \ , \tag{2.141}$$

$$\frac{\mathrm{d}A_3}{\mathrm{d}z} = -\mathrm{j}\sigma_3 A_1^2 \exp(+\mathrm{j}\Delta kz) \ . \tag{2.142}$$

Let us differentiate eqs. (2.141–142) with respect to z and introduce the *wave intensities* $I_1 = I_2$ and $I_3(\omega_3 = 2\omega_1)$:

$$I_1 = A_1 A_1^* = a_1^2; \quad I_3 = A_3 A_3^* = a_3^2 \ . \tag{2.143}$$

As a result, we will have instead of (2.141–142):

$$\frac{\mathrm{d}^2 A_1}{\mathrm{d}z^2} + \mathrm{j}\Delta k \frac{\mathrm{d}A_1}{\mathrm{d}z} - \sigma_1(\sigma_1 I_3 - \sigma_3 I_1) = 0 \ , \tag{2.144}$$

$$\frac{d^2 A_3}{dz^2} - j\Delta k \frac{dA_3}{dz} - 2\sigma_1 \sigma_3 A_3 I_1 = 0 \ . \tag{2.145}$$

In the *fixed-intensity approximation* we must assume that $I_1(z) \cong I_1(0)$, and, instead of (2.145), we will have

$$\frac{d^2 A_3}{dz^2} - j\Delta k \frac{dA_3}{dz} + 2\sigma_1 \sigma_3 I_1(0) A_3 = 0 \ . \tag{2.146}$$

Equation (2.146) can be solved analytically with boundary conditions $A_3(0) = 0$, and $(dA_3/dz)_{z=0} = -j\sigma_3 A_1^2(0)$:

$$A_3(z) = -j\sigma_3 A_1^2(0)z \ \exp(j\Delta kz/2) \ \mathrm{sinc}(\Lambda, z), \tag{2.147}$$

where

$$\Lambda = \left[(\Delta k/2)^2 + 2\sigma_1 \sigma_3 I_1(0) \right]^{1/2}$$

$$= (\Delta k/2) \left[1 + 8(\Delta k L_{\mathrm{NL}})^{-2} \right]^{1/2} \tag{2.148}$$

In the *fixed-field approximation* the corresponding equation for A_3 consists of value $(\Delta k z/2)$ instead of Λ in eq. (2.147) [2.4], see also (2.120). Therefore, the mismatch factor $\mathrm{sinc}^2 x = \sin^2 x/x^2$ in the *fixed-intensity approximation* depends not only on mismatch, but also on input laser intensity. With enough large mismatch

$$\Delta k \gg 2[2\sigma_1 \sigma_3 I_1(0)]^{1/2} \tag{2.149}$$

the fixed-intensity approximation coincides with the fixed-field approximation.

Equation (2.148) can be rewritten for real amplitudes and phases:

$$a_3(z) = \sigma_3 a_1^2(0)z \ \mathrm{sinc}(\Lambda z) \ , \tag{2.150}$$

$$\phi_3(z) = 2\phi_1(0) - \pi/2 + \Delta kz/2 \ . \tag{2.151}$$

The results of SHG efficiency calculation in *fixed-intensity approximation* practically coincide with the rigorous consideration in the nonlinear regime (both in the *plane-wave approximation*) under condition $\Lambda > \sigma_3 a_1(0)$. This inequality is not valid with small mismatch, but in any case the *fixed-intensity approximation* gives more accurate analytical estimations. For example, in this case the equation for the phase of the fundamental (laser) wave $\phi_1(z)$ can be derived:

$$\phi_1(z) = \phi_1(0) + \frac{\Delta k z \left[1 - \mathrm{sinc}(\Lambda z) \right]}{8 + (\Delta k)^2 / [\sigma_1 \sigma_3 I_1(0)]} \ . \tag{2.152}$$

It is seen from (2.152) that the dependence of *phase velocity* of the fundamental wave on intensity occurs, i.e., the *nonlinear self-action* in media with square nonlinearity takes place. This very important phenomena is absent in the *fixed-field approximation*.

It was shown before [2.4], that in the *plane-wave approximation* the solutions of rigorous nonlinear equations (for SHG efficiency) are coincident with the solutions in the *fixed-intensity approximation* with an accuracy determined by terms that are proportional to $(L/L_{NL})^6$. For the *fixed-field approximation* it is true only with an accuracy of an order of $(L/L_{NL})^4$. Dependences of SHG efficiency on L/L_{NL} for the rigorous *nonlinear regime* and for *fixed-intensity approximation* even for case $\Delta k = 0$ are practically the same up to $L/L_{NL} \approx 1$, whereas the *fixed-field approximation* gives a correct result only for $L/L_{NL} < 0.3$.

2.14.6 Frequency Conversion of Ultrashort Laser Pulses

When a laser operates in pulse regime, it is necessary to take into account *unsteady (nonstationary) phenomena* in nonlinear interaction, i.e., the relationships between crystal length L and two effective lengths: *quasi-static length* L_{qs} (2.92) and *dispersive spreading length* L_{dis} (2.95). We shall consider as an example the SHG process with complex amplitudes $A_{1,2}$ and frequencies $\omega_{1,2}(\omega_2 = 2\omega_1)$ for the case of enough wide beams $(L_a, L_{dif} \gg L)$. In the first approximation of *dispersion theory* [2.32], i.e., with $L_{dis} \gg L$, the following *truncated equations* can be written:

$$\frac{\partial A_1}{\partial z} + \frac{1}{u_1}\frac{\partial A_1}{\partial t} = -j\sigma_1 A_1^* A_2 \exp(j\Delta kz) \ , \tag{2.153}$$

$$\frac{\partial A_2}{\partial z} + \frac{1}{u_2}\frac{\partial A_2}{\partial t} = -j\sigma_2 A_1^2 \exp(-j\Delta kz) \ , \tag{2.154}$$

where $u_{1,2}$ - group velocities, $\sigma_1 \approx \sigma_2 = \sigma$ - nonlinear coefficients. There are two different cases:

1. $u_1 = u_2 = u$, i.e., *group-velocity matching* (or *quasi-static regime*, [2.32,33]) takes place. In this case the eqs. (2.153–154) have an analytical solutions; specifically, for $\Delta k = 0$:

$$a_1(t,\eta) = |A_1(t,\eta)| = a_1(\eta,0)\operatorname{sech}[\sigma a_1(\eta,0)z] \ , \tag{2.155}$$

$$a_2(t,\eta) = |A_2(t,\eta)| = a_1(\eta,0)\tanh[\sigma a_1(\eta,0)z] \ , \tag{2.156}$$

$$\phi_1(t,\eta) \equiv \phi_1(\eta), \quad \phi_2(t,\eta) = 2\phi_1(\eta) - \pi/2 \ , \tag{2.157}$$

where $\eta = t - z/u$.

In the *fixed-field approximation* $a_2(\eta,z) = \sigma a_1^2(\eta,0)z$, therefore the Gaussian laser pulse $a_1(\eta,0) = a_1(0,0)\exp(-t^2/2\tau_1^2)$ transforms to the Gaussian second-harmonic pulse $a_2(\eta,0) = a_2(0,0)\exp(-t^2/2\tau_2^2)$, where τ_1 and τ_2 are the pulse durations of the fundamental laser pulse and second-harmonic pulse, respectively. At low conversion efficiencies $\tau_2 = \tau_1/\sqrt{(2)}$, whereas at high ones (*nonlinear regime*) we obtain $\tau_2 \to \tau_1$.

Note, that for the *quasi-static regime* it is possible to use the results of theory in *plane-wave approximation* with subsequent integrating of these results with respect to the time (see above). The numerical results are the same as they were obtained by using (2.155,156).

2. $u_1 \neq u_2$, i.e., *group-velocity mismatch* (or *unsteady regime*) takes place. In this case the *inverse group-velocity mismatch* $v = u_2^{-1} - u_1^{-1}$ must be taken into account, and it is necessary to consider the relationship between L and L_{qs}. Note that (2.92) gives the dependence of L_{qs} on pulse duration τ_1, and it is valid only for the so-called *spectral limited pulses* ($\tau_1 \Delta\omega_1 \approx 1$, where $\Delta\omega_1$ is the spectral width of the pulse). The ultrashort pulses in many cases are *frequency-* (or *phase-*) *modulated*, hence the equality $\tau_1 \Delta\omega_1 \approx 1$ is not valid anymore. Therefore, in the general case, instead of (2.92), it is necessary to introduce a new *quasi-static length*

$$L'_{qs} = (|v| \, \Delta\omega_1)^{-1} \tag{2.158}$$

that characterizes the real length where the *group-velocity mismatch* can be ignored. Therefore, if the initial (laser) pulse is a *spectral-limited pulse*, then $L'_{qs} < L_{qs}$, and it is necessary to take into account the shortest effective length. If the following inequality is possible: $L'_{qs} < L < L_{qs}$, then it is easily seen that the evaluation of effective length by taking into the account only the laser pulse duration (but not the spectral width of the laser pulse) will lead to the incorrect results.

Under the condition $L < L_{qs} < L'_{qs}$ we have a *quasi-static regime* that is equivalent to the case $u_1 = u_2$ (see above). With $L > L_{qs}, L'_{qs}$ the nonlinear interaction is *nonstationary*, and in the *fixed-field approximation* the solution of (2.153,154) is [2.32]:

$$A_2(t,z) = -\mathrm{j}\sigma_2 \int_0^z A_1^2(t - z/u_2 + vx, 0) \, \exp(-\mathrm{j}\Delta kx)\mathrm{d}x. \tag{2.158}$$

The *spectral density* of second-harmonic radiation is equal to

$$S_2(\Omega, z) = \mathrm{sinc}^2[(v\Omega - \Delta k)z/2]S_2(\Omega) , \tag{2.159}$$

where

$$S_2(\Omega) = (\sigma_2 z)^2 \left| \int_{-\infty}^{\infty} A_{10}(\Omega - \Omega') A_{10}(\Omega') \, \mathrm{d}\Omega' \right|^2 \tag{2.160}$$

and $A_{10}(\Omega)$ is the Fourier spectrum of complex amplitude $A_1(t,0)$, $\mathrm{sinc}\,x = (\sin x)/x$. Equations (2.159,160) are valid for arbitrary forms of frequency (phase) modulation, as well for arbitrary pulse forms.

With $L \gg L'_{qs}$ the spectral width of the second-harmonic pulse is lower then for fundamental radiation one, therefore, the nonlinear process is accompanied by the strong spreading of the second-harmonic pulse. For a spectral-limited laser pulse the duration of a second-harmonic pulse is $\tau_2(z) \approx vz$, i.e. τ_2 is practically independent on τ_1. The second harmonic-spectrum has a maximum at $\omega_{max} = 2\omega_1 - v^{-1}\Delta k$, and its spectral width is equal to $\Delta\omega_2 = 2\pi(vz)^{-1}$. Therefore, we have a possibility to *tune* the average second-harmonic frequency by changing the Δk.

In *nonlinear regime* the situation is more complicated (see [2.32,34,35]). Some analytical results for the pulse shape, spectral width, and conversion efficiency can be successfully obtained in *fixed-intensity approximation* [2.32]. The existence of phase-modulation of the laser pulse prevents to optimal conversion due to the distortion of phase relations. In the general case (2.153,154) can be solved only by numerical calculation.

When $u_1 = u_2$, the influence of nonlinear media dispersion on conversion efficiency is connected with *dispersive spreading of pulses*, $L \approx L_{dis}$ (the *second approximation of dispersion theory* [2.32]). However, this situation usually doesn't take place in practice (may be only for $\tau_1 < 10^{-14}$s) because of the small value of *dispersive spreading* parameter g. It's typical value is about 10^{-27} $s^2 cm^{-1}$ [2.4] and even for $\tau = 10^{-13}$s we have $L_{dis} = \tau^2/g = 10$ cm.

Note that for effective frequency conversion of ultrashort laser pulses both the *group-velocity* and *phase matching* must be fulfilled together, but this is a very seldom case. It was found [2.33] that the *group-velocity matching* (simultaneously with collinear type $I^{(-)}$ *phase matching*) for the SHG process occurs only for the specific wavelengths of fundamental radiation: 1.04 µm for KDP, 2.04 µm for $LiNbO_3$ and 2.16 µm for $LiIO_3$.

The similar consideration of frequency conversion of laser ultrashort pulses can be done for SFG, DFG, optical parametric oscillators (OPO), Raman scattering, and so on [2.32,34,35].

For more detailed consideration of unsteady phenomena, taking into account both the diffraction and anisotropy, it is convenient to introduce the Fourier representations of the complex field amplitudes, i.e., to calculate the *spectra* of *spatial frequencies*. The rigorous calculation shows that the *spatial* (diffraction and anisotropy) and *temporal*, or *unsteady* (*group-velocity mismatch*) *phenomena* exert control over the total process of the frequency conversion and in this case we have the *nonlinear superposition* of both these effects.

2.14.7 Frequency Conversion of Laser Beams with Limited Aperture in the Stationary Regime

When the laser operates in the *stationary* (*continuous wave* or cw) *regime*, it is necessary first of all to take into account the limiting factors which are connected with *spatial beam modulation* and corresponding phenomena such as *diffraction* and *anisotropy*. Therefore, in this case $L_{qs} = L_{dis} = \infty$, $L_a \sim L$, and $L_{diff} \sim L$. Taking as an example the type $I^{(-)}$ SHG process in uniaxial crystal, let us write the corresponding *truncated equations*:

$$\frac{\partial A_1}{\partial z} + j\frac{1}{2k_1}\left(\frac{\partial A_1}{\partial x^2} + \frac{\partial A_1}{\partial y^2}\right) = -j\sigma_1 A_1^* A_2 \exp(-j\Delta kz) , \qquad (2.161)$$

$$\frac{\partial A_2}{\partial z} + j\frac{1}{2k_2}\left(\frac{\partial A_2}{\partial x^2} + \frac{\partial A_2}{\partial y^2}\right) + \rho\frac{\partial A_2}{\partial x} = -j A_1^2 \exp(+j\Delta kz) . \qquad (2.162)$$

There are two possible situations:

1. $L_{\text{dif}} \gg L$, therefore the diffraction and corresponding terms with second-order derivatives in (2.161,162) can be ignored, but the birefringence should be taken into account ($L_{\text{a}} \sim L$). The theoretical consideration can be done by two different ways: whether by direct solving of (2.161,162) without diffraction terms, or using the limited laser beam representation by partial beams with constant transverse intensity distribution with subsequent integrating with respect to beam square. The numerical results in both cases are of course the same. Note that the possibility to use the second way is connected with the absence of diffraction and with the corresponding possibility to consider the spatially-modulated beam in the *approximation of geometrical optics* [2.4].

Table 2.31 illustrates the expressions for SHG efficiency $\eta = P_2(L)/P_1(0)$ in the above mentioned *geometrical optics* or *quasi-plane-wave approximation* together with the *field-fixed approximation* for two types of phase matching ($\text{I}^{(-)}$ and $\text{II}^{(-)}$) in uniaxial crystals. Remember, that for type $\text{II}^{(-)}$ phase matching it is necessary to include in the *truncated equations* (2.161,162) two (instead of one) equations for o-and e-waves of the fundamental radiation. We have used in the equations of Table 2.31 the following designations for the case of type $\text{II}^{(-)}$ phase matching: $P_1^o(0)$ and $P_1^e(0)$ – the incident laser powers of o- and e-waves, respectively; $P_1(0) = P_1^o + P_1^e$ – total incident laser power (all for $z = 0$); ρ_1 and ρ_2 – *"walk-off" angles* for extraordinary beams at fundamental and second-harmonic frequencies, respectively; $L_{\text{a}1} = 2w_0/\rho_1$ and $L_{\text{a}2} = 2w_0/\rho_2$ – corresponding *aperture lengths* for e-beams.

The formulae of Table 2.31 show that the existence of the anisotropy leads to the sufficient decrease of SHG efficiency, especially for type $\text{II}^{(-)}$ phase matching. The physical sense of this fact is clear: for the ooe case the e-wave of second-harmonic radiation continuously escapes from the o-wave of

Table 2.31. Equations for calculating the SHG conversion efficiency in the case of exact phase matching and quasi-plane wave fixed-field approximation for spatially-limited laser beam neglecting diffraction, but taking into account crystal anisotropy (in the CGS system)

Type of interaction and relation between L and L_{a}	Conversion efficiency $\eta = P_2(L)/P_1(0)$
ooe $L \leq L_{\text{a}}$	$\eta = \dfrac{2^7 \pi^5 d_{\text{eff}}^2 L^2 P_1(0)}{c n_{o1}^2 n_{e2} \lambda_2^2 \pi w_0^2}\left(1 - \dfrac{L}{3L_{\text{a}}}\right)$
ooe $L > L_{\text{a}}$	$\eta = \dfrac{2^7 \pi^5 d_{\text{eff}}^2 P_1(0)}{c n_{o1}^2 n_{e2} \lambda_2^2 \rho^2}\left(\dfrac{L}{L_{\text{a}}} - \dfrac{1}{3}\right)$
oee $L \leq L_{\text{a}}$	$\eta = \dfrac{2^7 \pi^5 d_{\text{eff}}^2 L^2 P_1^o(0) P_1^e(0)}{c n_{o1} n_{e1} n_{e2} \lambda_2^2 \pi w_0^2 P_1(0)} \times \left(1 - \dfrac{L}{3L_{\text{a}1}} - \dfrac{L}{3L_{\text{a}2}}\right)$
oee $L > L_{\text{a}}$	$\eta = \dfrac{2^7 \pi^5 d_{\text{eff}}^2 P_1^o(0) P_1^e(0)}{3 c n_{o1} n_{e1} n_{e2} \lambda_2^2 \pi \rho_1 \rho_2 P_1(0)} \times \left(1 - \left(1 - \dfrac{L}{L_{\text{a}1}}\right)^3 \dfrac{\rho_2}{|\rho_2 - \rho_1|}\right)$

fundamental radiation, not interrupting the SHG process; to the contrary, for the oee case, as soon as the e-wave of fundamental radiation escapes from the o-wave of the same fundamental radiation, the SHG-process interrupts immediately. The cardinal way to eliminate this phenomena in the case of narrow beams is the use of 90° phase matching (in accordance with (2.21) $\rho = 0$ at $\theta = 90°$, and $L_a \to \infty$), but it is true only for type $I^{(-)}$ phase matching, because d_{eff} for type $II^{(-)}$ phase matching with $\theta = 90°$ is equal to zero (Table 2.3). The second way is to use the special schemes for SHG, namely, SHG in two crystals that are placed *in tandem* and have the opposite directions of the optical axis [2.4]; in this case the *"walk-off"* in the first crystal is compensated by opposite *"walk-off"* in the second one. The third way is to use the wide laser beams ($L_a \gg L$). Note that SHG in the crystals with large *"walk-off"* angle (LiIO$_3$) is accompanied by strong distortion of the spatial distribution of the output SH beam.

2. With $L_a \gg L$, the main limited factor is *diffraction*, which should be taken into account for narrow and especially for *focused* laser beams (for example, at $\theta_{pm} = 90°$). In this case the term $\rho \partial A_2 / \partial z$ in (2.162) can be ignored but not the terms with second-order derivatives. Note that if $\theta_{pm} \neq 90°$, all the terms in (2.161,162) should be conserved.

When the focused laser beams are used, there are two opposite processes: first, due to focusing of laser radiation into the nonlinear crystal, the laser power density in crystal increases, as well as the *vector phase matching* should be taken into account; second, the influence of the laser beam *divergence* and crystal *birefringence* increases also, hence the *phase mismatch* appears. The first process leads to the increase of SHG efficiency, the second to the decrease of SHG efficiency. Therefore, there is an *optimum* of the focusing (see below). In all practically important cases the *fixed-field approximation* can be used for the Gaussian beams, hence, the nonlinear term in the right-hand side of (2.161) can be ignored, and (2.161) can be solved in this case independently in (2.162) in *linear approximation*. The resulting solution for $A_1(x, y, z)$ should be substituted in (2.162). As a result the following expression for output second-harmonic power can be deduced [2.4]:

$$P_2(L) = CkLP_1^2(0)\,h(v, \alpha, \xi, \mu) \;, \tag{2.163}$$

where $h(v, \alpha, \xi, \mu)$, is the so-called *aperture function*, introduced by *Boyd* and *Kleinman* [2.36]; see also [2.4]:

$$h(v, \alpha, \xi, \mu) = \frac{2\pi\sqrt{\pi}}{\xi} \int_{-\infty}^{\infty} \exp(-4s^2)|H(v', \alpha, \xi, \mu)|^2 ds \;, \tag{2.164}$$

$$H(v', \alpha, \xi, \mu) = \frac{1}{2\pi} \int_{-\xi(1-\mu)}^{\xi(1+\mu)} \frac{\exp(jv'T)}{1 + jT} dT \;. \tag{2.165}$$

$v = kw_0^2\Delta k/2$, $v' = v + 4\alpha s$, $\alpha = \rho k w_0/2$, $\xi = L/kw_0^2$ – the ratio of nonlinear crystal length L to the *focal length* kw_0^2, or the *parameter of focussing*, $\mu = (L - 2z_0)/L$ – the *relative position of focus* inside the crystal, z_0 – the coordinate of the minimum beam radius $w(z)$, i.e., $w(z_0) = w_0$; with $z_0 = L/2$, i.e., when the focus is in the middle of crystal, $\mu = 0$; s – the *variable of integration*, $C = 2^7 \pi^4 d_{\mathrm{eff}}^2 / c n_1^2 n_2 \lambda_2^2$. The phase mismatch Δk is introduced here as a mismatch from the beam axis.

For the 90° phase matching ($\theta_{\mathrm{pm}} = 90°$, $\rho = 0$, $\alpha = 0$) there are two extreme cases (usually $\mu = 0$) which permit us to obtain the simple analytical expressions:

a) *Weak focusing*, $\xi \ll 1$. In this case the *aperture function* (2.164) has the following view:

$$h(v, 0, \xi, 0) = \xi \operatorname{sinc}^2(\Delta k L/2) \tag{2.166}$$

and for the SHG efficiency we have

$$\eta(L) = P_2(L)/P_1(0) = C P_1(0) L^2 \frac{\operatorname{sinc}^2(\Delta k L/2)}{w_0^2} \,. \tag{2.167}$$

Hence, the expressions for $\eta(L)$ in this case is the same as for SHG by Gaussian laser beam with $\Delta k \neq 0$ in the *plane-wave fixed-field approximation* [2.4].

b) *Strong focusing*, $\xi \gg 1$. In this case for $\Delta k < 0$:

$$h(v, 0, \xi, 0) = \xi^{-1} |\pi \exp(v) + \operatorname{Si}(-\Delta k L/2) - \pi/2|^2 \tag{2.168}$$

and for $\Delta k > 0$:

$$h(v, 0, \xi, 0) = \xi^{-1} |\operatorname{Si}(\Delta k L/2) - \pi/2|^2 \,, \tag{2.169}$$

where Si x is *integral sine* (2.134).

It is seen from eqs. (2.168,169) that functions h is asymmetrical with respect to phase-mismatch sign. As an example, for $\xi = 10$ the maximum value of h (and hence of second-harmonic power) occurs not at $\Delta k = 0$, but for $\Delta k L/2 \approx -3$, i.e., for negative phase mismatch. This effect is due to *vector (noncollinear) phase matching* that can be fulfilled only in the region of negative mismatch, i.e., in the region of *anomalous dispersion* (remember, that for $\Delta k < 0$, or for $\theta > \theta_{\mathrm{pm}}$, the inequality $n_2^e < n_{\mathrm{o}1}$ is valid which corresponds to *anomalous dispersion*). In region $\theta < \theta_{\mathrm{pm}}$ neither *scalar (collinear)* nor *vector (noncollinear)* phase matching are fulfilled, and the SHG efficiency decreases with increase of the *focusing parameter*.

In order to obtain the *optimal focusing* and hence the maximum efficiency it is necessary to optimize the function h (2.164) on the both parameters: v (mismatch) and ξ (focusing). Numerical calculation shows that the maximum h takes place at $\xi_{\mathrm{opt}} = 2.84$, $v_{\mathrm{opt}} = -0.55$ (it corresponds to $\Delta k_{\mathrm{opt}} L/2 = -1.6$), and $h(v_{\mathrm{opt}}, \xi_{\mathrm{opt}}) = 1.07$ [2.4].

At $\theta_{\mathrm{pm}} < 90°$ ($\rho \neq 0$) the calculation is more complicated, see [2.4,36]; note that the crystal *anisotropy* leads to the decrease of SHG efficiency in the case of non-focused narrow laser beams.

2.14.8 Linear Absorption

The absorption parameter ($\delta \neq 0$) in the absence of heat effects can be taken into account by multiplying the conversion efficiency calculated for $\delta = 0$ by the factor $\exp(-2\delta L)$. Since the nonlinear crystals are generally transparent for the interacting waves, the following expansion can be used:

$$\exp(-2\delta L) \approx 1 - 2\delta L \ . \tag{2.170}$$

Note that the linear absorption coefficient α (for intensity absorption) widely used in the literature is equal to 2δ. More rigorous inclusion of the absorption can be done by substitution of expression $L' = (\delta)^{-1}[1 - \exp(-\delta L)]$ instead of L in the analytical expressions for conversion efficiency. Note that these corrections are usually less than the experimental measurement accuracy, therefore they have a meaning only in the case of high conversion efficiency (near 100%) and allow us to define the *limiting* efficiency (for example, for $\delta \approx 0.01$ cm^{-1} and $L \approx 3$ cm, $\eta_{\lim} \approx 94\%$). A stronger influence of the linear absorption takes place when the *heating phenomena* are taken into account, which leads to the appearance of inhomogeneous *wave mismatch, hysteresis phenomena*, and finally to the decrease and temperature instabilities of the conversion efficiency. In powerful pulse regime it is necessary to also take into account the *nonlinear* (generally *two-photon*) *absorption* of the second-harmonic radiation (2.12). It should be emphasized that the absorption coefficients of the crystal surface and of the crystal volume are not the same: due to the existance of the *disrupted layer* on the crystal surface and of corresponding *surface defects*, the surface absorption may be several orders greater than the volume absorption. Due to very low thickness of the *disrupted surface layer* (2–10 nm), this absorption cannot affect the conversion efficiency, but it can exert the primary influence on the *surface-damage threshold*. Generally, if the *disrupted surface layer* is small (≤ 2nm), then both *surface-* and *bulk-damage thresholds* are the same. These comments show that in nonlinear optics of the powerful laser radiation it is necessary to take into account the *surface processing quality*.

2.15 Additional Comments

As it was already mentioned, *sum-frequency generation* is used in the so-called *up-converters*, where infrared signals are transformed to visible ones or visible signals to ultraviolet. The up-converter efficiency can be evaluated by the equations for SFG, see Tables 2.28–30 [2.3].

Parametric luminescence and *optical parametric oscillation* are widely used for designing the sources of tunable laser radiation (including IR region). For evaluation of the OPO conversion efficiency see [2.4, 33, 37].

In the last few years the lasers with *high* and *superhigh intensities* has been extensively developed; these lasers have the electric light fields which are much higher than *atomic* fields. In these fields the *cooperative nonlinear phenomena* may be observed [2.32,34,35]. Note that the *damage threshold* of a nonlinear crystal increases with decrease of the pulse duration (Chap. 3).

Note that recently great progress was achieved in the field of *intracavity* SHG (ICSHG), which employs the nonlinear crystal placed inside the laser cavity. In modern lasers with ICSHG the so-called "100% conversion efficiency" is achieved. It means that the real conversion efficiency of internal laser radiation to the second harmonic is optimum for this laser, and therefore the output second-harmonic power is the same as for this laser on fundamental frequency *without phase matching in nonlinear crystal* and with *optimum transparency* of the output laser mirror (remember, that with ICSHG the transparency of the output mirror on the laser frequency is equal to zero).

A very interesting direction in modern nonlinear optics is the development of a quantum theory of optical media nonlinearity, which relates the nonlinearity of the given crystal to its molecular structure [2.38,39]. This theory may become a tool for predicting and directing the search for new nonlinear optical materials with preset properties.

Finally, note that in the last years the so-called *active nonlinear* crystals were successfully developed. In these crystals the *active* (laser) and *nonlinear* properties are combined in the same crystal. For example, $LiNbO_3 : MgO : Nd$ crystal ($\lambda_1 = 1.085$ μm, $\lambda_2 = 0.542$ μm), $KTiOPO_4 : Cr$ crystal (tuned radiation $\lambda_1 = 800 \dots 850$ nm, $\lambda_2 = 400 \dots 425$ nm), $BBO : Nd$ crystal, $YVO_4 : Nd$ crystal, and so on. The fundamentals of theory for these crystals are given in [2.4].

3 Properties of Nonlinear Optical Crystals

This chapter contains the main reference material, namely, optical and non-linear optical properties of 77 crystals which are used in applied nonlinear optics for frequency conversion of laser light. All nonlinear optical crystals described in this chapter are divided into four groups: basic, frequently used, other inorganic, and other organic. The properties of crystalline quartz which is not phase-matchable but nevertheless is often used as a standard in the **d**-tensor coefficients' measurements are given in a separate section. Inside every group of crystals an ordering of crystals according to the wavelength of the short-wave absorption edge was done.

For each crystal all properties related to nonlinear frequency conversion are given, such as point group, transparency range, coefficients of linear and two-photon absorption, refractive indices and their temperature derivatives, experimental values of the phase-matching angle and temperature of non-critical phase matching, effective nonlinearity expressions in phase-matching direction, nonlinear coefficients, laser-induced surface- and bulk-damage thresholds, thermal conductivity coefficient, etc.

We have performed the comparison between numerous sets of Sellmeier equations (dispersion relations), in the existing literature. For each crystal using dispersion relations from different sources and formulae given above in Sect. 2.5, 9 we have calculated the theoretical values of the phase-matching angle. Then comparing these values with the experimental ones we have determined the best set of dispersion relations. Using this set for each crystal we have calculated the theoretical values of the phase-matching angle for the most frequently used laser frequencies and corresponding values of the birefringence ("walk-off") angle. For some basic nonlinear optical crystals we have also determined the inverse group-velocity mismatch for the SHG process.

In the shortened notation of polarization of interacting waves, everywhere the first symbol refers to the longest wavelength radiation and the third one represents the wave with the shortest wavelength. All given experimental values of internal angular, temperature and spectral bandwidths correspond to 1 cm length crystal. If the literature source contained the value of the external angular bandwidth then in calculating the value of the internal one for the case of type II interaction we have used the minimal value of the refraction index. Kleinman symmetry relations are assumed to be valid unless otherwise specified. The components of **d**-tensor are given everywhere in the system of principal crystallographic axes a, b, c (crystallographic reference frame) with the exception of crystals belonging to point group 2 for which the system of principal dielectric axes (X, Y, Z) or dielectric reference frame was used. All the reference material is given in the SI system. For the conversion of nonlinear coefficients' values to the CGS system one should bear in mind that $1 \text{ m/V} = (3/4\pi) \times 10^4$ esu.

The chapter contains about 650 references updated to the end of 1994. In Sect. 3.6 "New Developments" some important new references are given. The index of crystals is provided at the end of the handbook.

3.1 Basic Nonlinear Optical Crystals

3.1.1 LiB₃O₅, Lithium Triborate (LBO)

Negative biaxial crystal: $2V_Z = 109.2°$ at $\lambda = 0.5321 \mu m$ [3.1];
Point group: mm2
Assignment of dielectric and crystallographic axes:
$X, Y, Z \rightarrow a, c, b$ (Fig. 3.1);
Mass density: 2.47 g/cm³;

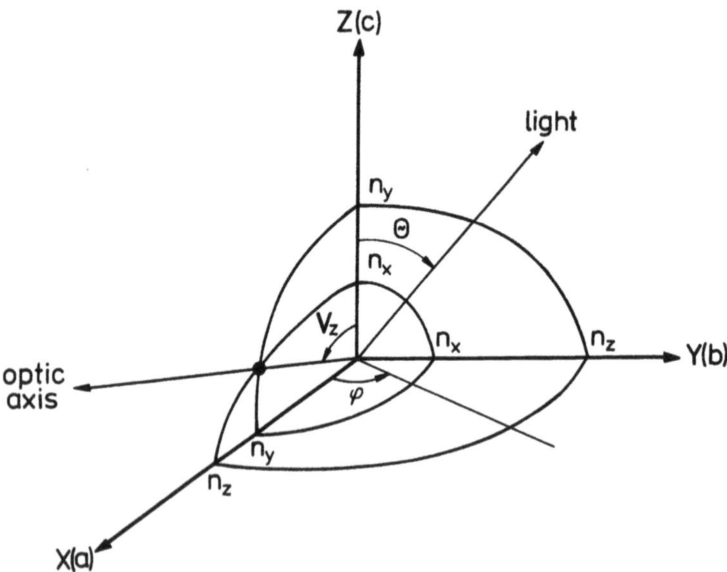

Fig. 3.1. Dependence of refractive index on light propagation direction and polarization (index surface) in the first octant of dielectric reference frame (X, Y, Z) of LBO crystal. Designations: θ is the polar angle, ϕ is the asimuthal angle, V_Z is the angle between one of the optical axes and the Z axis

Mohs hardness: 6;
Transparency range at "0" transmittance level: $0.155-3.2$ μm [3.1,2];
Linear absorption coefficient α [3.3]:

λ [μm]	α [cm^{-1}]
0.35–0.36	0.0031
1.0642	0.00035

Experimental values of refractive indices:

λ [μm]	n_X	n_Y	n_Z	Ref.
0.2537	1.6335	1.6582	1.6792	3.1
0.2894	1.6209	1.6467	1.6681	3.1
0.2968	1.6182	1.6450	1.6674	3.1
0.3125	1.6097	1.6415	1.6588	3.1
0.3341	1.6043	1.6346	1.6509	3.1
0.3650	1.59523	1.62518	1.64025	3.3
	1.5954	1.6250	1.6407	3.1
0.4000	1.58995	1.61918		3.3
0.4047	1.5907	1.6216	1.6353	3.1
0.4358	1.5859	1.6148	1.6297	3.1
0.4500	1.58449	1.61301	1.62793	3.3
0.4861	1.5817	1.6099	1.6248	3.1
0.5000	1.58059	1.60862	1.62348	3.3
0.5250	1.57906	1.60686		3.3
0.5321	1.57868	1.60642	1.62122	3.3
	1.5785	1.6065	1.6212	3.1
0.5398	1.5782	1.6212	1.6063	3.4
0.5461	1.5780	1.6057	1.6206	3.1
0.5500	1.57772	1.60535	1.62014	3.3
0.5780	1.5765	1.6039	1.6187	3.1
0.5893	1.5760	1.6035	1.6183	3.1
0.6000	1.57541	1.60276	1.61753	3.3
0.6328	1.5742	1.6014	1.6163	3.1
0.6563	1.5734	1.6006	1.6154	3.1
0.7000		1.59893	1.61363	3.3
0.8000	1.56959	1.59615	1.61078	3.3
0.9000	1.56764	1.59386	1.60843	3.3
1.0000	1.56586	1.59187	1.60637	3.3
1.0642	1.56487	1.59072	1.60515	3.3
	1.5656	1.5905	1.6055	3.1
1.0796	1.5655	1.6053	1.5902	3.4
1.1000	1.56432	1.59005	1.60449	3.3

Temperature derivative of refractive indices within interval 20–65 °C for spectral range $0.4 - 1.0$ µm [3.3]:

$$\mathrm{d}n_X/\mathrm{d}T \times 10^6 = -1.8 \ ,$$

$$\mathrm{d}n_Y/\mathrm{d}T \times 10^6 = -13.6 \ ,$$

$$\mathrm{d}n_Z/\mathrm{d}T \times 10^6 = -6.3 - 2.1\,\lambda \ ,$$

where λ in µm and $\mathrm{d}n_X/\mathrm{d}T$, $\mathrm{d}n_Y/\mathrm{d}T$, and $\mathrm{d}n_Z/\mathrm{d}T$ are in K^{-1}.

Experimental values of phase-matching angle ($T=293$ K) and comparison between different sets of dispersion relations:

XY plane, $\theta = 90°$

Interacting wavelengths [µm]	$\phi_{\exp}$ [deg]	ϕ_{theor} [deg]		
		[3.2]	[3.5]	[3.6]
SHG, o + o $\Rightarrow$ e				
1.908 $\Rightarrow$ 0.954	23.8 [3.5]	24.04	23.98	31.32
1.5 $\Rightarrow$ 0.75	7 [3.5]	7.03	6.81	10.18
1.0796 $\Rightarrow$ 0.5398	10.6 [3.5]	10.64	10.39	10.42
	10.7 [3.1]			
	10.7 [3.4]			
1.0642 $\Rightarrow$ 0.5321	11.3 [3.3]	11.60	11.36	11.39
	11.4 [3.7]			
	11.4 [3.8]			
	11.6 [3.2]			
	11.6 [3.5]			
	11.6 [3.9]			
	11.8 [3.10]			
0.896 $\Rightarrow$ 0.448	23.3 [3.11]	23.25	23.01	23.14
0.88 $\Rightarrow$ 0.44	24.5 [3.11]	24.53	24.29	24.44
0.84 $\Rightarrow$ 0.42	27.9 [3.11]	27.94	27.70	27.88
0.80 $\Rightarrow$ 0.40	31.7 [3.11]	31.69	31.47	31.66
0.78 $\Rightarrow$ 0.39	33.7 [3.11]	33.72	33.51	33.71
0.75 $\Rightarrow$ 0.375	37.1 [3.11]	37.02	36.83	37.03
0.7094 $\Rightarrow$ 0.3547	41.8 [3.5]	42.09	41.94	42.12
	41.9 [3.12]			
	42 [3.13]			
	43.5 [3.14]			
0.63 $\Rightarrow$ 0.315	55.6 [3.15]	55.32	55.29	55.42
0.555 $\Rightarrow$ 0.2775	86 [3.5]	85.75	85.74	85.97
0.554 $\Rightarrow$ 0.277	90 [3.16]	88.97	88.87	no pm

SFG, o + o ⇒ e				
1.0642 + 0.5321 ⇒				
⇒ 0.35473	37.1 [3.6]	37.21	36.86	37.30
	37.2 [3.2]			
	37.2 [3.3]			
1.0642 + 0.35473 ⇒				
⇒ 0.26605	60.7 [3.2]	60.63	61.35	61.02
	61 [3.5]			
1.3188 + 0.26605 ⇒				
⇒ 0.22139	70.2 [3.2]	70.13	78.47	71.32
1.3414 + 0.6707 ⇒				
⇒ 0.44713	20 [3.6]	20.02	19.41	20.18
0.21284 + 2.35524 ⇒				
⇒ 0.1952	50.3 [3.6]	48.41	68.17	52.48
0.21284 + 1.90007 ⇒				
⇒ 0.1914	63.8 [3.6]	60.99	no pm	64.07
0.21284 + 1.58910 ⇒				
⇒ 0.1877	88 [3.6]	81.21	no pm	no pm

YZ plane, $\phi = 90°$

Interacting wavelengths [μm]	θ_{exp} [deg]	θ_{theor} [deg]		
		[3.2]	[3.5]	[3.6]
SHG, o + e ⇒ o				
1.908 ⇒ 0.954	46.2 [3.5]	49.00	46.24	60.13
1.5 ⇒ 0.75	14.7 [3.5]	14.19	14.18	12.78
1.0796 ⇒ 0.5398	19.2 [3.5]	18.52	19.04	18.94
1.0642 ⇒ 0.5321	19.9 [3.3]	20.45	20.94	20.85
	20.6 [3.2]			
	20.5 [3.17]			
	21.0 [3.5]			
SFG, o + e ⇒ o				
1.0642 + 0.5321 ⇒				
⇒ 0.35473	42.2 [3.2]	42.19	42.16	42.63
	42.5 [3.6]			
	43.2 [3.3]			

XZ plane, $\phi = 0°, \theta < V_Z$

Interacting wavelengths [μm]	θ_{exp} [deg]	θ_{theor} [deg]		
		[3.2]	[3.5]	[3.6]
SHG, *o + o ⇒ e*				
1.3414 ⇒ 0.6707	4.2 [3.5]	4.67	4.17	5.00
	5.0 [3.9]			

| 1.3188 $\Rightarrow$ 0.6594 | 5.2 [3.2] | 5.10 | 4.62 | 5.29 |
| 1.3 $\Rightarrow$ 0.65 | 5.4 [3.9] | 5.26 | 4.78 | 5.36 |

XZ plane, $\phi = 0°, \theta > V_Z$

Interacting wavelengths [µm]	θ_{exp} [deg]	θ_{theor} [deg]		
		[3.2]	[3.5]	[3.6]
SHG, e + e $\Rightarrow$ o				
1.3414 $\Rightarrow$ 0.6707	86.3 [3.5] 86.6 [3.9]	86.47	86.22	88.93
1.3188 $\Rightarrow$ 0.6594	86.0 [3.2]	86.26	86.03	87.79
1.3 $\Rightarrow$ 0.65	86.1 [3.9]	86.25	86.01	87.41

Note: The other sets of dispersion relations from [3.1, 18, 3, 19, 20, 8, 21, 22, 23] show worse agreement with the experiment

Best set of dispersion relations (λ in µm, $T = 20$ °C) [3.2]:

$$n_X^2 = 2.4542 + \frac{0.01125}{\lambda^2 - 0.01135} - 0.01388\,\lambda^2,$$

$$n_Y^2 = 2.5390 + \frac{0.01277}{\lambda^2 - 0.01189} - 0.01848\,\lambda^2,$$

$$n_Z^2 = 2.5865 + \frac{0.01310}{\lambda^2 - 0.01223} - 0.01861\,\lambda^2.$$

Calculated values of phase-matching and "walk-off" angles:
XY plane, $\theta = 90°$

Interacting wavelengths [µm]	θ_{pm} [deg]	ρ_3 [deg]
SHG, o + o $\Rightarrow$ e		
2.098 $\Rightarrow$ 1.049	31.61	0.840
1.1523 $\Rightarrow$ 0.57615	6.06	0.213
1.0642 $\Rightarrow$ 0.5321	11.60	0.403
0.6943 $\Rightarrow$ 0.34715	44.19	1.086
0.5782 $\Rightarrow$ 0.2891	69.91	0.730
SFG, o + o $\Rightarrow$ e		
1.0642 + 0.5321 $\Rightarrow$ $\Rightarrow$ 0.35473	37.21	1.046
1.0642 + 0.35473 $\Rightarrow$ $\Rightarrow$ 0.26605	60.63	1.006
1.3188 + 0.6594 $\Rightarrow$ $\Rightarrow$ 0.4396	21.11	0.705

YZ plane, $\phi = 90°$

Interacting wavelengths [μm]	θ_{pm} [deg]	ρ_3 [deg]
SHG, o + e $\Rightarrow$ o		
2.098 $\Rightarrow$ 1.049	72.90	0.307
1.1523 $\Rightarrow$ 0.57615	9.28	0.169
1.0642 $\Rightarrow$ 0.5321	20.45	0.348
SFG, o + e $\Rightarrow$ o		
1.0642 + 0.5321 $\Rightarrow$		
$\Rightarrow$ 0.35473	42.19	0.533

XZ plane $\phi = 0°, \theta < V_Z$

Interacting wavelengths [μm]	θ_{pm} [deg]	ρ_1 [deg]	ρ_3 [deg]
SHG, e + o $\Rightarrow$ e			
1.3188 $\Rightarrow$ 0.6594	5.10	0.248	0.262

XZ plane, $\phi = 0°, \theta > V_Z$

Interacting wavelengths [μm]	θ_{pm}[deg]	ρ_1[deg]
SHG, e + e $\Rightarrow$ o		
1.3188 $\Rightarrow$ 0.6594	86.26	0.191

Calculated values of inverse group-velocity mismatch for the SHG process in LBO:

XY plane, $\theta = 90°$

Interacting wavelengths [μm]	ϕ_{pm} [deg]	β [fs/mm]
SHG, o + o $\Rightarrow$ e		
1.2 $\Rightarrow$ 0.6	2.36	18
1.1 $\Rightarrow$ 0.55	9.37	37
1.0 $\Rightarrow$ 0.5	15.74	59
0.9 $\Rightarrow$ 0.45	22.94	86
0.8 $\Rightarrow$ 0.4	31.69	123
0.7 $\Rightarrow$ 0.35	43.38	175
0.6 $\Rightarrow$ 0.3	62.63	257

YZ plane, $\phi = 90°$

Interacting wavelengths [μm]	θ_{pm} [deg]	β [fs/mm]
SHG, o + e ⇒ o		
$1.1 \Rightarrow 0.55$	15.98	82
$1.0 \Rightarrow 0.5$	28.96	106
$0.9 \Rightarrow 0.45$	45.36	139
$0.8 \Rightarrow 0.4$	76.88	186

Experimental values of NCPM temperature: along *X* axis

Interacting wavelengths [μm]	T [°C]	Ref
SHG, type I		
$1.25 \Rightarrow 0.625$	–2.9	3.7, 8
$1.215 \Rightarrow 0.6075$	21	3.8
$1.211 \Rightarrow 0.6055$	20	3.2
$1.2 \Rightarrow 0.6$	24.3	3.7, 8
$1.15 \Rightarrow 0.575$	61.1	3.7, 8
$1.135 \Rightarrow 0.5675$	77.4	3.10
$1.11 \Rightarrow 0.555$	108.2	3.7, 8
$1.0796 \Rightarrow 0.5398$	112	3.1
$1.0642 \Rightarrow 0.5321$	148	3.7, 8
	148.5	3.24, 25
	149	3.10
	149.5	3.26
	151	3.17
$1.047 \Rightarrow 0.5235$	166.5	3.27
	167	3.28
	172	3.29
	175	3.30
	176.5	3.31
	180	3.32
$1.025 \Rightarrow 0.5125$	190.3	3.7, 8
SFG, type I		
$1.908 + 1.0642 \Rightarrow$		
$\Rightarrow 0.6832$	81	3.10
$1.135 + 1.0642 \Rightarrow$		
$\Rightarrow 0.5491$	112	3.10

Experimental values of internal angular, temperature and spectral bandwidths: along X axis

Interacting wavelengths [µm]	T [°C]	$\Delta\phi^{\text{int}}$ [deg]	$\Delta\theta^{\text{int}}$ [deg]	ΔT [°C]	Ref.
SHG, type I					
$1.135 \Rightarrow 0.5675$	77.4			4.7	3.10
$1.0642 \Rightarrow 0.5321$	148	3.54	2.57	3.9	3.7
	148.5			2.7	3.24
	148.5			4.2	3.25
	149	2.3	1.9	4.0	3.10
	149.5			4.1	3.26
	151	2.1	2.1	2.9	3.17
$1.047 \Rightarrow 0.5235$	175			3.5	3.30
	176.5			3.5	3.31
SFG, type I					
$1.908 + 1.0642 \Rightarrow$					
$\Rightarrow 0.6832$	81			7.4	3.10
$1.135 + 1.0642 \Rightarrow$					
$\Rightarrow 0.5491$	112			5.0	3.10

XY plane, $\theta = 90°(T = 20\ °C)$

Interacting wavelengths [µm]	ϕ_{pm} [deg]	$\Delta\phi^{\text{int}}$ [deg]	$\Delta\theta^{\text{int}}$ [deg]	ΔT [°C]	$\Delta\nu$ [cm^{-1}]	Ref.
SHG, $o + o \Rightarrow e$						
$1.0796 \Rightarrow 0.5398$	10.7	0.31				3.4
$1.0642 \Rightarrow 0.5321$	10.8(?)	0.27	2.63			3.17
	11.4	0.24	1.79			3.8
	11.6			5.8		3.2
		0.34	2.64	6.7	8.8	3.33
$0.886 \Rightarrow 0.443$	24.1			7.8	15.9	3.11
$0.870 \Rightarrow 0.435$	25.4	0.12				3.34
		0.10				3.11
$0.78 \Rightarrow 0.39$	33.7	0.08				3.34
		0.07				3.11
$0.7605 \Rightarrow 0.38025$	35.9			15.3	10.5	3.11
$0.715 \Rightarrow 0.3575$	41	0.06				3.34
SFG, $o + o \Rightarrow e$						
$1.0642 + 0.3547 \Rightarrow$						
$\Rightarrow 0.2661$	60.7			3.8		3.2

YZ plane, $\phi = 90°(T = 20\ °C)$

Interacting wavelengths [μm]	ϕ_{pm} [deg]	$\Delta\phi^{int}$ [deg]	$\Delta\theta^{int}$ [deg]	ΔT [°C]	$\Delta\nu$ [cm^{-1}]	Ref.
SHG, $o + e \Rightarrow o$						
$1.0642 \Rightarrow 0.5321$	20.6	0.77	3.20			3.17
		0.81	3.00		11.5	3.33
				6.2		3.2
SFG, $o + o \Rightarrow e$						
$1.0642 + 0.5321 \Rightarrow$						
$\Rightarrow 0.3547$	42.2	0.18				3.2
		0.18	3.11			3.18

Effective nonlinearity expressions in the phase-matching direction for three-wave interactions in the principal planes of LBO crystal [3.35], [3.36]:

XY plane

$$d_{ooe} = d_{32} \cos \phi \ ;$$

YZ plane

$$d_{oeo} = d_{oeo} = d_{31} \cos \theta$$

XZ plane, $\theta < V_Z$

$$d_{eoe} = d_{oee} = d_{32} \sin^2 \theta + d_{31} \cos^2 \theta \ ;$$

XZ plane, $\theta > V_Z$

$$d_{eeo} = d_{32} \sin^2 \theta + d_{31} \cos^2 \theta \ .$$

Effective nonlinearity expressions for three-wave interactions in the arbitrary direction of LBO crystal are given in [3.36].

Nonlinear coefficients [3.37]:

$$d_{31}(1.0642\ \mu m) = \mp\ 0.67\ pm/V$$

$$d_{32}(1.0642\ \mu m) = \pm\ 0.85\ pm/V$$

$$d_{33}(1.0642\ \mu m) = \pm\ 0.04\ pm/V$$

Laser-induced surface-damage threshold:

λ [μm]	τ_p [ns]	$I_{thr} \times 10^{-12}$ [W/m^2]	Ref.	Note
0.2661	12	> 0.4	3.38	
0.3078	17	> 0.6	3.39	
	17	> 0.6	3.40	
	10	> 1.0	3.41	
	0.0003	470 000(?)	3.42	
0.3547	18	> 1.8	3.43	10 Hz

λ [μm]	τ_p [ns]	$I_{thr} \times 10^{-12}$ [W/m^2]	Ref.	Note
0.3547	10	> 0.4	3.12	
	10	> 2.0	3.44	
	8	> 1.3	3.19	
	7	> 1.4	3.45	
	0.03	> 94	3.46	10 Hz
	0.03	> 180	3.47	10 Hz
	0.015	> 28	3.14	
	0.018	> 50	3.13	
	0.025	> 60	3.48	10 Hz
0.5145	cw	> 0.0003	3.49	
0.5235	0.055	> 11	3.32	500 Hz
	0.055	> 50	3.50	500 Hz
0.5321	cw	> 0.004	3.26	
	60	> 0.7	3.51	900 Hz
	10	> 2.2	3.9	
	0.1	> 45	3.52	500 Hz
	0.035	> 31	3.24	
	0.015	> 44	3.20	
0.605	0.0002	> 250	3.53	
0.616	0.0004	310000 (?)	3.42	
	0.0004	350000 (?)	3.54	
	0.0004	380000 (?)	3.55	
0.652	0.02	> 8.1	3.21	
0.7–0.9	10	> 0.3	3.11	10 Hz
0.71–0.87	25	11–14	3.34	25 Hz
0.72–0.85	0.001	> 80	3.56	
1.0642	cw	> 0.01	3.26	
	60	> 0.6	3.51	1333 Hz
	18	> 6	3.43	10 Hz
	9	> 9	3.57	10 Hz
	8	> 5	3.17	
	1.3	190	3.33	
	0.1	250	3.1	
	0.035	> 48	3.24	
	0.025	> 33	3.48	10 Hz
1.0796	0.04	300	3.42	

Thermal conductivity coefficient [3.58]:

$$\kappa = 3.5 \, \text{W/mK} \, .$$

3.1.2 KH₂PO₄, Potassium Dihydrogen Phosphate (KDP)

Negative uniaxial crystal: $n_o > n_e$;
Point group: $\bar{4}2m$;
Mass density: 2.3383 g/cm^3 at 293 K [3.59];
Mohs hardness: 2.5;
Transparency range at "0" transmittance level: 0.174 – 1.57 μm [3.60, 59];
Transparency range at 0.5 transmittance level for a 0.8 cm long crystal:
0.178 – 1.45 μm [3.60, 59];
Linear absorption coefficient α:

λ [μm]	α [cm^{-1}]	Ref.	Note
0.212	0.2	3.61	
0.25725	0.01–0.2	3.62	e – wave, $\perp c$
	0.007	3.63	e – wave, $\perp c$
0.3–1.15	< 0.07	3.64	
0.3513	0.003	3.65	e – wave, $\perp c$
0.5145	0.00005	3.62	o – wave
0.5265	0.01	3.66	o – wave
0.94	0.01	3.67	
1.053	0.05	3.66	o – wave
1.054	0.058	3.65	o – wave
	0.02	3.65	e – wave, $\perp c$
1.22	0.1	3.68	o – wave
1.3152	0.3	3.69	
1.32	0.1	3.68	e – wave, $\perp c$

Two-photon absorption coefficient β:

λ[μm]	$\beta \times 10^{13}$ [m/W]	Ref.	Note
0.216	60 ± 5	3.70	
0.2661	27 ± 8.1	3.71	$\theta = 41°, \phi = 45°$
	40–80	3.72	
0.3547	0.59 ± 0.21	3.71	e – wave, $\perp c$

Experimental values of refractive indices at T = 298 K [3.73]:

λ [μm]	n_o	n_e	λ [μm]	n_o	n_e
0.2138560	1.60177	1.54615	0.2980628	1.54618	1.49824
0.2288018	1.58546		0.3021499	1.54433	1.49708
0.2446905	1.57228		0.3035781		1.49667
0.2464068	1.57105		0.3125663	1.54117	1.49434
0.2536519	1.56631	1.51586	0.3131545	1.54098	1.49419
0.2800869	1.55263	1.50416	0.3341478		1.48954

0.3650146	1.52932	1.48432	0.5769580	1.50987	
0.3654833	1.52923	1.48423	0.5790654	1.50977	1.46856
0.3662878	1.52909	1.48409	0.6328160	1.50737	1.46685
0.3906410		1.48089	1.0139750	1.49535	1.46041
0.4046561	1.52341	1.47927	1.1287040	1.49205	1.45917
0.4077811	1.52301	1.47898	1.1522760	1.49135	1.45893
0.4358350	1.51990	1.47640	1.3570700	1.48455	
0.4916036		1.47254	1.5231000		1.45521
0.5460740	1.51152	1.46982	1.5295250		1.45512

Temperature derivative of refractive indices [3.74]:

λ [μm]	$dn_{\mathrm{o}}/dT \times 10^5$ [K^{-1}]	$dn_{\mathrm{e}}/dT \times 10^5$ [K^{-1}]
0.405	−3.27	−3.15
0.436	−3.27	−2.88
0.546	−3.28	−2.90
0.578	−3.25	−2.87
0.633	−3.94	−2.54

Temperature dependences of refractive indices upon cooling from room temperature to T [K].

for the spectral range 0.365 – 0.690 μm [3.75]:

$$n_{\mathrm{o}}(T) = n_{\mathrm{o}}(298) + 0.402 \times 10^{-4}\left\{[n_{\mathrm{o}}(298)]^2 - 1.432\right\}(298 - T) \; ;$$

$$n_{\mathrm{e}}(T) = n_{\mathrm{e}}(298) + 0.221 \times 10^{-4}\left\{[n_{\mathrm{e}}(298)]^2 - 1.105\right\}(298 - T) \; ;$$

for the spectral range 0.436 – 0.589 μm [3.76]:

$$n_{\mathrm{o}}(T) = n_{\mathrm{o}}(300) + 10^{-4}(143.3 - 0.618T + 4.81 \times 10^{-4} \, T^2) \; ,$$

$$n_{\mathrm{e}}(T) = n_{\mathrm{e}}(300) + 10^{-4}(153.3 - 0.969T + 1.57 \times 10^{-3} \, T^2) \; .$$

Experimental values of phase-matching angle ($T = 293$ K) and comparison between different sets of dispersion relations:

Interacting wavelengths [μm]	θ_{exp} [deg]	θ_{theor} [deg]		
		[3.73]	[3.77]	[3.78]K
SHG, o + o ⇒ e				
0.517 ⇒ 0.2585	90 [3.74]	no pm	no pm	73.6
0.6576 ⇒ 0.3288	53.6 [3.69]	53.6	53.6	53.2
0.6943 ⇒ 0.34715	50.4 [3.79]	50.6	50.6	50.4
0.8707 ⇒ 0.43535	42.4 [3.80]	42.8	42.7	42.8
1.06 ⇒ 0.53	41 [3.81]	41.2	41.0	40.9
	41 [3.82]			

1.3152 ⇒ 0.6576	44.3 [3.69]	44.6	44.7	44.1
SFG, o + o ⇒ e				
1.415 + 0.22027 ⇒				
⇒ 0.1906	88.7 [3.83]	83.7	83.6	54.3
1.3648 + 0.6943 ⇒				
⇒ 0.46019	40.9 [3.80]	41.7	41.7	41.6
1.3152 + 0.6576 ⇒				
⇒ 0.4384	42.2 [3.69]	42.1	42.1	42.0
1.0642 + 0.2707 ⇒				
⇒ 0.21581	87.6 [3.84]	87.5	87.3	62.9
1.0642 + 0.5321 ⇒				
⇒ 0.35473	47.3 [3.85]	47.3	47.3	47.1
1.06 + 0.53 ⇒				
⇒ 0.35333	47.5 [3.82]	47.4	47.4	47.3
0.6576 + 0.4384 ⇒				
⇒ 0.26304	74 [3.86]	75.2	75.4	68.6
SHG, e + o ⇒ e				
1.3152 ⇒ 0.6576	61.4 [3.69]	61.8	61.8	60.7
1.06 ⇒ 0.53	59 [3.82]	59.0	58.8	58.6
SFG, e + o ⇒ e				
1.0642 + 0.5321 ⇒				
⇒ 0.35473	58.3 [3.85]	58.2	58.3	57.9
1.06 + 0.53 ⇒				
⇒ 0.35333	59.3 [3.82]	58.5	58.5	58.1

Note: The other sets of dispersion relations from [3.74] and [3, 78]E show worse agreement with the experiment.
[3.78]K ⇒ see [3.78], data of *Kirby* et al.;
[3.78]E ⇒ see [3.78], data of *Eimerl*.

Experimental values of NCPM temperature:

Interacting wavelengths [μm]	T [°C]	Ref.
SHG, o + o ⇒ e		
0.5145 ⇒ 0.25725	−13.7	3.63
	−11	3.62
0.517 ⇒ 0.2585	20	3.74
0.5321 ⇒ 0.26605	177	3.87
	177	3.88
SFG, o + o ⇒ e		
1.06 + 0.265 ⇒ 0.212	−70	3.61
1.0642 + 0.26605 ⇒ 0.21284	−40	3.89
	−35	3.90

Best set of dispersion relations (λ in µm, $T = 20\,°\mathrm{C}$) [3.74] :

$$n_\mathrm{o}^2 = 2.259276 + \frac{13.00522\lambda^2}{\lambda^2 - 400} + \frac{0.01008956}{\lambda^2 - (77.26408)^{-1}} \ ,$$

$$n_\mathrm{e}^2 = 2.132668 + \frac{3.2279924\lambda^2}{\lambda^2 - 400} + \frac{0.008637494}{\lambda^2 - (81.42631)^{-1}} \ .$$

Temperature-dependent Sellmeier equations (λ in µm, T in K) [3.77] :

$$n_\mathrm{o}^2 = (1.44896 + 3.185 \times 10^{-5}T) + \frac{(0.84181 - 1.4114 \times 10^{-4}\ T)\lambda^2}{\lambda^2 - (0.0128 - 2.13 \times 10^{-7}T)}$$
$$+ \frac{(0.90793 + 5.75 \times 10^{-7}\ T)\lambda^2}{\lambda^2 - 30} \ ,$$

$$n_\mathrm{e}^2 = (1.42691 - 1.152 \times 10^{-5}\ T) + \frac{(0.72722 - 6.139 \times 10^{-5}\ T)\lambda^2}{\lambda^2 - (0.01213 + 3.104 \times 10^{-7}\ T)}$$
$$+ \frac{(0.22543 - 1.98 \times 10^{-7}\ T)\lambda^2}{\lambda^2 - 30} \ .$$

Calculated values of phase-matching and "walk-off" angles:

Interacting wavelengths [µm]	θ_pm [deg]	ρ_1 [deg]	ρ_3 [deg]
SHG, $\mathrm{o} + \mathrm{o} \Rightarrow \mathrm{e}$			
$0.5321 \Rightarrow 0.26605$	76.60		0.808
$0.5782 \Rightarrow 0.2891$	64.03		1.391
$0.6328 \Rightarrow 0.3164$	56.15		1.611
$0.6594 \Rightarrow 0.3297$	53.43		1.657
$0.6943 \Rightarrow 0.34715$	50.55		1.687
$1.0642 \Rightarrow 0.5321$	41.21		1.603
$1.3188 \Rightarrow 0.6594$	44.70		1.549
SFG, $\mathrm{o} + \mathrm{o} \Rightarrow \mathrm{e}$			
$0.5782 + 0.5105 \Rightarrow 0.27112$	72.46		1.025
$1.0642 + 0.5321 \Rightarrow 0.35473$	47.28		1.712
$1.3188 + 0.6594 \Rightarrow 0.4396$	42.05		1.657
SHG, $\mathrm{e} + \mathrm{o} \Rightarrow \mathrm{e}$			
$1.0642 \Rightarrow 0.5321$	58.98	1.149	1.404
$1.3188 \Rightarrow 0.6594$	61.85	0.922	1.269
SFG, $\mathrm{e} + \mathrm{o} \Rightarrow \mathrm{e}$			
$1.0642 + 0.5321 \Rightarrow 0.35473$	58.23	1.166	1.521
$1.3188 + 0.6594 \Rightarrow 0.4396$	49.42	1.104	1.634

Calculated values of inverse group-velocity mismatch for SHG process in KDP:

Interacting wavelengths [μm]	θ_{pm} [deg]	β [fs/mm]
SHG, $o + o \Rightarrow e$		
$1.2 \Rightarrow 0.6$	42.45	42
$1.1 \Rightarrow 0.55$	41.38	17
$1.0 \Rightarrow 0.5$	41.22	9
$0.9 \Rightarrow 0.45$	42.24	40
$0.8 \Rightarrow 0.4$	44.91	77
$0.7 \Rightarrow 0.35$	50.14	128
$0.6 \Rightarrow 0.3$	60.40	208
SHG, $e + o \Rightarrow e$		
$1.2 \Rightarrow 0.6$	59.54	89
$1.1 \Rightarrow 0.55$	58.87	67
$1.0 \Rightarrow 0.5$	59.75	89
$0.9 \Rightarrow 0.45$	62.97	118
$0.8 \Rightarrow 0.4$	70.71	158

Experimental values of internal angular and temperature bandwidths:

Interacting wavelengths [μm]	T [°C]	θ_{pm} [deg]	$\Delta\theta^{int}$ [deg]	ΔT [°C]	Ref.
SHG, $o + o \Rightarrow e$					
$1.1523 \Rightarrow 0.57615$	20	41	0.074		3.91
$1.0642 \Rightarrow 0.5321$	20	41	0.070		3.92
	25			23	3.93
$1.064 \Rightarrow 0.532$	20	41	0.069		3.94
$1.06 \Rightarrow 0.53$	20	41	0.063		3.81
$1.054 \Rightarrow 0.527$	25	41	0.060		3.95
$0.5321 \Rightarrow 0.26605$	25			1.7	3.93
	177	90		1.9	3.87
	177	90		2	3.88
$0.53 \Rightarrow 0.265$	20	77	0.059		3.96
	20	77	0.066		3.97
SFG, $o + o \Rightarrow e$					
$1.0642 + 0.5321 \Rightarrow$					
$\Rightarrow 0.35473$	25			5.5	3.93
$1.054 + 0.527 \Rightarrow$					
$\Rightarrow 0.35133$	25	48	0.046		3.95
SHG, $e + o \Rightarrow e$					
$1.0642 \Rightarrow 0.5321$	25			18.3	3.93
$1.06 \Rightarrow 0.53$	20	59	0.129		3.96

1.054 ⇒ 0.527	25	59	0.126	3.95

SFG, e + o ⇒ e

1.0642 + 0.5321 ⇒				
⇒ 0.35473	25		5.2	3.93
1.06 + 0.53 ⇒				
⇒ 0.35333	20	59	0.062	3.97
1.054 + 0.527 ⇒				
⇒ 0.35133	25	59	0.059	3.95

Experimental values of spectral bandwidth:

Interacting wavelengths [µm]	T [°C]	θ_{pm} [deg]	$\Delta \nu$ [cm^{-1}]	Ref.
SHG, o + o ⇒ e				
1.06 ⇒ 0.53	20	41	178	3.81
0.53 ⇒ 0.265	20	77	1.2	3.96
SHG, e + o ⇒ e				
1.06 ⇒ 0.53	20	59	101.5	3.96

Temperature variation of phase-matching angle:

Interacting wavelengths [µm]	T [°C]	θ_{pm} [deg]	$d\theta_{pm}/dT$ [deg/K]	Ref.
SHG, o + o ⇒ e				
1.0642 ⇒ 0.5321	25		0.0028	3.93
1.054 ⇒ 0.527	25	41	0.0046	3.95
0.5321 ⇒ 0.26605	25		0.0382	3.93
SFG, o + o ⇒ e				
1.0642 + 0.5321 ⇒ 0.35473	25		0.0073	3.93
1.054 + 0.527 ⇒ 0.35133	25	59	0.0046	3.95
SHG, e + o ⇒ e				
1.0642 ⇒ 0.5321	25	59	0.0069	3.98
	25		0.0069	3.93
1.06 ⇒ 0.53	20	59	0.0057	3.96
1.054 ⇒ 0.527	25	59	0.0086	3.95
	20	59	0.0069	3.65
SHG, e + o ⇒ e				
1.0642 + 0.5321 ⇒ 0.35473	25	58	0.0106	3.98
	25		0.0117	3.93
1.054 + 0.527 ⇒ 0.35133	25	59	0.0152	3.95
	20	59	0.0075	3.65

Temperature tuning of noncritical SHG [3.74]:

Interacting wavelengths [μm]	$d\lambda_1/dT$ [nm/K]
SHG, $o + o \Rightarrow e$	
$0.517 \Rightarrow 0.2585$	0.048

Temperature variation of birefringence for noncritical SHG process:

Interacting wavelengths [μm]	$d(n_2^e - n_1^o)/dT \times 10^5$ [K^{-1}]	Ref.
$0.5145 \Rightarrow 0.25725$	1.745	3.99
$0.5321 \Rightarrow 0.26605$	1.2	3.87

Effective nonlinearity expressions in the phase-matching direction [3.100]:

$$d_{\mathrm{ooe}} = d_{36} \sin\theta \sin 2\phi \ ,$$
$$d_{\mathrm{eoe}} = d_{\mathrm{oee}} = d_{36} \sin 2\theta \cos 2\phi \ .$$

Nonlinear coefficient [3.37]:

$$d_{36}(1.064 \ \mu\mathrm{m}) = 0.39 \ \mathrm{pm/V} \ ,$$

Laser-induced bulk-damage threshold:

λ [μm]	τ_{p} [ns]	$I_{\mathrm{thr}} \times 10^{-12}$ [W/m^2]	Ref.
0.52	330	2	3.101
0.5265	20	30	3.66
	0.6	90	3.66
0.527	0.5	> 140	3.102
0.53	0.2	170	3.103
	0.005	100 00(?)	3.104
0.5321	0.6	> 80	3.72
	0.03	300	3.105
0.596	330	2.4	3.101
	20	30	3.101
0.6943	20	> 4	3.106
1.053	25	40	3.66
	1	180	3.66
	1	200	3.107
1.054	0.14	> 70	3.108
1.06	60	2	3.109
	12–25	2.5	3.81
	0.5	> 30	3.110
	0.2	230	3.103
1.064	20	3–6	3.111
	1.3	80	3.33

λ [µm]	τ_{p} [ns]	$I_{\mathrm{thr}} \times 10^{-12}$ [W/m^2]	Ref.
1.064	1	30–70	3.111
	1	50	3.112
	0.1	70	3.1

Thermal conductivity coefficient [3.59]:

T[K]	κ [W/mK], $\parallel c$	κ [W/mK], $\perp c$
302	1.21	
319		1.34

3.1.3 KD$_2$PO$_4$, Deuterated Potassium Dihydrogen Phosphate (DKDP)

Negative uniaxial crystal: $n_{\mathrm{o}} > n_{\mathrm{e}}$;
Point group: $\bar{4}$2m;
Mass density: 2.355 g/cm^3;
Mohs hardness: 2.5;
Transparency range at "0" transmittance level: $0.2 - 2.1$ µm [3.113, 114];
Linear absorption coefficient α:

λ [µm]	α [cm^{-1}]	Ref.	Note
0.266	0.035	3.115	
0.5321	0.004–0.005	3.116	98–99% deuteration
0.82–1.21	< 0.015	3.67	
0.94	0.005	3.67	
1.0642	0.004–0.005	3.116	98–99% deuteration
1.315	0.025	3.117	
1.57	0.1	3.68	o – wave, 95% deuteration
1.74	0.1	3.68	e – wave, 95% deuteration

Two-photon absorption coefficient β:

λ [µm]	$\beta \times 10^{13}$ [m/W]	Ref.	Note
0.2661	2.0 ± 1.0	3.118	
	2.7 ± 0.7	3.115	
0.3547	0.54 ± 0.19	3.71	e – wave, $\perp c$

Experimental values of refractive indices at $T = 298$ K [3.95]:

λ [µm]	n_{o}	n_{e}
0.4047	1.5189	1.4776
0.4078	1.5185	1.4772

λ [μm]	n_o	n_e
0.4358	1.5155	1.4747
0.4916	1.5111	1.4710
0.5461	1.5079	1.4683
0.5779	1.5063	1.4670
0.6234	1.5044	1.4656
0.6907	1.5022	1.4639

Temperature derivative of refractive indices [3.74]:

λ [μm]	$dn_o/dT \times 10^5$ [K^{-1}]	$dn_e/dT \times 10^5$ [K^{-1}]
0.405	−3.00	−1.86
0.436	−3.37	−2.13
0.546	−2.99	−1.95
0.578	−3.00	−2.52
0.633	−3.16	−2.03

Temperature dependences of refractive indices upon cooling from room temperature to T [K]
for the spectral range 0.365 − 0.690 μm [3.75] :

$$n_o(T) = n_o(298) + 0.228 \times 10^{-4}\left\{[n_o(298)]^2 - 1.047\right\}(298 - T) \;;$$

$$n_e(T) = n_e(298) + 0.955 \times 10^{-5}[n_e(298)]^2(298 - T) \;;$$

for the spectral range 0.436 − 0.589 μm [3.76]:

$$n_o(T) = n_o(300) + 10^{-4}(85.2 - 0.0695\,T - 7.25 \times 10^{-4}T^2) \;,$$

$$n_e(T) = n_e(300) + 10^{-4}(21.8 - 0.445\,T - 1.24 \times 10^{-3}T^2) \;.$$

Experimental values of phase-matching angle (T = 293 K) and comparison between different sets of dispersion relations:

Interacting wavelengths [μm]	θ_{exp} [deg]	θ_{theor} [deg]		
		[3.77]	[3.78]K	[3.78]E
SHG, o + o $\Rightarrow$ e				
0.530 $\Rightarrow$ 0.265	90 [3.119]	no pm	no pm	87.4
0.6943 $\Rightarrow$ 0.34715	52 [3.79]	50.6	50.9	51.0
1.062 $\Rightarrow$ 0.531	37.1 [3.120]	38.6	36.6	36.6
SHG, e + o $\Rightarrow$ e				
1.3152 $\Rightarrow$ 0.6576	51.3 [3.69]	63.2	51.7	49.4

Note: The set of dispersion relations from [3.74] shows worse agreement with the experiment.
[3.78]K $\Rightarrow$ see [3.78], data of *Kirby* et al.;

[3.78]E $\Rightarrow$ see [3.78], data of *Eimerl*.
Experimental values of NCPM temperature:

Interacting wavelengths [µm]	T [°C]	Ref.	Note
SHG, o + o $\Rightarrow$ e			
0.528 $\Rightarrow$ 0.264	−30	3.119	
0.5321 $\Rightarrow$ 0.26605	42	3.89	99% deuteration
	45	3.87	95% deuteration
	46	3.90	99% deuteration
	49.8	3.121	> 95% deuteration
	60.8	3.122	90% deuteration
0.536 $\Rightarrow$ 0.268	100	3.119	

Best set of dispersion relations (λ in µm, $T = 20$ °C) [3.78]K :

$$n_\text{o}^2 = 2.240921 + \frac{2.246956\lambda^2}{\lambda^2 - (11.26591)^2} + \frac{0.009676}{\lambda^2 - (0.124981)^2} \ ,$$

$$n_\text{e}^2 = 2.126019 + \frac{0.784404\lambda^2}{\lambda^2 - (11.10871)^2} + \frac{0.008578}{\lambda^2 - (0.109505)^2} \ .$$

Temperature-dependent Sellmeier equations (λ in µm, T in K) [3.77] :

$$n_\text{o}^2 = (1.55934 + 3.3935 \times 10^{-4} \ T) + \frac{(0.71098 - 4.1655 \times 10^{-4} \ T)\lambda^2}{\lambda^2 - (0.01407 + 6.4904 \times 10^{-6} \ T)}$$
$$+ \frac{(0.67671 + 4.8281 \times 10^{-5} \ T)\lambda^2}{\lambda^2 - 30} \ ,$$

$$n_\text{e}^2 = (1.68647 + 3.43 \times 10^{-6} \ T) + \frac{(0.46629 - 6.26 \times 10^{-5} \ T)\lambda^2}{\lambda^2 - (0.01663 + 1.3626 \times 10^{-6} \ T)}$$
$$+ \frac{(0.59614 + 2.41 \times 10^{-7} \ T)\lambda^2}{\lambda^2 - 30} \ .$$

Calculated values of phase-matching and "walk-off" angles:

Interacting wavelengths [µm]	θ_pm [deg]	ρ_1 [deg]	ρ_3 [deg]
SHG, o + o $\Rightarrow$ e			
0.5321 $\Rightarrow$ 0.26605	86.20		0.225
0.5782 $\Rightarrow$ 0.2891	66.87		1.197
0.6328 $\Rightarrow$ 0.3164	57.53		1.467
0.6594 $\Rightarrow$ 0.3297	54.31		1.522
1.6943 $\Rightarrow$ 0.34715	50.86		1.558
1.0642 $\Rightarrow$ 0.5321	36.60		1.450
1.3188 $\Rightarrow$ 0.6594	36.36		1.412

SFG, o + o ⇒ e

0.5782 + 0.5105 ⇒ 0.27112	77.88		0.595
1.0642 + 0.5321 ⇒ 0.35473	46.82		1.580
1.3188 + 0.6594 ⇒ 0.4396	39.18		1.515

SFG, e + o ⇒ e

1.0642 ⇒ 0.5321	53.47	1.286	1.427
1.3188 ⇒ 0.6594	51.70	1.222	1.420

SFG, e + o ⇒ e

1.0642 + 0.5321 ⇒ 0.35473	59.38	1.174	1.378
1.3188 + 0.6594 ⇒ 0.4396	47.70	1.254	1.527

Calculated values of inverse group-velocity mismatch for SHG process in DKDP:

Interacting wavelengths [μm]	θ_{pm} [deg]	β [fs/mm]
SHG, o + o ⇒ e		
1.2 ⇒ 0.6	35.94	< 1
1.1 ⇒ 0.55	36.28	18
1.0 ⇒ 0.5	37.47	38
0.9 ⇒ 0.45	39.79	63
0.8 ⇒ 0.4	43.75	96
0.7 ⇒ 0.35	50.37	143
0.6 ⇒ 0.3	62.54	218
SHG, e + o ⇒ e		
1.2 ⇒ 0.6	51.62	55
1.1 ⇒ 0.55	52.73	71
1.0 ⇒ 0.5	55.37	92
0.9 ⇒ 0.45	60.41	120
0.8 ⇒ 0.4	70.43	159

Experimental values of internal angular and temperature bandwidths:

Interacting wavelengths [μm]	T [°C]	θ_{pm} [deg]	$\Delta\theta^{int}$ [deg]	ΔT [°C]	Ref.
SHG, o + o ⇒ e					
1.0642 ⇒ 0.5321	20	37	0.081		3.92
0.5321 ⇒ 0.26605	60.8	90		1.8	3.122
	45	90		1.9	3.87
SHG, e + o ⇒ e					
1.0642 ⇒ 0.5321	20	54	0.131		3.123
	20		0.126		3.124
1.06 ⇒ 0.53	20	60	0.143		3.96

Experimental value of spectral bandwidth [3.96]:

Interacting wavelengths [μm]	T [°C]	θ_{pm} [deg]	$\Delta\nu$ [cm^{-1}]
SHG, $e + o \Rightarrow e$			
$1.06 \Rightarrow 0.53$	20	60	74.8

Temperature variation of phase-matching angle [3.96]:

Interacting wavelengths [μm]	T [°C]	θ_{pm} [deg]	$d\theta pm/dT$ [deg/K]
SHG, $e + o \Rightarrow e$			
$1.06 \Rightarrow 0.53$	20	60	0.0063

Temperature tuning of noncritical SHG [3.74]:

Interacting wavelengths [μm]	$d\lambda_1/dT$ [nm/K]
SHG, $o + o \Rightarrow e$	
$0.519 \Rightarrow 0.2595$	0.068

Effective nonlinearity in the phase-matching direction [3.100]:

$$d_{ooe} = d_{36} \sin\theta \sin 2\phi \; ,$$

$$d_{eoe} = d_{oee} = d_{36} \sin 2\theta \cos 2\phi \; .$$

Nonlinear coefficient [3.37]:

$$d_{36}(1.064\mu m) = 0.37 \text{ pm/V} \; .$$

Laser-induced bulk-damage threshold:

λ [μm]	τ_p [ns]	$I_{thr} \times 10^{-12}$ [W/m^2]	Ref.
0.266	0.03	> 100	3.115
0.532	30	> 0.5	3.122
	8	170	3.125
	0.6	> 80	3.72
	0.03	> 80	3.118
0.6	330	3	3.101
1.062	0.007	> 10	3.120
1.064	40	> 2.5	3.122
	18	> 1.0	3.116
	14	80	3.125
	1	60	3.124
	0.25	> 30	3.116
1.315	1	15	3.69

Thermal conductivity coefficient [3.78]:

$$\kappa = 1.86 \text{ Wm/K } (\parallel c) \ ,$$

$$\kappa = 2.09 \text{ Wm/K } (\perp c) \ .$$

3.1.4 $NH_4H_2PO_4$, Ammonium Dihydrogen Phosphate (ADP)

Negative uniaxial crystal : $n_o > n_e$;
Point group: $\bar{4}$2m;
Mass density: 1.803 g/cm^3 at 293 K [3.59];
Mohs hardness: 2;
Transparency range at "0" transmittance level: 0.18 – 1.53 µm [3.60, 126];
Transparency range at 0.5 transmittance level for a 0.8 cm long crystal:
0.185 – 1.45 µm [3.60, 59]

Linear absorption coefficient α:

λ [µm]	α [cm^{-1}]	Ref.	Note
0.25725	0.002	3.62	e – wave, $\perp c$
0.265	0.07	3.127	e – wave, $\perp c$
0.266	0.035	3.115	
0.3–1.15	< 0.07	3.64	
0.5145	0.00005	3.62	o – wave, $\perp c$
1.027	0.086	3.67	
1.083	0.208	3.67	
1.144	0.150	3.67	

Two-photon absorption coefficient β:

λ [µm]	$\beta \times 10^{13}$ [m/W]	Ref.	Note
0.2661	6 ± 1	3.118	
	11 ± 3	3.115	
	24 ± 7	3.71	$\theta = 42°$, $\phi = 45°$
0.3078	23 ± 5	3.128	
0.3547	0.68 ± 0.24	3.71	e – wave, $\perp c$

Experimental values of refractive indices at $T = 298$ K [3.73, 129]:

λ [µm]	n_o	n_e	λ [µm]	n_o	n_e
0.2138560	1.62598	1.56738	0.3021499	1.56270	1.51163
0.2288018	1.60785	1.55138	0.3125663	1.55917	1.50853
0.2536519	1.58688	1.53289	0.3131545	1.55897	1.50832
0.2967278	1.56462	1.51339	0.3341478	1.55300	1.50313

λ [μm]	n_o	n_e	λ [μm]	n_o	n_e
0.3650146	1.54615	1.49720	0.5460740	1.52662	1.48079
0.3654833	1.54608	1.49712	0.5769590	1.52478	1.47939
0.3662878	1.54592	1.49698	0.5790654	1.52466	1.47930
0.3906410	1.54174		0.6328160	1.52195	1.47727
0.4046561	1.53969	1.49159	1.0139750	1.50835	1.46895
0.4077811	1.53925	1.49123	1.1287040	1.50446	1.46704
0.4358350	1.53578	1.48831	1.1522760	1.50364	1.46666
0.4916036		1.48390			

Temperature derivative of refractive indices [3.74]:

λ [μm]	$dn_\mathrm{o}/dT \times 10^5$ [K^{-1}]	$dn_\mathrm{e}/dT \times 10^5$ [K^{-1}]
0.405	−4.78	≈ 0
0.436	−4.94	≈ 0
0.546	−5.23	≈ 0
0.578	−4.60	≈ 0
0.633	−5.08	≈ 0

Temperature dependences of refractive indices upon cooling from room temperature to T [K].
for the spectral range 0.365 – 0.690 μm [3.75]:

$$n_\mathrm{o}(T) = n_\mathrm{o}(298) + 0.713 \times 10^{-2}\{[n_\mathrm{o}(298)]^2$$
$$- 3.0297\, n_\mathrm{o}(298) + 2.3004\}(298 - T) \ ,$$

$$n_\mathrm{e}(T) = n_\mathrm{e}(298) + 0.675 \times 10^{-6}(298 - T) \ ;$$

for the spectral range 0.436 – 0.589 μm [3.76]:

$$n_\mathrm{o}(T) = n_\mathrm{o}(300) + 10^{-4}(141.8 - 0.322\, T - 5.02 \times 10^{-4}\, T^2) \ ,$$

$$n_\mathrm{e}(T) = n_\mathrm{e}(300) + 10^{-4}(2.5 - 0.01763\, T + 2.901 \times 10^{-5}\, T^2) \ .$$

Experimental values of phase-matching angle (T = 293 K) and comparison between different sets of dispersion relations:

Interacting wavelengths [μm]	θ_exp [deg]	θ_theor [deg]		
		[3.73] [3.129]	[3.77]	[3.78]K
SHG, o + o ⇒ e				
0.524 ⇒ 0.262	90 [3.74]	no pm	no pm	83.6
0.530 ⇒ 0.265	81.7 [3.97]	81.6	82.2	79.6
0.6943 ⇒ 0.34715	51.9 [3.79]	51.1	51.1	51.5
0.7035 ⇒0.35175	50.5 [3.130]	50.4	50.5	50.8

$1.06 \Rightarrow 0.53$	41.9 [3.79]	41.7	41.7	42.2
	42 [3.81]			
SFG, $o + o \Rightarrow e$				
$1.0642 + 0.5321 \Rightarrow$				
$\Rightarrow 0.35473$	46.9 [3.85]	47.8	47.9	48.3
$1.0642 + 0.2810 \Rightarrow$				
$\Rightarrow 0.22230$	90 [3.84]	89.0	no pm	74.7
$0.81219 + 0.34715 \Rightarrow$				
$\Rightarrow 0.24320$	90 [3.131]	no pm	no pm	81.1
SFG $e + o \Rightarrow e$				
$1.0642 + 0.5321 \Rightarrow$				
$\Rightarrow 0.35473$	60.2 [3.85]	59.9	60.0	60.4

Note: The other sets of dispersion relations from [3.74] and [3,78]E show worse agreement with the experiment.

[3.78]K $\Rightarrow$ see [3.78], data of *Kirby* et al.:

[3.78]E $\Rightarrow$ see [3.78], data of *Eimerl*.

Experimental values of NCPM temperature:

Interacting wavelengths [µm]	T [°C]	Ref.	Note
SHG, $o + o \Rightarrow e$			
$0.4920 \Rightarrow 0.2460$	−116	3.132	
$0.4965 \Rightarrow 0.24825$	−93.2	3.133	
$0.5017 \Rightarrow 0.25085$	−68.4	3.133	
$0.5145 \Rightarrow 0.25725$	−11.7	3.99	
	−10.2	3.133	
	−9.2	3.62	
$0.524 \Rightarrow 0.262$	20	3.74	
$0.52534 \Rightarrow 0.26267$	30	3.134	
$0.53 \Rightarrow 0.265$	43	3.127	
	47	3.97	
	48	3.135	
	49.6	3.136	
$0.5321 \Rightarrow 0.26605$	47.1	3.90	
	49.5	3.137	
	50	3.138	
	51.2	3.139	0.1–1 Hz
	44.6	3.139	20 Hz
	51–52	3.140	
$0.548 \Rightarrow 0.274$	100	3.134	
$0.557 \Rightarrow 0.2785$	120	3.119	
SFG, $o + o \Rightarrow e$			
$1.0642 + 0.26605 \Rightarrow 0.21284$	−55	3.141	

Best set of dispersion relations (λ in μm, $T = 20$ °C) [3.73], [3.129] :

$$n_{\mathrm{o}}^2 = 2.302842 + \frac{15.102464\lambda^2}{\lambda^2 - 400} + \frac{0.011125165}{\lambda^2 - (75.450861)^{-1}} \; ,$$

$$n_{\mathrm{e}}^2 = 2.163510 + \frac{5.919896\lambda^2}{\lambda^2 - 400} + \frac{0.009616676}{\lambda^2 - (76.98751)^{-1}} \; .$$

Temperature-dependent Sellmeier equations (λ in μm, T in K) [3.77] :

$$n_{\mathrm{o}}^2 = (1.6996 - 8.7835 \times 10^{-4}\ T) + \frac{(0.64955 + 7.2007 \times 10^{-4}\ T)\lambda^2}{\lambda^2 - (0.01723 - 1.40526 \times 10^{-5}\ T)}$$
$$+ \frac{(1.10624 - 1.179 \times 10^{-4}\ T)\lambda^2}{\lambda^2 - 30} \; ,$$

$$n_{\mathrm{e}}^2 = (1.42036 - 1.089 \times 10^{-5}\ T) + \frac{(0.74453 + 5.14 \times 10^{-6}\ T)\lambda^2}{\lambda^2 - (0.013 - 2.471 \times 10^{-7}\ T)}$$
$$+ \frac{(0.42033 - 9.99 \times 10^{-7}\ T)\lambda^2}{\lambda^2 - 30} \; .$$

Calculated values of phase-matching and "walk-off" angles:

Interacting wavelengths [μm]	θ_{pm} [deg]	ρ_1 [deg]	ρ_3 [deg]
SHG o + o $\Rightarrow$ e			
0.5321 $\Rightarrow$ 0.26605	80.15		0.639
0.5782 $\Rightarrow$ 0.2891	65.28		1.427
0.6328 $\Rightarrow$ 0.3164	56.91		1.703
0.6594 $\Rightarrow$ 0.3297	54.07		1.762
0.6943 $\Rightarrow$ 0.34715	51.09		1.803
1.0642 $\Rightarrow$ 0.5321	41.74		1.746
1.3188 $\Rightarrow$ 0.6594	45.55		1.694
SFG o + o $\Rightarrow$ e			
0.5782 + 0.5105 $\Rightarrow$ 0.27112	74.84		0.955
1.0642 + 0.5321 $\Rightarrow$ 0.35473	47.82		1.836
1.3188 + 0.6594 $\Rightarrow$ 0.4396	42.56		1.794
SHG e + o $\Rightarrow$ e			
1.0642 $\Rightarrow$ 0.5321	61.39	1.230	1.449
1.3188 $\Rightarrow$ 0.6594	65.63	0.968	1.250
SFG e + e $\Rightarrow$ e			
1.0642 + 0.5321 $\Rightarrow$ 0.35473	59.85	1.272	1.582
1.3188 + 0.6594 $\Rightarrow$ 0.4396	50.86	1.274	1.748

Calculated values of inverse group-velocity mismatch for SHG process in ADP:

Interacting wavelengths [μm]	θ_{pm} [deg]	β [fs/mm]
SHG, o + o $\Rightarrow$ e		
1.2 $\Rightarrow$ 0.6	43.10	49
1.1 $\Rightarrow$ 0.55	41.94	21
1.0 $\Rightarrow$ 0.5	41.71	8
0.9 $\Rightarrow$ 0.45	42.68	42
0.8 $\Rightarrow$ 0.4	45.34	85
0.7 $\Rightarrow$ 0.35	50.67	142
0.6 $\Rightarrow$ 0.3	61.39	233
SHG, e + o $\Rightarrow$ e		
1.2 $\Rightarrow$ 0.6	62.50	105
1.1 $\Rightarrow$ 0.55	61.39	78
1.0 $\Rightarrow$ 0.5	62.02	95
0.9 $\Rightarrow$ 0.45	65.24	127
0.8 $\Rightarrow$ 0.4	73.80	173

Experimental values of internal angular and temperature bandwidths:

Interacting wavelengths [μm]	T [°C]	θ_{pm} [deg]	$\Delta\theta^{int}$ [deg]	ΔT [°C]	Ref.
SHG, o + o $\Rightarrow$ e					
1.06 $\Rightarrow$ 0.53	20	42	0.057		3.81
0.5321 $\Rightarrow$ 0.26605	49.5	90		0.60	3.137
	51	90	1.086	0.53	3.139
0.53 $\Rightarrow$ 0.265	20	82	0.118		3.103
	20	82	0.088		3.96
	20	82	0.089	0.63	3.97

Experimental values of spectral bandwidth:

Interacting wavelengths [μm]	T [°C]	θ_{pm} [deg]	$\Delta\nu$ [cm^{-1}]	Ref.
SHG, o + o $\Rightarrow$ e				
1.06 $\Rightarrow$ 0.53	20	42	178	3.81
0.53 $\Rightarrow$ 0.265	20	82	1.2	3.96

Temperature variation of phase-matching angle [3.97]:

Interacting wavelengths [μm]	T [°C]	θ_{pm} [deg]	$d\theta_{pm}/dT$ [deg/K]
SHG, o + o $\Rightarrow$ e			
0.53 $\Rightarrow$ 0.265	20	82	0.1418
	47	90	1.1020

Temperature tuning of noncritical SHG [3.74]:

Interacting wavelengths [μm]	$d\lambda_1/dT$ [nm/K]
SHG, o + o $\Rightarrow$ e	
0.524 $\Rightarrow$ 0.262	0.306

Temperature tuning of noncritical SFG [3.142]:

Interacting wavelengths [μm]	$d\lambda_3/dT$ [nm/K]
SFG, o + o $\Rightarrow$ e	
0.6943 + 0.39961 $\Rightarrow$ 0.25363	0.171

Temperature variation of birefringence for noncritical SHG process ($0.5145\,\mu m \Rightarrow 0.25725\,\mu m$, o + o $\Rightarrow$ e):

$$d(n_2^e - n_1^o)/dT = 5.65 \times 10^{-5} K^{-1} [3.99].$$

Effective nonlinearity expressions in the phase-matching direction [3.100]:

$$d_{ooe} = d_{36} \sin\theta \sin 2\phi \ ,$$

$$d_{eoe} = d_{oee} = d_{36} \sin 2\theta \cos 2\phi \ .$$

Nonlinear coefficient [3.37]:

$$d_{36}(1.064\,\mu m) = 0.47 \text{ pm/V} \ .$$

Laser-induced bulk-damage threshold:

λ [μm]	τ_p [ns]	$I_{thr} \times 10^{-12}$ [W/m²]	Ref.	Note
0.265	30	> 10	3.127	
0.266	0.03	> 100	3.120	
0.53	0.5	> 130	3.110	
0.5321	3	> 7.5	3.140	30 Hz
	0.6	> 80	3.72	
	0.03	> 80	3.118	
0.6	330	18	3.101	
1.06	60	5	3.109	

Thermal conductivity coefficient [3.59]:

T [K]	κ [W/mK], $\parallel c$	κ [W/mK], $\perp c$
315	0.71	1.26
340	0.71	1.34

3.1.5 β-BaB$_2$O$_4$, Beta-Barium Borate (BBO)

Negative uniaxial crystal: $n_\mathrm{o} > n_\mathrm{e}$;
Point group: 3m;
Mass density: 3.85 g/cm^3;
Mohs hardness: 4 [3.124];
Transparency range at "0" transmittance level: $0.189 - 3.5$ µm [3.143, 144];
Transparency range at 0.5 transmittance level for a 0.8 cm long crystal: 0.198-2.6 µm [3.145];
Linear absorption coefficient α:

λ [µm]	α [cm^{-1}]	Ref.
0.532	0.01	3.146
2.55	0.5	3.147

Experimental values of refractive indices [3.148]:

λ [µm]	n_o	n_e
0.40466	1.69267	1.56796
0.43583	1.68679	1.56376
0.46782	1.68198	1.56024
0.47999	1.68044	1.55914
0.50858	1.67722	1.55691
0.54607	1.67376	1.55465
0.57907	1.67131	1.55298
0.58930	1.67049	1.55247
0.64385	1.66736	1.55012
0.81890	1.66066	1.54589
0.85212	1.65969	1.54542
0.89435	1.65862	1.54469
1.01400	1.65608	1.54333

Temperature derivative of refractive indices at $\lambda = 0.4 - 1.0$ µm [3.148]:

$$dn_\mathrm{o}/dT = -16.6 \times 10^{-6} \text{ K}^{-1} \, ,$$

$$dn_\mathrm{e}/dT = -9.3 \times 10^{-6} \text{ K}^{-1} \, .$$

Experimental values of phase-matching angle ($T = 293$ K) and comparison between different sets of dispersion relations:

Interacting wavelengths [μm]	θ_{exp} [deg]	θ_{theor} [deg]		
		[3.149]	[3.148]	[3.145]
SHG, $o + o \Rightarrow e$				
$0.4096 \Rightarrow 0.2048$	90 [3.145]	89.36	86.51	88.82
$0.41 \Rightarrow 0.205$	90 [3.150]	87.25	85.54	86.97
$0.41152 \Rightarrow 0.20576$	82.8 [3.145]	84.11	82.99	83.77
$0.41546 \Rightarrow 0.20773$	79.2 [3.145]	79.80	78.87	79.31
$0.4765 \Rightarrow 0.23825$	57 [3.151]	57.79	56.57	56.73
$0.488 \Rightarrow 0.244$	54.5 [3.151]	55.53	54.29	54.46
$0.4965 \Rightarrow 0.24825$	52.5 [3.151]	54.00	52.76	52.94
$0.5145 \Rightarrow 0.25725$	49.5 [3.151]	51.13	49.87	50.06
$0.5321 \Rightarrow 0.26605$	47.3 [3.148]	48.67	47.42	47.62
	47.5 [3.145]			
	47.5 [3.152]			
	47.6 [3.153]			
	47.6 [3.45]			
	48 [3.154]			
$0.604 \Rightarrow 0.302$	40 [3.155]	41.00	39.89	40.13
$0.6156 \Rightarrow 0.3078$	39 [3.156]	40.02	38.95	39.18
$0.70946 \Rightarrow 0.35473$	32.9 [3.157]	33.65	32.94	33.15
	32.9 [3.158]			
	33 [3.159]			
	33 [3.152]			
	33 [3.160]			
	33.1 [3.45]			
	33.3 [3.147]			
	33.7 [3.161]			
$1.0642 \Rightarrow 0.5321$	22.7 [3.148]	21.42	22.88	22.78
	22.8 [3.145]			
	22.8 [3.152]			
	22.8 [3.33]			
	22.8 [3.162]			
	22.8 [3.45]			
	22.8 [3.163]			
SFG, $o + o \Rightarrow e$				
$0.73865 + 0.25725 \Rightarrow$ $\Rightarrow 0.1908$	81.7 [3.164]	72.94	75.27	76.11
$0.72747 + 0.26325 \Rightarrow$ $\Rightarrow 0.1933$	76 [3.165]	71.79	73.59	74.22
$0.5922 + 0.2961 \Rightarrow$ $\Rightarrow 0.1974$	88 [3.166]	80.44	82.13	83.22

$0.5964 + 0.2982 \Rightarrow$				
$\Rightarrow 0.1988$	82.5 [3.167]	78.02	79.11	79.81
$0.5991 + 0.29955 \Rightarrow$				
$\Rightarrow 0.1997$	80 [3.166]	76.71	77.57	78.14
$0.60465 + 0.30233 \Rightarrow$				
$\Rightarrow 0.20155$	76.2 [3.167]	74.41	74.92	75.34
$0.5321 + 0.32561 \Rightarrow$				
$\Rightarrow 0.202$	83.9 [3.145]	80.88	81.22	81.95
$0.6099 + 0.30495 \Rightarrow$				
$\Rightarrow 0.2033$	73.5 [3.166]	72.51	72.82	73.16
$0.5321 + 0.34691 \Rightarrow$				
$\Rightarrow 0.21$	71.9 [3.145]	72.11	71.60	71.84
$1.0642 + 0.26605 \Rightarrow$				
$\Rightarrow 0.21284$	51.1 [3.145]	50.69	51.04	51.12
$1.0642 + 0.35473 \Rightarrow$				
$\Rightarrow 0.26605$	40.2 [3.145]	40.75	40.19	40.31
$1.0642 + 0.5321 \Rightarrow$				
$\Rightarrow 0.35473$	31.1 [3.148]	31.52	31.12	31.28
	31.3 [3.145]			
	31.4 [3.161]			
$0.5782 + 0.5106 \Rightarrow$				
$\Rightarrow 0.27115$	46 [3.168]	45.23	46.03	46.24
$0.59099 + 0.5321 \Rightarrow$				
$\Rightarrow 0.28$	44.7 [3.169]	45.23	44.03	44.25
$2.68823 + 0.5712 \Rightarrow$				
$\Rightarrow 0.4711$	21.8 [3.170]	18.37	21.73	21.39
$1.41831 + 1.0642 \Rightarrow$				
$\Rightarrow 0.608$	21 [3.171]	18.40	21.26	20.96
SHG, $e + o \Rightarrow e$				
$0.5321 \Rightarrow 0.26605$	81 [3.145]	no pm	82.03	80.78
$0.70946 \Rightarrow 0.35473$	48 [3.159]	48.72	47.61	47.92
	48.1 [3.152]			
$1.0642 \Rightarrow 0.5321$	31.6 [3.172]	30.00	31.94	32.18
	32.4 [3.148]			
	32.7 [3.152]			
	32.7 [3.33]			
	32.9 [3.145]			
SFG, $e + o \Rightarrow e$				
$1.0642 + 0.35473 \Rightarrow$				
$\Rightarrow 0.26605$	46.6 [3.145]	46.81	46.11	46.31
$1.0642 + 0.5321 \Rightarrow$				
$\Rightarrow 0.35473$	38.4 [3.148]	38.39	37.77	38.15
	38.5 [3.145]			

SFG, $o + e \Rightarrow e$
$1.0642 + 0.5321 \Rightarrow$

$\Rightarrow 0.35473$	59.8 [3.145]	59.46	58.91	58.89

Note: The sets of dispersion relations from [3.143, 154, 170] show worse agreement with the experiment.

Best set of dispersion relations (λ in µm, $T = 20\ °\text{C}$) [3.145]:

$$n_o^2 = 2.7359 + \frac{0.01878}{\lambda^2 - 0.01822} - 0.01354\lambda^2 \ ,$$

$$n_e^2 = 2.3753 + \frac{0.01224}{\lambda^2 - 0.01667} - 0.01516\lambda^2 \ .$$

Calculated values of phase-matching and "walk-off" angles:

Interacting wavelengths [µm]	θ_{pm} [deg]	ρ_3 [deg]
SHG, $o + o \Rightarrow e$		
$0.4880 \Rightarrow 0.2440$	54.46	4.757
$0.5105 \Rightarrow 0.25525$	50.66	4.861
$0.5145 \Rightarrow 0.25725$	50.06	4.869
$0.5321 \Rightarrow 0.26605$	47.62	4.879
$0.5782 \Rightarrow 0.2891$	42.46	4.782
$0.6328 \Rightarrow 0.3164$	37.87	4.571
$0.6594 \Rightarrow 0.3297$	36.05	4.457
$0.6943 \Rightarrow 0.34715$	33.96	4.306
$1.0642 \Rightarrow 0.5321$	22.78	3.189
$1.3188 \Rightarrow 0.6594$	20.36	2.881
SFG, $o + o \Rightarrow e$		
$1.3188 + 0.6594 \Rightarrow 0.4396$	25.39	3.515
$1.3188 + 0.4396 \Rightarrow 0.3297$	31.19	4.205
$1.3188 + 0.3297 \Rightarrow 0.26376$	37.40	4.897
$1.3188 + 0.26376 \Rightarrow 0.2198$	44.52	5.588
$1.0642 + 0.5321 \Rightarrow 0.35473$	31.28	4.132
$1.0642 + 0.35473 \Rightarrow 0.26605$	40.31	4.941
$1.0642 + 0.26605 \Rightarrow 0.21284$	51.12	5.497
$0.6943 + 0.34715 \Rightarrow 0.23143$	55.00	4.882
$0.5782 + 0.5105 \Rightarrow 0.27112$	46.12	4.872
$0.5145 + 0.4880 \Rightarrow 0.25045$	52.17	4.831

Interacting wavelengths [µm]	θ_{pm} [deg]	ρ_1 [deg]	ρ_2 [deg]	ρ_3 [deg]
SHG, e + o $\Rightarrow$ e				
0.5321 $\Rightarrow$ 0.26605	80.78	1.252	1.252	1.446
0.5782 $\Rightarrow$ 0.2891	65.08	3.068	3.068	3.460
0.6328 $\Rightarrow$ 0.3164	55.98	3.773	3.773	4.163
0.6594 $\Rightarrow$ 0.3297	52.77	3.941	3.941	4.310
0.6943 $\Rightarrow$ 0.34715	49.25	4.070	4.070	4.408
1.0642 $\Rightarrow$ 0.5321	32.18	3.840	3.840	3.940
1.3188 $\Rightarrow$ 0.6594	28.77	3.632	3.632	3.663
SFG, e + o $\Rightarrow$ e				
1.3188 + 0.6594 $\Rightarrow$ 0.4396	30.88	3.773		3.947
1.3188 + 0.4396 $\Rightarrow$ 0.3297	35.71	4.013		4.444
1.3188 + 0.3297 $\Rightarrow$ 0.26376	41.38	4.140		4.973
1.0642 + 0.5321 $\Rightarrow$ 0.35473	38.15	4.078		4.441
1.0642 + 0.35473 $\Rightarrow$ 0.26605	46.31	4.108		4.913
1.0642 + 0.26605 $\Rightarrow$ 0.21284	56.96	3.666		5.048
0.6943 + 0.34715 $\Rightarrow$ 0.23143	72.50	2.254		2.860
0.5782 + 0.5105 $\Rightarrow$ 0.27112	70.05	2.555		2.951
SFG, o + e $\Rightarrow$ e				
1.3188 + 0.6594 $\Rightarrow$ 0.4396	45.50		4.164	4.312
1.3188 + 0.4396 $\Rightarrow$ 0.3297	78.68		1.556	1.640
1.0642 + 0.5321 $\Rightarrow$ 0.35473	58.89		3.619	3.831
0.5782 + 0.5105 $\Rightarrow$ 0.27112	84.64		0.737	0.842

Calculated values of inverse group-velocity mismatch for SHG process in BBO:

Interacting wavelengths [µm]	θ_{pm} [deg]	β [fs/mm]
SHG, o + o $\Rightarrow$ e		
1.2 $\Rightarrow$ 0.6	21.18	54
1.1 $\Rightarrow$ 0.55	22.28	76
1.0 $\Rightarrow$ 0.5	23.85	104
0.9 $\Rightarrow$ 0.45	26.07	141
0.8 $\Rightarrow$ 0.4	29.18	194
0.7 $\Rightarrow$ 0.35	33.65	275
0.6 $\Rightarrow$ 0.3	40.47	415
0.5 $\Rightarrow$ 0.25	52.34	695
SHG, e + o $\Rightarrow$ e		
1.2 $\Rightarrow$ 0.6	29.91	103
1.1 $\Rightarrow$ 0.55	31.46	130
1.0 $\Rightarrow$ 0.5	33.73	164

0.9 ⇒ 0.45	36.98	210
0.8 ⇒ 0.4	41.67	276
0.7 ⇒ 0.35	48.74	373
0.6 ⇒ 0.3	60.91	531

Experimental values of internal angular, temperature and spectral bandwidths at $T = 293$ K:

Interacting wavelengths [µm]	θ_{pm} [deg]	$\Delta\theta^{\mathrm{int}}$ [deg]	ΔT [°C]	$\Delta\nu$ [cm^{-1}]	Ref.
SHG, o + o ⇒ e					
0.5321 ⇒ 0.26605	47.3	0.010	4		3.148
1.0642 ⇒ 0.5321	22.8	0.021	37	9.7	3.33
	21.9	0.028			3.154
	22.7	0.030	51		3.148
SFG, o + o ⇒ e					
1.0642 + 0.5321 ⇒ 0.35473	31.1	0.015	16		3.148
2.44702 + 0.5712 ⇒ 0.4631	22.1	0.026			3.170
2.68823 + 0.5712 ⇒ 0.4711	21.8	0.028			3.170
SHG, e + o ⇒ e					
1.0642 ⇒ 0.5321	32.7	0.034		8.8	3.33
	32.4	0.046	37		3.148
SFG, e + o ⇒ e					
1.0642 + 0.5321 ⇒ 0.35473	38.4	0.020	13		3.148
SFG, o + e ⇒ e					
1.0642 + 0.5321 ⇒ 0.35473	58.4	0.050	12		3.148

Temperature variation of phase-matching angle at $T = 293$ K [3.148]:

Interacting wavelengths [µm]	θ_{pm} [deg]	$\mathrm{d}\theta_{\mathrm{pm}}/\mathrm{d}T$ [deg/K]
SHG, o + o ⇒ e		
0.5321 ⇒ 0.26605	47.3	0.00250
1.0642 ⇒ 0.5321	22.7	0.00057
SFG, o + o ⇒ e		
1.0642 + 0.5321 ⇒ 0.35473	31.1	0.00099
SHG, e + o ⇒ e		
1.0642 ⇒ 0.5321	32.4	0.00120
SFG, e + o ⇒ e		
1.0642 + 0.5321 ⇒ 0.35473	38.4	0.00150
SFG, o + e ⇒ e		
1.0642 + 0.5321 ⇒ 0.35473	58.4	0.00421

Effective nonlinearity expressions in the phase-matching direction [3.100]:

$$d_{ooe} = d_{31} \sin\theta - d_{22} \cos\theta \sin 3\phi \ ,$$

$$d_{eoe} = d_{oee} = d_{22} \cos^2\theta \cos 3\phi \ .$$

Nonlinear coefficients [3.37, 143, 170]:

$$d_{22}(1.0642\,\mu m) = \pm\, 2.3 \text{ pm/V} \ ,$$
$$d_{31}(1.0642\,\mu m) = \mp\, 0.16 \text{ pm/V} \ .$$

Laser-induced damage threshold:

λ [μm]	τ_p [ns]	$I_{thr} \times 10^{-12}$ [W/m^2]	Ref.	Note
0.266	8	> 1.2	3.153	
0.3078	12	> 2.0	3.173	
0.3547	10	50	3.147	
	8	250	3.125	1 pulse
	8	190	3.125	1800 pulses
	0.03	> 4	3.46	10 Hz
	0.015	> 30	3.157	
0.5106	20	> 2.5	3.174	4 kHz
0.51–0.58	20	10	3.175	4–14 kHz, surface damage
	20	100	3.176	bulk damage
0.5145	cw	> 0.004	3.177	
0.5321	8	480	3.125	1 pulse
	8	320	3.125	1800 pulses
	1	70	3.154	
	0.25	100	3.124	
	0.075	> 70	3.144	
	0.025	> 42	3.158	10 Hz
	0.025	> 40	3.163	
0.5398	0.015	1200–1500 (?)	3.178	1 pulse, surface damage
0.62	0.0002	> 500	3.179	
	0.0001	10 000 (?)	3.180	
0.6943	0.02	100	3.143	
1.054	0.005	500	3.181	
1.0642	14	500	3.125	1 pulse
	14	230	3.125	1800 pulses
	10	50	3.124	
	1.3	100	3.33	
	1.0	135	3.124	
	0.1	100	3.124	
	0.035	> 50	3.158	
1.0796	0.015	2500–3500 (?)	3.178	1 pulse, surface damage

Thermal conductivity coefficient:

κ [W/mK], $\parallel c$	κ [W/mK], $\perp c$	Ref.
0.8	0.08	3.148
1.6	1.2	3.58

3.1.6 LiIO$_3$, Lithium Iodate

Negative uniaxial crystal: $n_\mathrm{o} > n_\mathrm{e}$;
Point group: 6;
Mass density: 4.49 g/cm^3 [3.182];
Mohs hardness: 3.5 − 4.0;
Transparency range at "0" transmittance level: 0.28 − 6 µm [3.183, 184];
Linear absorption coefficient α:

λ [µm]	α [cm^{-1}]	Ref.	Note
0.34715	0.1	3.185	$\parallel c$
	0.3	3.185	e − wave, $\perp c$
0.5145	0.0024	3.186	$\parallel c$
	0.0025	3.186	e − wave, $\perp c$
0.5321	0.3	3.187	e − wave
0.5422	0.37	3.184	
0.6594	0.0007–0.0023	3.186	$\parallel c$
	0.0006–0.0017	3.186	e − wave, $\perp c$
1.0642	0.1	3.187	o − wave
	0.25	3.187	e − wave
	< 0.0002	3.186	$\parallel c$
	0.0008	3.186	e − wave, $\perp c$
1.0845	0.06	3.184	
1.315	0.0005	3.182	
1.3188	0.0008–0.0036	3.186	$\parallel c$
	0.0007–0.0010	3.186	e − wave, $\perp c$

Two-photon absorption coefficient at $\lambda = 0.532$ µm:
$\beta < 4 \times 10^{-12}$ m/W [3.188].

Experimental values of refractive indices:

λ [µm]	n_o	n_e	Ref.	λ [µm]	n_o	n_e	Ref.
0.3547	1.9822	1.8113	3.189	0.3996	1.9464	1.7842	3.189
0.3669	1.9706	1.8026	3.189	0.4047	1.9443	1.7826	3.190
0.3712	1.9671	1.8000	3.189	0.4358	1.9275	1.7702	3.189
0.3795	1.9600	1.7947	3.189	0.4545	1.9184	1.7638	3.191
0.3877	1.9544	1.7905	3.189	0.4579	1.9170	1.7630	3.191

λ [µm]	n_o	n_e	Ref.	λ [µm]	n_o	n_e	Ref.
0.4658	1.9141	1.7611	3.191	0.7000	8746	1.7300	3.96
0.4727	1.9122	1.7600	3.191	0.7660	1.8694	1.7261	3.96
0.4765	1.9100	1.7583	3.191	0.8000	1.8673	1.7245	3.96
0.4800	1.9109	1.7579	3.189	0.8630	1.8640	1.7220	3.96
0.4880	1.9083	1.7556	3.191	0.9000	1.8623	1.7207	3.96
0.5017	1.9053	1.7537	3.191	1.0000	1.8587	1.7180	3.96
0.5086	1.9031	1.7514	3.189	1.1000	1.8559	1.7160	3.96
0.5145	1.9012	1.7487	3.191	1.2000	1.8536	1.7143	3.96
0.5320	1.8975	1.7475	3.189	1.3000	1.8517	1.7130	3.96
0.5461	1.8950	1.7455	3.96	1.3674	1.8508	1.7122	3.190
0.5600	1.8921	1.7433	3.189	1.5296	1.8482	1.7101	3.190
0.5791	1.8894	1.7413	3.190	1.6920	1.8464	1.7089	3.190
0.5800	1.8889	1.7403	3.189	1.9701	1.8431	1.7072	3.190
0.5896	1.8875	1.7400	3.190	2.2493	1.8385	1.7050	3.190
0.6000	1.8859	1.7383	3.189	2.5000	1.8378	1.7037	3.192
0.6200	1.8828	1.7361	3.189	3.0000	1.8319	1.7001	3.192
0.6328	1.8815	1.7351	3.192	3.5000	1.8266	1.6971	3.192
0.6438	1.8807	1.7346	3.190	4.0000	1.8140	1.6897	3.192
0.6560	1.8789	1.7332	3.96	5.0000	1.7940	1.6783	3.192

Optical activity at $T = 300$ K :

λ [µm]	ρ [deg/mm]	Ref.	λ [µm]	ρ [deg/mm]	Ref.
0.286	1052.9	3.193	0.429	222.46	3.193
0.290	964.99	3.193	0.448	198.72	3.193
0.295	886.65	3.193	0.470	175.75	3.193
0.299	814.39	3.193	0.492	153.61	3.193
0.304	748.76	3.193	0.520	133.02	3.193
0.310	687.46	3.193	0.546	117.42	3.193
0.317	630.44	3.193	0.551	113.36	3.193
0.324	579.01	3.193	0.600	95.27	3.193
0.331	532.44	3.193	0.628	86.80	3.193
0.339	489.47	3.193	1.084	25.0	3.184
0.347	448.42	3.193	1.1	23.83	3.194
0.355	410.37	3.193	1.6	11.00	3.194
0.363	374.34	3.193	2.1	6.33	3.194
0.374	340.18	3.193	2.6	4.12	3.194
0.386	308.07	3.193	3.1	2.89	3.194
0.399	277.45	3.193	3.6	2.32	3.194
0.412	249.32	3.193			

Temperature derivative of refractive indices [3.186]:

λ [μm]	$dn_{\mathrm{o}}/dT \times 10^5$ [K^{-1}]	$dn_{\mathrm{e}}/dT \times 10^5$ [K^{-1}]
0.5321	−9.64	−8.61
0.6594	−9.49	−8.39
1.0642	−8.93	−7.52
1.3188	−9.44	−8.49

Experimental values of phase-matching angle ($T = 293$ K) and comparison between different sets of dispersion relations:

Interacting wavelengths [μm]	θ_{exp} [deg]	θ_{theor} [deg]		
		[3.195]	[3.192]	[3.196]
SHG, o + o ⇒ e				
0.586 ⇒ 0.293	90 [3.195]	87.7	81.0	83.6
0.5863 ⇒ 0.29315	90 [3.196]	86.9	80.7	83.2
0.6 ⇒ 0.3	75.6 [3.196]	75.5	73.5	73.7
0.62 ⇒ 0.31	68.2 [3.196]	68.0	67.0	66.3
0.6943 ⇒ 0.34715	52 [3.197]	53.5	53.4	52.0
	52 [3.198]			
1.06 ⇒ 0.53	30 [3.199]	30.2	30.3	29.5
1.0642 ⇒ 0.5321	30.2 [3.200]	30.1	30.2	29.4
	30.2 [3.92]			
	30 [3.201]			
	30 [3.202]			
1.0845 ⇒ 0.54225	28.9 [3.184]	29.5	29.6	28.8
1.1523 ⇒ 0.57615	27.2 [3.184]	27.6	27.7	27.1
1.3886 ⇒ 0.6943	23.1 [3.203]	23.2	23.2	23.1
1.746 ⇒ 0.873	20 [3.204]	19.9	19.6	20.3
SFG, o + o ⇒ e				
5.33 + 1.32969 ⇒ 1.0642	21 [3.202]	20.1	20.0	22.1
4.44 + 1.39968 ⇒ 1.0642	20.2 [3.202]	19.4	19.0	21.2
5.2 + 0.80129 ⇒ 0.6943	19.5 [3.205]	19.6	19.6	20.6
2.5 + 0.96126 ⇒ 0.6943	21 [3.206]	21.5	21.4	21.7
5.0 + 0.66251 ⇒ 0.585	20.3 [3.207]	20.3	20.3	20.9
2.0 + 0.82686 ⇒ 0.585	25.1 [3.207]	25.1	25.1	24.8
4.16 + 0.61015 ⇒ 0.5321	21.6 [3.208]	21.7	21.7	22.0
2.66 + 0.66514 ⇒ 0.5321	24.5 [3.208]	24.6	24.6	24.4
0.946 + 0.5484 ⇒ 0.34715	50 [3.209]	51.0	50.9	49.6
2.67 + 0.6943 ⇒ 0.55102	24.4 [3.210]	24.1	24.1	23.9
1.98 + 0.6943 ⇒ 0.51405	27.4 [3.210]	27.6	27.6	27.1
1.2013 + 0.6943 ⇒ 0.44	35.1 [3.203]	36.2	36.4	35.2
3.3913 + 0.5145 ⇒ 0.44673	24 [3.211]	25.4	25.4	25.1
2.38 + 0.4880 ⇒ 0.40497	30.5 [3.212]	31.1	31.1	30.5
1.0642 + 0.5321 ⇒ 0.35473	47.5 [3.200]	47.8	47.8	46.5

Note: the other sets of dispersion relations from [3.213, 214, 215, 189, 202] show worse agreement with the experiment.

Best set of dispersion relations (λ in μm, $T = 20\ °C$) [3.295] (a corrected set from [3.213]):

$$n_o^2 = 3.4132 + \frac{0.0476}{\lambda^2 - 0.0338} - 0.0077\,\lambda^2\ ,$$

$$n_e^2 = 2.9211 + \frac{0.0346}{\lambda^2 - 0.0320} - 0.0042\,\lambda^2\ .$$

Calculated values of phase-matching and "walk-off" angles:

Interacting wavelengths [μm]	θ_{pm}[deg]	ρ_3[deg]
SHG, o + o $\Rightarrow$ e		
0.6328 $\Rightarrow$ 0.3164	64.52	3.90
0.6943 $\Rightarrow$ 0.34715	53.48	4.76
1.0642 $\Rightarrow$ 0.5321	30.08	4.23
1.3188 $\Rightarrow$ 0.6594	24.27	3.63
2.9365 $\Rightarrow$ 1.46825	20.15	3.04
SFG, o + o $\Rightarrow$ e		
1.0642 + 0.5321 $\Rightarrow$ 0.354733	47.81	5.00
1.3188 + 0.6594 $\Rightarrow$ 0.4396	35.42	4.67

Experimental values of internal angular, temperature and spectral bandwidths ($T = 293$K):

Interacting wavelengths [μm]	θ_{pm} [deg]	$\Delta\theta^{int}$ [deg]	ΔT [°C]	$\Delta\nu$ [cm^{-1}]	Ref.
SHG, o + o $\Rightarrow$ e					
0.586 $\Rightarrow$ 0.293	90	0.5–0.58		2.04	3.195
0.6943 $\Rightarrow$ 0.34715	52	0.018			3.197
1.06 $\Rightarrow$ 0.53	30	0.019		6.27	3.96
1.0642 $\Rightarrow$ 0.5321	30	0.022			3.201
	30	0.022	40		3.216
	30	0.024	52.4		3.217
	30	0.026			3.92
1.0845 $\Rightarrow$ 0.54225	29	0.020			3.184

Temperature variation of phase-matching angle:

Interacting wavelengths [μm]	θ_{pm}[deg]	dθ_{pm}/dT [deg/K]	Ref.
SHG, o + o $\Rightarrow$ e			
1.0845 $\Rightarrow$ 0.54225	29	$< -1.3 \times 10^{-3}$	3.184
1.0642 $\Rightarrow$ 0.5321	30	-8.4×10^{-4}	3.217

Effective nonlinearity expression in the phase-matching direction:

$$d_{\text{ooe}} = d_{31} \sin \theta$$

Nonlinear coefficient [3.37]:

$$d_{31}(1.064 \,\mu\text{m}) = 4.4 \text{ pm/V} \ ,$$

$$d_{33}(1.064 \,\mu\text{m}) = 4.5 \text{ pm/V} \ .$$

Laser-induced bulk-damage threshold:

λ [μm]	τ_p [ns]	$I_{\text{thr}} \times 10^{-12}$ [W/m^2]	Ref.	Note
0.44–0.62	200–300	0.1	3.218	
0.53	15	0.4–0.5	3.199	
	20	0.7–0.8	3.219	
0.5321	0.031	50	3.220	
	0.032	100–120	3.221	25 Hz
	0.035	80–100	3.222	1 Hz
	0.035	40–50	3.222	12.5 Hz
	0.1	10	3.220	
	12	0.3	3.202	
0.64	330	0.04	3.101	
0.6943	10	1.2	3.206	
	20	1.3	3.185	10 pulses
	20	0.25	3.203	500 pulses
1.0642	0.045	190	3.220	
	0.13	80	3.220	
	10	1.2	3.223	100 Hz
	12	1.2	3.202	
	300	0.02	3.201	1 kHz
	1.8×10^5	> 0.5	3.224	50 Hz

Thermal conductivity coefficient [3.182]:

$$\kappa = 1.47 \text{ W/mk}$$

3.1.7 KTiOPO$_4$, Potassium Titanyl Phosphate (KTP)

Positive biaxial crystal: $2V_Z = 37.4°$ at $\lambda = 0.5461 \,\mu\text{m}$ [3.225];
Point group: mm2
Assignment of dielectric and crystallographic axes:
$X, Y, Z \Rightarrow a, b, c$ (Fig. 3.2);
Mass density: 2.945 g/cm^3 [3.226, 227];
3.023 g/cm^3 [3.228]; 3.03 g/cm^3 [3.229];
Mohs hardness: 5 [3.227];
Vickers hardness: 531 [3.228], 566 [3.230];
Knoop hardness: 702 [3.228];

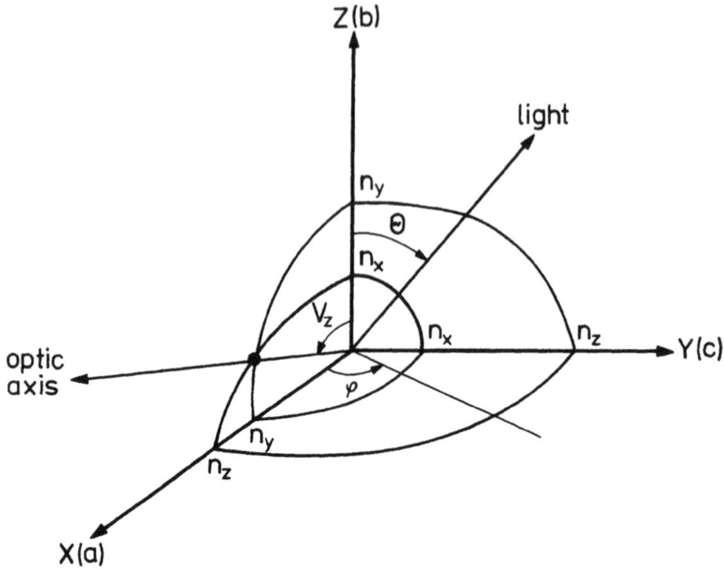

Fig. 3.2. Dependence of refractive index on light propagation direction and polarization (index surface) in the first octant of dielectric reference frame (X, Y, Z) of KTP crystal. Designations: θ is the polar angle, ϕ is the asimuthal angle, V_Z is the angle between one of the optical axes and the Z axis

Transparency range at "0" transmittance level: $0.35 - 4.5\,\mu m$ [3.231, 232]; Linear absorption coefficient α :

λ [μm]	α [cm^{-1}]	Ref.	Note
0.43-0.78	< 0.004	2.233	oxygen annealing + cerium doping
0.5145	0.013	3.186	along a axis
	0.027	3.186	along b axis
	0.026	3.186	along c axis
0.53-0.78	< 0.005	2.233	oxygen annealing
0.5321	0.04	3.234	along SHG direction
	< 0.02	3.235	
0.6594	0.0065	3.186	along a axis
	0.0087	3.186	along b axis
	0.0065	3.186	along c axis
1.06	< 0.01	3.229	
1.0642	< 0.006	3.235	
	0.005	3.234	along SHG direction
	0.0002	3.186	along a axis
	0.0005	3.186	along b axis
	0.0004	3.186	along c axis

λ [μm]	α [cm^{-1}]	Ref.	Note
1.3188	0.0015	3.186	along a axis
	0.0004	3.186	along b axis
	0.001	3.186	along c axis

Experimental values of refractive indices:
hydrothermally grown KTP [3.229]

λ[μm]	n_X	n_Y	n_Z
0.53	1.7787	1.7924	1.8873
1.06	1.7400	1.7469	1.8304

flux-grown KTP

λ [μm]	n_X	n_Y	n_Z	Ref.
0.4047	1.8249	1.8410	1.9629	3.225
0.4358	1.8082	1.8222	1.9359	3.225
0.4916	1.7883	1.8000	1.9044	3.225
0.5343	1.7780	1.7887	1.8888	3.225
0.53975	1.7764	1.7869	1.8863	3.236
0.5410	1.7767	1.7873	1.8869	3.225
0.5461	1.7756	1.7860	1.8850	3.225
0.5770	1.7703	1.7803	1.8769	3.225
0.5790	1.7699	1.7798	1.8764	3.225
0.5853	1.7689	1.7787	1.8749	3.225
0.5893	1.7684	1.7780	1.8740	3.225
0.6234	1.7637	1.7732	1.8672	3.225
0.6328	1.7622	1.7714	1.8649	3.236
0.6410	1.7617	1.7709	1.8641	3.225
0.6939	1.7565	1.7652	1.8564	3.225
0.6943	1.7564	1.7652	1.8564	3.225
0.7050	1.7555	1.7642	1.8550	3.225
1.0640	1.7381	1.7458	1.8302	3.225
1.0642	1.7379	1.7454	1.8297	3.236
1.0795	1.7375	1.7450	1.8291	3.236
1.3414	1.7314	1.7387	1.8211	3.236

Temperature derivative of refractive indices [3.237] :

$$\mathrm{d}n_X/\mathrm{d}T \times 10^5 = 0.1323\,\lambda^{-3} - 0.4385\,\lambda^{-2} + 1.2307\,\lambda^{-1} + 0.7709\ ,$$
$$\mathrm{d}n_Y/\mathrm{d}T \times 10^5 = 0.5014\,\lambda^{-3} - 2.0030\,\lambda^{-2} + 3.3016\,\lambda^{-1} + 0.7498\ ,$$
$$\mathrm{d}n_Z/\mathrm{d}T \times 10^5 = 0.3896\,\lambda^{-3} - 1.3332\,\lambda^{-2} + 2.2762\,\lambda^{-1} + 2.1151\ ,$$

where λ *in* μm and $\mathrm{d}n_X/\mathrm{d}T$, $\mathrm{d}n_Y/\mathrm{d}T$, and $\mathrm{d}n_Z/\mathrm{d}T$ are in K^{-1}.

Temperature derivative of refractive indices [3.237] :

λ [μm]	$dn_X/dT \times 10^5$ [K^{-1}]	$dn_Y/dT \times 10^5$ [K^{-1}]	$dn_Z dT \times 10^5$ [K^{-1}]
0.5321	2.41	3.21	4.27
1.0642	1.65	2.50	3.40

Experimental values of phase-matching angle ($T = 293$ K) and comparison between different sets of dispersion relations:
hydrothermally grown KTP
XY plane, $\theta = 90°$

Interacting wavelengths [μm]	ϕ_{exp} [deg]	ϕ_{theor} [deg] [3.238]	[3.239]
SHG, e + o ⇒ e			
1.053 ⇒ 0.5265	34 [3.240]	32.01	33.94
1.062 ⇒ 0.531	25 [3.229]	24.54	27.52
1.0642 ⇒ 0.5321	24 [3.238]	22.46	25.81
	26 [3.121]		
	26 [3.226]		
SFG, e + o ⇒ e			
1.3188 + 0.6594 ⇒			
⇒ 0.4396	3.8 [3.241]	no pm	13.64

YZ plane, $\phi = 90°$

Interacting wavelengths [μm]	θ_{exp} [deg]	θ_{theor} [deg] [3.238]	[3.239]
SFG, o + e ⇒ o			
1.3188 + 0.6594 ⇒			
⇒ 0.4396	65.1 [3.241]	65.01	65.33
1.338 + 0.669 ⇒			
⇒ 0.446	63.2 [3.241]	63.17	63.39

XZ plane, $\phi = 0°, \theta > V_z$

Interacting wavelengths [μm]	θ_{exp} [deg]	θ_{theor} [deg] [3.238]	[3.239]
SHG, o + e ⇒ o			
1.3188 + 0.6594 ⇒			
⇒ 0.4396	87.7 [3.241]	88.10	no pm

$1.338 + 0.669 \Rightarrow$			
$\Rightarrow 0.446$	79.9 [3.241]	80.31	82.33
$1.0642 + 1.4581 \Rightarrow$			
$\Rightarrow 0.6152$	78 [3.171]	74.20	76.84
$1.0642 + 1.4762 \Rightarrow$			
$\Rightarrow 0.6184$	76.6 [3.171]	73.95	76.66
$1.0642 + 1.5918 \Rightarrow$			
$\Rightarrow 0.6378$	75.8 [3.171]	72.77	75.91

flux-grown KTP
XY plane, $\theta = 90°$

Interacting wavelengths [μm]	ϕ_{exp} [deg]	ϕ_{theor} [deg]		
		[3.242]	[3.232]	[3.236]
SHG, e + o $\Rightarrow$ e				
$1.0642 \Rightarrow 0.5321$	23.0 [3.243]	21.12	24.59	22.89
	23.2 [3.225]			
	23.3 [3.244]			
	24.1 [3.245]			
	25.0 [3.227]			
	25.2 [3.231]			
	25.2 [3.238]			
	25.2 [3.246]			
	25.3 [3.230]			

YZ plane, $\phi = 90°$

Interacting wavelengths [μm]	θ_{exp} [deg]	θ_{theor} [deg]		
		[3.242]	[3.232]	[3.236]
SHG, o + e $\Rightarrow$ o				
$1.0642 \Rightarrow 0.5321$	69.0 [3.247]	68.03	68.67	68.83
	69.2 [3.238]			
$1.068 \Rightarrow 0.534$	67.8 [3.247]	67.52	68.16	68.32
$1.182 \Rightarrow 0.591$	57.4 [3.247]	56.77	57.41	57.64
$1.3188 \Rightarrow 0.6594$	50.0 [3.238]	49.42	50.25	50.38
$1.5 \Rightarrow 0.75$	44.6 [3.247]	43.80	45.02	44.87

XZ plane, $\phi = 0°, \theta > V_z$

Interacting wavelengths [μm]	$\theta_{\exp}$ [deg]	θ_{theor} [deg]		
		[3.242]	[3.232]	[3.236]
SHG, o + e ⇒ o				
1.0796 ⇒ 0.5398	85.3 [3.248]	85.68	no pm	86.94
	86.7 [3.236]			
1.3188 ⇒ 0.6594	58.3 [3.238]	59.03	60.38	59.58
	58.9 [3.249]			
1.3414 ⇒ 0.6707	58.7 [3.236]	58.02	59.42	58.58
1.54 ⇒ 0.77	53 [3.250]	52.02	53.93	52.64
1.90768 ⇒ 0.95384	51.1 [3.249]	48.33	51.32	49.07
2.05 ⇒ 1.025	50.8 [3.249]	48.6	51.82	48.82
2.1284 ⇒ 1.0642	53.7 [3.251]	48.63	52.36	49.15
	54 [3.249]			
SFG, o + e ⇒ o				
1.3188 + 0.6594 ⇒				
⇒ 0.4396	87.6 [3.238]	84.76	86.84	83.14
	87.1 [3.241]			
1.338 + 0.669 ⇒				
⇒ 0.446	79.8 [3.241]	79.21	80.23	78.53
1.3414 + 0.6707 ⇒				
⇒ 0.44713	78.1 [3.252]	78.52	79.50	77.91
1.0642 + 1.90768 ⇒				
⇒ 0.68333	77.2 [3.249]	72.47	75.21	72.73
1.0796 + 1.3414 ⇒				
⇒ 0.59817	74.9 [3.236]	75.03	76.49	74.48
1.54 + 0.78 ⇒				
⇒ 0.51776	61 [3.253]	59.87	60.79	60.17
1.90768 + 2.40688 ⇒				
⇒ 1.0642	58.6 [3.249]	52.79	57.08	53.37
1.58053 + 1.54 ⇒				
⇒ 0.78	52.1 [3.253]	51.21	53.15	51.83
1.90768 + 1.0642 ⇒				
⇒ 0.68333	48.7 [3.249]	46.70	48.17	47.22

Note: The other sets of dispersion relations from [3.225, 254, 255, 238] show worse agreement with the experiment.

Best sets of dispersion relations (λ in μm, $T = 20°$) hydrothermally grown KTP [3.239] :

$$n_x^2 = 2.1146 + \frac{0.89188\lambda^2}{\lambda^2 - (0.20861)^2} - 0.01320\lambda^2 \ ,$$

$$n_y^2 = 2.1518 + \frac{0.87862\lambda^2}{\lambda^2 - (0.21801)^2} - 0.01327\lambda^2 \ ,$$

$$n_z^2 = 2.3136 + \frac{1.00012\lambda^2}{\lambda^2 - (0.23831)^2} - 0.01679\lambda^2 \ .$$

flux-grown KTP [3.232] :

$$n_x^2 = 3.0065 + \frac{0.03901}{\lambda^2 - 0.04251} - 0.01327\lambda^2 \ ,$$

$$n_y^2 = 3.0333 + \frac{0.04154}{\lambda^2 - 0.04547} - 0.01408\lambda^2 \ ,$$

$$n_z^2 = 3.3134 + \frac{0.05694}{\lambda^2 - 0.05658} - 0.01682\lambda^2 \ .$$

Calculated values of phase-matching and "walk-off" angles for flux-grown KTP:

XY plane, $\theta = 90°$

Interacting wavelengths [μm]	θ_{pm} [deg]	ρ_1 [deg]	ρ_3 [deg]
SHG, e + o $\Rightarrow$ e			
1.0642 $\Rightarrow$ 0.5321	24.59	0.202	0.268

YZ plane, $\phi = 90°$

Interacting wavelengths [μm]	θ_{pm} [deg]	ρ_2 [deg]
SHG, o + e $\Rightarrow$ o		
1.0642 $\Rightarrow$ 0.5321	68.67	1.829
1.1523 $\Rightarrow$ 0.57615	59.59	2.314
1.3188 $\Rightarrow$ 0.6594	50.25	2.544
2.098 $\Rightarrow$ 1.049	43.01	2.481
2.9365 $\Rightarrow$ 1.46825	57.95	2.225
SFG, o + e $\Rightarrow$ o		
1.3188 + 0.6594 $\Rightarrow$		
$\Rightarrow$ 0.4396	65.14	2.210

XZ plane, $\phi = 0°, \theta > V_z$

Interacting wavelengths [μm]	θ_{pm} [deg]	ρ_2 [deg]
SHG, o + e ⇒ o		
1.1523 ⇒ 0.57615	72.01	1.747
1.3188 ⇒ 0.6594	60.38	2.487
2.098 ⇒ 1.049	52.13	2.671
2.9365 ⇒ 1.46825	67.36	1.928
SFG, o + e ⇒ o		
1.3188 + 0.6594 ⇒		
⇒ 0.4396	86.84	0.362

Calculated values of inverse group-velocity mismatch for SHG process in flux-grown KTP:

XY plane, $\theta = 90°$

Interacting wavelengths [μm]	ϕ_{pm} [deg]	β [fs/mm]
SHG, e + o ⇒ e		
1.0 ⇒ 0.5	73.18	475
1.05 ⇒ 0.525	35.03	434

YZ plane, $\phi = 90°$

Interacting wavelengths [μm]	θ_{pm} [deg]	β [fs/mm]
SHG, o + e ⇒ o		
1.0 ⇒ 0.5	83.17	490
1.1 ⇒ 0.55	64.36	361
1.2 ⇒ 0.6	56.22	329
1.3 ⇒ 0.65	51.02	228
1.4 ⇒ 0.7	47.46	186
1.5 ⇒ 0.75	45.02	153
1.6 ⇒ 0.8	43.40	126
1.7 ⇒ 0.85	42.44	103
1.8 ⇒ 0.9	41.99	84
1.9 ⇒ 0.95	41.98	83
2.0 ⇒ 1.0	42.35	100

XZ plane, $\phi = 0°, \theta > V_z$

Interacting wavelengths [μm]	θ_{pm} [deg]	β [fs/mm]
SHG, o + e ⇒ o		
$1.1 \Rightarrow 0.55$	80.31	391
$1.2 \Rightarrow 0.6$	67.47	307
$1.3 \Rightarrow 0.65$	61.25	246
$1.4 \Rightarrow 0.7$	57.32	200
$1.5 \Rightarrow 0.75$	54.70	164
$1.6 \Rightarrow 0.8$	52.99	135
$1.7 \Rightarrow 0.85$	51.94	111
$1.8 \Rightarrow 0.9$	51.42	90
$1.9 \Rightarrow 0.95$	51.32	81
$2.0 \Rightarrow 1.0$	51.57	98

Experimental values of NCPM temperature and corresponding temperature bandwidth:
hydrothermally grown KTP
along X axis

Interacting wavelengths [μm]	T [°C]	ΔT [°C]	Ref.
SFG, type II			
$1.3188^y + 0.6594^z \Rightarrow 0.4396^y$	47	8.5	3.241
$1.338^y + 0.669^z \Rightarrow 0.446^y$	463	8.5	3.241

along Y axis

Interacting wavelengths [μm]	T [°C]	ΔT [°C]	Ref.
SHG, type II			
$0.9943^x + 0.9943^z \Rightarrow 0.49715^x$	20	175	3.256
SFG, type II			
$1.0642^x + 0.809^z \Rightarrow 0.45961^x$	20	122	3.257

flux-grown KTP
along X axis

Interacting wavelengths [μm]	T [°C]	ΔT [°C]	Ref.
SHG, type II			
$1.0796^y + 1.0796^z \Rightarrow 0.5398^y$	153(?)	20	3.248
	63	30	3.258
SFG, type II			
$1.090^y + 1.039^z \Rightarrow 0.5321^y$	20		3.259
	20		3.260

$2.15^y + 1.04^z \Rightarrow 0.70094^y$	20		3.261
$3.09^y + 1.38^z \Rightarrow 0.95396^y$	20		3.261
$3.297^y + 1.571^z \Rightarrow 1.047^y$	20		3.262
$3.276^y + 1.539^z \Rightarrow 1.0642^y$	20		3.262
$1.3188^y + 0.6594^z \Rightarrow 0.4396^y$	60.2	8.5	3.241
$1.338^y + 0.669^z \Rightarrow 0.446^y$	484	8.5	3.241

along Y axis

Interacting wavelengths [µm]	T [°C]	Ref.
SHG, type II		
$0.99^x + 0.99^z \Rightarrow 0.495^x$	20	3.254
SFG, type II		
$1.0642^x + 0.8068^z \Rightarrow 0.4589^x$	20	3.254
$1.0642^x + 0.808^z \Rightarrow 0.45929^x$	20	3.238
$1.0642^x + 0.9691^z \Rightarrow 0.5072^x$	20	3.254

Note: Superscripts of interacting wavelengths represent polarization directions

Experimental values of internal angular, temperature, and spectral bandwidths:

XY plane, $\theta = 90°(T = 20\ °C)$

Interacting wavelengths [µm]	ϕ_{pm} [deg]	$\Delta\phi^{int}$ [deg]	$\Delta\theta^{int}$ [deg]	ΔT [°C]	Δv [cm^{-1}]	Ref.
SHG, $e + o \Rightarrow e$						
$1.0582 \Rightarrow 0.5921$		0.43	2.01			3.263
$1.062 \Rightarrow 0.531$	25	0.49	2.23	25	4.9	3.229
$1.0642 \Rightarrow 0.5321$	23	0.53		20		3.243
	23.2	0.58	1.82	24		3.225
	23.3	0.43		20	4.0	3.244
	25				6.2	3.227
	25.2			25		3.231
	25.2	0.42		17.5		3.246
	25.2	0.52	2.52	25.7		3.230

YZ plane, $\phi = 90°(T = 20\ °C)$

Interacting wavelengths [µm]	θ_{pm} [deg]	$\Delta\theta^{int}$ [deg]	$\Delta\phi^{int}$ [deg]	ΔT [°C]	Δv [cm^{-1}]	Ref.
SHG, $o + e \Rightarrow o$						
$0.9943 \Rightarrow 0.49715$	90	2.96	5.70	175	7.1	3.256

1.0642 $\Rightarrow$ 0.5321	69			100	3.237
	69	0.11		47	3.264
2.532 $\Rightarrow$ 1.266	56	0.20		30.7	3.264
SFG, type II					
$1.0642^x + 0.809^z \Rightarrow$					
$\Rightarrow 0.45961^x$	90	2.72	6.13	17.6(Δv_2)	3.257

XZ plane, $\phi = 0°, \theta > V_z$

Interacting wavelengths [μm]	T [°C]	θ_{pm} [deg]	$\Delta\theta^{int}$ [deg]	Ref.
SHG, o + e $\Rightarrow$ o				
1.0796 $\Rightarrow$ 0.5398	20	85.3	0.34	3.248
	153	90	1.70	3.248

Note: Superscripts of interacting wavelengths represent polarization directions

Effective nonlinearity in the phase-matching direction for three-wave interactions in the principal planes of KTP crystal [3.35, 36]:

XY plane

$$d_{eoe} = d_{oee} = d_{31} \sin^2 \phi + d_{32} \cos^2 \phi \ ,$$

YZ plane

$$d_{oeo} = d_{eoo} = d_{31} \sin \theta \ ,$$

XZ plane, $\theta < V_z$

$$d_{ooe} = d_{32} \sin \theta \ ,$$

XZ plane, $\theta > V_z$

$$d_{oeo} = d_{eoo} = d_{32} \sin \theta \ .$$

Effective nonlinearity for three-wave interactions in the arbitary direction of KTP crystal are given in [3.36]

Nonlinear coefficients [3.265] :

$$d_{31}(1.0642 \, \mu m) = 1.4 \ pm/V \ ,$$

$$d_{32}(1.0642 \, \mu m) = 2.65 \ pm/V \ ,$$

$$d_{33}(1.0642 \, \mu m) = 10.7 \ pm/V \ .$$

Laser-induced damage threshold:
hydrothermally grown KTP

λ [μm]	τ_p [ns]	$I_{thr} \times 10^{-12}$ [W/m^2]	Ref.	Note
0.526	0.03	300	3.239	
	0.03	300	3.235	10 Hz
1.0642	125000	0.01	3.266	
	30	1.5	3.267	
	20	> 1.5	3.268	
	11	20–30	3.269	10 Hz

flux-grown KTP

λ [μm]	τ_p [ns]	$I_{thr} \times 10^{-12}$ [W/m^2]	Ref.	Note
0.526	0.03	100	3.235	10 Hz
0.5291	18	0.8–1.0	3.263	surface damage
0.5321	14	0.5	3.246	60 pulses
	8	14–22	3.270	2 Hz, surface damage
	8	20–32	3.270	2 Hz, bulk damage
	0.06	> 18	3.245	5 Hz
1.0582	25	1.8–2.2	3.263	surface damage
1.0642	30	> 3.3	3.249	
	25	> 6	3.271	250 000 pulses, bulk darkening
	25	> 3	3.271	3 500 000 pulses, bulk darkening
	20	1.5	3.246	60 pulses
	11	15–22	3.270	2 Hz, surface damage
	11	24–35	3.270	2 Hz, bulk damage
	10	9–10	3.243	
	9	310	3.272	1 pulse, bulk damage
	1.3	46	3.33	surface damage
	1	150	3.225	1 pulse
	1	> 150	3.112	

Thermal conductivity coefficient [3.235] :

κ [W/mK], along a	κ [W/mK], along b	κ [W/mK], along c
2	3	3.3

3.1.8 LiNbO$_3$, Lithium Niobate

Negative uniaxial crystal: $n_o > n_e$;
Point group: 3m;
Mass density: 4.628 g/cm^3 [3.273];
Mohs hardness: 5 − 5.5;
Transparency range at "0" transmittance level: 0.4 − 5.5 µm [3.274, 275];
Linear absorption coefficient α:

λ [µm]	α [cm^{-1}]	Ref.	Note
0.5145	0.025	3.276	
	0.019–0.025	3.186	$\parallel c$
	0.035–0.045	3.186	e − wave, $\perp c$
0.6594	0.0021–0.0044	3.186	$\parallel c$
	0.0085–0.0096	3.186	e − wave, $\perp c$
1.0642	0.0019–0.0023	3.186	$\parallel c$
	0.0014–0.0019	3.186	e − wave, $\perp c$
	0.0042	3.277	$\parallel c$
	0.0028	3.277	$\perp c$
1.3188	0.0018–0.0044	3.186	$\parallel c$
	0.0017–0.0110	3.186	e − wave, $\perp c$

Two-photon absorption coefficient β:

λ [µm]	$\beta \times 10^{11}$ [m/W]	Ref.	Note
0.5288	0.15 (?)	3.278	
0.53	5.0	3.279	
0.5321	2.90	3.188	o − wave
	1.57	3.188	e − wave

Experimental values of refractive indices for
lithium-rich lithium niobate, $T = 293$ K [3.280] :

λ [µm]	n_o	n_e	λ [µm]	n_o	n_e
0.3250	2.6360	2.4670	0.4880	2.3495	2.2398
0.4545	2.3751	2.2608	0.4965	2.3437	2.2352
0.4579	2.3719	2.2584	0.5017	2.3405	2.2329
0.4658	2.3658	2.2530	0.5145	2.3334	2.2270
0.4727	2.3604	2.2489	0.6328	2.2878	2.1890
0.4765	2.3573	2.2465	1.0642	2.2339	2.1440

lithium niobate grown from stoichiometric melt (mole ratio Li/Nb = 1.000), $T = 293$ K [3.274] :

λ [μm]	n_o	n_e	λ [μm]	n_o	n_e
0.42	2.4089	2.3025	1.80	2.2049	2.1306
0.45	2.3780	2.2772	2.00	2.1974	2.1250
0.50	2.3410	2.2457	2.20	2.1909	2.1183
0.55	2.3132	2.2237	2.40	2.1850	2.1129
0.60	2.2967	2.2082	2.60	2.1778	2.1071
0.70	2.2716	2.1874	2.80	2.1703	2.1009
0.80	2.2571	2.1745	3.00	2.1625	2.0945
0.90	2.2448	2.1641	3.20	2.1543	2.0871
1.00	2.2370	2.1567	3.40	2.1456	2.0804
1.20	2.2269	2.1478	3.60	2.1363	2.0725
1.40	2.2184	2.1417	3.80	2.1263	2.0642
1.60	2.2113	2.1361	4.00	2.1155	2.0553

lithium niobate grown from congruent melt (mole ratio Li/Nb = 0.946), $T = 293$ K [3.281] :

λ [μm]	n_o	n_e
0.43584	2.39276	2.29278
0.54608	2.31657	2.22816
0.63282	2.28647	2.20240
1.1523	2.2273	2.1515
3.3913	2.1451	2.0822

$T = 297.5$ K [3.282]:

λ [μm]	n_o	n_e	λ [μm]	n_o	n_e
0.40463	2.4317	2.3260	0.66782	2.2778	2.1953
0.43584	2.3928	2.2932	0.70652	2.2699	2.1886
0.46782	2.3634	2.2683	0.80926	2.2541	2.1749
0.47999	2.3541	2.2605	0.87168	2.2471	2.1688
0.50858	2.3356	2.2448	0.93564	2.2412	2.1639
0.54607	2.3165	2.2285	0.95998	2.2393	2.1622
0.57696	2.3040	2.2178	1.01400	2.2351	2.1584
0.57897	2.3032	2.2171	1.09214	2.2304	2.1545
0.58756	2.3002	2.2147	1.15392	2.2271	2.1517
0.64385	2.2835	2.2002	1.15794	2.2269	2.1515

λ [μm]	n_o	n_e	λ [μm]	n_o	n_e
1.28770	2.2211	2.1464	2.39995	2.1840	2.1151
1.43997	2.2151	2.1413	2.61504	2.1765	2.1087
1.63821	2.2083	2.1356	2.73035	2.1724	2.1053
1.91125	2.1994	2.1280	2.89733	2.1657	2.0999
2.18428	2.1912	2.1211	3.05148	2.1594	2.0946

Temperature derivative of refractive indices for lithium-rich niobate, $T = 298$ K [3.280] :

λ [μm]	$\mathrm{d}n_o/\mathrm{d}T \times 10^5$ [K^{-1}]	$\mathrm{d}n_e/\mathrm{d}T \times 10^5$ [K^{-1}]
0.3250	8.71	12.9
0.4545	1.93	6.22
0.6328	0.522	4.31
1.0642	0.141	3.85

stoichiometric melt (mole ratio Li/Nb = 1.000), $\lambda = 0.45 - 0.70\,\mu$m, $T = 293$ K) [3.283] :

$$\mathrm{d}n_\mathrm{o}/\mathrm{d}T = 2.0 \times 10^{-5}\ \mathrm{K}^{-1}\ ,$$

$$\mathrm{d}n_\mathrm{e}/\mathrm{d}T = 7.6 \times 10^{-5}\ \mathrm{K}^{-1}\ ;$$

Sellmeier equations (λ in μm, $T = 20$ °C) for lithium-rich niobate [3.280] :

$$n_\mathrm{o}^2 = 4.91296 + \frac{0.116275}{\lambda^2 - 0.048398} - 0.0273\lambda^2\ ,$$

$$n_\mathrm{e}^2 = 4.54528 + \frac{0.091649}{\lambda^2 - 0.046079} - 0.0303\lambda^2\ ;$$

stoichiometric melt (mole ratio Li/Nb = 1.000) [3.284] :

$$n_\mathrm{o}^2 = 4.91300 + \frac{0.118717}{\lambda^2 - 0.045932} - 0.0278\lambda^2\ ,$$

$$n_\mathrm{e}^2 = 4.57906 + \frac{0.099318}{\lambda^2 - 0.042286} - 0.0224\lambda^2\ ;$$

congruent melt (mole ratio Li/Nb = 0.946) [3.281] :

$$n_\mathrm{o}^2 = 4.9048 + \frac{0.117680}{\lambda^2 - 0.047500} - 0.027169\lambda^2\ ,$$

$$n_\mathrm{e}^2 = 4.5820 + \frac{0.099169}{\lambda^2 - 0.044432} - 0.021950\lambda^2\ .$$

Temperature-dependent Sellmeier equations (λ in µm, T in K) for lithium-rich lithium niobate [3.280]

$$n_\mathrm{o}^2 = 4.913 + 1.6 \times 10^{-8}(T^2 - 88506.25)$$

$$+ \frac{0.1163 + 0.94 \times 10^{-8}(T^2 - 88506.25)}{\lambda^2 - [0.2201 + 3.98 \times 10^{-8}(T^2 - 88506.25)]^2} - 0.0273\lambda^2 \ ,$$

$$n_\mathrm{e}^2 = 4.546 + 2.72 \times 10^{-7}(T^2 - 88506.25)$$

$$+ \frac{0.0917 + 1.93 \times 10^{-8}(T^2 - 88506.25)}{\lambda^2 - [0.2148 + 5.3 \times 10^{-8}(T^2 - 88506.25)]^2} - 0.0303\lambda^2 \ .$$

stoichiometric melt (mole ratio Li/Nb = 1.000) [3.284] :

$$n_\mathrm{o}^2 = 4.9130 + \frac{0.1173 + 1.65 \times 10^{-8}\ T^2}{\lambda^2 - (0.212 + 2.7 \times 10^{-8}\ T^2)^2} - 0.0278\lambda^2 \ ,$$

$$n_\mathrm{e}^2 = 4.5567 + 2.605 \times 10^{-7}T^2 + \frac{0.0970 + 2.70 \times 10^{-8}T^2}{\lambda^2 - (0.201 + 5.4 \times 10^{-8}T^2)^2} - 0.0224\lambda^2 \ ,$$

congruent melt (mole ratio Li/Nb = 0.946) [3.285] :

$$n_\mathrm{o}^2 = 4.9048 + 2.1429 \times 10^{-8}(T^2 - 88506.25)$$

$$+ \frac{0.11775 + 2.2314 \times 10^{-8}(T^2 - 88506.25)}{\lambda^2 - [0.21802 - 2.9671 \times 10^{-8}(T^2 - 88506.25)]^2} - 0.027153\lambda^2 \ ,$$

$$n_\mathrm{e}^2 = 4.5820 + 2.2971 \times 10^{-7}(T^2 - 88506.25)$$

$$+ \frac{0.09921 + 5.2716 \times 10^{-8}(T^2 - 88506.25)}{\lambda^2 - [0.21090 - 4.9143 \times 10^{-8}(T^2 - 88506.25)]^2} - 0.021940\lambda^2 \ .$$

Experimental and theoretical values of phase-matching angle and calculated values of "walk-off" angle:
lithium-rich lithium niobate, $T = 295$ K

Interacting wavelengths [µm]	θ_exp [deg]	θ_theor [deg] [3.280]	ρ_3 [deg]
SHG, o + o $\Rightarrow$ e 1.0642 $\Rightarrow$ 0.5321	67.5 [3.280]	66.76	1.776

stoichiometric melt (mole ratio Li/Nb = 1.000), $T = 293$ K

Interacting wavelengths [μm]	θ_{exp} [deg]	θ_{theor} [deg] [3.284]	ρ_3 [deg]
SHG, o + o ⇒ e			
1.118 ⇒ 0.559	71.7 [3.284]	71.80	1.312
1.1523 ⇒ 0.57615	67.6 [3.284]	67.74	1.543
	68 [3.274]		
	69 [3.286]		
SFG, o + o ⇒ e			
2.17933 + 0.8529 ⇒ 0.613	55 [3.287]	54.75	2.073
4.0 + 0.72394 ⇒ 0.613	47.5 [3.287]	47.48	2.212

congruent melt (mole ratio Li/Nb = 0.946), $T = 293$ K

Interacting wavelengths [μm]	θ_{exp} [deg]	θ_{theor} [deg] [3.284]	ρ_3 [deg]
SHG, o + o ⇒ e			
1.1523 ⇒ 0.57615	72 [3.286]	70.39	1.341
2.12 ⇒ 1.06	43.8 [3.288]	45.25	1.988
2.1284 ⇒ 1.0642	44.6 [3.289]	45.28	1.987
	47 [3.290]		
SFG, o + o ⇒ e			
1.95160 + 1.0642 ⇒ 0.68867	52.7 [3.291]	52.86	2.000
2.57887 + 1.0642 ⇒ 0.75333	48.1 [3.291]	48.13	2.047
3.22241 + 1.0642 ⇒ 0.80000	46.5 [3.291]	46.50	2.044
4.19039 + 1.0642 ⇒ 0.84867	47 [3.291]	46.90	2.026

Note: The PM angle values are strongly dependent on melt stoichiometry

Experimental values of NCPM temperature:
lithium-rich lithium niobate

Interacting wavelengths [μm]	T [°C]	Ref.
SHG, o + o ⇒ e		
0.954 ⇒ 0.477	−62.5	3.280
1.0642 ⇒ 0.5321	233.7	3.277
	238	3.280
1.3188 ⇒ 0.6594	520	3.280

stoichiometric melt (mole ratio Li/Nb = 1.000)

Interacting wagelengths [μm]	T [°C]	Ref.
SHG, $o + o \Rightarrow e$		
1.029 $\Rightarrow$ 0.5145	15	3.292
1.058 $\Rightarrow$ 0.529	0	3.293
1.0642 $\Rightarrow$ 0.5321	43	3.294
	72	3.295
1.084 $\Rightarrow$ 0.542	97	3.296
1.118 $\Rightarrow$ 0.559	153.5	3.284
1.1523 $\Rightarrow$ 0.57615	193	3.293
	208	3.284
	211	3.295

congruent melt (mole ratio Li/Nb = 0.946)

Interacting wavelengths [μm]	T [°C]	Ref.
SHG, $o + o \Rightarrow e$		
1.029 $\Rightarrow$ 0.5145	−66	3.292
1.0576 $\Rightarrow$ 0.5288	−14	3.278
1.0642 $\Rightarrow$ 0.5321	−8	3.297
	6	3.298
	11.5	3.294
1.084 $\Rightarrow$ 0.542	38	3.299
	42	3.297
	46	3.292
1.1523 $\Rightarrow$ 0.57615	172	3.297
	174	3.282

Note: The NCPM temperature values are strongly dependent on melt stoichiometry

Experimental value of internal angular bandwidth [3.81]:

Interacting wagelengths [μm]	$\Delta\theta^{\mathrm{int}}$[deg]
SHG, $o + o \Rightarrow e$	
1.06 $\Rightarrow$ 0.53	0.040

Experimental values of temperature and spectral bandwidths:

Interacting wavelengths [μm]	T [°C]	θ_{pm} [deg]	ΔT [°C]	$\Delta\nu_1$ [cm^{-1}]	Ref.
SHG, $o + o \Rightarrow e$					
1.06 $\Rightarrow$ 0.53	20	68		3.2	3.81

$1.0642 \Rightarrow 0.5321$	51	90	0.72		3.300
	234	90	0.52		3.277
$1.084 \Rightarrow 0.542$	38	90	0.74		3.292
	46	90	0.74		3.299
$1.1523 \Rightarrow 0.57615$	172	90	0.66		3.297
SFG, o + o + e					
$1.7 + 0.6943 \Rightarrow 0.493$	70	90	1.6	7.9	3.301
$2.65 + 0.488 \Rightarrow 0.4115$	90	90		2.9	3.302

Effective nonlinearity expressions in the phase-matching direction [3.100]:

$$d_{\mathrm{ooe}} = d_{31} \sin \theta - d_{22} \cos \theta \sin 3\phi \ ,$$

$$d_{\mathrm{eoe}} = d_{\mathrm{oee}} = d_{22} \cos^2 \theta \cos 3\phi \ .$$

Nonlinear coefficients:

stoichiometric melt (mole ratio Li/Nb = 1.000)

$$d_{22}(1.058\,\mu\mathrm{m}) = 2.46 \pm 0.23 \ \mathrm{pm/V} \ [3.274, 37] \ ,$$
$$d_{31}(1.058\,\mu\mathrm{m}) = -4.64 \pm 0.66 \ \mathrm{pm/V} \ [3.274, 37] \ ,$$
$$d_{33}(1.058\,\mu\mathrm{m}) = -41.7 \pm 7.8 \ \mathrm{pm/V} \ [3.274, 37] \ .$$

congruent melt (mole ratio Li/Nb = 0.946)

$$d_{22}(1.06\,\mu\mathrm{m}) = 2.10 \pm 0.21 \ \mathrm{pm/V} \ [3.303, 37] \ ,$$
$$d_{31}(1.06\,\mu\mathrm{m}) = -4.35 \pm 0.44 \ \mathrm{pm/V} \ [3.303, 37],$$
$$d_{33}(1.06\,\mu\mathrm{m}) = -27.2 \pm 2.7 \ \mathrm{pm/V} \ [3.303, 37] \ .$$

Laser-induced surface-damage threshold:

λ [μm]	τ_{p} [ns]	$I_{\mathrm{thr}} \times 10^{-12}$ [W/m^2]	Ref.	Note
0.53	0.007	> 100	3.304	
0.5321	0.002	> 700	3.305	10 Hz
0.59–0.596	≈ 10	> 3.5	3.305	10 Hz
0.6943	25	1.5	3.306	1 pulse
1.06	30	1.2	3.307	
	30	1.7	3.308	
	10–30	3.0	3.309	
	30	12	3.307	bulk damage
	0.006	> 100	3.288	
1.0642	20	> 1	3.289	
	30	150–200	3.310	with coating

Thermal conductivity coefficient [3.64]:

$$\kappa = 4.6 \ \mathrm{W/mK} \ \mathrm{at} \ T = 300 \ \mathrm{K} \ .$$

3.1.9 KNbO₃, Potassium Niobate

Negative biaxial crystal: $2V_Z = 66.78°$ at $\lambda = 0.5321$ μm [3.311];
Point group: mm2;
Assignment of dielectric and crystallographic axes:
$X,Y,Z \Rightarrow$ b,a,c (Fig. 3.3)
Transparency range at "0" transmittance level: $\approx 0.4 - > 4$μm [3.312, 313];
Linear absorption coefficient α:

λ [μm]	α [cm⁻¹]	Ref.	Note
0.42–1.06	< 0.05	3.314	
0.82	0.015	3.315	
1.0642	0.0018–0.0025	3.316	along b axis

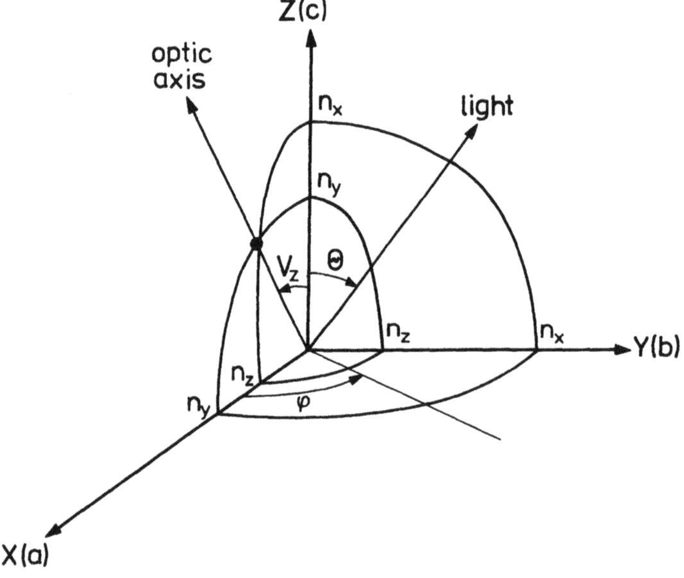

Fig. 3.3. Dependence of refractive index on light propagation direction and polarization (index surface) in the first octant of dielectric reference frame (X, Y, Z) of KNbO₃ crystal. Designations: θ is the polar angle, ϕ is the asimuthal angle, V_Z is the angle between one of the optical axes and the Z axis

Experimental values of refractive indices at $T = 295$ K [3.312]:

λ [μm]	n_X	n_Y	n_Z
0.430	2.4974	2.4145	2.2771
0.488	2.4187	2.3527	2.2274
0.514	2.3951	2.3337	2.2121
0.633	2.3296	2.2801	2.1687

λ [μm]	n_X	n_Y	n_Z
0.860	2.2784	2.2372	2.1338
1.064	2.2576	2.2195	2.1194
1.500	2.2341	2.1992	2.1029
2.000	2.2159	2.1832	2.0899
2.500	2.1981	2.1674	2.0771
3.000	2.1785	2.1498	2.0630

Experimental values of phase-matching angle ($T = 293$ K) and comparison between different sets of dispersion relations:

XY plane, $\theta = 90°$

Interacting wavelengths [μm]	$\phi_{\exp}$ [deg]	ϕ_{theor} [deg]		
		[3.311]	[3.312]	[3.317]
SHG, e + e $\Rightarrow$ o				
0.946 $\Rightarrow$ 0.473	$\approx$ 30 [3.318]	26.88	29.97	30.43

YZ plane, $\phi = 90°$

Interacting wavelengths [μm]	$\theta_{\exp}$ [deg]	θ_{theor} [deg]		
		[3.311]	[3.312]	[3.317]
SHG, o + o $\Rightarrow$ e				
0.86 $\Rightarrow$ 0.43	83.5 [3.319]	77.37	83.13	87.98
0.89 $\Rightarrow$ 0.445	70.7 [3.319]	69.03	70.67	71.92
0.92 $\Rightarrow$ 0.46	64 [3.319]	63.36	63.92	64.94
0.94 $\Rightarrow$ 0.47	60.5 [3.319]	60.27	60.43	61.37
1.0642 $\Rightarrow$ 0.5321	$\approx$ 47 [3.311]	46.57	45.95	46.52

XZ plane, $\phi = 0°, \theta > V_Z$

Interacting wavelengths [μm]	$\theta_{\exp}$ [deg]	θ_{theor} [deg]		
		[3.311]	[3.312]	[3.317]
SHG, o + o $\Rightarrow$ e				
1.0642 $\Rightarrow$ 0.5321	70.4 [3.320]	71.05	71.85	71.16
	$\approx$ 71[3.311]			
	71 [3.314]			
	71 [3.317]			

Note: The dispersion relations given in [3.321] show worse agreement with the experiment

Experimental values of NCPM temperature:
along X axis

Interacting wavelengths [μm]	T [°C]	Ref.
SHG, type I		
$0.972 \Rightarrow 0.486$	−20	3.322
$0.982 \Rightarrow 0.491$	20	3.323
$0.986 \Rightarrow 0.493$	20	3.324
$0.988 \Rightarrow 0.494$	20	3.314
$1.047 \Rightarrow 0.5235$	162	3.325
$1.0642 \Rightarrow 0.5321$	178	3.326
	181	3.311
	182	3.320
	184	3.300
	188	3.327

along Y axis

Interacting wavelengths [μm]	T [° C]	Ref.
SHG, type I		
$0.8385 \Rightarrow 0.41925$	−34.2	3.328
$0.8406 \Rightarrow 0.4203$	−28.3	3.329
$0.842 \Rightarrow 0.421$	−22.8	3.330
$0.856 \Rightarrow 0.428$	15	3.331
$0.857 \Rightarrow 0.4285$	20	3.332
$0.8593 \Rightarrow 0.42965$	20	3.328
$0.86 \Rightarrow 0.43$	22	3.324
$0.8615 \Rightarrow 0.43075$	30	3.333
$0.862 \Rightarrow 0.431$	34	3.334
$0.879 \Rightarrow 0.4395$	70	3.334
$0.9289 \Rightarrow 0.46445$	158	3.328
$0.95 \Rightarrow 0.475$	180	3.324
SFG, type I		
$0.6764 + 1.0642 \Rightarrow 0.41355$	−4	3.335
$0.6943 + 1.0642 \Rightarrow 0.42017$	27.2	3.335

Best set of dispersion relations (λ in μm, $T = 22$ °C) [3.312]:

$$n_x^2 = 1 + \frac{1.44121874\lambda^2}{\lambda^2 - 0.07439136} + \frac{2.54336918\lambda^2}{\lambda^2 - 0.01877036} - 0.02845018\lambda^2 \ ,$$

$$n_y^2 = 1 + \frac{1.33660410\lambda^2}{\lambda^2 - 0.06664629} + \frac{2.49710396\lambda^2}{\lambda^2 - 0.01666505} - 0.02517432\lambda^2 \ ,$$

$$n_z^2 = 1 + \frac{1.04824955\lambda^2}{\lambda^2 - 0.06514225} + \frac{2.37108379\lambda^2}{\lambda^2 - 0.01433172} - 0.01943289\lambda^2 \ .$$

Temperature-dependent dispersion relations (λ in µm, T in K) [3.336]:

$$n_x^2 = 1 + \frac{(2.5389409 + 3.8636303 \times 10^{-6}\, F)\lambda^2}{\lambda^2 - (0.1371639 + 1.767 \times 10^{-7}\, F)^2}$$

$$+ \frac{(1.4451842 - 3.909336 \times 10^{-6}F - 1.2256136 \times 10^{-4}\, G)\lambda^2}{\lambda^2 - (0.2725429 + 2.38 \times 10^{-7}F - 6.78 \times 10^{-5}\, G)^2}$$

$$- (2.837 \times 10^{-2} - 1.22 \times 10^{-8}F)\lambda^2 - 3.3 \times 10^{-10}F\,\lambda^4 \ ,$$

$$n_y^2 = 1 + \frac{(2.6386669 + 1.6708469 \times 10^{-6}\, F)\lambda^2}{\lambda^2 - (0.1361248 + 0.796 \times 10^{-7}\, F)^2}$$

$$+ \frac{(1.1948477 - 1.3872635 \times 10^{-6}\, F - 0.90742707 \times 10^{-4}\, G)\lambda^2}{\lambda^2 - (0.2621917 + 1.231 \times 10^{-7}\, F - 1.82 \times 10^{-5}\, G)^2}$$

$$- (2.513 \times 10^{-2} - 0.558 \times 10^{-8}\, F)\lambda^2 - 4.4 \times 10^{-10}F\,\lambda^4 \ ,$$

$$n_z^2 = 1 + \frac{(2.370517 + 2.8373545 \times 10^{-6}\, F)\lambda^2}{\lambda^2 - (0.1194071 + 1.75 \times 10^{-7}\, F)^2}$$

$$+ \frac{(1.048952 - 2.1303781 \times 10^{-6}\, F - 1.8258521 \times 10^{-4}\, G)\lambda^2}{\lambda^2 - (0.2553605 + 1.89 \times 10^{-7}\, F - 2.48 \times 10^{-5}\, G)^2}$$

$$- (1.939 \times 10^{-2} - 0.27 \times 10^{-8}\, F)\lambda^2 - 5.7 \times 10^{-10}\, F\lambda^4 \ ,$$

where $F = T^2 - 295.15^2$, and $G = T - 293.15$.

Calculated values of phase-matching and "walk-off" angles:
YZ plane, $\phi = 90°$

Interacting wavelengths [µm]	θ_{pm} [deg]	ρ_3 [deg]
SHG, o + o $\Rightarrow$ e		
1.0642 $\Rightarrow$ 0.5321	45.95	3.009
1.3188 $\Rightarrow$ 0.6594	29.87	2.507

XZ plane, $\phi = 0°$ $\theta > V_z$

Interacting wavelengths [μm]	θ_{pm} [deg]	ρ_3 [deg]
SHG, o + o ⇒ e		
1.0642 ⇒ 0.5321	71.85	2.479
1.3188 ⇒ 0.6594	57.47	3.553

Experimental values of the internal angular bandwidth:
XZ plane, $\phi = 0°$

Interacting wavelengths [μm]	T [°C]	θ_{pm} [deg]	$\Delta\theta^{int}$ [deg]	Ref.
SHG, o + o ⇒ e				
1.0642 ⇒ 0.5321	20	71	0.013-0.014	3.323

along Y axis

Interacting wavelengths [μm]	T [°C]	θ_{pm} [deg]	$\Delta\theta^{int}$ [deg]	$\Delta\phi^{int}$ [deg]	Ref.
SHG, type I					
0.857 ⇒ 0.4285	20	90	0.659	1.117	3.323

Experimental values of temperature bandwidth:
along X axis

Interacting wavelengths [μm]	T [°C]	θ_{pm} [deg]	ΔT [°C]	Ref.
SHG, type I				
1.0642 ⇒ 0.5321	181	90	0.27–0.32	3.311
	182	90	0.28	3.320
	184	90	0.28–0.29	3.300
	188	90	0.34	3.327

along Y axis

Interacting wavelengths [μm]	T [°C]	θ_{pm} [deg]	ΔT [°C]	Ref.
SHG, type I				
0.8385 ⇒ 0.41925	−34.2	90	0.27	3.328
0.842 ⇒ 0.421	−22.8	90	0.30	3.330
0.855 ⇒ 0.4275	26.4 (?)	90	0.265	3.314
0.92 ⇒ 0.46	163.5 (?)	90	0.285	3.314
SFG, type I				
0.6764 + 1.0642 ⇒ 0.41355	−4	90	0.35	3.335

Temperature of noncritical SHG [3.323]
along X axis

$$\lambda_1 = 0.97604 + 2.53 \times 10^{-4}\, T + 1.146 \times 10^{-6}\, T^2 \; ;$$

along Y axis

$$\lambda_1 = 0.85040 + 2.94 \times 10^{-4}\, T + 1.234 \times 10^{-6}\, T^2 \; ;$$

where λ_1 in μm, and T in °C.
Temperature variation of birefringence for noncritical SHG process [3.314]:
along X axis $(1.0642\,\mu\text{m} \Rightarrow 0.5321\mu\text{m})$

$$\mathrm{d}[n_Z(2\omega) - n_Y(\omega)]/\mathrm{d}T = 1.10 \times 10^{-4}\ \text{K}^{-1} \; ;$$

along Y axis $(0.92\,\mu\text{m} \Rightarrow 0.46\,\mu\text{m})$

$$\mathrm{d}[n_Z(2\omega) - n_X(\omega)]/\mathrm{d}T = 1.43 \times 10^{-4}\ \text{K}^{-1} \; .$$

Effective nonlinearity expressions in the phase-matching direction for three-wave interactions in the principal planes of $KNbO_3$ crystal [3.35], [3.36]:
XY plane

$$d_{\text{eeo}} = d_{32} \sin^2 \phi + d_{31} \cos^2 \phi \, ;$$

YZ plane

$$d_{\text{ooe}} = d_{32} \sin \theta \, ;$$

XY plane, $\theta < V_Z$

$$d_{\text{oeo}} = d_{\text{eoo}} = d_{31} \sin \theta \, ;$$

XZ plane, $\theta > V_Z$

$$d_{\text{ooe}} = d_{31} \sin \theta \, .$$

Effective nonlinearity expressions for three-wave interactions in the arbitrary direction of $KNbO_3$ crystal are given in [3.36].
Nonlinear coefficients [3.323, 37, 313]:

$$d_{31}(1.0642\,\mu\text{m}) = -11.9\ \text{pm/V} \; ,$$
$$d_{32}(1.0642\,\mu\text{m}) = -13.7\ \text{pm/V} \; ,$$
$$d_{33}(1.0642\,\mu\text{m}) = -20.6\ \text{pm/V} \; .$$

Laser-induced surface-damage threshold:

λ [μm]	τ_p [ns]	$I_{\text{thr}} \times 10^{-12}$ [W/m^2]	Ref.	Note
0.527	0.5	88–94	3.337	along b axis, $\mathbf{E} \parallel c$
	0.5	120–150	3.337	along b axis, $\mathbf{E} \perp c$
0.5321	10	0.55	3.326	
	25	1.5–1.8	3.300	
1.047	11	> 0.3	3.325	4 kHz, 2000 hours

λ [μm]	τ_p [ns]	$I_{\text{thr}} \times 10^{-12}$ [W/m^2]	Ref.	Note
1.054	0.7	110	3.337	along a axis, $\mathbf{E} \perp c$
	0.7	180	3.337	along b axis, $\mathbf{E} \perp c$
	0.7	370	3.337	along b axis, $\mathbf{E} \perp c$
1.0642	25	1.5–1.8	3.300	
	0.1	> 1000	3.323	

Thermal conductivity coefficient:

$$\kappa > 3.5 \text{ W/mK } [3.316] \ .$$

3.1.10 AgGaS$_2$, Silver Thiogallate

Negative uniaxial crystal: $n_o > n_e$ (at $\lambda < 0.497\,\mu\text{m}$ $n_e > n_o$);
Point group: $\bar{4}2m$;
Mass density: 4.58 g/cm^3 [3.338] ;
Mohs hardness: 3 – 3.5 ;
Transparency range at "0" transmittance level: 0.47 – 13 μm [3.339] ;
Linear absorption coefficient α:

λ [μm]	α [cm^{-1}]	Ref.
0.5–13	< 0.1	3.340
0.6–0.65	0.04	3.341
0.6–12	< 0.09	3.339
0.633	0.05	3.342
0.9–8.5	< 0.9	3.343
1.064	0.01	3.342
4–8.5	< 0.04	3.341

Experimental values of refractive indices [3.344]:

λ [μm]	n_o	n_e	λ [μm]	n_o	n_e	λ [μm]	n_o	n_e
0.490	2.7148	2.7287	0.850	2.4802	2.4279	2.200	2.4142	2.3684
0.500	2.6916	2.6867	0.900	2.4716	2.4192	2.400	2.4119	2.3583
0.525	2.6503	2.6239	0.950	2.4644	2.4118	2.600	2.4102	2.3567
0.550	2.6190	2.5834	1.000	2.4582	2.4053	2.800	2.4094	2.3559
0.575	2.5944	2.5537	1.100	2.4486	2.3954	3.000	2.4080	2.3545
0.600	2.5748	2.5303	1.200	2.4414	2.3881	3.200	2.4068	2.3534
0.625	2.5577	2.5116	1.300	2.4359	2.3819	3.400	2.4062	2.3522
0.650	2.5437	2.4961	1.400	2.4315	2.3781	3.600	2.4046	2.3511
0.675	2.5310	2.4824	1.500	2.4280	2.3745	3.800	2.4024	2.3491
0.700	2.5205	2.4706	1.600	2.4252	2.3716	4.000	2.4024	2.3488
0.750	2.5049	2.4540	1.800	2.4206	2.3670	4.500	2.4003	2.3461
0.800	2.4909	2.4395	2.000	2.4164	2.3637	5.000	2.3955	2.3419

λ [µm]	n_{o}	n_{e}	λ [µm]	n_{o}	n_{e}	λ [µm]	n_{o}	n_{e}
5.500	2.3938	2.3401	8.000	2.3757	2.3219	10.50	2.3486	2.2948
6.000	2.3908	2.3369	8.500	2.3699	2.3163	11.00	2.3417	2.2880
6.500	2.3874	2.3334	9.000	2.3663	2.3121	11.50	2.3329	2.2789
7.000	2.3827	2.3291	9.500	2.3606	2.3064	12.00	2.3266	2.2716
7.500	2.3787	2.3252	10.00	2.3548	2.3012	12.50	2.3177	

Optical activity [3.339, 345]:

$\rho = 522$ deg/mm at isotropic point ($n_{\mathrm{o}} = n_{\mathrm{e}}$, $\lambda = 0.4973\,\mu\mathrm{m}$)

Temperature dependences of refractive indices (λ in µm)[3.346] :

$$\mathrm{d}n_{\mathrm{o}}/\mathrm{d}T = \frac{10^{-5}}{2n_{\mathrm{o}}} \times \left[-\frac{39.88\lambda^2}{\lambda^2 - 0.0676} + \frac{112.20\lambda^4}{(\lambda^2 - 0.0676)^2} \right] ,$$

$$\mathrm{d}n_{\mathrm{e}}/\mathrm{d}T = \frac{10^{-5}}{2n_{\mathrm{e}}} \times \left[+\frac{25.50\lambda^2}{\lambda^2 - 0.107584} + \frac{45.72\lambda^4}{(\lambda^2 - 0.107584)^2} \right] .$$

Note: *Canarelli* et al. [3.347] observed the discrepancy between these dispersion formulas and the experiment

Experimental values of phase-matching angle ($T = 293$ K) and comparison between different sets of dispersion relations:

Interacting wavelengths [µm]	θ_{exp} [deg]	θ_{theor} [deg]		
		[3.348]	[3.349]	[3.350]
SHG, o + o ⇒ e				
3.3913 ⇒ 1.69565	33 [3.339]	34.1	33.2	33.5
10.6 ⇒ 5.3	67 [3.351]	70.7	73.3	71.7
	67.5 [3.352]			
	68 [3.339]			
	70.8 [3.344]			
SFG, o + o ⇒ e				
11.538 + 1.17233 ⇒ 1.0642	34.7 [3.48]	35.9	35.3	35.7
9.9 + 1.19237 ⇒ 1.0642	35.9 [3.353]	36.4	35.6	36.2
8.7 + 1.21252 ⇒ 1.0642	37 [3.354]	37.3	36.4	37.0
6.24 + 1.28301 ⇒ 1.0642	41.1 [3.355]	40.9	39.8	40.4
5.89 + 1.29888 ⇒ 1.0642	42.1 [3.353]	41.7	40.5	41.2
4.8 + 1.36735 ⇒ 1.0642	44 [3.354]	44.7	43.4	44.1
4.0 + 1.44996 ⇒ 1.0642	47.7 [3.355]	47.7	46.1	46.9
3.09 + 1.62325 ⇒ 1.0642	51 [3.350]	51.9	50.0	50.9
2.53 + 1.83683 ⇒ 1.0642	53.4 [3.350]	54.4	52.4	53.4
6.85 + 1.0642 ⇒ 0.92110	42 [3.356]	43.9	42.7	43.6
4.43 + 1.0642 ⇒ 0.85807	55 [3.356]	57.1	55.3	56.7
6.6 + 0.77593 ⇒ 0.6943	60 [3.357]	60.5	60.4	61.8

$4.8 + 0.81171 \Rightarrow 0.6943$	75.5 [3.357]	79.5	79.0	83.9
$11.66329 + 0.617 \Rightarrow 0.586$	64 [3.358]	58.9	67.0	63.4
$10.12478 + 0.622 \Rightarrow 0.586$	70 [3.358]	64.2	75.4	70.1
SFG, $e + o \Rightarrow e$				
$10.9 + 1.17934 \Rightarrow 1.0642$	38.3 [3.359]	38.3	37.5	38.0
$8.8 + 1.21060 \Rightarrow 1.0642$	40.3 [3.359]	40.2	39.1	39.9
$7.0 + 1.25500 \Rightarrow 1.0642$	43.6 [3.359]	43.7	42.4	43.2
$5.2 + 1.33803 \Rightarrow 1.0642$	50.6 [3.359]	50.6	48.7	49.9
$10.6 + 1.0642 \Rightarrow 0.96711$	39.8 [3.360]	39.7	38.8	39.5
$9.6 + 1.0642 \Rightarrow 0.95800$	41.5 [3.360]	41.0	40.0	40.8
$10.6 + 0.6943 \Rightarrow 0.65162$	55 [3.361]	54.0	55.3	55.8

Note: The other sets of dispersion relations from [3.348, 362, 48] show worse agreement with the experiment

Best of dispersion relations (λ in μm, $T = 20$ °C) [3.350].

$$n_o^2 = 3.3970 + \frac{2.3982\lambda^2}{\lambda^2 - 0.09311} + \frac{2.1640\lambda^2}{\lambda^2 - 950.0} \; ,$$

$$n_e^2 = 3.5873 + \frac{1.9533\lambda^2}{\lambda^2 - 0.11066} + \frac{2.3391\lambda^2}{\lambda^2 - 1030.7} \; .$$

Calculated values of phase-matching and "walk-off" angles:

Interacting wavelengths [μm]	θ_{pm} [deg]	ρ_1 [deg]	ρ_3 [deg]
SHG, $o + o \Rightarrow e$			
$10.6 \Rightarrow 5.3$	71.68		0.76
$9.6 \Rightarrow 4.8$	58.15		1.15
$5.3 \Rightarrow 2.65$	32.00		1.17
$4.8 \Rightarrow 2.4$	31.04		1.15
$2.9365 \Rightarrow 1.46825$	37.27		1.24
$2.1284 \Rightarrow 1.0642$	54.23		1.18
SFG, $o + o \Rightarrow e$			
$10.6 + 3.533 \Rightarrow 2.65$	37.40		1.25
$10.6 + 2.65 \Rightarrow 2.12$	34.79		1.21
$10.6 + 1.0642 \Rightarrow 0.96711$	37.31		1.21
$10.6 + 0.6943 \Rightarrow 0.65162$	52.85		1.04
SFG, $e + o \Rightarrow e$			
$10.6 + 5.3 \Rightarrow 3.533$	58.15	1.18	1.15
$10.6 + 1.0642 \Rightarrow 0.96711$	39.52	1.32	1.23
$10.6 + 0.6943 \Rightarrow 0.65162$	55.76	1.23	1.00

Experimental values of internal angular and spectral bandwidths at $T = 293$ K:

Interacting wavelengths [μm]	θ_{pm} [deg]	$\Delta\theta^{int}$ [deg]	$\Delta\nu_1$ [cm^{-1}]	Ref.
SHG, o + o $\Rightarrow$ e				
10.6 $\Rightarrow$ 5.3	67.5	0.41		3.339
SFG, o + o $\Rightarrow$ e				
4.6 + 0.8177 $\Rightarrow$ 0.6943	82.7	0.42		3.357
10.53 + 0.589 $\Rightarrow$ 0.56589	90	2.34		3.349
6.24 + 1.283 $\Rightarrow$ 1.0642	41.1		9.8	3.355
4.817 + 1.0642 $\Rightarrow$ 0.87163	52		5.9	3.356
10.619 + 0.634 $\Rightarrow$ 0.598	90		1.73	3.341
10.6 + 0.598 $\Rightarrow$ 0.566	90		1.5	3.363
10.6 + 0.5968 $\Rightarrow$ 0.565	90		1.44	3.364

Temperature variation of phase-matching angle [3.360]:

Interacting wavelengths [μm]	T [°C]	θ_{pm} [deg]	$d\theta_{pm}/dT$ [deg/K]
SFG, e + o $\Rightarrow$ e			
10.6 + 1.0642 $\Rightarrow$ 0.9671	20	39.8	0.03

Temperature tuning of noncritical SFG [3.347]:

Interacting wavelengths [μm]	$d\lambda_1/dT$ [nm/K]
SHG, o + o $\Rightarrow$ e	
7.8 + 0.65 $\Rightarrow$ 0.6	≈ 4

Experimental value of temperature bandwidth for the noncritical SFG process (10.6 μm + 0.598 μm $\Rightarrow$ 0.566 μm, o + o $\Rightarrow$ e):

$$\Delta T = 2.5 \ ^\circ C \ [3.346] \ .$$

Effective nonlinearity expressions in the phase-matching direction [3.100]:

$$d_{ooe} = d_{36} \sin\theta \sin 2\phi \ ,$$
$$d_{eoe} = d_{oee} = d_{36} \sin 2\theta \cos 2\phi \ .$$

Nonlinear coefficient:

$$d_{36}(10.6\,\mu m) = 0.134 \times d_{36}(\text{GaAs}) \pm 15\% =$$
$$11.1 \pm 1.7 \ \text{pm/V} \ [3.344], [3.37] \ ,$$
$$d_{36}(10.6\,\mu m) = 0.15 \times d_{36}(\text{GaAs}) \pm 20\% =$$
$$12.5 \pm 2.5 \ \text{pm/V} \ [3.351], [3.37] \ .$$

Laser-induced surface-damage threshold:

λ [µm]	τ_p [ns]	$I_{thr} \times 10^{-12}$ [W/m²]	Ref.	Note
0.59	500	0.2	3.358	10 pulses
0.598	3	0.15	3.363	
0.625	500	0.25–0.36	3.358	10 pulses
0.6943	30	0.006	3.361	1 Hz, 1000 pulses
	10	0.1	3.357	100 pulses
	10	0.2	3.348	
1.06	35	0.2–0.25	3.348	
1.0642	20	0.1	3.350	10 Hz
	17.5	> 0.12	3.365	1000 pulses
	15	0.2	3.352	
	12	0.35	3.359	10 Hz
	0.023	> 0.75	3.366	10 Hz
	0.025	> 7	3.48	10 Hz
	0.002	> 10	3.367	
	0.021	> 20	3.355	
	0.020	30	3.353	
10.6	150	0.1	3.349	
	150	0.2	3.368	
	220	0.25	3.365	1000 pulses

Thermal conductivity coefficient at $T = 293$ K [3.58]:

κ [W/mK], $\parallel c$	κ [W/mK], $\perp c$
1.4	1.5

3.1.11 ZnGeP₂, Zinc Germanium Phosphide

Positive uniaxial crystal: $n_e > n_o$;
Point group: $\bar{4}2m$;
Mass density: 4.12 g/cm³ [3.338] ;
Mohs hardness: 5.5 ;
Transparency range at "0" transmittance level: $0.74 - 12$ µm [3.369, 370]
Linear absorption coefficient α:

λ [µm]	α [cm⁻¹]	Ref.	Note
1.9	0.8–0.95	3.371	
2.15	0.6	3.372	
2.5–8	< 0.1	3.373	
2.5–8.3	< 0.2	3.374	

λ [μm]	α [cm^{-1}]	Ref.	Note
2.5–8.5	< 0.1	3.375	
2.8–8.3	< 0.1	3.376	
3–8	< 0.1	3.377	
3.5–3.9	0.41	3.378	o – wave, SFG direction
3.5	0.4	3.379	
3.8	0.1–0.18	3.371	
4.5–8	0.03	3.380	best samples
4.65	0.4	3.381	
	0.1–0.2	3.382	
4.8	0.16	3.383	
5.3–6.1	0.32	3.378	e – wave, SFG direction
8.3–9.5	< 0.3	3.374	
9	≈ 1	3.379	
9.28	0.4	3.373	
9.3	0.8	3.381	
	0.4–0.5	3.382	
9.6	0.56	3.383	
10.3	0.42	3.384	
10.4	0.6	3.372	
10.6	0.9	3.379	
	0.83	3.378	e – wave, SFG direction

Experimental values of refractive indices [3.369]:

λ [μm]	n_o	n_e	λ [μm]	n_o	n_e
0.64	3.5052	3.5802	2.40	3.1388	3.1780
0.66	3.4756	3.5467	2.60	3.1357	3.1745
0.68	3.4477	3.5160	2.80	3.1327	3.1717
0.70	3.4233	3.4885	3.00	3.1304	3.1693
0.75	3.3730	3.4324	3.20	3.1284	3.1671
0.80	3.3357	3.3915	3.40	3.1263	3.1647
0.85	3.3063	3.3593	3.60	3.1257	3.1632
0.90	3.2830	3.3336	3.80	3.1237	3.1616
0.95	3.2638	3.3124	4.00	3.1223	3.1608
1.00	3.2478	3.2954	4.20	3.1209	3.1595
1.10	3.2232	3.2688	4.50	3.1186	3.1561
1.20	3.2054	3.2493	4.70	3.1174	3.1549
1.30	3.1924	3.2346	5.00	3.1149	3.1533
1.40	3.1820	3.2244	5.50	3.1131	3.1518
1.60	3.1666	3.2077	6.00	3.1101	3.1480
1.80	3.1562	3.1965	6.50	3.1057	3.1445
2.00	3.1490	3.1889	7.00	3.1040	3.1420
2.20	3.1433	3.1829	7.50	3.0994	3.1378

λ [μm]	n_o	n_e	λ [μm]	n_o	n_e
8.00	3.0961	3.1350	10.50	3.0738	3.1137
8.50	3.0919	3.1311	11.00	3.0689	3.1087
9.00	3.0880	3.1272	11.50	3.0623	3.1008
9.50	3.0836	3.1231	12.00	3.0552	3.0949
10.00	3.0788	3.1183			

Temperature derivative of refractive indices [3.369]:

λ [μm]	$dn_o/dT \times 10^5$ [K^{-1}]	$dn_e/dT \times 10^5$ [K^{-1}]	λ [μm]	$dn_o/dT \times 10^5$ [K^{-1}]	$dn_e/dT \times 10^5$ [K^{-1}]
0.64	35.94	37.58	3.40	14.40	15.46
0.66	31.23	37.34	3.60	15.58	16.29
0.68	29.52	32.53	3.80	14.58	16.53
0.70	28.63	31.82	4.00	14.26	15.02
0.75	26.22	28.26	4.20	13.57	15.14
0.80	24.69	26.43	4.50	15.31	16.60
0.85	24.12	25.39	4.70	15.51	16.71
0.90	22.34	24.61	5.00	15.05	16.43
0.95	21.32	24.26	5.50	14.49	15.42
1.00	21.18	23.01	6.00	14.58	16.30
1.10	20.11	22.08	6.50	15.60	16.13
1.20	18.63	20.51	7.00	12.85	15.01
1.30	16.84	20.12	7.50	18.15	18.59
1.40	15.34	16.55	8.00	16.10	17.43
1.60	15.10	16.75	8.50	15.16	17.37
1.80	13.20	14.40	9.00	15.56	17.50
2.00	14.19	15.29	9.50	16.27	17.11
2.20	14.60	15.28	10.00	16.53	18.41
2.40	14.14	15.49	10.50	15.40	16.84
2.60	15.13	16.80	11.00	15.25	16.34
2.80	15.48	16.05	11.50	14.74	18.32
3.00	13.26	13.96	12.00	14.24	16.59
3.20	14.94	16.28			

Experimental values of phase-matching angle ($T = 293$ K) and comparison between different sets of dispersion relations:

Interacting wavelengths [μm]	θ_{exp} [deg]	θ_{theor} [deg]	
		[3.362]	[3.385]
SHG, e + e $\Rightarrow$ o			
3.8 $\Rightarrow$ 1.9	57.8 ±0.3 [3.371]	59.8	59.7
4.34 $\Rightarrow$ 2.17	55.8 ±0.2 [3.372]	52.5	52.4
4.64 $\Rightarrow$ 2.32	47.5 [3.386]	50.1	49.9

$9.2 \Rightarrow 4.6$	63.8 [3.387]	64.4	64.0
$9.3 \Rightarrow 4.65$	61.3 [3.375]	65.5	65.1
	61.3 [3.385]		
	62.7–64.4 [3.382]		
	64 [3.381]		
$9.5 \Rightarrow 4.75$	62.1 [3.375]	67.9	67.6
	62.1 [3.385]		
	66.8 [3.387]		
$9.6 \Rightarrow 4.8$	64.9 [3.382]	69.3	69.0
$10.2 \Rightarrow 5.1$	72 [3.375]	81.6	81.3
$10.3 \Rightarrow 5.15$	74.3 [3.384]	86.9	86.4
SFG, e + e $\Rightarrow$ o			
$10.668 + 4.34 \Rightarrow 3.085$	54.3 ± 0.2 [3.372]	51.5	51.3
$9.74 + 4.2039 \Rightarrow 2.9365$	49.6 [3.370]	49.5	49.3
SFG, o + e $\Rightarrow$ o			
$6.74 + 5.2036 \Rightarrow 2.9365$	76 [3.388]	75.9	74.9
$6.45 + 5.3908 \Rightarrow 2.9365$	79.2 [3.374]	80.1	78.8
$6.25 + 5.5389 \Rightarrow 2.9365$	84.0 [3.374]	85.9	83.3
$6.15 + 5.6199 \Rightarrow 2.9365$	85.5 [3.374]	no pm	89.0
$6.29 + 5.0173 \Rightarrow 2.791$	76 [3.376]	77.4	76.5
$6.19 + 5.0828 \Rightarrow 2.791$	77.6 [3.376]	79.1	78.0
$6.06 + 5.1739 \Rightarrow 2.791$	80.5 [3.376]	82.0	80.5
$6.015 + 5.207 \Rightarrow 2.791$	84 [3.389]	83.3	81.6
$5.95 + 5.2569 \Rightarrow 2.791$	83.4 [3.376]	86.1	83.6
$5.90 + 5.2965 \Rightarrow 2.791$	87 [3.376]	no pm	85.8
$10.6 + 1.0642 \Rightarrow 0.9671$	84 [3.379]	83.0	83.4

Best set of dispersion relations (λ in μm, $T = 20\ °$C) [3.385]:

$$n_o^2 = 4.47330 + \frac{5.26576\lambda^2}{\lambda^2 - 0.13381} + \frac{1.49085\lambda^2}{\lambda^2 - 662.55}\ ,$$

$$n_e^2 = 4.63318 + \frac{5.34215\lambda^2}{\lambda^2 - 0.14255} + \frac{1.45795\lambda^2}{\lambda^2 - 662.55}\ .$$

dispersion relation for $T = 93$ K, 173 K, 373 K, 473 K, and 673 K are given in [3.390], for $T = 343$ K in [3.391].

Calculated values of phase-matching and "walk-off" angles:

Interacting wavelengths [μm]	θ_{pm} [deg]	ρ_1 [deg]	ρ_2 [deg]
SHG, e + e $\Rightarrow$ o			
$9.6 \Rightarrow 4.8$	68.95	0.49	0.49
$5.3 \Rightarrow 2.65$	47.08	0.70	0.70

4.8 ⇒ 2.4	48.97	0.69	0.69
SFG, e + e ⇒ o			
10.6 + 2.65 ⇒ 2.12	50.11	0.72	0.66
9.6 + 2.4 ⇒ 1.92	51.08	0.71	0.69
10.6 + 1.0642 ⇒ 0.96711	72.54	0.42	0.47
9.6 + 1.0642 ⇒ 0.958	82.66	0.18	0.21
SFG, o + e ⇒ o			
10.6 + 5.3 ⇒ 3.533	81.66		0.20
9.6 + 4.8 ⇒ 3.2	69.74		0.46
10.6 + 1.0642 ⇒ 0.96711	83.31		0.19

Experimental values of internal angular bandwidth:

Interacting wavelengths [μm]	$\Delta\theta^{int}$ [deg]	Ref.
SHG, e + e ⇒ o		
3.8 ⇒ 1.9	1.33	3.371
4.34 ⇒ 2.17	1.05	3.372
5.3 ⇒ 2.65	0.69	3.386
9.3 ⇒ 4.65	0.74–0.80	3.382
	1.15	3.381
9.6 ⇒ 4.8	0.8	3.382
10.2 ⇒ 5.1	1.35	3.375
10.3 ⇒ 5.15	1.20	3.384
SFG, e + e ⇒ o		
10.668 + 4.34 ⇒ 3.085	1.23	3.372
SFG, o + e ⇒ o		
10.6 + 1.064 ⇒ 0.967	0.55	3.379

Experimental values of spectral bandwidth:

Interacting wavelength [μm]	$\Delta\nu$ [cm^{-1}]	Ref.
SHG, e + e ⇒ o		
4.34 ⇒ 2.17	7.9	3.372
10.2 ⇒ 5.1	4.9	3.375

Experimental value of temperature bandwidth for SHG process
(10.2 μm ⇒ 5.1 μm, e + e ⇒ o);

$$\Delta T = 50 \,°C \,[3.375] \;.$$

Temperature variation of phase-matching angle:

Interacting wavelengths [µm]	$\mathrm{d}\theta_{\mathrm{pm}}/\mathrm{d}T$ [deg/K]	Ref.
SHG, e + e $\Rightarrow$ o		
9.2 $\Rightarrow$ 4.6	0.014	3.387
10.3 $\Rightarrow$ 5.15	0.072	3.375
10.6 $\Rightarrow$ 5.3	0.107	3.375
SFG, o + e $\Rightarrow$ o		
10.6 + 1.0642 $\Rightarrow$ 0.9671	0.007	3.379

Effective nonlinearity expressions in the phase-matching direction [3.100]:

$$d_{\mathrm{ooe}} = d_{36} \sin\theta \sin 2\phi \ ,$$
$$d_{\mathrm{eoe}} = d_{\mathrm{oee}} = d_{36} \sin 2\theta \cos 2\phi \ .$$

Nonlinear coefficient:

$$d_{36}(10.6\,\mu\mathrm{m}) = 0.83 \times d_{36}(\mathrm{GaAs}) \pm 15\% =$$
$$68.9 \pm 10.3 \ \mathrm{pm/V} \ [3.369], [3.37] \ ,$$
$$d_{36}(9.6\,\mu\mathrm{m}) = 75 \pm 8 \ \mathrm{pm/V} \ [3.383] \ .$$

Laser-induced surface-damage threshold:

λ [µm]	τ_{p} [ns]	$I_{\mathrm{thr}} \times 10^{-12}$ [W/m^2]	Ref.	Note
1.064	30	> 0.03	3.392	12.5 Hz
	10	0.03	3.369	
2.79	0.15	300	3.376	
	0.1	350	3.389	
2.94	0.11	300	3.388	
	0.11	300	3.370	
5.3–6.1	cw	> 0.0001	3.386	
	cw	0.0025	3.378	
9.28	2	12.5	3.373	
9.3–10.6	125	0.3-0.4	3.384	2 Hz
	125	0.25	3.384	20 Hz
9.3	100	0.12	3.381	100 Hz
9.6	120	0.78	3.383	
10.2–10.8	$10^5 - 10^7$	0.6	3.375	1500 Hz
	cw	> 0.00001	3.375	
10.6	cw	> 0.0000001	3.392	
	cw	0.002	3.378	

Thermal conductivity coefficient at $T = 293$ K [3.58]:

κ [W/mK], $\parallel c$	κ [W/mK], $\perp c$
36	35

3.2 Frequently Used Nonlinear Optical Crystals

3.2.1 KB$_5$O$_8 \cdot$ 4H$_2$O, Potassium Pentaborate Tetrahydrate (KB5)

Positive biaxial crystal: $2V_Z = 126.3°$ at $\lambda = 0.5461$ μm [3.393];
Point group: mm2;
Assignment of dielectric and crystallographic axes:
$X,Y,Z \Rightarrow a,b,c$ (Fig. 3.4);
Molecular mass: 1.74 g/cm^3 [3.394];
Mohs hardness: 2.5 [3.394];
Transparency range at "0" transmittance level: $0.162 - 1.5$ μm [3.395];

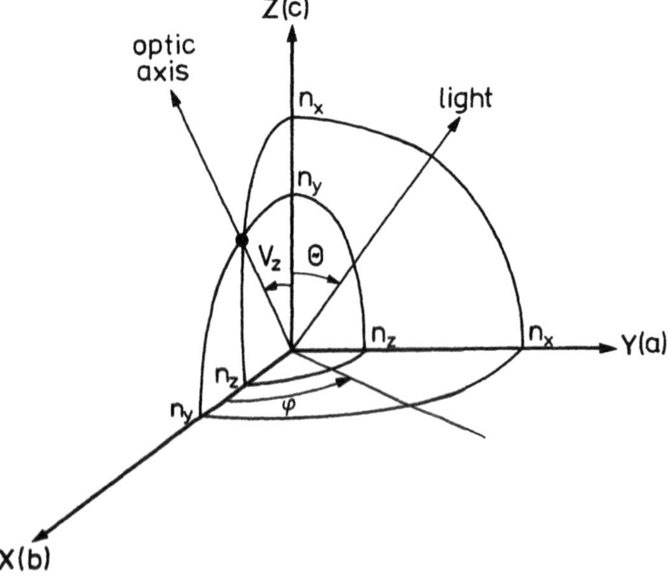

Fig. 3.4. Dependence of refractive index on light propagation direction and polarization (index surface) in the first octant of dielectric reference frame (X, Y, Z) of KB5 crystal. Designations: θ is the polar angle, ϕ is the asimuthal angle, V_Z is the angle between one of the optical axes and the Z axis

Linear absorption coefficient α:

λ [μm]	α [cm^{-1}]	Ref.	Note
0.2128	0.18	3.396	o – wave, XY plane, FIHG direction
	0.14	3.397	o – wave, XY plane, FIHG direction
0.2314	0.12	3.393	o – wave, XY plane, THG direction
0.2661	0.12	3.396	e – wave, XY plane, FIHG direction
	0.06	3.397	e – wave, XY plane, FIHG direction
0.3472	0.04	3.393	e – wave, XY plane, THG direction
0.3547	< 0.01	3.398	along Y

λ [µm]	α [cm^{-1}]	Ref.	Note
0.5321	0.02	3.397	XY plane, FIHG direction
	< 0.01	3.398	along Y
0.6943	0.03	3.393	e – wave, XY plane, THG direction
1.0642	0.06	3.397	e – wave, XY plane, FIHG direction

Two-photon absorption coefficient β (along b axis) [3.399]:

λ[µm]	$\beta \times 10^{12}$ [m/W]
0.216	6.5 ± 1.0
0.270	3.5 ± 0.5

Experimental values of refractive indices:

λ [µm]	n_X	n_Y	n_Z	Ref.
0.217			1.4969	3.400
0.220			1.4938	3.400
0.225			1.4891	3.400
0.230			1.4848	3.400
0.2345		1.4930		3.401
0.235			1.4809	3.400
0.240			1.4774	3.400
0.245			1.4740	3.400
0.250			1.4708	3.400
0.390	1.5021	1.4457	1.4327	3.401
0.400	1.5005	1.4453	1.4320	3.401
0.420	1.4984	1.4438	1.4303	3.401
0.450	1.4956	1.4414	1.4280	3.401
0.500	1.4917	1.4380	1.4251	3.401
0.546	1.4888	1.4357	1.4230	3.401
0.600	1.4859	1.4334	1.4211	3.401
0.650	1.4839	1.4319	1.4196	3.401
0.700	1.4823	1.4306	1.4182	3.401
0.730	1.4815	1.4297	1.4176	3.401
0.765	1.4813	1.4292	1.4171	3.401

Experimental values of phase-matching angle (T = 293 K) and comparison between different sets of dispersion relations:
XY plane, $\theta = 90°$

Interacting wavelengths [µm]	ϕ_{exp} [deg]	ϕ_{theor} [deg]	
		[3.401]	[3.402]
SHG, e + e $\Rightarrow$ o			
0.434 $\Rightarrow$ 0.217	90 [3.400]	81.6	no pm

$0.4342 \Rightarrow 0.2171$	90 [3.403]	81.3	no pm
$0.4384 \Rightarrow 0.2192$	80.5 [3.86]	77.3	80.4
$0.50 \Rightarrow 0.25$	52.8 [3.400]	53.8	54.1
$0.630 \Rightarrow 0.315$	31 [3.403]	33.0	32.8
SFG, e + e $\Rightarrow$ o			
$0.5435 + 0.3511 \Rightarrow 0.2133$	90 [3.404]	78.9	87.3
$0.6943 + 0.3472 \Rightarrow 0.2314$	57 [3.405]	56.3	57.9
$0.5737 + 0.3345 \Rightarrow 0.2113$	90 [3.404]	77.9	87.2
$0.6522 + 0.3261 \Rightarrow 0.2174$	68 [3.398]	65.7	68.8
$0.6219 + 0.3110 \Rightarrow 0.2073$	90 [3.398]	76.9	no pm
$0.6943 + 0.30519 \Rightarrow 0.2120$	70 [3.406]	66.2	70.5
$0.6943 + 0.28409 \Rightarrow 0.2016$	90 [3.406]	74.2	no pm
$1.06415 + 0.26604 \Rightarrow 0.2128$	53 [3.397]	48.5	52.1
$0.78971 + 0.26604 \Rightarrow 0.1990$	75 [3.407]	67.5	76.1
$0.75322 + 0.26604 \Rightarrow 0.1966$	90 [3.407]	72.5	no pm
$0.79737 + 0.25725 \Rightarrow 0.1945$	84 [3.408]	70.0	83.3
$0.79235 + 0.25725 \Rightarrow 0.1942$	90 [3.408]	70.7	85.6
$0.9 + 0.23287 \Rightarrow 0.185$	90 [3.409]	68.4	no pm

YZ plane, $\phi = 90°$

Interacting wavelengths [µm]	θ_{exp} [deg]	θ_{theor} [deg] [3.401]	[3.402]
SHG, o + o $\Rightarrow$ e			
$0.4346 \Rightarrow 0.2173$	90 [3.405]	69.1	83.4
$0.4690 \Rightarrow 0.2345$	17 [3.405]	no pm	12.8
$0.4796 \Rightarrow 0.2398$	0 [3.403]	no pm	no pm
SFG, o + o $\Rightarrow$ e			
$0.5634 + 0.3511 \Rightarrow 0.2163$	63 [3.404]	49.2	59.9
$0.5948 + 0.3345 \Rightarrow 0.2141$	63 [3.404]	47.0	59.8
$0.6264 + 0.3132 \Rightarrow 0.2088$	68 [3.398]	52.2	72.0
$0.7621 + 0.26604 \Rightarrow 0.1972$	68 [3.407]	38.5	75.3

Best set of dispersion relations (revised data of [3.401], given in [3.402], λ in µm, $T = 293$ K):

$$n_X^2 = 1 + \frac{\lambda^2}{0.848117\lambda^2 - 0.0074477} \; ,$$

$$n_Y^2 = 1 + \frac{\lambda^2}{0.972682\lambda^2 - 0.0087757} \; ,$$

$$n_Z^2 = 1 + \frac{\lambda^2}{1.008157\lambda^2 - 0.0094050} \; .$$

Calculated values of phase-matching and "walk-off" angles:
XY plane, $\theta = 90°$

Interacting wagelengths [µm]	θ_{pm} [deg]	ρ_1 [deg]	ρ_2 [deg]
SHG, e + e ⇒ o			
$0.5105 \Rightarrow 0.25525$	51.62	2.037	2.037
$0.532075 \Rightarrow 0.26604$	47.19	2.073	2.073
$0.5782 \Rightarrow 0.2891$	39.57	2.020	2.020
$0.6973 \Rightarrow 0.34715$	25.83	1.585	1.585
SFG, e + e ⇒ o			
$1.06415 + 0.532075 \Rightarrow 0.35473$	20.65	1.324	1.332
$1.06415 + 0.35473 \Rightarrow 0.26604$	36.35	1.946	1.979
$1.06415 + 0.26604 \Rightarrow 0.21283$	52.12	2.015	2.078
$0.6943 + 0.34715 \Rightarrow 0.23143$	57.93	1.889	1.918
$0.5782 + 0.5105 \Rightarrow 0.27112$	45.17	2.017	2.075

Experimental values of NCPM temperature:
along b axis

Interacting wavelengths [µm]	T [°C]	Ref.
SFG, type I		
$0.6943 + 0.28334 \Rightarrow 0.20122$	−15	3.406
$0.6943 + 0.28361 \Rightarrow 0.20136$	0	3.406
$0.6943 + 0.28405 \Rightarrow 0.20158$	20	3.406
$0.6943 + 0.28449 \Rightarrow 0.20180$	35	3.406
$0.79202 + 0.25725 \Rightarrow 0.19418$	25	3.408
$0.79344 + 0.25725 \Rightarrow 0.19427$	40	3.408

Effective nonlinearity expressions in the phase-matching direction for three-wave interactions in the principal planes of KB5 crystal; [3.35, 36]:
XY plane

$$d_{eeo} = d_{31} \sin^2 \phi + d_{32} \cos^2 \phi \; ;$$

YZ plane

$$d_{ooe} = d_{31} \sin \theta \; ;$$

XZ plane, $\theta < V_Z$

$$d_{oeo} = d_{eoo} = d_{32} \sin \theta \; ;$$

XZ plane, $\theta > V_Z$

$$d_{ooe} = d_{32} \sin \theta \; .$$

Effective nonlinearity expressions for three-wave interactions in the arbitrary direction of KB5 crystal are given in [3.36].

Nonlinear coefficients [3.37]:

$$d_{31}(0.5321\ \mu m) = 0.04\ \text{pm/V}\ ,$$
$$d_{32}(0.5321\ \mu m) = 0.003\ \text{pm/V}\ ,$$
$$d_{33}(0.5321\ \mu m) = 0.05\ \text{pm/V}\ .$$

Laser-induced surface-damage threshold:

λ [μm]	τ_p [ns]	$I_{thr} \times 10^{-12}$ [W/m^2]	Ref.	Note
0.2661	8	> 0.43	3.397	10 Hz
	0.03	> 4.8	3.410	1 Hz
0.311	10	> 0.13	3.398	10 Hz
0.3472	8	> 0.9	3.393	
0.45	7	10	3.405	15 Hz
0.622	10	> 0.4	3.398	10 Hz
0.6943	10	> 0.8	3.393	
0.74–0.91	30	> 0.5	3.409	
1.0642	12	> 0.85	3.397	10 Hz

3.2.2 CO(NH$_2$)$_2$, Urea

Positive uniaxial crystal: $n_e > n_o$;
Point group: $\bar{4}$2m;
Mass density: 1.318 g/cm^3;
Mohs hardness: < 2.5 ;
Transparency range at 0.5 transmittance level for a 0.5 cm long crystal cut at
$\theta = 74°$: 0.2 – 1.43 μm [3.411];
Linear absorption coefficient α [3.411]:

λ [μm]	α [cm^{-1}]	Note
0.213	0.10	o – wave, FIHG direction
0.266	0.04	e – wave, FIHG direction
1.064	0.02	e – wave, FIHG direction

The graph of n_o and n_e dependences versus wavelength is given in [3.412, 413].
Experimental values of phase-matching angle ($T = 293$ K) and comparison
between different sets of dispersion relations:

Interacting wavelengths [μm]	θ_{exp} [deg]	θ_{theor} [deg]		
		[3.414]	[3.415]*	[3.416]
SHG, e + e $\Rightarrow$ o				
0.476 $\Rightarrow$ 0.238	90 [3.414]	82.2	no pm	no pm
0.500 $\Rightarrow$ 0.250	67.6 [3.414]	67.5	76.7	72.2
0.550 $\Rightarrow$ 0.275	54 [3.414]	54.2	55.9	55.2
0.600 $\Rightarrow$ 0.300	46.6 [3.414]	46.5	46.5	46.5

SFG, e + e ⇒ o

0.6943 + 0.34715 ⇒				
⇒ 0.23143	77 [3.411]	81.5	no pm	no pm
1.0642 + 0.26605 ⇒				
⇒ 0.21284	72 [3.411]	86.7	no pm	no pm

SHG, o + e ⇒ o

0.597 ⇒ 0.2985	90 [3.414]	no pm	no pm	no pm
0.650 ⇒ 0.325	63.6 [3.414]	65.4	63.5	64.6
0.700 ⇒ 0.350	55.6 [3.414]	56.6	54.6	55.6

SFG, o + e ⇒ o

1.0642 + 0.29146 ⇒ 0.2288	90 [3.414]	no pm	no pm	76.6
1.0642 + 0.29668 ⇒ 0.2320	80 [3.414]	83.6	80.9	72.8
1.0642 + 0.30656 ⇒ 0.2380	70.4 [3.414]	75.0	70.0	67.8
1.0642 + 0.42792 ⇒ 0.3052	47.5 [3.414]	49.9	48.8	47.0
1.0642 + 0.63501 ⇒ 0.3977	37.7 [3.414]	39.1	37.1	37.6
0.720 + 0.53764 ⇒ 0.3078	63 [3.417]	64.7	63.1	62.7
0.646 + 0.58793 ⇒ 0.3078	69 [3.418]	71.6	70.0	70.3
0.62875 + 0.5321 ⇒ 0.2882	90 [3.414]	no pm	no pm	no pm
0.63980 + 0.5321 ⇒ 0.2905	80.5 [3.414]	85.1	84.3	81.5
0.66406 + 0.5321 ⇒ 0.2954	73.4 [3.414]	75.3	74.3	73.1

SFG, e + o ⇒ o

1.0642 + 0.50787 ⇒ 0.3438	90 [3.414]	79.2	84.5	no pm
1.0642 + 0.53 ⇒ 0.3538	72.2 [3.414]	70.9	72.3	74.8
1.0642 + 0.575 ⇒ 0.3733	62.5 [3.414]	61.4	61.5	63.0
1.0642 + 0.63195 ⇒ 0.3965	53.5 [3.414]	53.9	53.4	54.5

*with correction given in [3.419].

Best set of dispersion relations (λ in μm, $T = 293$ K) [3.415, 419]:

$$n_o^2 = 2.1548 + \frac{0.01310}{\lambda^2 - 0.0318},$$

$$n_e^2 = 2.5527 + \frac{0.01784}{\lambda^2 - 0.0294} + \frac{0.0288(\lambda - 1.5)}{(\lambda - 1.5)^2 + 0.03371}.$$

Calculated values of phase-matching and "walk-off" angles:

Interacting wavelengths [μm]	θ_{pm} [deg]	ρ_1 [deg]	ρ_2 [deg]
SHG, o + e ⇒ o			
0.6118 ⇒ 0.3059	75.95		2.31
0.6328 ⇒ 0.3164	68.01		3.35
0.6594 ⇒ 0.3297	61.49		3.98
0.6943 ⇒ 0.34715	55.40		4.34

SFG, o + e ⇒ o

$1.0642 + 0.5321 \Rightarrow 0.35473$	41.10		4.52
$1.3188 + 0.6594 \Rightarrow 0.4396$	30.46		3.82

SFG, e + o ⇒ o

$1.0642 + 0.5321 \Rightarrow 0.35473$	71.63	2.69
$1.3188 + 0.6594 \Rightarrow 0.4396$	48.98	3.97

Experimental value of internal angular bandwidth [3.411]:

Interacting wavelengths [μm]	$\Delta\theta^{\text{int}}$ [deg]
FIHG, e + e ⇒ o	
$1.064 + 0.266 \Rightarrow 0.213$	0.017

Temperature tuning for noncritical SHG [3.414]:

Interacting wavelengths [μm]	$d\lambda_1/T$ [nm/K]
SHG, e + o ⇒ e	
$0.597 \Rightarrow 0.2985$	−0.013

Effective nonlinearity expressions in the phase-matching direction [3.100]:

$$d_{\text{ooe}} = d_{36} \sin\theta \sin 2\phi \ ,$$
$$d_{\text{eoe}} = d_{\text{oee}} = d_{36} \sin 2\theta \cos 2\phi \ .$$

Nonlinear coefficient:

$$d_{36}(0.6\,\mu m) \approx 3 \times d_{36}(\text{KDP}) = 1.17 \text{ pm/V} \ [3.412, 37],$$
$$d_{36}(0.597\,\mu m) = 2.4 \times d_{36}(\text{ADP}) \pm 8\% =$$
$$1.13 \pm 0.09 \text{ pm/V} \ [3.419, 37] \ .$$

Laser-induced bulk-damage threshold:

λ [μm]	τ_p [ns]	$I_{\text{thr}} \times 10^{-12}$ [W/m²]	Ref.	Note
0.266	10	5	3.420	single pulse
0.355	10	14	3.420	single pulse
	10	1.5	3.421	3000 pulses
0.532	10	30	3.420	single pulse
1.064	10	50	3.420	single pulse

3.2.3 CsH_2AsO_4, Cesium Dihydrogen Arsenate (CDA)

Negative uniaxial crystal: $n_o > n_e$;
Point group: $\bar{4}2m$;
Mass density: 3.53 g/cm^3 ;
Transparency range at 0.5 transmittance level for a 17.5 mm long crystal cut at
$\theta = 90°, \phi = 45° : 0.26 - 1.43\,\mu m$ [3.422] ;
UV edge of transmission spectrum at "0" transmittance level:
$0.216\,\mu m$ [3.113] ;
IR edge of transmission spectrum at "0" transmittance level:
$1.87\,\mu m$ for o – wave, $1.67\,\mu m$ for e – wave [3.78] ;
Linear absorption coefficient α :

λ [μm]	α [cm^{-1}]	Ref.
0.35–1.4	0.6	3.113
1.062	0.041	3.120
1.064	0.041	3.422

Two-photon absorption coefficient $\beta(\theta = 90°, \phi = 45°)$ [3.71]:

λ[μm]	$\beta \times 10^{13}$ [m/W]	Note
0.355	2.81	e – wave

Experimental values of refractive indices [3.422]:

λ [μm]	n_o	n_e
0.3472	1.6027	1.5722
0.5321	1.5733	1.5514
0.6943	1.5632	1.5429
1.0642	1.5516	1.5330

Temperature derivative of refractive indices [3.74]:

λ [μm]	$dn_o/dT \times 10^5$ [K^{-1}]	$dn_e/dT \times 10^5$ [K^{-1}]
0.405	−3.15	−1.89
0.436	−3.05	−2.09
0.546	−2.59	−2.12
0.578	−2.76	−2.39
0.633	−2.80	−2.56

Experimental values of phase-matching angle ($T = 293$ K) and comparison between different sets of dispersion relations:

Interacting wavelengths [μm]	θ_{exp} [deg]	θ_{theor} [deg]		
		[3.74]	[3.78]K	[3.78] E
SHG, o + o ⇒ e				
1.05 ⇒ 0.525	90 [3.119]	59.8	no pm	no pm
1.052 ⇒ 0.526	90 [3.74]	59.7	no pm	no pm
1.06 ⇒ 0.53	87 [3.423]	59.0	no pm	no pm
	87 [3.96]			
1.0642 ⇒ 0.5321	83.5 [3.424]	58.6	no pm	88.7
	83.5 [3.425]			
	84.2 [3.422]			
	84.4 [3.426]			
1.068 ⇒ 0.534	–	58.3	88.2	86.5

Note: [3.78] K ⇒ see [3.78], set of *Kirby* et al. ;
　　　[3.78] E ⇒ see [3.78], set of *Eimerl*.

Experimental values of NCPM temperature:

Interacting wavelengths [μm]	T [°C]	Ref.	Note
SHG, o + o ⇒ e			
1.05 ⇒ 0.525	20	3.119	
1.052 ⇒ 0.526	20	3.74	
1.06 ⇒ 0.53	31	3.423	
1.0642 ⇒ 0.5321	40.3	3.427	10 Hz
	41	3.425	
	42	3.428	
	43	3.426	
	44.5	3.90	
	45	3.120	
	46	3.424	12.5 Hz
	48	3.422	0.1–1 Hz
	39.6	3.422	20 Hz
	49.2	3.429	10 Hz
1.073 ⇒ 0.5365	61	3.428	
1.078 ⇒ 0.539	100	3.119	

Best set of dispersion relations (λ in μm, $T = 293$ K) [3.78]E:

$$n_o^2 = 1.8776328 - 0.03602222\lambda^2 + 0.005234121\lambda^4 + \frac{0.5503951\lambda^2}{\lambda^2 - (0.1625700)^2} \, ,$$

$$n_\text{e}^2 = 1.6862889 - 0.01372244\lambda^2 + 0.003948463\lambda^4 + \frac{0.6694571\lambda^2}{\lambda^2 - (0.1464712)^2} \quad .$$

Calculated values of phase-matching and "walk-off" angles:

Interacting wavelengths [μm]	θ_pm [deg]	ρ_3 [deg]
SHG, o + o $\Rightarrow$ e		
1.0642 $\Rightarrow$ 0.5321	88.72	0.035
1.3188 $\Rightarrow$ 0.6594	74.52	0.384

Experimental values of internal angular and temperature bandwidths:

Interacting wavelengths [μm]	$T[^\circ\text{C}]$	θ_pm [deg]	$\Delta\theta^\text{int}$ [deg]	$\Delta T[^\circ\text{C}]$	Ref.
SHG, o + o $\Rightarrow$ e					
1.06 $\Rightarrow$ 0.53	22	87	≈ 0.4		3.423
	31	90	≈ 3.8	≈ 3	3.423
	20	87	0.43		3.96
	63 (?)	90	3.03		3.96
1.062 $\Rightarrow$ 0.531	45	90	2.85	6.5	3.120
1.0642 $\Rightarrow$ 0.5321	40.3	90		6.8	3.427
	24	83.5	0.86	~ 8	3.424
	46	90	3.2		3.424
	20	84.2	0.70		3.422
	48	90	2.91	6	3.422
	20	84.4	0.70		3.426
	43	90	≈ 3		3.426

Temperature variation of phase-matching angle:

Interacting wavelengths [μm]	T [$^\circ$C]	θ_pm [deg]	$d\theta_\text{pm}/dT$ [deg/K]	Ref.
SHG, o + o $\Rightarrow$ e				
1.06 $\Rightarrow$ 0.53	20	87	0.085	3.96
	63(?)	90	0.481	3.96
1.0642 $\Rightarrow$ 0.5321	24	83.5	0.129	3.424
	20	84.4	0.131	3.426
	35	86.5	0.194	3.426
	39	87.6	0.251	3.426
	41	88.3	0.537	3.426

Temperature tuning of noncritical SHG [3.74]:

Interacting wavelengths [μm]	$d\lambda_1/dT$ [nm/K]
SHG, $o + o \Rightarrow e$ $1.052 \Rightarrow 0.526$	0.308

Temperature variation of birefringence for noncritical SHG process ($1.0642\,\mu m \Rightarrow 0.5321\,\mu m, o + o \Rightarrow e$):

$$d(n_2^e - n_1^o)/dT = 7.2 \times 10^{-6} \text{ K}^{-1} \text{ [3.427]} ,$$
$$d(n_2^e - n_1^o)/dT = (8.0 \pm 0.2) \times 10^{-6} \text{ K}^{-1} \text{ [3.422]} .$$

Effective nonlinearity expressions in the phase-matching direction [3.100]:

$$d_{ooe} = d_{36} \sin \theta \sin 2\phi,$$
$$d_{eoe} = d_{oee} = d_{36} \sin 2\theta \cos 2\phi .$$

Nonlinear coefficient:

$$d_{36}(1.0642\,\mu m) = 0.40 \pm 0.05 \text{ pm/V [3.422]} .$$

Laser-induced bulk-damage threshold:

λ [μm]	τ_p [ns]	$I_{thr} \times 10^{-12}$ [W/m^2]	Ref.	Note
0.532	10	> 3	3.429	
1.062	0.007	> 40	3.120	
1.064	12	> 2.6	3.422	10–20 Hz
	10	3.5	3.424	12.5 Hz
	18	4	3.427	2–50 Hz

3.2.4 CsD$_2$AsO$_4$, Deuterated Cesium Dihydrogen Arsenate (DCDA)

Negative uniaxial crystal: $n_o > n_e$:
Point group: $\bar{4}2m$;
Transparency range at 0.5 transmittance level for a 13.5 mm long crystal cut at $\theta = 90°, \phi = 45° : 0.27 - 1.66\,\mu m$ [3.422]
IR edge of transmission spectrum at "0" transmittance level:
$2.03\,\mu m$ for o − wave, $1.78\,\mu m$ for e − wave [3.78] ;

Linear absorption coefficient α :

λ [μm]	α [cm^{-1}]	Ref.
1.062	0.01	3.120
1.064	0.02	3.422

Two-photon absorption coefficient $\beta(\theta = 90°, \phi = 45°)$ [3.71]:

λ [μm]	$\beta \times 10^{13}$ [m/W]	Note
0.355	8.0	o − wave
	5.1	e − wave

Experimental values of refractive indices [3.422]:

λ [μm]	n_o	n_e
0.3472	1.5895	1.5685
0.5321	1.5681	1.5495
0.6943	1.5596	1.5418
1.0642	1.5503	1.5326

Temperature derivative of refractive indices [3.74]:

λ [μm]	$dn_\mathrm{o}/dT \times 10^5$ [K^{-1}]	$dn_\mathrm{e}/dT \times 10^5$ [K^{-1}]
0.405	−2.26	−1.77
0.436	−2.26	−1.51
0.546	−2.47	−1.64
0.578	−2.31	−1.71
0.633		−1.70

Experimental values of phase-matching angle ($T = 293$ K) and comparison between different sets of dispersion relations:

Interacting wavelengths [μm]	θ_exp [deg]	θ_theor [deg]		
		[3.74]	[3.78]K	[3.78]E
SHG, o + o ⇒ e				
1.034 ⇒ 0.517	90 [3.119]	65.2	no pm	no pm
1.037 ⇒ 0.5185	90 [3.74]	64.8	no pm	no pm
1.046 ⇒ 0.523		63.7	88.4	88.1

| 1.0642 ⇒ 0.5321 | 79.35 [3.422]
80.8 [3.426] | 61.8 | 82.4 | 82.3 |

Note: [3.78]K ⇒ see [3.78], data of *Kirby* et al. ;
    [3.78]E ⇒ see [3.78], data of *Eimerl*.

Experimental values of NCPM temperature:

Interacting wavelengths [μm]	T [°C]	Ref.	Note
SHG, o + o ⇒ e			
1.034 ⇒ 0.517	20	3.119	
1.037 ⇒ 0.5185	20	3.74	
1.0642 ⇒ 0.5321	102	3.428	
	102	3.425	
	112.3	3.422	90% deuteration, < 1 Hz
	109.8	3.422	90% deuteration, 20 Hz
	96.4	3.426	70% deuteration
	108	3.119	

Best set of dispersion relations (λ in μm, $T = 293$ K) [3.78]E :

$$n_o^2 = 1.6278496 - 0.018220310\lambda^2$$

$$+ 0.00028133331\lambda^4 + \frac{0.7808170\lambda^2}{\lambda^2 - (0.1407699)^2} \, ,$$

$$n_e^2 = 1.6236063 - 0.009338692\lambda^2$$

$$+ 0.00196541330\lambda^4 + \frac{0.7249589\lambda^2}{\lambda^2 - (0.1414850)^2} \, .$$

Calculated values of phase-matching and "walk-off" angles:

Interacting wavelengths [μm]	θ_{pm} [deg]	ρ_3 [deg]
SHG, o + o ⇒ e		
1.0642 ⇒ 0.5321	82.32	0.188
1.3188 ⇒ 0.6594	69.54	0.449

Experimental values of internal angular and temperature bandwidths:

Interacting wavelengths [µm]	T [°C]	θ_{pm} [deg]	$\Delta\theta^{int}$ [deg]	ΔT [°C]	Ref.
SHG, $o + o \Rightarrow e$					
$1.0642 \Rightarrow 0.5321$	20	79.35	0.41		3.422
	112.3	90	2.90	6.1	3.422
	20	80.8	0.50		3.426
	96.4	90	≈ 3.5		3.426

Temperature variation of the phase-matching angle [3.426]:

Interacting wavelengths [µm]	T [°C]	θ_{pm} [deg]	$d\theta_{pm}/dT$ [deg /K]
SHG, $o + o \Rightarrow e$			
$1.0642 \Rightarrow 0.5321$	20	80.8	0.042
	66.3	84.3	0.081
	80	86.4	0.270
	87.7	88.1	0.533

Temperature tuning of noncritical SHG [3.74]:

Interacting wavelengths [µm]	$d\lambda_1/dT$ [nm/K]
SHG, $o + o \Rightarrow e$	
$1.037 \Rightarrow 0.5185$	0.317

Temperature variation of birefringence for noncritical SHG process $(1.0642\,\mu m \Rightarrow 0.5321\,\mu m, \; o + o \Rightarrow e)$:

$$\frac{d(n_2^e - n_1^o)}{dT} = (7.8 \pm 0.2) \times 10^{-6}\,\text{K}^{-1}\ [3.422]\ .$$

Effective nonlinearity expressions in the phase-matching direction [3.100]:

$$d_{ooe} = d_{36} \sin\theta \sin 2\phi \ ,$$
$$d_{eoe} = d_{oee} = d_{36} \sin 2\theta \cos 2\phi \ .$$

Nonlinear coefficient:

$$d_{36}(1.0642\,\mu m) = 0.40 \pm 0.05\,\text{pm/V}\ [3.422]\ .$$

Laser-induced bulk-damage threshold:

λ [µm]	τ_p [ns]	$I_{thr} \times 10^{-12}$ [W/m^2]	Ref.	Note
1.064	12	>2.6	3.422	10–20 Hz
	12	>2.5	3.139	0.1–20 Hz

3.2.5 KTiOAsO$_4$, Potassium Titanyl Arsenate (KTA)

Positive biaxial crystal: $2V_Z = 34.5°$ at $\lambda = 0.5321\,\mu\text{m}$;
Point group: mm2;
Assignment of dielectric and crystallographic axes:
$X, Y, Z \Rightarrow a, b, c$;
Transparency range at "0" transmittance level: $0.35 - 5.3\,\mu\text{m}$ [3.430, 431];
Linear absorption coefficient α [3.432] :

λ [μm]	α [cm^{-1}]
4.0	0.2
5.0	1.0

Experimental values of refractive indices [3.433]:

λ [μm]	n_X	n_Y	n_Z
0.6328	1.8083	1.8142	1.9048

Experimental values of phase-matching angle ($T = 293\,\text{K}$) and comparison between different sets of dispersion relations:

XY plane, $\theta = 90°$

Interacting wavelengths [μm]	ϕ_{exp} [deg]	ϕ_{theor} [deg]	
		[3.433]	[3.434]
SHG, e + o $\Rightarrow$ e			
1.053 $\Rightarrow$ 0.5265	65 [3.434]	no pm	64.97
1.0642 $\Rightarrow$ 0.5321	57.8 [3.434]	no pm	57.58
SFG, e + o $\Rightarrow$ e			
1.3188 + 0.6594 $\Rightarrow$			
$\Rightarrow$ 0.4396	47.8 [3.434]	68.84	47.79
1.0642 + 1.5791 $\Rightarrow$			
$\Rightarrow$ 0.6358	19.8 [3.434]	16.64	19.63

YZ plane, $\phi = 90°$

Interacting wavelengths [μm]	θ_{exp} [deg]	θ_{theor} [deg]	
		[3.433]	[3.434]
SHG, o + e $\Rightarrow$ o			
1.0642 $\Rightarrow$ 0.5321	76.3 [3.434]	no pm	76.28
1.1523 $\Rightarrow$ 0.57615	64 [3.434]	69.30	63.94

1.3188 $\Rightarrow$ 0.6594	55.9 [3.433]	56.22	53.09
SFG, o + e $\Rightarrow$ o			
1.3188 + 0.6594 $\Rightarrow$			
$\Rightarrow$ 0.4396	71.2 [3.434]	82.37	71.15
1.0642 + 1.5791 $\Rightarrow$			
$\Rightarrow$ 0.6358	67.3 [3.434]	73.04	67.29
4.15 + 1.0642 $\Rightarrow$			
$\Rightarrow$ 0.847	30.3 [3.431]	31.19	31.87

XZ plane, $\phi = 0°$, $\theta > V_Z$

Interacting wavelengths [μm]	θ_{exp} [deg]	θ_{theor} [deg]	
		[3.433]	[3.434]
SHG, o + e $\Rightarrow$ o			
1.1523 $\Rightarrow$ 0.57615	82.9 [3.434]	80.61	83.00
1.3188 $\Rightarrow$ 0.6594	64.2 [3.434]	63.28	64.25
SFG, o + e $\Rightarrow$ o			
1.5791 + 0.6358 $\Rightarrow$			
$\Rightarrow$ 0.4533	73.7 [3.434]	72.82	73.74

Best set of dispersion relations (λ in μm) [3.434]:

$$n_X^2 = 3.1533 + \frac{0.04029}{\lambda^2 - 0.04932} - 0.01320\,\lambda^2 \,,$$

$$n_Y^2 = 3.1775 + \frac{0.04353}{\lambda^2 - 0.05640} - 0.01444\,\lambda^2 \,,$$

$$n_Z^2 = 3.4487 + \frac{0.06334}{\lambda^2 - 0.05887} - 0.01646\,\lambda^2 \,.$$

Calculated values of phase-matching and "walk-off" angles:

XY plane, $\theta = 90°$

Interacting wavelengths [μm]	θ_{pm} [deg]	ρ_1 [deg]	ρ_3 [deg]
SHG, e + o $\Rightarrow$ e			
1.0642 $\Rightarrow$ 0.5321	57.58	0.211	0.337
SFG, e + o $\Rightarrow$ e			
1.3188 + 0.6594 $\Rightarrow$			
$\Rightarrow$ 0.4396	47.79	0.217	0.511

YZ plane, $\phi = 90°$

Interacting wavelengths [μm]	θ_{pm} [deg]	ρ_2 [deg]
SHG, o + e ⟹ o		
1.0642 ⟹ 0.5321	76.28	1.179
1.1523 ⟹ 0.57615	63.94	1.978
1.3188 ⟹ 0.6594	53.09	2.344
2.098 ⟹ 1.049	44.71	2.345
2.9365 ⟹ 1.46825	59.80	2.042
SFG, o + e ⟹ o		
1.3188 + 0.6594 ⟹		
⟹ 0.4396	71.15	1.708

XZ plane, $\phi = 0°, \theta > V_Z$

Interacting wavelengths [μm]	θ_{pm} [deg]	ρ_2 [deg]
SHG, o + e ⟹ o		
1.1523 ⟹ 0.57615	83.00	0.676
1.3188 ⟹ 0.6594	64.25	2.119
2.098 ⟹ 1.049	53.50	2.445
2.9365 ⟹ 1.46825	69.37	1.657

Experimental values of internal angular and temperature bandwidths:
XY plane, $\theta = 90°$

Interacting wavelengths [μm]	ϕ_{pm} [deg]	$\Delta\phi^{int}$ [deg]	ΔT [°C]	Ref.
SHG, e + o ⟹ e				
1.053 ⟹ 0.57615	65	0.4		3.430
1.0642 ⟹ 0.5321	57.8	0.37	10.4	3.434

YZ plane, $\phi = 90°$

Interacting wavelengths [μm]	θ_{pm} [deg]	$\Delta\theta^{int}$ [deg]	Ref.
SHG, o + e ⟹ o			
1.3188 ⟹ 0.6594	55.9	0.093	3.433

Effective nonlinearity expressions in the phase-matching direction for three-wave interactions in the principal planes of KTA crystal [3.35, 36]:

XY plane

$$d_{\mathrm{eoe}} = d_{\mathrm{oee}} = d_{31} \sin^2 \phi + d_{32} \cos^2 \phi \; ;$$

YZ plane

$$d_{\mathrm{oeo}} = d_{\mathrm{eoo}} = d_{31} \sin \theta \; ;$$

XZ plane, $\theta < V_Z$

$$d_{\mathrm{ooe}} = d_{32} \sin \theta \; ;$$

XZ plane, $\theta > V_Z$

$$d_{\mathrm{oeo}} = d_{\mathrm{eoo}} = d_{32} \sin \theta \; .$$

Effective nonlinearity expressions for three-wave interactions in the arbitrary direction of KTA crystal are given in [3.36]

Nonlinear coefficients:

$$d_{31}(1.0642\,\mu\mathrm{m}) = 2.5 \pm 0.3 \text{ pm/V [3.434]} , \; 2.8 \pm 0.3 \text{ pm/V [3.433]} \; ;$$

$$d_{32}(1.0642\,\mu\mathrm{m}) = 4.2 \pm 0.4 \text{ pm/V [3.433]} , \; 4.5 \pm 0.5 \text{ pm/V [3.434]} \; ;$$

$$d_{33}(1.0642\,\mu\mathrm{m}) = 16.2 \pm 1.0 \text{ pm/V [3.433]} \; .$$

Laser-induced surface-damage threshold:

λ [μm]	τ_{p} [ns]	$I_{\mathrm{thr}} \times 10^{-12}$ [W/m^2]	Ref.	Note
0.85	2	>10	3.431	
1.0642	8	>12	3.432	20 Hz, 1000 pulses

3.2.6 MgO : LiNbO$_3$, Magnesium-Oxide-Doped Lithium Niobate (5 mole % MgO)

Negative uniaxial crystal: $n_{\mathrm{o}} > n_{\mathrm{e}}$;
Point group: 3m ;
Transparency range at "0" transmittance level: $\approx 0.4 - \approx 5\,\mu$m [3.435], [3.436]
Linear absorption coefficient α:

λ [μm]	α [cm^{-1}]	Ref.
0.5321	0.02	3.437
1.0642	<0.01	3.437
	0.003	3.438

Experimental values of refractive indices:

5 mole % MgO, mole ratio Li/Nb = 0.97 [3.435]

λ [μm]	n_{o}	n_{e}
0.4358	2.3863	2.2802
0.4916	2.3403	2.2416
0.5461	2.3114	2.2172
0.5770	2.2988	2.2068
0.5790	2.2980	2.2062
0.6328	2.2816	2.1922
0.6943	2.2678	2.1805
0.8400	2.2460	2.1622
1.0642	2.2272	2.1463

5 mole % MgO, mole ratio Li/Nb = 0.946 (congruent melt) [3.436]

λ [μm]	n_{o}	n_{e}	λ [μm]	n_{o}	n_{e}
0.4047	2.4247	2.3111	0.5790	2.2982	2.2056
0.4078	2.4202	2.3073	0.5893	2.2945	2.2027
0.4358	2.3863	2.2795	0.6234	2.2840	2.1938
0.4861	2.3441	2.2444	0.6563	2.2756	2.1867
0.4916	2.3404	2.2412	0.6907	2.2681	2.1802
0.4962	2.3376	2.2389	0.6943	2.2669	2.1793
0.5461	2.3112	2.2167	1.0640	2.2237	2.1456
0.5770	2.2989	2.2063			

Experimental values of phase-matching angle ($T = 293$ K) and comparison between different sets of dispersion relations:

Interacting wavelengths [μm]	θ_{exp} [deg]	θ_{theor} [deg]	
		[3.435]	[3.436]
SHG, o + o $\Rightarrow$ e			
1.0642 $\Rightarrow$ 0.5321	74.5 [3.436]	76.83	82.45
	76 [3.437]		
	76.5 [3.435]		
1.0796 $\Rightarrow$ 0.5398	74 [3.435]	74.08	78.62

Note: The PM angle values are strongly dependent on melt stoichiometry

Experimental values of NCPM temperature:

Interacting wavelengths [μm]	T [°C]	Ref.
SHG, o + o ⇒ e		
1.047 ⇒ 0.5235	75.3	3.439
1.0642 ⇒ 0.5321	107	3.437
	107	3.438
	107	3.440
	107	3.441
	110	3.442
	110.6	3.92
	113	3.443
	116	3.444

Note: The PM temperature values are strongly dependent on melt stoichiometry

Best set of dispersion relations (λ in μm, $T = 20$ °C) [3.435]:

$$n_o^2 = 4.9017 + \frac{0.112280}{\lambda^2 - 0.049656} - 0.039636\,\lambda^2\,,$$

$$n_e^2 = 4.5583 + \frac{0.091806}{\lambda^2 - 0.048086} - 0.032068\,\lambda^2\,.$$

Calculated values of phase-matching and "walk-off" angles:

Interacting wavelengths [μm]	θ_{pm} [deg]	ρ_3 [deg]
SHG, o + o ⇒ e		
1.0642 ⇒ 0.5321	76.83	1.025
1.3188 ⇒ 0.6594	55.87	2.076
2.9365 ⇒ 1.46825	69.09	1.333

Experimental values of angular and temperature bandwidths:

Interacting wavelengths [μm]	T [°C]	θ_{pm} [deg]	$\Delta\theta^{int}$ [deg]	ΔT [°C]	Ref.
SHG, o + o ⇒ e					
1.0642 ⇒ 0.5321	20	76	0.063		3.437
	107	90	2.160	0.73	3.437
	110.6	90		0.73	3.92

Effective nonlinearity expressions in the phase-matching direction [3.100]:

$$d_{\text{ooe}} = d_{31} \sin\theta - d_{22}\cos\theta\sin 3\phi \; ,$$

$$d_{\text{eoe}} = d_{\text{oee}} = d_{22}\cos^2\theta\cos 3\phi \; .$$

Nonlinear coefficient [3.92]:

$$d_{31}(1.0642\,\mu\text{m}) = -4.69 \pm 0.13\,\text{pm/V} \; .$$

Laser-induced surface-damage threshold [3.437]:

λ [µm]	τ_{p} [ns]	$I_{\text{thr}} \times 10^{-12}$ [W/m^2]
0.5321	≈ 20	3.4
1.0642	≈ 20	6.1

3.2.7 Ag$_3$AsS$_3$, Proustite

Negative uniaxial crystal: $n_{\text{o}} > n_{\text{e}}$;
Point group: 3m;
Mass density: 5.63 g/cm^3 at 293 K [3.64];
Mohs hardness: 2 – 2.5 [3.64];
Transparency range at "0" transmittance level: 0.6 – 13 µm [3.445];
Linear absorption coefficient α:

λ [µm]	α [cm^{-1}]	T [K]	Ref.	Note
0.593	0.89	77	3.446	e – wave, SFG direction
0.6328	0.81	77	3.446	o – wave, SFG direction
	0.64	77	3.446	e – wave, DFG direction
0.6789	0.64	77	3.446	o – wave, DFG direction
9.31	0.25	77	3.446	e – wave, SFG and DFG directions
0.576	36	300	3.445	e – wave, $\perp c$
0.593	16.1	300	3.446	e – wave, SFG direction
0.6328	1.83	300	3.446	o – wave, SFG direction
	1.59	300	3.446	e – wave, DFG direction
0.6358	1.88	300	3.447	e – wave, SFG direction
0.6764	0.95	300	3.447	o – wave, SFG direction
0.6789	0.83	300	3.446	o – wave, DFG direction
0.6943	0.1	300	3.448	
	0.2	300	3.449	o – wave, $\parallel c$
1.06	0.1	300	3.449	o – wave, $\parallel c$
1.064	0.02	300	3.448	
0.61–13.3	<0.1	300	3.445	e – wave, $\perp c$
0.63–12.5	<0.1	300	3.445	o – wave, $\perp c$
5.3	0.3	300	3.450	e – wave, SHG direction

$\lambda[\mu m]$	$\alpha[cm^{-1}]$	T [K]	Ref.	Note
	0.32	300	3.451	e − wave, SHG direction
9.2	0.29	300	3.452	o − wave, SHG direction
9.3	0.53	300	3.446	e − wave, DFG direction
10.2	1.2	300	3.445	o − wave, $\perp c$
	1.3	300	3.453	o − wave
10.6	0.16	300	3.454	o − wave
	0.38	300	3.451	o − wave
	0.45	300	3.447	o − wave, SFG direction
	0.6	300	3.450	o − wave, SHG direction
	0.8	300	3.455	o − wave
	1	300	3.351	o − wave
11.6	0.5	300	3.453	o − wave
14.5	≈ 70	300	3.453	o − wave
15.2–20.8	<20	300	3.456	o − wave

Two-photon absorption coefficient β (along c axis):

λ [μm]	$\beta \times 10^{10}$ [m/W]	Ref.
0.6943	10	3.451
	2	3.448
1.06	3	3.451
1.064	<0.3	3.448

Experimental values of refractive indices at $T = 293$ K [3.445] :

λ [μm]	n_o	n_e
0.5876		2.7896
0.6328	3.0190	2.7391
0.6678	2.9804	2.7094
1.014	2.8264	2.5901
1.129	2.8067	2.5756
1.367	2.7833	2.5570
1.530	2.7728	2.5485
1.709	2.7654	2.5423
2.50	2.7478	2.5282
3.56	2.7379	2.5213
4.62	2.7318	2.5178

Experimental values of phase-matching angle ($T = 293$ K) and comparison between different sets of dispersion relations:

Interacting wavelengths [µm]	θ_{exp} [deg]	θ_{theor} [deg]		
		[3.445]	[3.457]	[3.458]
SHG, $o + o \Rightarrow e$				
$10.6 \Rightarrow 5.3$	23.6 [3.459]	21.50	22.13	19.75
$10.59 \Rightarrow 5.295$	21.5 [3.455]	21.48	22.11	19.74
$9.2 \Rightarrow 4.6$	19.9 [3.452]	18.96	19.09	17.51
$2.13 \Rightarrow 1.065$	29.5 [3.460]	29.78	29.44	29.83
$2.1284 \Rightarrow 1.0642$	29.4 [3.461]	29.80	29.46	29.86
SFG, $o + o \Rightarrow e$				
$12.2 + 1.064 \Rightarrow 0.9786$	17.2 [3.462]	18.51	18.64	18.44
$8.9 + 1.064 \Rightarrow 0.9504$	20.0 [3.462]	20.96	21.00	21.04
$6.3 + 1.064 \Rightarrow 0.9103$	23.5 [3.462]	24.60	24.68	24.83
$10.57 + 0.6943 \Rightarrow 0.6515$	25.3 [3.463]	24.15	25.31	25.08
$10.6 + 0.6764 \Rightarrow 0.6358$	25.7 [3.447]	24.50	25.79	25.52
$10.6935 + 0.6726 \Rightarrow 0.6328$	25.8 [3.464]	24.47	25.79	25.51
$10.5881 + 0.6730 \Rightarrow 0.6328$	25.9 [3.464]	24.55	25.87	25.60
$10.3006 + 0.6742 \Rightarrow 0.6328$	26.1 [3.464]	24.87	26.20	25.93
$10.1918 + 0.6747 \Rightarrow 0.6328$	26.4 [3.464]	25.03	26.37	26.10
$9.5333 + 0.6778 \Rightarrow 0.6328$	27.1 [3.464]	25.81	27.18	26.93
$9.2688 + 0.6792 \Rightarrow 0.6328$	27.4 [3.464]	26.17	27.56	27.30
$6.3552 + 0.7028 \Rightarrow 0.6328$	32.9 [3.465]	31.26	32.91	32.69
$6.2571 + 0.7040 \Rightarrow 0.6328$	33.2 [3.465]	31.47	33.14	32.91
$6.1629 + 0.7052 \Rightarrow 0.6328$	33.4 [3.465]	31.69	33.37	33.14
$5.9079 + 0.7087 \Rightarrow 0.6328$	34.1 [3.465]	32.34	34.05	33.83
$5.7375 + 0.7112 \Rightarrow 0.6328$	34.5 [3.465]	32.77	34.50	34.28
$5.5393 + 0.7144 \Rightarrow 0.6328$	35.1 [3.465]	33.35	35.11	34.90
$5.2578 + 0.7194 \Rightarrow 0.6328$	36.0 [3.465]	34.23	36.03	35.82
SFG, $e + o \Rightarrow e$				
$10.59 + 1.064 \Rightarrow 0.967$	20.0 [3.466]	20.27	20.14	20.26
$10.59 + 0.6943 \Rightarrow 0.6516$	27.7 [3.467]	24.69	25.85	25.62
$9.31 + 0.6789 \Rightarrow 0.6328$	29.0 [3.446]	26.71	28.11	27.87
SFG, $o + e \Rightarrow e$				
$7.8 + 2.47 \Rightarrow 1.8759$	33 [3.468]	33.44	32.57	32.33

Best set of dispersion relations (λ in µm, $T = 293$ K) [3.457]:

$$n_o^2 = 9.220 + \frac{0.4454}{\lambda^2 - 0.1264} + \frac{1733}{\lambda^2 - 1000} \; ,$$

$$n_e^2 = 7.007 + \frac{0.3230}{\lambda^2 - 0.1192} + \frac{660}{\lambda^2 - 1000} \; .$$

Calculated values of phase-matching and "walk-off" angles:

Interacting wavelengths [µm]	θ_{pm} [deg]	ρ_1 [deg]	ρ_2 [deg]	ρ_3 [deg]
SHG, o + o $\Rightarrow$ e				
10.6 $\Rightarrow$ 5.3	22.13			3.42
9.6 $\Rightarrow$ 4.8	19.90			3.17
5.3 $\Rightarrow$ 2.65	14.71			2.50
4.8 $\Rightarrow$ 2.4	15.01			2.56
2.9365 $\Rightarrow$ 1.46825	21.02			3.45
2.1284 $\Rightarrow$ 1.0642	29.46			4.44
SFG, o + o $\Rightarrow$ e				
10.6 + 1.0642 $\Rightarrow$ 0.96711	19.57			3.39
10.6 + 0.6943 $\Rightarrow$ 0.65162	25.28			4.52
SHG, e + o $\Rightarrow$ e				
10.6 $\Rightarrow$ 5.3	30.28	3.84		4.19
9.6 $\Rightarrow$ 4.8	27.27	3.70		3.97
2.9365 $\Rightarrow$ 1.46825	29.45	4.21		4.32
2.1284 $\Rightarrow$ 1.0642	42.46	4.77		4.96
SFG, e + o $\Rightarrow$ e				
10.6 + 5.3 $\Rightarrow$ 3.533	19.78	2.87		3.19
9.6 + 4.8 $\Rightarrow$ 3.2	18.70	2.81		3.06
10.6 + 1.0642 $\Rightarrow$ 0.96711	20.29	2.93		3.49
10.6 + 0.6943 $\Rightarrow$ 0.65162	25.83	3.49		4.59
SFG, o + e $\Rightarrow$ e				
10.6 + 5.3 $\Rightarrow$ 3.533	28.65		4.06	4.13
9.6 + 4.8 $\Rightarrow$ 3.2	26.89		3.93	3.99

Experimental values of internal angular bandwidth:

Interacting wavelengths [µm]	$\Delta\theta^{int}$ [deg]	Ref.
SHG, o + o $\Rightarrow$ e		
10.6 $\Rightarrow$ 5.3	0.098	3.450
9.2 $\Rightarrow$ 4.6	0.082	3.452
SFG, e + o $\Rightarrow$ e		
10.6 + 0.6943 $\Rightarrow$ 0.6516	0.031	3.467

Effective nonlinearity expressions in the phase-matching direction [3.100]:

$$d_{ooe} = d_{31} \sin\theta - d_{22} \cos\theta \sin 3\phi \ ,$$

$$d_{eoe} = d_{oee} = d_{22} \cos^2\theta \cos 3\phi \ .$$

Nonlinear coefficients [3.455, 37]:

$$|d_{22}(10.6\,\mu m)| = (0.2 \pm 0.03) \times |d_{36}(GaAs)|$$
$$= 16.6 \pm 2.5\,pm/V \ ,$$

$$|d_{31}(10.6\,\mu\text{m})| = (1.6 \pm 0.1)^{-1} \times |d_{22}(\text{Ag}_3\text{AsS}_3)|$$
$$= 10.4 \pm 2.2\,\text{pm/V} \,.$$

Laser-induced surface-damage threshold:

λ [μm]	τ_p [ns]	$I_{thr} \times 10^{-12}$ [W/m^2]	Ref.
0.6943	10^6	0.00006	3.469
	14	0.03	3.365
1.0642	cw	0.000001	3.469
	18	>0.12	3.365
2.098	200	>0.1	3.365
10.6	190	>0.46	3.365
	150	0.53	3.450

3.2.8 GaSe, Gallium Selenide

Negative uniaxial crystal: $n_o > n_e$;
Point group: $\bar{6}2m$;
Mass density: 5.03 g/cm^3 [3.338];
Mohs hardness: ≈ 0;
Transparency range at "0" transmittance level: $0.62 - 20\,\mu\text{m}$ [3.388];
Linear absorption coefficient α:

λ [μm]	α [cm^{-1}]	Ref.
0.65–18	<1	3.470
0.7	<0.3	3.471
1.06	0.45	3.472
	<0.25	3.473
	<0.1	3.474
1.9	0.1	3.472
2	<0.1	3.474
9.3–10.6	<0.05	3.384
10	<0.1	3.474
10.6	0.081	3.454

Experimental values of refractive indices [3.470]:

λ[μm]	n_o	n_e
0.6328	2.97	2.74
1.1523	2.90	2.54
3.3913	2.81	2.46

Experimental values of phase-matching angle ($T = 293$ K) and comparison between different sets of dispersion relations:

Interacting wavelengths [μm]	θ_{exp} [deg]	θ_{theor} [deg]		
		[3.470]	[3.473]	[3.475]
SHG, o + o ⇒ e				
2.36 ⇒ 1.18	18.7 [3.470]	18.48	19.22	18.92
5.30 ⇒ 2.65	10.2 [3.470]	9.83	9.77	10.70
9.30 ⇒ 4.65	12.8 [3.384]	11.14	10.75	13.25
9.60 ⇒ 4.80	13.2 [3.384]	11.46	11.04	13.61
10.3 ⇒ 5.15	14.0 [3.384]	12.28	11.81	14.49
10.6 ⇒ 5.3	12.7 [3.470]	12.67	12.16	14.89
	14.4 [3.384]			
SFG, o + o ⇒ e				
17.4 + 3.5327 ⇒ 2.9365	13 [3.388]	12.41	12.03	13.02
	13 [3.370]			
11.6 + 3.9318 ⇒ 2.9365	10 [3.388]	10.16	9.94	11.36
	10 [3.370]			
10.8 + 2.3611 ⇒ 1.9375	10.7 [3.472]	10.46	10.46	11.18
7.4 + 2.4859 ⇒ 1.8608	11.2 [3.472]	11.14	11.23	11.70
5 + 2.7039 ⇒ 1.7549	12.4 [3.472]	12.43	12.60	12.81
10.1 + 1.1895 ⇒ 1.0642	13.3 [3.476]	12.68	13.21	13.51
7.15 + 1.2503 ⇒ 1.0642	15 [3.476]	14.41	15.06	15.20
19.1 + 1.1144 ⇒ 1.053	11.5 [3.367]	11.51	11.86	11.88
12 + 1.1543 ⇒ 1.053	12 [3.367]	12.10	12.60	12.89
5.8 + 1.2866 ⇒ 1.053	15.7 [3.367]	15.70	16.44	16.48
10.6 + 1.0642 ⇒				
⇒ 0.96711	13.6 [3.473]	12.94	13.59	13.88
4.9 + 1.0642 ⇒ 0.8743	18.8 [3.473]	18.48	19.63	19.81
17.17 + 0.7235 ⇒				
⇒ 0.6943	15.2 [3.471]	12.99	13.63	15.19
9.99 + 0.7462 ⇒ 0.6943	18.3 [3.471]	15.68	16.56	18.73
SFG, e + o ⇒ e				
15.5 + 1.1427 ⇒ 1.0642	12.4 [3.476]	11.87	12.32	12.45
12.0 + 1.1678 ⇒ 1.0642	13.3 [3.476]	12.58	13.14	13.32
9.4 + 1.2001 ⇒ 1.0642	14.4 [3.476]	13.75	14.42	14.52
7.4 + 1.2430 ⇒ 1.0642	16.4 [3.476]	15.27	16.07	16.01
10.6 + 1.0642 ⇒				
⇒ 0.96711	14.4 [3.473]	13.51	14.25	14.42
18.28 + 0.7217 ⇒				
⇒ 0.6943	15.2 [3.471]	13.22	13.89	15.15
11.10 + 0.7406 ⇒				
⇒ 0.6943	18.6 [3.471]	15.49	16.44	18.46

Best set of dispersion relations (λ in µm, $T = 293$ K) [3.475]

$$n_{\mathrm{o}}^2 = 7.443 + \frac{0.4050}{\lambda^2} + \frac{0.0186}{\lambda^4} + \frac{0.0061}{\lambda^6} + \frac{3.1485\,\lambda^2}{\lambda^2 - 2194} \; ,$$

$$n_{\mathrm{e}}^2 = 5.76 + \frac{0.3879}{\lambda^2} - \frac{0.2288}{\lambda^4} + \frac{0.1223}{\lambda^6} + \frac{1.8550\,\lambda^2}{\lambda^2 - 1780} \; .$$

Calculated values of phase-matching and "walk-off" angles:

Interacting wavelengths [µm]	θ_{pm} [deg]	ρ_1 [deg]	ρ_2 [deg]	ρ_3 [deg]
SHG, o + o $\Rightarrow$ e				
10.6 $\Rightarrow$ 5.3	14.89			4.059
9.6 $\Rightarrow$ 4.8	13.61			3.750
5.3 $\Rightarrow$ 2.65	10.70			3.007
4.8 $\Rightarrow$ 2.4	10.99			3.081
2.9365 $\Rightarrow$ 1.46825	15.39			4.208
SFG, o + o $\Rightarrow$ e				
10.6 + 2.65 $\Rightarrow$ 2.12	11.07			3.102
9.6 + 2.4 $\Rightarrow$ 1.92	11.26			3.156
10.6 + 1.0642 $\Rightarrow$ 0.96711	13.88			3.978
9.6 + 1.0642 $\Rightarrow$ 0.95800	14.44			4.121
SHG, e + o $\Rightarrow$ e				
10.6 $\Rightarrow$ 5.3	20.47	5.252		5.264
9.6 $\Rightarrow$ 4.8	18.65	4.891		4.898
5.3 $\Rightarrow$ 2.65	14.60	3.991		3.984
4.8 $\Rightarrow$ 2.4	15.00	4.084		4.078
SFG, e + o $\Rightarrow$ e				
10.6 + 5.3 $\Rightarrow$ 3.53333	13.92	3.817		3.823
9.6 + 4.8 $\Rightarrow$ 3.2	13.28	3.664		3.666
10.6 + 1.0642 $\Rightarrow$ 0.96711	14.42	3.938		4.114
9.6 + 1.0642 $\Rightarrow$ 0.95800	15.06	4.092		4.274
SFG, o + e $\Rightarrow$ e				
10.6 + 5.3 $\Rightarrow$ 3.53333	19.51		5.074	5.070
9.6 + 4.8 $\Rightarrow$ 3.2	18.57		4.881	4.876

Experimental values of internal angular bandwidth:

Interacting wavelengths [µm]	$\Delta\theta^{\mathrm{int}}$ [deg]	Ref.
SHG, o + o $\Rightarrow$ e		
10.3 $\Rightarrow$ 5.15	0.146	3.384
SFG, o + o $\Rightarrow$ e		
7 + 2.51 $\Rightarrow$ 1.8475	0.086	3.472
12.5 + 0.7351 $\Rightarrow$ 0.6943	0.021	3.471

Effective nonlinearity expressions in the phase-matching direction [3.100]:

$$d_{\text{ooe}} = d_{36} \sin\theta \sin 2\phi \ ,$$
$$d_{\text{eoe}} = d_{\text{oee}} = d_{36} \sin 2\theta \cos 2\phi \ .$$

Nonlinear coefficient:

$$|d_{22}(10.6\,\mu\text{m})| = 3 \times |d_{31}(\text{CdSe})| \pm 20\% = 54 \pm 10.8\,\text{pm/V} \ [3.470, 37].$$

Laser-induced surface-damage threshold:

λ [µm]	τ_{p} [ns]	$I_{\text{thr}} \times 10^{-12}$ [W/m^2]	Ref.	Note
0.6943	30	0.2	3.471	
1.053	0.002	>10	3.367	1 Hz
1.064	10	0.3	3.473	
2.36	40	>0.05	3.470	
2.94	0.11	300	3.388	1 Hz
	0.11	300	3.370	1 Hz
10.6	125	0.3	3.384	2–20 Hz

3.2.9 AgGaSe$_2$, Silver Gallium Selenide

Negative uniaxial crystal: $n_{\text{o}} > n_{\text{e}}$ $(at\ \lambda < 0.804\,\mu\text{m}\ n_{\text{e}} > n_{\text{o}})$;
Point group: $\bar{4}$2m ;
Molecular mass: 5.71 g/cm^3 [3.338] ;
Mohs hardness: $3 - 3.5$;
Transparency range at "0" transmittance level: $0.71 - 19\,\mu\text{m}$ [3.477, 478];
Linear absorption coefficient α :

λ [µm]	α [cm^{-1}]	Ref.	Note
1	<0.02	3.479	
1.3	0.002	3.480	o – wave, OPO direction
	0.002	3.480	e – wave, OPO direction
2.0	0.012	3.481	o – wave, OPO direction
	0.030	3.481	e – wave, OPO direction
	0.004	3.482	
2.05	<0.01	3.483	
	0.015–0.058	3.484	
2.1	0.06–0.07	3.478	
	0.012–0.072	3.485	
2.2	0.002–0.004	3.480	o – wave, OPO direction
	0.02–0.05	3.480	e – wave, OPO direction
5-11	<0.02	3.479	
10.6	0.089	3.454	
	0.01–0.06	3.486	
	0.002	3.477	

Experimental values of refractive indices [3.487]:

λ [μm]	n_o	n_e	λ [μm]	n_o	n_e
0.725	2.8452	2.8932	3.800	2.6203	2.5876
0.750	2.8191	2.8415	4.000	2.6189	2.5863
0.800	2.7849	2.7866	4.500	2.6166	2.5840
0.850	2.7598	2.7522	5.000	2.6144	2.5819
0.900	2.7406	2.7275	5.500	2.6128	2.5800
0.950	2.7252	2.7085	6.000	2.6113	2.5784
1.000	2.7132	2.6934	6.500	2.6094	2.5765
1.100	2.6942	2.6712	7.000	2.6070	2.5743
1.200	2.6806	2.6554	7.500	2.6049	2.5723
1.300	2.6705	2.6438	8.000	2.6032	2.5704
1.400	2.6624	2.6347	8.500	2.6009	2.5681
1.600	2.6516	2.6224	9.000	2.5988	2.5659
1.800	2.6432	2.6131	9.500	2.5964	2.5635
2.000	2.6376	2.6071	10.00	2.5939	2.5608
2.200	2.6336	2.6027	10.50	2.5917	2.5585
2.400	2.6304	2.5992	11.00	2.5890	2.5555
2.600	2.6286	2.5968	11.50	2.5868	2.5536
2.800	2.6261	2.5943	12.00	2.5837	2.5505
3.000	2.6245	2.5925	12.50	2.5805	2.5473
3.200	2.6231	2.5912	13.00	2.5771	2.5439
3.400	2.6221	2.5899	13.50	2.5731	2.5404
3.600	2.6213	2.5889			

Optical activity [3.345]:

$\rho = 7 \deg /\text{mm}$ at isotropic point $(n_o = n_e,\ \lambda = 0.804\,\mu\text{m})$.

Temperature derivative of refractive indices at $\lambda = 3.3913\,\mu\text{m}$ $(T = 35\,°\text{C})$ [3.478]:

$$\frac{dn_o}{dT} = 4.5 \times 10^{-5}\,\text{K}^{-1}\ ,$$

$$\frac{dn_e}{dT} = 7.6 \times 10^{-5}\,\text{K}^{-1}\ .$$

Experimental values of phase-matching angle $(T = 293\,\text{K})$ and comparison between different sets of dispersion relations:

Interacting wavelengths [μm]	θ_{exp} [deg]	θ_{theor} [deg]		
		[3.488]	[3.362]	[3.479]
SHG, o + o ⇒ e				
10.63 ⇒ 5.315	55.9 [3.488]	55.2	57.3	58.0
10.6 ⇒ 5.3	57.5 [3.477]	55.0	57.1	57.7

$10.55 \Rightarrow 5.275$	55.3 [3.488]	54.7	56.7	57.4
$10.3 \Rightarrow 5.15$	53.7 [3.488]	53.1	54.9	55.7
$10.21 \Rightarrow 5.15$	53.1 [3.488]	52.5	54.3	55.1
$6 \Rightarrow 3$	42.2 [3.488]	39.5	39.4	40.1
$5.2 \Rightarrow 2.6$	40.3 [3.488]	41.5	40.8	41.3
$4.1 \Rightarrow 2.05$	49.7 [3.483]	50.6	48.3	48.6
SFG, $o + o \Rightarrow e$				
$12.15 + 10.63 \Rightarrow 5.67$	61 [3.488]	60.7	63.5	63.6
$10.63 + 5.33 \Rightarrow 3.55$	42.7 [3.488]	42.1	42.7	43.3
$5.515 + 3.3913 \Rightarrow 2.1$	$\approx$48 [3.478]	48.1	46.2	46.5
$4.84 + 3.55 \Rightarrow 2.0479$	49.2 [3.483]	50.1	48.0	48.2
$5.13 + 2.685 \Rightarrow 1.763$	61.3 [3.474]	57.1	53.3	53.5
$6.00 + 2.586 \Rightarrow 1.807$	56 [3.474]	54.9	51.7	51.9
$7.43 + 2.484 \Rightarrow 1.862$	49.5 [3.474]	49.0	46.6	46.9
$9.93 + 2.384 \Rightarrow 1.923$	45.8 [3.474]	44.6	42.9	43.1
$6.95 + 1.66 \Rightarrow 1.34$	$\approx$78 [3.483]	83.1	68.6	69.2
$7.4 + 1.604 \Rightarrow 1.318$	80 [3.477]	no pm	69.8	70.4
$8.8 + 1.550 \Rightarrow 1.318$	70 [3.477]	69.0	61.2	61.7
$12.3 + 1.476 \Rightarrow 1.318$	60 [3.477]	58.2	53.1	53.4

Best set of Sellmeier equations (λ in μm, $T = 293\,\mathrm{K}$) [3.488]:

$$n_o^2 = 3.9362 + \frac{2.9113\,\lambda^2}{\lambda^2 - (0.38821)^2} + \frac{1.7954\,\lambda^2}{\lambda^2 - 1600} \; ,$$

$$n_e^2 = 3.3132 + \frac{3.3616\,\lambda^2}{\lambda^2 - (0.38201)^2} + \frac{1.7677\,\lambda^2}{\lambda^2 - 1600} \; .$$

Calculated values of phase-matching and "walk-off" angles:

Interacting wavelengths [μm]	θ_{pm} [deg]	ρ_1 [deg]	ρ_3 [deg]
SHG, $o + o \Rightarrow e$			
$10.6 \Rightarrow 5.3$	55.02		0.68
$9.6 \Rightarrow 4.8$	49.00		0.71
$5.3 \Rightarrow 2.65$	41.10		0.69
$4.8 \Rightarrow 2.4$	43.63		0.68
SFG, $o + o \Rightarrow e$			
$10.6 + 2.65 \Rightarrow 2.12$	43.71		0.67
$9.6 + 2.4 \Rightarrow 1.92$	46.36		0.66
SHG, $e + o \Rightarrow e$			
$5.3 \Rightarrow 2.65$	72.03	0.42	0.40

SFG, e + o ⇒ e

10.6 + 5.3 ⇒ 3.533	55.60	0.68	0.66
9.6 + 4.8 ⇒ 3.2	55.36	0.70	0.67

Experimental values of internal angular bandwidth

Interacting wavelengths [µm]	$\Delta\theta^{\text{int}}$ [deg]	Ref.
SHG, o + o ⇒ e		
10.25 ⇒ 5.125	0.84	3.486
SFG, o + o ⇒ e		
5.515 + 3.3913 ⇒ 2.1	0.54	3.478

Effective nonlinearity expressions in the phase-matching direction [3.100]:

$$d_{\text{ooe}} = d_{36} \sin\theta \sin 2\phi \; ,$$

$$d_{\text{eoe}} = d_{\text{oee}} = d_{36} \sin 2\theta \cos 2\phi \; .$$

Nonlinear coefficient:

$$d_{36}(10.6\,\mu\text{m}) = 33\,\text{pm/V} \; [3.37] \; ,$$

$$d_{36}(9.5\,\mu\text{m}) = 32 \pm 4\,\text{pm/V} \; [3.489] \; .$$

Laser-induced surface-damage threshold:

λ [µm]	τ_{p} [ns]	$I_{\text{thr}} \times 10^{-12}$ [W/m^2]	Ref.	Note
1.064	23	0.13–0.4	3.483	
	35	0.3	3.488	single pulse
	35	0.11	3.488	1000 pulses
2.0	30	0.083	3.481	5 kHz, uncoated crystal
	30	<0.13	3.481	5 kHz, coated crystal
	20–30	0.2–0.3	3.482	
2.05	50	0.25	3.483	
2.1	50	0.13	3.478	
	180	0.094	3.485	uncoated crystal
	180	0.17	3.485	coated crystal
9.5	30	0.33	3.489	
10.25	75	0.12	3.486	10 pulses
10.6	150	0.1–0.2	3.368	

Thermal conductivity coefficient at $T = 293\,\text{K}$ [3.58]:

κ [W/mK], $\parallel c$	κ [W/mK], $\perp c$
1.0	1.1

3.2.10 CdSe, Cadmium Selenide

Positive uniaxial crystal: $n_e > n_o$;
Point group: 6mm ;
Mass density: $5.81 \, \text{g/cm}^3$ [3.338] ;
Mohs hardness: 3.25 [3.59];
Transparency range at "0" transmittance level: $0.75 - 25 \, \mu\text{m}$ [3.490, 59];
Linear absorption coefficient α :

λ [μm]	α [cm^{-1}]	Ref.
0.75–20	<0.1	3.490
1.064	0.02	3.448
1.32	0.01	3.448
3.39	0.01	3.491
4	0.04	3.492
10.6	0.0005	3.493
	0.016	3.492
	0.032	3.494

Two-photon absorption coefficient β :

λ [μm]	$\beta \times 10^{11}$ [m/W]	Ref.	Note
1.06	140	3.495	o – wave, $\perp c$ and $\parallel c$
	60	3.495	e – wave, $\perp c$
1.064	208	3.496	o – wave, $\parallel c$
	<20	3.448	
1.32	2	3.448	

Experimental values of refractive indices:

λ [μm]	n_o	n_e	Ref.	λ [μm]	n_o	n_e	Ref.
0.8	2.6448	2.6607	3.497	2.2	2.4642	2.4840	3.497
0.9	2.5826	2.6027	3.497	2.3253	2.4627	2.4823	3.494
1.0	2.5502	2.5696	3.497	2.4	2.4612	2.4798	3.497
1.0139	2.5481	2.5677	3.494	2.6	2.4590	2.4784	3.497
1.1287	2.5246	2.5444	3.494	2.8	2.4562	2.4757	3.497
1.2	2.5132	2.5331	3.497	3.0	2.4553	2.4748	3.494
1.3673	2.4971	2.5170	3.494	3.2	2.4532	2.4726	3.497
1.4	2.4929	2.5133	3.497	3.4	2.4518	2.4714	3.497
1.5295	2.4861	2.5059	3.494	3.6	2.4509	2.4702	3.497
1.6	2.4818	2.5008	3.497	3.8	2.4498	2.4694	3.497
1.7109	2.4776	2.4974	3.494	4.0	2.4491	2.4685	3.497
1.8	2.4732	2.4930	3.497	5.0	2.4464	2.4657	3.494
2.0	2.4682	2.4873	3.497	6.0	2.4434	2.4625	3.494

λ [μm]	n_o	n_e	Ref.	λ [μm]	n_o	n_e	Ref.
7.0	2.4398	2.4586	3.494	10.0	2.4294	2.4475	3.494
8.0	2.4367	2.4552	3.494	11.0	2.4252	2.4430	3.494
9.0	2.4333	2.4514	3.494	12.0	2.4204	2.4379	3.494

Experimental values of phase-matching angle ($T = 293$ K) and comparison between different sets of dispersion relations:

Interacting wavelengths [μm]	θ_{exp} [deg]	θ_{theor} [deg]	
		[3.468]	[3.362]
SFG, o + e ⇒ o			
$16.4 + 3.479 \Rightarrow 2.87$	73.7 [3.493]	71.3	72.4
$15.96 + 2.28 \Rightarrow 1.995$	62.2 [3.498]	64.2	64.6
$14.1 + 3.604 \Rightarrow 2.87$	70.9 [3.493]	68.9	69.7
$13.7 + 2.8492 \Rightarrow 2.3587$	65 [3.499]	65.2	65.5
$10.6 + 2.72 \Rightarrow 2.1646$	70.5 [3.491]	70.4	70.5
$10.361 + 2.227 \Rightarrow 1.833$	78 [3.500]	78.7	78.5
$9.871 + 2.251 \Rightarrow 1.833$	84 [3.500]	83.9	83.5
$9.776 + 2.256 \Rightarrow 1.833$	90 [3.500]	85.8	85.1
$8.278 + 4.3 \Rightarrow 2.83$	84 [3.492]	81.4	81.9
$8.253 + 4.4 \Rightarrow 2.87$	84 [3.492]	82.4	83.0
$8.236 + 4.5 \Rightarrow 2.91$	84 [3.492]	83.6	84.5
$7.88 + 3.36 \Rightarrow 2.3587$	90 [3.499]	no pm	no pm
$7.86 + 3.37 \Rightarrow 2.3587$	90 [3.490]	no pm	no pm

Best set of dispersion relations (λ in μm, $T = 20$ °C) [3.362]:

$$n_o^2 = 4.2243 + \frac{1.7680\,\lambda^2}{\lambda^2 - 0.2270} + \frac{3.1200\,\lambda^2}{\lambda^2 - 3380} \ ,$$

$$n_e^2 = 4.2009 + \frac{1.8875\,\lambda^2}{\lambda^2 - 0.2171} + \frac{3.6461\,\lambda^2}{\lambda^2 - 3629} \ .$$

dispersion relations for the temperatures 73 K, 173 K, 373 K, 573 K are given in [3.390].

Calculated values of phase-matching and "walk-off" angles:

Interacting wavelengths [μm]	θ_{pm} [deg]	ρ_2 [deg]
SFG, o + e ⇒ o		
$22 + 2.9365 \Rightarrow 2.5907$	83.2	0.11
$20 + 2.9365 \Rightarrow 2.5605$	74.0	0.24
$15 + 2.9365 \Rightarrow 2.4557$	66.0	0.34
$10 + 2.9365 \Rightarrow 2.2699$	71.6	0.27
$9 + 2.9365 \Rightarrow 2.2141$	77.6	0.19

$15 + 3.6513 \Rightarrow 2.9365$	71.5	0.27
$10 + 4.1573 \Rightarrow 2.9365$	73.3	0.25
$20 + 3.2437 \Rightarrow 2.791$	80.9	0.14
$15 + 3.4290 \Rightarrow 2.791$	69.3	0.30
$10 + 3.8715 \Rightarrow 2.791$	71.8	0.27

Experimental value of internal angular bandwidth [3.491]:

Interacting wavelengths [μm]	$\Delta\theta^{\text{int}}$ [deg]
SFG, $o + e \Rightarrow o$	
$10.6 + 2.72 \Rightarrow 2.1646$	1.24

Experimental value of spectral bandwidth [3.491]:

Interacting wavelengths [μm]	Δv [cm^{-1}]
SFG, $o + e \Rightarrow o$	
$10.6 + 2.72 \Rightarrow 2.1646$	15

Effective nonlinearity expression in the phase-matching direction [3.100]:

$$d_{\text{oeo}} = d_{\text{eoo}} = d_{31} \sin\theta .$$

Nonlinear coefficients [3.37]:

$$d_{31}(10.6\,\mu\text{m}) = -18\,\text{pm/V} ,$$

$$d_{33}(10.6\,\mu\text{m}) = 36\,\text{pm/V} .$$

Laser-induced surface-damage threshold:

λ [μm]	τ_{p} [ns]	$I_{\text{thr}} \times 10^{-12}$ [W/m^2]	Ref.
1.833	200	0.3	3.494
1.995	20	>0.5	3.498
2.36	35	0.5	3.490
10.6	200	0.6	3.365

Thermal conductivity coefficient at $T = 293$ K [3.58]:

κ [W/mK], $\parallel c$	κ [W/mK], $\perp c$
6.9	6.2

3.2.11 CdGeAs$_2$, Cadmium Germanium Arsenide

Positive uniaxial crystal: $n_e > n_o$;
Point group: $\bar{4}2m$;
Mass density : 5.60 g/cm^3 [3.338] ;
Mohs hardness: 3.5–4 ;
Transparency range at "0" transmittance level: 2.4 − 18 μm [3.501];
Linear absorption coefficient α :

λ [μm]	T [K]	α [cm^{-1}]	Ref.
3.39	300	5.7	3.502
4–18	300	<0.9	3.503
5.3	77	0.4	3.504
	300	1.3	3.502
5.85	77	0.42	3.505
	300	1.5	3.505
8.6-12	77	<0.2	3.380
	300	<0.5	3.380
9–11	300	0.23	3.503
10.6	77	0.1	3.504
	300	0.4	3.501
	300	0.5	3.502
	300	2.4	3.454
10.6–11.7	77	0.14	3.505
	300	0.5	3.505

Experimental values of refractive indices [3.506]:

λ [μm]	n_o	n_e	λ [μm]	n_o	n_e
2.3	3.6076		4.8	3.5354	3.6273
2.4	3.5973	3.7545	5.0	3.5336	3.6249
2.5	3.5895	3.7316	5.5	3.5285	3.6178
2.6	3.5823	3.7156	6.0	3.5251	3.6134
2.7	3.5773	3.7030	6.5	3.5223	3.6104
2.8	3.5721	3.6926	7.0	3.5200	3.6073
2.9	3.5684	3.6846	7.5	3.5175	3.6050
3.0	3.5645	3.6775	8.0	3.5157	3.6030
3.1	3.5615	3.6714	8.5	3.5140	3.6009
3.2	3.5581	3.6661	9.0	3.5120	3.5988
3.4	3.5536	3.6574	9.5	3.5098	3.5966
3.6	3.5503	3.6508	10.0	3.5078	3.5942
3.8	3.5468	3.6454	10.5	3.5054	3.5922
4.0	3.5440	3.6402	11.0	3.5031	3.5896
4.2	3.5415	3.6368	11.5	3.5004	3.5871
4.4	3.5391	3.6329	12.0	3.4977	
4.6	3.5372	3.6299	12.5	3.4950	

Experimental values of phase-matching angle ($T = 293\,\mathrm{K}$) and comparison between different sets of dispersion relations:

Interacting wavelengths [μm]	$\theta_{\exp}$ [deg]	θ_{theor} [deg]		
		[3.501]	[3.503]	[3.362]
SHG, e + e ⇒ o				
10.6 ⇒ 5.3	32 [3.502]	35.6	33.5	33.6
	35 [3.506]			
	33.8 [3.505]			
	34 [3.380]			
11.7 ⇒ 5.85	35.7 [3.505]	37.7	34.6	35.1
SHG, o + e ⇒ o				
10.6 ⇒ 5.3	51.6 [3.506]	54.9	50.8	51.1
	52 [3.501]			
	50.7 [3.503]			
	48.4 [3.503]			
	49 [3.502]			
SFG, o + e ⇒ o				
16.4 + 9.54 ⇒ 6.03	47 [3.503]	55.4	49.8	50.4
12.9 + 9.59 ⇒ 5.5	46.1[3.503]	51.7	47.8	48.0

Note: The dispersion relations in [3.503] are given with a mistake. The first term in the equation for n_o^2 should be 3.4141 instead of 6.4141

Best set of dispersion relations (λ in μm, $T = 293\,\mathrm{K}$) [3.362]:

$$n_o^2 = 10.1064 + \frac{2.2998\,\lambda^2}{\lambda^2 - 1.0872} + \frac{1.6247\,\lambda^2}{\lambda^2 - 1370} \, ,$$

$$n_e^2 = 11.8018 + \frac{1.2152\,\lambda^2}{\lambda^2 - 2.6971} + \frac{1.6922\,\lambda^2}{\lambda^2 - 1370} \, .$$

Calculated values of phase-matching and "walk-off" angles:

Interacting wavelengths [μm]	θ_{pm} [deg]	ρ_1 [deg]	ρ_2 [deg]
SHG, e + e ⇒ o			
10.6 ⇒ 5.3	33.57	1.27	1.27
9.6 ⇒ 4.8	32.92	1.26	1.26
5.3 ⇒ 2.65	48.45	1.44	1.44
4.8 ⇒ 2.4	55.73	1.38	1.38
SFG, e + e ⇒ o			
10.6 + 5.3 ⇒ 3.5333	35.41	1.31	1.36
9.6 + 4.8 ⇒ 3.2	37.72	1.34	1.41
SHG, o + e ⇒ o			

$10.6 \Rightarrow 5.3$	51.06	1.37
$9.6 \Rightarrow 4.8$	49.87	1.38
SFG, $o + e \Rightarrow o$		
$10.6 + 5.3 \Rightarrow 3.533$	44.62	1.45
$9.6 + 4.8 \Rightarrow 3.2$	47.73	1.47

Experimental values of internal angular bandwidth:

Interacting wavelengths [μm]	$\Delta\theta^{\text{int}}$ [deg]	Ref.
SHG, $e + e \Rightarrow o$		
$10.6 \Rightarrow 5.3$	0.84	3.380
SHG, $o + e \Rightarrow o$		
$10.6 \Rightarrow 5.3$	0.29	3.501

Effective nonlinearity expressions in the phase-matching direction [3.100]:

$$d_{\text{ooe}} = d_{36} \sin\theta \sin 2\phi \,,$$
$$d_{\text{eoe}} = d_{\text{oee}} = d_{36} \sin 2\theta \cos 2\phi \,.$$

Nonlinear coefficient:

$$|d_{36}(10.6\,\mu\text{m})| = 3.4 \times |d_{36}(\text{GaAs})| \pm 20\%$$

$$= 282 \pm 56\,\text{pm/V} \,[3.501, 37] \,,$$

$$|d_{36}(10.6\,\mu\text{m})| = 2.62 \times |d_{36}(\text{GaAs})| \pm 15\%$$

$$= 217 \pm 33\,\text{pm/V} \,[3.506, 37] \,.$$

Laser-induced surface-damage threshold:

λ [μm]	τ_{p} [ns]	$I_{\text{thr}} \times 10^{-12}$ [W/m^2]	Ref.	Note
10.6	cw	>0.0013	3.504	$T = 77\,\text{K}$
	cw	>0.0001	3.501	
	160	>0.04	3.501	
	160	0.38	3.503	
	150	0.33–0.4	3.368	

Thermal conductivity coefficient [3.338]:

$$\kappa = 4.18 \text{ W/mK} \text{ or } 6.69 \text{ W/mK} \,(?) \,.$$

3.3 Other Inorganic Nonlinear Optical Crystals

3.3.1 KB$_5$O$_8$ · 4D$_2$O, Deuterated Potassium Pentaborate Tetrahydrate (DKB5)

Positive biaxial crystal;
Point group: mm2 ;
Assignment of dielectric and crystallographic axes:
$X, Y, Z \Rightarrow a, b, c$;
Transparency range at "0" transmittance level: $0.16 - 2.1\ \mu m$ [3.395];
Experimental values of phase-matching angle $(T = 295\,\mathrm{K})$:
XY plane $\theta = 90°$

Interacting wavelengths [µm]	θ_{exp} [deg]
SHG, e + e $\Rightarrow$ o	
$0.4323 \Rightarrow 0.21615$	90 [3.395]
SFG, e + e $\Rightarrow$ o	
$0.74806 + 0.26604 \Rightarrow 0.19264$	90 [3.395]

Dispersion relations   (λ in µm, $T = 22\ °C$) [3.395]:

$$n_X^2 = 1 + \frac{\lambda^2}{0.84857\,\lambda^2 - 0.0075428} \ ,$$

$$n_Z^2 = 1 + \frac{\lambda^2}{1.01230\,\lambda^2 - 0.0095376} \ .$$

Effective nonlinearity expressions in the phase-matching direction for three-wave interactions in the principal planes of DKB5 crystal [3.35, 36]:
XY plane

$$d_{\mathrm{eeo}} = d_{31} \sin^2 \phi + d_{32} \cos^2 \phi \ ;$$

YZ plane

$$d_{\mathrm{ooe}} = d_{31} \sin \theta \ ;$$

XZ plane, $\theta < V_Z$

$$d_{\mathrm{oeo}} = d_{\mathrm{eoo}} = d_{32} \sin \theta \ ;$$

XZ plane, $\theta > V_Z$

$$d_{\mathrm{ooe}} = d_{32} \sin \theta \ .$$

Effective nonlinearity expressions for three-wave interactions in the arbitrary direction of DKB5 crystal are given in [3.36].
Nonlinear coefficients [3.395, 37]:

$$d_{31} \geq d_{31}(\mathrm{KB5}) = 0.04\,\mathrm{pm/V} \ ,$$

$$d_{32} \geq d_{32}(\text{KB5}) = 0.003 \, \text{pm/V} \; .$$

Laser-induced bulk-damage threshold [3.395, 405]:

λ [μm]	τ_p [ns]	$I_{\text{thr}} \times 10^{-12}$ [W/m^2]
0.43	7	10

3.3.2 CsB$_3$O$_5$, Cesium Triborate (CBO)

Negative biaxial crystal: $2V_Z = 97.3°$ at $\lambda = 0.5321 \, \mu$m [3.507];
Point group: 222 ;
Mass density: 3.357 g/cm^3 ;
Transparency range at "0" transmittance level: $0.167 - 3.0 \, \mu$m [3.507] ;
Experimental values of refractive indices [3.507]:

λ [μm]	n_X	n_Y	n_Z
0.3547	1.5499	1.5849	1.6145
0.4765	1.5370	1.5758	1.6031
0.4880	1.5367	1.5736	1.6009
0.4965	1.5362	1.5716	1.5996
0.5145	1.5349	1.5690	1.5974
0.5321	1.5328	1.5662	1.5936
0.6328	1.5294	1.5588	1.5864
1.0642	1.5194	1.5505	1.5781

Dispersion relations (λ in μm, $T = 20$ °C) [3.507]:

$$n_X^2 = 2.2916 + \frac{0.02105}{\lambda^2 + 0.06525} - 3.1848 \times 10^{-5} \, \lambda^2 \; ;$$

$$n_Y^2 = 2.3731 + \frac{0.03437}{\lambda^2 + 0.11600} - 7.2632 \times 10^{-5} \, \lambda^2 \; ;$$

$$n_Z^2 = 2.4607 + \frac{0.03202}{\lambda^2 + 0.08961} - 5.6332 \times 10^{-5} \, \lambda^2 \; .$$

Note: The dispersion relations in [3.507] are given with a mistake. The numerator of the second term in the equation for n_Y^2 should be 0.03437 instead of 0.3437

Experimental and theoretical values of phase-matching angle and calculated values of "walk-off" angles:
XZ plane, $\phi = 0°$, $\theta > V_Z$

Interacting wavelengths [μm]	θ_{exp} [deg]	θ_{theor} [deg] [3.507]	ρ_1 [deg]	ρ_2 [deg]
SHG, e + e $\Rightarrow$ o				
$1.0642 \Rightarrow 0.5321$	62 [3.507]	67.53	1.54	1.54
SFG, e + e $\Rightarrow$ o				
$1.0642 + 0.5321 \Rightarrow 0.35473$	76 [3.507]	76.31	1.01	1.08

Experimental value of internal angular bandwidth [3.507]:
XZ plane, $\phi = 0°$

Interacting wavelengths [μm]	θ_{pm} [deg]	$\Delta\theta^{int}$ [deg]
SHG, e + e $\Rightarrow$ o		
$1.0642 \Rightarrow 0.5321$	62	0.064

Effective nonlinearity expressions in the phase-matching direction for three-wave interactions in the principal planes of CsB_3O_5, crystal [3.35]:
XY plane

$$d_{eoe} = d_{oee} = d_{14} \sin 2\phi \;;$$

YZ plane

$$d_{eeo} = d_{14} \sin 2\theta \;;$$

XZ plane, $\theta < V_Z$

$$d_{eoe} = d_{oee} = d_{14} \sin 2\theta \;;$$

XZ plane, $\theta > V_Z$

$$d_{eeo} = d_{14} \sin 2\theta \;.$$

Nonlinear coefficient:

$$d_{14}(1.064\,\mu m) = 0.468 \times d_{22}\,(BBO) = 1.08\,pm/V \;[3.507, 37] \;.$$

Laser-induced damage threshold [3.507]:

λ [μm]	τ_p [ns]	$I_{thr} \times 10^{-12}$ [W/m^2]
1.053	1	260

3.3.3 BeSO$_4$ · 4H$_2$O, Beryllium Sulfate

Negative uniaxial crystal: $n_o > n_e$;
Point group: $\bar{4}$2m ;
Mass density: 1.713 g/cm^3 [3.508] ;
Mohs hardness: > 2.5 [3.509] ;
Transparency range at "0" transmittance level: 0.17 − 1.58 μm [3.508, 510]
Linear absorption coefficient α :

λ [μm]	α [cm^{-1}]	Ref.	Note
0.3164	0.6	3.508	e − wave, SHG direction
0.6328	0.17	3.508	o − wave, SHG direction
0.187–1.3	<0.01	3.511	

Experimental values of refractive indices [3.510]:

λ [μm]	n_o	n_e
0.4154	1.4847	1.4431
0.4825	1.4782	1.4379
0.5321	1.4749	1.4348
0.6328	1.4701	1.4315
0.6471	1.4692	1.4312
0.6764	1.4681	1.4304
0.7525	1.4668	1.4292

Experimental values of phase-matching angle ($T = 293$ K) and comparison between different sets of dispersion relations:

Interacting wavelengths [μm]	θ_{exp} [deg]	θ_{theor} [deg] [3.510]	θ_{theor} [deg] [3.511]
SHG, o + o ⇒ e			
1.1523 ⇒ 0.5762	42 [3.508]	30.4	42.9
0.6328 ⇒ 0.3164	55 [3.508]	59.9	56.2
	60 [3.509]		
0.5400 ⇒ 0.2700	77 [3.511]	79.0	76.7
0.5340 ⇒ 0.2670	80 [3.511]	81.9	80.1
0.5321 ⇒ 0.2661	81.5 [3.510]	83.1	81.5
	81.6 [3.511]		
0.5266 ⇒ 0.2633	90 [3.511]	no pm*	no pm#
SHG, e + o ⇒ e			
1.1523 ⇒ 0.5762	64 [3.508]	43.7	65.3
0.7606 ⇒ 0.3803	90 [3.511]	78.3	89.3

SFG, $o + o \Rightarrow e$

$1.0642 + 0.5321 \Rightarrow 0.3547$	47.4 [3.511]	47.8	47.4
$1.0642 + 0.3547 \Rightarrow 0.2661$	62.4 [3.511]	59.4	62.5
$0.9070 + 0.3547 \Rightarrow 0.2550$	72.3 [3.511]	67.3	72.4
$0.8468 + 0.3547 \Rightarrow 0.2500$	80 [3.511]	71.8	80.0
$0.8209 + 0.3547 \Rightarrow 0.2477$	90 [3.511]	74.2	89.5

* NCPM corresponds to the SHG with $\lambda_1 = 0.5271\ \mu m$;
# NCPM corresponds to the SHG with $\lambda_1 = 0.5268\ \mu m$.

Best set of dispersion relations (λ in μm, $T = 20\ °C$) [3.511]:

$$n_o^2 = 2.1545 + \frac{0.00835}{\lambda^2 - 0.01606} - 0.03573\,\lambda^2\ ,$$

$$n_e^2 = 2.0335 + \frac{0.00806}{\lambda^2 - 0.01354} - 0.01970\,\lambda^2\ .$$

Calculated values of phase-matching and "walk-off" angles:

Interacting wavelengths [μm]	θ_{pm} [deg]	ρ_1 [deg]	ρ_3 [deg]
SHG, $o + o \Rightarrow e$			
$1.0642 \Rightarrow 0.5321$	41.88		1.59
$0.6943 \Rightarrow 0.34715$	50.32		1.60
$0.5782 \Rightarrow 0.2891$	64.99		1.25
$0.5321 \Rightarrow 0.26605$	81.46		0.48
SFG, $o + o \Rightarrow e$			
$1.0642 + 0.3547 \Rightarrow 0.26605$	62.50		1.36
$0.5782 + 0.5105 \Rightarrow 0.2711$	75.34		0.80
SHG, $e + o \Rightarrow e$			
$1.0642 \Rightarrow 0.5321$	64.07	1.11	1.23
SFG, $e + o \Rightarrow e$			
$1.0642 + 0.5321 \Rightarrow 0.3547$	60.87	1.20	1.37

Experimental values of internal angular, temperature, and spectral bandwidths at $T = 293\ K$:

Interacting wavelengths [μm]	θ_{pm} [deg]	$\Delta\theta^{int}$ [deg]	ΔT [°C]	$\Delta\nu$ [cm^{-1}]	Ref.
SHG, $o + o \Rightarrow e$					
$0.5321 \Rightarrow 0.2661$	81.5	0.09			3.510
	81.6	0.11	1.45	4.9	3.511

Temperature variation of phase-matching angle [3.511]:

Interacting wavelengths [μm]	$T[°C]$	θ_{pm} [deg]	$d\theta_{pm}/dT$ [deg /K]
SHG, $o + o \Rightarrow e$			
$0.5321 \Rightarrow 0.2661$	20	81.6	0.077

Effective nonlinearity expressions in the phase-matching direction [3.100]:

$$d_{ooe} = d_{36} \sin\theta \sin 2\phi ,$$

$$d_{eoe} = d_{oee} = d_{36} \sin 2\theta \cos 2\phi .$$

Nonlinear coefficient:

$$d_{36}(0.5321\,\mu m) = 0.62 \times d_{36}(DKDP) \pm 10\%$$
$$= 0.23 \pm 0.02\,pm/V\,[3.510, 37] .$$

Laser-induced surface-damage threshold:

λ [μm]	τ_P [ns]	$I_{thr} \times 10^{-12}$ [W/m^2]	Ref.	Note
0.2661	8	1	3.510	10 Hz
0.5321	8	>2.2	3.511	3 Hz

3.3.4 MgBaF$_4$, Magnesium Barium Fluoride

Negative biaxial crystal: $2V_Z = 117.5°$ at $\lambda = 0.5321\,\mu m$ [3.512];
Point group: mm2;
Assignment of dielectric and crystallographic axes:
$X, Y, Z \Rightarrow b, c, a$;
Transparency range: $0.17 - 8\,\mu m$ [3.513];
Experimental values of refractive indices [3.512]:

λ [μm]	n_X	n_Y	n_Z
0.5321	1.4508	1.4678	1.4742
1.0642	1.4436	1.4604	1.4674

Sellmeier equations (λ in μm, $T = 20$ °C) [3.512]:

$$n_X^2 = 2.0770 + \frac{0.00760}{\lambda^2 - 0.0079} ,$$

$$n_Y^2 = 2.1238 + \frac{0.00860}{\lambda^2} ,$$

$$n_Z^2 = 2.1462 + \frac{0.00736}{\lambda^2 - 0.0090} .$$

Experimental and theoretical values of phase-matching angle and calculated values of "walk-off" angle:

XY plane, $\theta = 90°$

Interacting wavelengths [μm]	$\phi_{\exp}$ [deg]	ϕ_{theor} [deg] [3.512]	ρ_3 [deg]
SHG, $o + o \Rightarrow e$ 1.0642 $\Rightarrow$ 0.5321	9.2 [3.512]	9.65	0.223

XZ plane, $\phi = 0°$, $\theta < V_z$

Interacting wavelengths [μm]	$\theta_{\exp}$ [deg]	θ_{theor} [deg] [3.512]	ρ_1 [deg]	ρ_3 [deg]
SHG, $e + o \Rightarrow e$ 1.0642 $\Rightarrow$ 0.5321	18.9 [3.512]	17.39	0.525	0.516

Effective nonlinearity expressions in the phase-matching direction for three-wave interactions in the principal planes of $MgBaF_4$ crystal [3.35], [3.36]:
XY plane

$$d_{ooe} = d_{31} \cos \phi \; ;$$

YZ plane

$$d_{oeo} = d_{eoo} = d_{32} \cos \theta \; ;$$

XZ plane, $\theta < V_z$

$$d_{oee} = d_{eoe} = d_{31} \sin^2 \theta + d_{32} \cos^2 \theta \; ;$$

XZ plane, $\theta > V_z$

$$d_{eeo} = d_{31} \sin^2 \theta + d_{32} \cos^2 \theta \; .$$

Effective nonlinearity expressions for three-wave interactions in the arbitrary direction of $MgBaF_4$ crystal are given in [3.36].
Nonlinear coefficient:

$$d_{31}(1.0642 \, \mu m) = \pm 0.057 \times d_{36} \, (\mathbf{KDP}) \pm 23\%$$

$$= \pm 0.022 \pm 0.005 \, pm/V \, [3.512, 37] \; ,$$

$$d_{32}(1.0642 \, \mu m) = \pm 0.085 \times d_{36} \, (\mathbf{KDP}) \pm 12\%$$

$$= \pm 0.033 \pm 0.012 \, pm/V \, [3.512, 37] \; ,$$

$$d_{33}(1.0642\,\mu m) = \pm\, 0.023 \times d_{36}\,(KDP) \pm 14\%$$
$$= \pm\, 0.009 \pm 0.001\,pm/V\,[3.512, 37]\ .$$

Laser-induced surface-damage threshold [3.513]:

λ [μm]	τ_P [ns]	$I_{thr} \times 10^{-12}$ [W/m^2]
1.0642	≈ 20	>10

3.3.5 NH$_4$D$_2$PO$_4$, Deuterated Ammonium Dihydrogen Phosphate (DADP)

Negative uniaxial crystal: $n_o > n_e$;
Point group: $\bar{4}2m$;
IR edge of transmission spectrum (at "0" transmittance level): 1.9 μm [3.78];
Linear absorption coefficient:
$\alpha < 0.013\,cm^{-1}$ in the range $0.78 - 1.03\,\mu m$ [3.67];
Experimental values of refractive indices:

λ [μm]	n_o	n_e	Ref.
0.3472	1.5414	1.4923	3.126
0.4358	1.5278	1.4831	3.126
0.53	1.5198	1.4784	3.79
0.5461	1.5194	1.4759	3.126
0.6943	1.5142	1.4737	3.126
1.06	1.5088	1.4712	3.79

Experimental values of phase-matching angle ($T = 293$ K) and comparison between different sets of dispersion relations:

Interacting wavelengths [μm]	θ_{exp} [deg]	θ_{theor} [deg]	
		[3.78]K	[3.78]E
SHG, o + o $\Rightarrow$ e			
0.528 $\Rightarrow$ 0.264	90 [3.119]	82.2	no pm (?)
0.6943 $\Rightarrow$ 0.34715	47 [3.514]	50.3	no pm (?)

Note: [3.78]K $\Rightarrow$ see [3.78], data of *Kirby* et al.;
      [3.78]E $\Rightarrow$ see [3.78], data of *Eimerl*

Experimental values of NCPM temperature [3.119]:

Interacting wavelengths [μm]	T [$^\circ$C]
SHG, o + o $\Rightarrow$ e	
0.516 $\Rightarrow$ 0.258	-20

$0.524 \Rightarrow 0.262$	0
$0.528 \Rightarrow 0.264$	20
$0.554 \Rightarrow 0.277$	100

Best set of dispersion relations (λ in μm, $T = 20\ °C$) [3.78]K:

$$n_o^2 = 2.279481 + \frac{1.215879\,\lambda^2}{\lambda^2 - (7.614168)^2} + \frac{0.010761}{\lambda^2 - (0.115165)^2}\ ,$$

$$n_e^2 = 2.151161 + \frac{1.199009\,\lambda^2}{\lambda^2 - (11.25169)^2} + \frac{0.009652}{\lambda^2 - (0.098550)^2}\ .$$

Calculated values of phase-matching and "walk-off" angles:

Interacting wavelengths [μm]	θ_{pm} [deg]	ρ_1 [deg]	ρ_3 [deg]
SHG, $o + o \Rightarrow e$			
$0.5321 \Rightarrow 0.26605$	79.53		0.652
$0.5782 \Rightarrow 0.2891$	65.24		1.357
$0.6328 \Rightarrow 0.3164$	56.61		1.611
$0.6594 \Rightarrow 0.3297$	53.58		1.664
$0.6943 \Rightarrow 0.34715$	50.31		1.700
$1.0642 \Rightarrow 0.5321$	36.93		1.599
$1.3188 \Rightarrow 0.6594$	37.18		1.569
SFG, $o + o \Rightarrow e$			
$0.5782 + 0.5105 \Rightarrow 0.27112$	74.57		0.930
$1.0642 + 0.5321 \Rightarrow 0.35473$	46.44		1.728
$1.3188 + 0.6594 \Rightarrow 0.4396$	39.29		1.659
SHG, $e + o \Rightarrow e$			
$1.0642 \Rightarrow 0.5321$	54.47	1.411	1.547
$1.3188 \Rightarrow 0.6594$	53.55	1.339	1.533
SFG, $e + o \Rightarrow e$			
$1.0642 + 0.5321 \Rightarrow 0.35473$	59.17	1.308	1.504
$1.3188 + 0.6594 \Rightarrow 0.4396$	48.09	1.399	1.668

Effective nonlinearity expressions in the phase-matching direction [3.100]:

$$d_{\mathrm{ooe}} = d_{36} \sin\theta \sin 2\phi\ ,$$

$$d_{\mathrm{eoe}} = d_{\mathrm{oee}} = d_{36} \sin 2\theta \cos 2\phi\ .$$

Nonlinear coefficient:

$$d_{36}(0.6943\,\mu\mathrm{m}) = 1.10 \times d_{36}(\mathrm{KDP}) \pm 15\%$$

$$= 0.43 \pm 0.06\,\mathrm{pm/V}\ [3.514, 37]\ .$$

3.3.6 RbH_2PO_4, Rubidium Dihydrogen Phosphate (RDP)

Negative uniaxial crystal: $n_o > n_e$;
Point group: $\bar{4}2m$;
Mass density: 2.805 g/cm^3;
Transparency range at 0.5 transmittance level for a 15.3 mm long crystal cut at
$\theta = 50°, \phi = 45° : 0.19 - 1.38$ μm [3.515];
IR edge of transmission spectrum (at "0" transmittance level) :
1.65 μm for o − wave, 1.87 μm for e − wave [3.78] ;
Linear absorption coefficient α:

λ [μm]	α [cm^{-1}]	Ref.	Note
0.25–1.25	< 0.03	3.113	
0.3547	0.015	3.515	$\theta = 50°, \phi = 45°$
0.5321	0.01	3.515	$\theta = 50°, \phi = 45°$
1.0642	0.041	3.515	$\theta = 50°, \phi = 45°$

Two-photon absorption coefficient β ($\theta = 90°, \phi = 45°$) [3.71]:

λ [μm]	$\beta \times 10^{14}$ [m/W]	Note
0.355	5.9	e − wave

Experimental values of refractive indices:

λ [μm]	n_o	n_e	Ref.
0.3472	1.5284	1.4969	3.516
0.4358	1.5165	1.4857	3.516
0.4765	1.5140	1.4861	3.517
0.4880	1.5132	1.4832	3.517
0.4965	1.5126	1.4827	3.517
0.5017	1.5121	1.4825	3.517
0.5145	1.5116	1.4820	3.517
0.5321	1.5106	1.4811	3.517
0.5468	1.5082	1.4790	3.516
0.5893	1.5053	1.4765	3.516
0.6328	1.4976	1.4775	3.517
0.6943	1.5020	1.4735	3.516
1.0642	1.4926	1.4700	3.517

λ [μm]	n_o	Ref.	λ [μm]	n_e	Ref.
0.4699	1.5148	3.518	0.4658	1.4851	3.518
0.4950	1.5128	3.518	0.4780	1.4845	3.518

0.5120	1.5117	3.518	0.4950	1.4833	3.518
0.5329	1.5104	3.518	0.5324	1.4810	3.518
0.5851	1.5074	3.518	0.5577	1.4798	3.518
0.5980	1.5069	3.518	0.5878	1.4787	3.518
0.6245	1.5056	3.518	0.6165	1.4776	3.518
0.6474	1.5047	3.518	0.6521	1.4766	3.518
0.6662	1.5042	3.518	0.6640	1.4763	3.518

Temperature derivative of refractive indices [3.74]:

λ [µm]	$dn_\mathrm{o}/dT \times 10^5$ [K^{-1}]	$dn_\mathrm{e}/dT \times 10^5$ [K^{-1}]
0.405	−3.69	−2.67
0.436	−3.86	−2.76
0.546	−3.72	−2.54
0.578	−3.72	−2.80
0.633	−3.72	−2.89

Experimental values of phase-matching angle (T = 293 K) and comparison between different sets of dispersion relations:

Interacting wavelengths [µm]	θ_exp [deg]	θ_theor [deg]			
		[3.517]	[3.74]	[3.78]K	[3.78]E
SHG, o + o $\Rightarrow$ e					
0.626 $\Rightarrow$ 0.313	90 [3.74]	no pm	no pm	85.8	no pm
0.627 $\Rightarrow$ 0.3135	90 [3.119]	no pm	no pm	84.9	no pm
0.6275 $\Rightarrow$ 0.31375	90 [3.519]	no pm	no pm	84.5	no pm
0.6294 $\Rightarrow$ 0.3147	86.6 [3.519]	no pm	no pm	83.1	no pm
0.6328 $\Rightarrow$ 0.3164	83.2 [3.520]	no pm	no pm	81.3	no pm
0.6386 $\Rightarrow$ 0.3193	78.9 [3.519]	no pm	no pm	78.9	84.3
0.6550 $\Rightarrow$ 0.3275	73.9 [3.519]	no pm	no pm	74.1	76.5
0.6700 $\Rightarrow$ 0.3350	70.8 [3.519]	no pm	81.3	70.9	72.5
0.6943 $\Rightarrow$ 0.34715	66 [3.516]	no pm	72.9	67.0	67.9
1.0642 $\Rightarrow$ 0.5321	50.8 [3.521]	52.5	39.4	51.1	51.0
	50.8 [3.515]				
	50.9 [3.425]				
1.1523 $\Rightarrow$ 0.57615	51 [3.520]	48.4	36.0	51.7	51.3
SHG, e + o $\Rightarrow$ e					
1.0642 $\Rightarrow$ 0.5321	83.1 [3.521]	no pm	61.3	85.3	84.6
1.1523 $\Rightarrow$ 0.57615	77.1 [3.520]	74.6	54.4	82.0	80.0
THG, o + o $\Rightarrow$ e					
1.0642 + 0.5321					
$\Rightarrow$ 0.3547	61.2 [3.515]	75.5	62.1	60.9	61.4

Note: [3.78]K $\rightarrow$ see [3.78], data of *Kirby* et al.;
[3.78]E $\Rightarrow$ see [3.78], data of *Eimerl*

Experimental values of NCPM temperature:

Interacting wavelengths [μm]	T [°C]	Ref.
SGH, o + o $\Rightarrow$ e		
0.627 $\Rightarrow$ 0.3135	20	[3.425, 119]
0.6275 $\Rightarrow$ 0.31375	20	[3.519]
0.635 $\Rightarrow$ 0.3175	100	[3.425, 119]
0.637 $\Rightarrow$ 0.3185	98	[3.519]

Best set of dispersion relations (λ in μm, $T = 20\ °C$) [3.78]K:

$$n_\mathrm{o}^2 = 2.249885 + \frac{3.688005\ \lambda^2}{\lambda^2 - (11.27829)^2} + \frac{0.010560}{\lambda^2 - (0.088207)^2}\ ,$$

$$n_\mathrm{e}^2 = 2.159913 + \frac{0.988431\ \lambda^2}{\lambda^2 - (11.30013)^2} + \frac{0.009515}{\lambda^2 - (0.092076)^2}\ .$$

Calculated values of phase-matching and "walk-off" angles:

Interacting wavelengths [μm]	θ_pm [deg]	ρ_1 [deg]	ρ_3 [deg]
SHG, o + o $\Rightarrow$ e			
0.6328 $\Rightarrow$ 0.3164	81.31		0.357
0.6594 $\Rightarrow$ 0.3297	73.05		0.664
0.6943 $\Rightarrow$ 0.34715	66.96		0.853
1.0642 $\Rightarrow$ 0.5321	51.08		1.093
1.3188 $\Rightarrow$ 0.6594	55.49		0.994
SFG, o + o $\Rightarrow$ e			
1.0642 + 0.5321 $\Rightarrow$ 0.35473	60.86		1.008
1.3188 + 0.6594 $\Rightarrow$ 0.4396	52.53		1.114
SHG, e + o $\Rightarrow$ e			
1.0642 $\Rightarrow$ 0.5321	85.26	0.141	0.182
SFG, e + o $\Rightarrow$ e			
1.3188 + 0.6594 $\Rightarrow$ 0.4396	62.54	0.567	0.938

Experimental values of internal angular bandwidth at $T = 293$ K:

Interacting wavelengths [μm]	θ_pm [deg]	$\Delta\theta^\mathrm{int}$ [deg]	Ref.
SHG, o + o $\Rightarrow$ e			
0.6275 $\Rightarrow$ 0.31375	90	1.73	3.519
0.6943 $\Rightarrow$ 0.34715	66	0.14	3.522
1.0642 $\Rightarrow$ 0.5321	50.8	0.10	3.521

$1.0642 \Rightarrow 0.5321$	50.8	0.11	3.515
SHG, $e + o \Rightarrow e$			
$1.0642 \Rightarrow 0.5321$		0.40	3.523
	83.1	0.54	3.521
THG, $o + o \Rightarrow e$			
$1.0642 + 0.5321 \Rightarrow 0.3547$	61.2	0.08	3.515

Temperature tuning of noncritical SHG:

Interacting wavelengths [μm]	$d\lambda_1/dT$ [nm/K]	Ref.
SHG, $o + o \Rightarrow e$		
$0.626 \Rightarrow 0.313$	0.12	3.74
$0.6275 \Rightarrow 0.31375$	0.123	3.519

Experimental value of temperature bandwidth for noncritical SHG process ($0.6275\ \mu m \Rightarrow 0.31375\ \mu m$, $o + o \Rightarrow e$):

$$\Delta T = 2.5 \pm 0.3\ °C\ [3.519].$$

Temperature variation of birefringence for noncritical SHG process ($0.6275\ \mu m \Rightarrow 0.31375\ \mu m$, $o + o \Rightarrow e$):

$$d(n_2^e - n_1^o)/dT = (1.1 \pm 0.1) \times 10^{-5} K^{-1}\ [3.519].$$

Effective nonlinearity expressions in the phase-matching direction [3.100]:

$$d_{ooe} = d_{36} \sin\theta\ \sin 2\phi,$$

$$d_{eoe} = d_{oee} = d_{36} \sin 2\theta\ \cos 2\phi.$$

Nonlinear coefficient:

$$d_{36}(0.6943\ \mu m) = 1.04 \times d_{36}(KDP) \pm 15\%$$
$$= 0.41 \pm 0.06\ pm/V\ [3.514, 37],$$

$$d_{36}(0.6943\ \mu m) = 0.92 \times d_{36}\ (KDP) \pm 10\%$$
$$= 0.36 \pm 0.04\ pm/V\ [3.198, 37].$$

Laser-induced bulk-damage threshold:

λ [μm]	τ_p [ns]	$I_{thr} \times 10^{-12}$ [W/m^2]	Ref.	Note
0.6281	330	5.5	3.101	
0.6943	10	> 1.8	3.522	
1.0642	12	> 2.6	3.521	10–20 Hz

3.3.7 RbD$_2$PO$_4$, Deuterated Rubidium Dihydrogen Phosphate (DRDP)

Negative uniaxial crystal: $n_o > n_e$;
Point group: $\bar{4}$2m;
IR edge of transmission spectrum (at "0" transmittance level):
1.66 µm [3.78];
Best set of dispersion relations (λ in µm, $T = 20$ °C) [3.78]K:

$$n_o^2 = 2.235596 + \frac{2.355322\,\lambda^2}{\lambda^2 - (11.26298)^2} + \frac{0.010929}{\lambda^2 - (0.0376136)^2}\;,$$

$$n_e^2 = 2.152727 + \frac{0.691253\,\lambda^2}{\lambda^2 - (11.27007)^2} + \frac{0.010022}{\lambda^2 - (0.037137)^2}\;.$$

Calculated values of phase-matching and "walk-off" angles:

Interacting wavelengths [µm]	θ_{pm} [deg]	ρ_1 [deg]	ρ_3 [deg]
SHG, o + o $\Rightarrow$ e			
0.6328 $\Rightarrow$ 0.3164	81.66		0.319
0.6594 $\Rightarrow$ 0.3297	73.26		0.610
0.6943 $\Rightarrow$ 0.34715	66.98		0.793
1.0642 $\Rightarrow$ 0.5321	47.19		1.054
1.3188 $\Rightarrow$ 0.6594	47.35		1.021
SFG, o + o $\Rightarrow$ e			
1.0642 + 0.5321 $\Rightarrow$ 0.35473	60.01		0.955
1.3188 + 0.6594 $\Rightarrow$ 0.4396	50.09		1.064
SHG, e + o $\Rightarrow$ e			
1.0642 $\Rightarrow$ 0.5321	75.61	0.427	0.502
1.3188 $\Rightarrow$ 0.6594	70.09	0.502	0.648
SFG, e + o $\Rightarrow$ e			
1.3188 + 0.6594 $\Rightarrow$ 0.4396	61.81	0.654	0.894

Effective nonlinearity expressions in the phase-matching direction [3.100]:

$$d_{ooe} = d_{36} \sin\theta \sin 2\phi\,.$$

$$d_{eoe} = d_{oee} = d_{36} \sin\theta \cos 2\phi\,.$$

Nonlinear coefficient:

$$d_{36} \approx 0.38 \text{ pm/V } [3.78]\,.$$

3.3.8 KH$_2$AsO$_4$, Potassium Dihydrogen Arsenate (KDA)

Negative uniaxial crystal: $n_o > n_e$;
Point group: $\bar{4}$2m;
Calculated mass density: 2.872 g/cm^3;

Transparency range at "0" transmittance level:
0.213–1.82 μm [3.113, 524, 78];
Linear absorption coefficient α:

λ [μm]	α [cm^{-1}]	Ref.
0.35–1.45	0.3–0.9	3.113

Two-photon absorption coefficient β ($\theta = 90°, \phi = 45°$) [3.71]:

λ [μm]	$\beta \times 10^{13}$ [m/W]	Note
0.355	4.84	e – wave

Experimental values of refractive indices [3.517]:

λ [μm]	n_o	n_e
0.4861	1.5762	1.5252
0.5460	1.5707	1.5206
0.5893	1.5674	1.5179
0.6563	1.5632	1.5146

Temperature derivative of refractive indices [3.74]:

λ [μm]	$dn_\mathrm{o}/dT \times 10^5$ [K^{-1}]	$dn_\mathrm{e}/dT \times 10^5$ [K^{-1}]
0.436	−3.64	−2.31
0.546	−4.07	−2.13
0.578	−3.98	−2.51
0.633	−4.09	−2.12

Experimental values of the phase-matching angle ($T = 293$ K) and comparison between different sets of dispersion relations:

Interacting wavelengths [μm]	θ_exp [deg]	θ_theor [deg]		
		[3.74]	[3.78]K	[3.78]E
SHG, o + o $\Rightarrow$ e				
0.596 $\Rightarrow$ 0.298	90 [3.74]	70.7	74.2	no pm
0.616 $\Rightarrow$ 0.308		65.0	68.7	88.3
0.6943 $\Rightarrow$ 0.34715	59 [3.514]	51.7	56.5	60.1
1.0642 $\Rightarrow$ 0.5321	40.5 [3.425]	29.2	40.0	41.9

Note: [3.78]K $\Rightarrow$ see [3.78], data of *Kirby* et al.;
  [3.78]E $\Rightarrow$ see [3.78], data of *Eimerl*

Experimental values of NCPM temperature [3.425]:

Interacting wavelengths [μm]	T [°C]
SHG, o + o $\Rightarrow$ e	
0.594 $\Rightarrow$ 0.297	20
0.601 $\Rightarrow$ 0.3005	100

Best set of dispersion relations (λ in μm, $T = 20$ °C) [3.78]E:

$$n_o^2 = 1.988413 - 0.05826141\,\lambda^2 + 0.01409368\,\lambda^4 + \frac{0.4430935\,\lambda^2}{\lambda^2 - (0.1710929)^2}\,,$$

$$n_e^2 = 2.011142 - 0.03195326\,\lambda^2 + 0.01217516\,\lambda^4 + \frac{0.2681806\,\lambda^2}{\lambda^2 - (0.1925064)^2}\,.$$

Calculated values of phase-matching and "walk-off" angles:

Interacting wavelengths [μm]	θ_{pm} [deg]	ρ_1 [deg]	ρ_3 [deg]
SHG, o + o $\Rightarrow$ e			
0.6328 $\Rightarrow$ 0.3164	74.64		0.986
0.6594 $\Rightarrow$ 0.3297	66.48		1.423
0.6943 $\Rightarrow$ 0.34715	60.09		1.688
1.0642 $\Rightarrow$ 0.5321	41.89		1.860
1.3188 $\Rightarrow$ 0.6594	38.82		1.762
SFG, o + o $\Rightarrow$ e			
1.0642 + 0.5321 $\Rightarrow$ 0.35473	54.29		1.859
1.3188 + 0.6594 $\Rightarrow$ 0.4396	43.32		1.926
SHG, e + o $\Rightarrow$ e			
1.0642 $\Rightarrow$ 0.5321	61.38	1.298	1.541
1.3188 $\Rightarrow$ 0.6594	53.50	1.334	1.698
SFG, e + o $\Rightarrow$ e			
1.0642 + 0.5321 $\Rightarrow$ 0.35473	71.12	0.939	1.182
1.3188 + 0.6594 $\Rightarrow$ 0.4396	51.93	1.356	1.855

Temperature tuning for noncritical SHG [3.74]:

Interacting wavelengths [μm]	$d\lambda_1/dT$ [nm/K]
SHG, o + o $\Rightarrow$ e	
0.596 $\Rightarrow$ 0.298	0.077

Effective nonlinearity expressions in the phase-matching direction [3.100]:

$$d_{ooe} = d_{36}\sin\theta\sin 2\phi\,,$$

$$d_{eoe} = d_{oee} = d_{36}\sin 2\theta\cos 2\phi\,.$$

Nonlinear coefficient:

$$d_{36}(0.6943 \ \mu m) = 0.70 \times d_{36} \ (KDP) \pm 15\%$$
$$= 0.27 \pm 0.04 \ pm/V \ [3.514, 37] \ ,$$

$$d_{36}(1.064 \ \mu m) = 1.06 \times d_{36} \ (KDP) \pm 5\% = 0.41 \pm 0.02 \ pm/V \ [3.525, 37] \ .$$

Laser-induced bulk-damage threshold [3.101]:

λ [μm]	τ_p [ns]	$I_{thr} \times 10^{-12}$ [W/m^2]
0.6	330	0.12

3.3.9 KD$_2$AsO$_4$, Deuterated Potassium Dihydrogen Arsenate (DKDA)

Negative uniaxial crystal: $n_o > n_e$;
Point group: $\bar{4}2m$;
Transparency range at "0" transmittance level: $0.22 - 2.3 \ \mu m$ [3.524];
Two-photon absorption coefficient β $(\theta = 90°, \phi = 45°)$

λ [μm]	$\beta \times 10^{13}$ [m/W]	Note	Ref.
0.355	2.66	e − wave	3.71

Experimental values of NCPM temperature [3.425]:

Interacting wavelengths [μm]	T [°C]
SHG, o + o $\Rightarrow$ e	
$0.609 \Rightarrow 0.3045$	20
$0.615 \Rightarrow 0.3075$	100

Effective nonlinearity expressions in the phase-matching direction [3.100]:

$$d_{ooe} = d_{36} \sin \theta \sin 2\phi \ ,$$

$$d_{eoe} = d_{oee} = d_{36} \sin 2\theta \cos 2\phi \ .$$

Nonlinear coefficient:

$$d_{36} \approx d_{36} \ (KDP) = 0.39 \ pm/V \ [3.78, 37] \ .$$

Laser-induced bulk-damage threshold [3.101]:

λ [μm]	τ_p [ns]	$I_{thr} \times 10^{-12}$ [W/m^2]
0.61	330	0.24

3.3.10 $NH_4H_2AsO_4$, Ammonium Dihydrogen Arsenate (ADA)

Negative uniaxial crystal: $n_o > n_e$;
Point group: $\bar{4}2m$;
Transparency range at "0" transmittance level: $0.218 - 1.53$ μm [3.526, 78]
Two-photon absorption coefficient β ($\theta = 90°, \phi = 45°$) [3.71]:

λ [μm]	$\beta \times 10^{13}$ [m /W]	Note
0.355	3.53	e − wave

Temperature derivative of refractive indices [3.74]:

λ [μm]	$dn_o/dT \times 10^5$ [K^{-1}]	$dn_e/dT \times 10^5$ [K^{-1}]
0.436	−4.85	+ 1.27
0.546	−4.39	+ 1.31
0.578	−4.53	+ 1.24
0.633	−4.45	+ 1.19

Experimental values of phase-matching angle ($T = 293$ K) and comparison between different sets of dispersion relations:

Interacting wavelengths [μm]	θ_{exp} [deg]	θ_{theor} [deg]		
		[3.74]	[3.78]K	[3.78]E
SHG, o + o ⇒ e				
0.58 ⇒ 0.29	90 [3.425]	76.5	no pm	no pm
0.582 ⇒ 0.291	90 [3.74]	75.8	no pm	no pm
0.584 ⇒ 0.292	–	75.1	87.3	no pm
1.0642 ⇒ 0.5321	41.3 [3.425]	32.8	41.7	41.7

Note: [3.78]K ⇒ see [3.78], data of *Kirby* et al.;
      [3.78]E ⇒ see [3.78], data of *Eimerl*

Experimental values of NCPM temperature:

Interacting wavelengths [μm]	T [°C]	Ref.
SHG, o + o ⇒ e		
0.568 ⇒ 0.284	−30	3.119
0.572 ⇒ 0.286	−10	3.425
0.58 ⇒ 0.29	20	3.425
0.586 ⇒ 0.293	25	3.527
0.606 ⇒ 0.303	80	3.101
0.611 ⇒ 0.3055	100	3.425
0.619 ⇒ 0.3095	120	3.119

Best set of dispersion relations (λ in μm, $T = 20$ °C) [3.78]K:

$$n_o^2 = 2.443449 + \frac{2.017752\,\lambda^2}{\lambda^2 - (7.604942)^2} + \frac{0.016757}{\lambda^2 - (0.135177)^2}\ ,$$

$$n_e^2 = 2.275962 + \frac{1.598260\,\lambda^2}{\lambda^2 - (11.26433)^2} + \frac{0.014296}{\lambda^2 - (0.128689)^2}\ .$$

Calculated values of phase-matching and "walk-off" angles:

Interacting wavelengths [μm]	θ_{pm} [deg]	ρ_1 [deg]	ρ_3 [deg]
SHG, $o + o \Rightarrow e$			
$0.6328 \Rightarrow 0.3164$	67.42		1.544
$0.6594 \Rightarrow 0.3297$	62.69		1.764
$0.6943 \Rightarrow 0.34715$	58.05		1.928
$1.0642 \Rightarrow 0.5321$	41.71		2.023
$1.3188 \Rightarrow 0.6594$	42.58		1.964
SFG, $o + o \Rightarrow e$			
$1.0642 + 0.5321 \Rightarrow 0.35473$	53.05		2.065
$1.3188 + 0.6594 \Rightarrow 0.4396$	44.31		2.087
SHG, $e + o \Rightarrow e$			
$1.0642 \Rightarrow 0.5321$	62.22	1.423	1.640
$1.3188 \Rightarrow 0.6594$	61.26	1.315	1.627
SFG, $e + o \Rightarrow e$			
$1.0642 + 0.5321 \Rightarrow 0.35473$	69.20	1.139	1.402
$1.3188 + 0.6594 \Rightarrow 0.4396$	53.77	1.497	1.968

Temperature tuning of noncritical SHG [3.74]:

Interacting wavelengths [μm]	$d\lambda_1/dT$ [nm/K]
SHG, $o + o \Rightarrow e$	
$0.582 \Rightarrow 0.291$	0.359

Effective nonlinearity expressions in the phase-matching direction [3.100]:

$$d_{ooe} = d_{36} \sin\theta \sin 2\phi\,,$$

$$d_{eoe} = d_{oee} = d_{36} \sin 2\theta \cos 2\phi\,.$$

Nonlinear coefficient:

$$d_{36}(\text{ADA}) = d_{36}(\text{ADP}) = 0.45 \text{ pm/V [3.414, 419, 37]}.$$

Laser-induced bulk-damage threshold [3.101]:

λ [μm]	τ_p [ns]	$I_{thr} \times 10^{-12}$ [W/m^2]
0.581	330	6.1
0.606	330	4.8

3.3.11 NH₄D₂AsO₄, Deuterated Ammonium Dihydrogen Arsenate (DADA)

Negative uniaxial crystal: $n_o > n_e$;
Point group: $\bar{4}2m$;
Experimental value of the phase-matching angle ($T = 293$ K) and comparison between different sets of dispersion relations:

Interacting wavelengths [μm]	θ_{exp} [deg]	θ_{theor} [deg]	
		[3.78]K	[3.78]E
SHG, o + o ⇒ e			
0.585 ⇒ 0.2925	90 [3.119]	no pm	86.6

Note: [3.78]K ⇒ see [3.78], data of *Kirby* et al.;
    [3.78]E ⇒ see [3.78], data of *Eimerl*

Experimental values of NCPM temperature:

Interacting wavelengths [μm]	T [°C]	Ref.
SHG, o + o ⇒ e		
0.585 ⇒ 0.2925	20	3.119
0.592 ⇒ 0.296	25	3.101

Best set of dispersion relations (λ in μm, $T = 20°C$) [3.78]E:

$$n_o^2 = 1.5985275 - 0.02238475\,\lambda^2$$
$$- 0.0003971065\,\lambda^4 + \frac{0.8226489\,\lambda^2}{\lambda^2 - (0.1402481)^2}\ ,$$

$$n_e^2 = 0.8036475 - 0.0002608396\,\lambda^2$$
$$+ 0.0037782240\,\lambda^4 + \frac{1.4554770\,\lambda^2}{\lambda^2 - (0.1025233)^2}\ .$$

Calculated values of phase-matching and "walk-off" angles:

Interacting wavlengths [μm]	θ_{pm} [deg]	ρ_1 [deg]	ρ_3 [deg]
SHG, o + o ⇒ e			
0.6328 ⇒ 0.3164	68.11		1.453
0.6594 ⇒ 0.3297	63.26		1.666
0.6943 ⇒ 0.34715	58.39		1.827
1.0642 ⇒ 0.5321	39.04		1.893
1.3188 ⇒ 0.6594	37.59		1.818
SFG, o + o ⇒ e			
1.0642 + 0.5321 ⇒ 0.35473	52.89		1.968

$1.3188 + 0.6594 \Rightarrow 0.4396$	42.71		1.971
SHG, $e + o \Rightarrow e$			
$1.0642 \Rightarrow 0.5321$	55.91	1.488	1.762
$1.3188 \Rightarrow 0.6594$	50.00	1.342	1.827
SFG, $e + o \Rightarrow e$			
$1.0642 + 0.5321 \Rightarrow 0.35473$	68.13	1.098	1.392
$1.3188 + 0.6594 \Rightarrow 0.4396$	50.56	1.336	1.923

Effective nonlinearity expressions in the phase-matching direction [3.100]:

$$d_{ooe} = d_{36}\ \sin\theta \sin 2\phi,$$

$$d_{eoe} = d_{oee} = d_{36} \sin 2\theta \cos 2\phi.$$

Laser-induced bulk-damage threshold [3.101]:

λ [μm]	τ_p [ns]	$I_{thr} \times 10^{-12}$ [W/m^2]
0.592	330	2.4

3.3.12 RbH$_2$AsO$_4$, Rubidium Dihydrogen Arsenate (RDA)

Negative uniaxial crystal: $n_o > n_e$;
Point group: $\bar{4}$2m;
Mass density: 3.28 g/cm^3;
Transparency range at "0" transmittance level: $0.22 - 1.82$ μm [3.528];
Transparency range at 0.5 transmittance level for a 14.8 mm long crystal cut at
$\theta = 50°, \phi = 45° : 0.26 - 46$ μm [3.529];
IR edge of transmission spectrum (at "0" transmittance level):
1.65 μm for o – wave, 1.87 μm for e – wave [3.78];
Linear absorption coefficient α:

λ [μm]	α [cm^{-1}]	Ref.	Note
0.3–1.4	0.1–0.2	3.113	
0.3547	0.051	3.529	$\theta = 50°, \phi = 45°$
0.5321	0.031	3.529	$\theta = 50°, \phi = 45°$
1.0642	0.036	3.529	$\theta = 50°, \phi = 45°$

Two-photon absorption coefficient β ($\theta = 90°, \phi = 45°$) [3.71]:

λ [μm]	$\beta \times 10^{13}$ [m/W]	Note
0.355	4.99	e – wave

Experimental values of refractive indices [3.530]:

λ [µm]	n_o	n_e
0.3472	1.5971	1.5531
0.6943	1.5543	

Temperature derivative of refractive indices [3.74]:

λ [µm]	$dn_o/dT \times 10^5$ [K^{-1}]	$dn_e/dT \times 10^5$ [K^{-1}]
0.436	−3.09	−1.97
0.546	−3.62	−2.34
0.578	−3.38	−2.17
0.633	−3.37	−2.35

Experimental values of phase-matching angle ($T = 293$ K) and comparison between different sets of dispersion relations:

Interacting wavelengths [µm]	θ_{exp} [deg]	θ_{theor} [deg]		
		[3.74]	[3.78]K	[3.78]E
SHG, o + o $\Rightarrow$ e				
0.684 $\Rightarrow$ 0.342	90 [3.74]	79.8	83.4	13.8(?)
0.6943 $\Rightarrow$ 0.34715	80 [3.514]	76.1	79.1	13.6(?)
	80.3 [3.530]			
1.0642 $\Rightarrow$ 0.5321	48.8 [3.425]	40.4	49.5	10.3(?)
	50.1* [3.529]			
TGH, o + o $\Rightarrow$ e				
1.0642 + 0.5321				
$\Rightarrow$ 0.3547	66.2* [3.529]	63.8	67.4	12.8(?)

*$T = 298$ K

Note: [3.78]K $\Rightarrow$ see [3.78], data of *Kirby* et al.;
        [3.78]E $\Rightarrow$ see [3.78], data of *Eimerl*.

Experimental values of NCPM temperature:

Interacting wavelengths [µm]	T [°C]	Ref.
SHG, o + o $\Rightarrow$ e		
0.679 $\Rightarrow$ 0.3395	−10	3.425
0.684 $\Rightarrow$ 0.342	20	3.425
0.6943 $\Rightarrow$ 0.34715	92	3.425
	92.6	3.531
	96.5	3.530
	97.4	3.198
0.695 $\Rightarrow$ 0.3475	100	3.119
0.698 $\Rightarrow$ 0.349	110	3.425

Best set of dispersion relations (λ in μm, $T = 293$ K) [3.78]K:

$$n_o^2 = 2.390661 + \frac{3.487176\,\lambda^2}{\lambda^2 - (11.25899)^2} + \frac{0.015513}{\lambda^2 - (0.134582)^2} \, ,$$

$$n_e^2 = 2.275570 + \frac{0.720099\,\lambda^2}{\lambda^2 - (11.25304)^2} + \frac{0.013915}{\lambda^2 - (0.120800)^2} \, .$$

Calculated values of phase-matching and "walk-off" angles:

Interacting wavelengths [μm]	θ_{pm} [deg]	ρ_1 [deg]	ρ_3 [deg]
SHG, $o + o \Rightarrow e$			
$0.6943 \Rightarrow 0.34715$	79.06		0.558
$1.0642 \Rightarrow 0.5321$	49.52		1.367
$1.3188 \Rightarrow 0.6594$	49.53		1.309
SFG, $o + o \Rightarrow e$			
$1.0642 + 0.5321 \Rightarrow 0.35473$	67.35		1.064
$1.3188 + 0.6594 \Rightarrow 0.4396$	53.38		1.372
SHG, $e + o \Rightarrow e$			
$1.0642 \Rightarrow 0.5321$	81.77	0.314	0.385
$1.3188 \Rightarrow 0.6594$	72.53	0.543	0.748
SFG, $e + o \Rightarrow e$			
$1.3188 + 0.6594 \Rightarrow 0.4396$	65.46	0.718	1.073

Experimental values of internal angular and temperature bandwidths:

Interacting wavelengths [μm]	T [°C]	θ_{pm} [deg]	$\Delta\theta^{int}$ [deg]	ΔT [°C]	Ref.
SHG, $o + o \Rightarrow e$					
$0.6943 \Rightarrow 0.34715$	20	80.3	0.126		3.530
	20	80	0.13		3.531
	92.6	90	≈ 2		3.531
	96.5	90	1.57	3.3	3.530
	97.4	90		3.4	3.198
$1.0642 \Rightarrow 0.5321$	25	50.1	0.08		3.529
THG, $o + o \Rightarrow e$					
$1.0642 + 0.5321$					
$\Rightarrow 0.3547$	25	66.2	0.057		3.529

Temperature tuning of noncritical SHG [3.74]:

Interacting wavelengths [μm]	$d\lambda_1/dT$ [nm/K]
SHG, $o + o \Rightarrow e$	
$0.684 \Rightarrow 0.342$	0.136

Temperature variation of birefringence for noncritical SHG process ($0.6943 \ \mu m \Rightarrow 0.3472 \ \mu m$, $o + o \Rightarrow e$):

$$d(n_2^e - n_1^o)/dT = (9.3 \pm 0.4) \times 10^{-6} \mathrm{K}^{-1} [3.530]$$

Effective nonlinearity in the phase-matching direction [3.100]:

$$d_{ooe} = d_{36} \sin \theta \sin 2\phi \,,$$

$$d_{eoe} = d_{oee} = d_{36} \sin 2\theta \cos 2\phi \,.$$

Nonlinear coefficient:

$$d_{36}(0.6943 \ \mu m) = 1.04 \times d_{36}(\mathrm{KDP}) \pm 10\%$$
$$= 0.41 \pm 0.04 \ \mathrm{pm/V} \ [3.198, 37] \,,$$

$$d_{36}(0.6943 \ \mu m) = 0.39 \pm 0.04 \ \mathrm{pm/V} \ [3.530] \,.$$

Laser-induced bulk-damage threshold:

λ [μm]	τ_p [ns]	$I_{thr} \times 10^{-2}$ [W/m^2]	Ref.
0.684	330	1.2	3.101
0.6943	20	3.5	3.530

3.3.13 RbD$_2$AsO$_4$, Deuterated Rubidium Dihydrogen Arsenate (DRDA)

Negative uniaxial crystal: $n_o > n_e$;
Point group: $\bar{4}2m$;
Transparency range at "0" transmittance level: $0.22 - 2.3 \ \mu m$ [3.528];
IR edge of transmission spectrum (at "0" transmittance level):
$2 \ \mu m$ for o – wave, $2.3 \ \mu m$ for e – wave [3.78];
Experimental value of phase-matching angle ($T = 293$ K) and comparison between different sets of dispersion relations:

Interacting wavelengths [μm]	θ_{exp} [deg]	θ_{theor} [deg]	
		[3.78]K	[3.78]E
SHG, $o + o \Rightarrow e$			
$0.698 \Rightarrow 0.349$	90 [3.425]	no pm	no pm
$0.700 \Rightarrow 0.350$		86.9	no pm

Note: [3.78]K $\Rightarrow$ see [3.78], data of *Kirby* et al.;
    [3.78]E $\Rightarrow$ see [3.78], data of *Eimerl*

Experimental values of NCPM temperature [3.425]:

Interacting wavelengths [μm]	T [°C]
SHG, $o + o \Rightarrow e$	
$0.698 \Rightarrow 0.349$	20
$0.714 \Rightarrow 0.357$	100

Best set of dispersion relations (λ in μm, $T = 20\ °C$) [3.78]K:

$$n_o^2 = 2.373255 + \frac{1.979528\ \lambda^2}{\lambda^2 - (11.26884)^2} + \frac{0.015430}{\lambda^2 - (0.125845)^2}\ ,$$

$$n_e^2 = 2.270806 + \frac{0.275372\ \lambda^2}{\lambda^2 - (7.621351)^2} + \frac{0.013592}{\lambda^2 - (0.126357)^2}\ .$$

Calculated values of phase-matching and "walk-off" angles:

Interacting wavelengths [μm]	θ_{pm} [deg]	ρ_1 [deg]	ρ_3 [deg]
SHG, $o + o \Rightarrow e$			
$1.0642 \Rightarrow 0.5321$	46.62		1.278
$1.3188 \Rightarrow 0.6594$	42.98		1.242
SFG, $o + o \Rightarrow e$			
$1.0642 + 0.5321 \Rightarrow 0.35473$	69.79		0.875
$1.3188 + 0.6594 \Rightarrow 0.4396$	52.14		1.272
SHG, $e + o \Rightarrow e$			
$1.0642 \Rightarrow 0.5321$	77.09	0.484	0.547
$1.3188 \Rightarrow 0.6594$	63.77	0.821	0.973
SFG, $e + o \Rightarrow e$			
$1.3188 + 0.6594 \Rightarrow 0.4396$	66.99	0.744	0.935

Effective nonlinearity expressions in the phase-matching direction [3.100]:

$$d_{ooe} = d_{36} \sin\theta \sin 2\phi\,,$$

$$d_{eoe} = d_{oee} = d_{36} \sin 2\theta \cos 2\phi\,.$$

Nonlinear coefficient:

$$d_{36} \approx 0.31\ \text{pm/V}\ [3.78]$$

Laser-induced bulk-damage threshold [3.101]:

λ [μm]	τ_p [ns]	$I_{thr} \times 10^{-12}$ [W/m^2]
0.7	330	0.21

3.3.14 LiCOOH · H₂O, Lithium Formate Monohydrate (LFM)

Negative biaxial crystal: $2V_Z = 123.8°$ at $\lambda = 0.5321$ μm [3.532];
Point group: mm2;
Assignment of dielectric and crystallographic axes:
$X, Y, Z \Rightarrow a, b, c;$
Mass density: 1.46 g/cm³ [3.532];
Transparency range at "0" transmittance level: 0.23 – 1.56 μm [3.532, 533];
Linear absorption coefficient α ($\theta = 90°, \phi = 10°$) [3.534]:

λ [μm]	α [cm⁻¹]
0.3547	0.025
0.5321	0.012
1.0642	0.017

Experimental values of refractive indices [3.535]:

λ [μm]	n_X	n_Y	n_Z	λ [μm]	n_X	n_Y	n_Z
0.35	1.3810	1.5073	1.5540	0.60	1.3643	1.4796	1.5174
0.36	1.3791	1.5051	1.5510	0.62	1.3638	1.4787	1.5161
0.37	1.3777	1.5034	1.5484	0.64	1.3633	1.4778	1.5152
0.38	1.3767	1.5017	1.5458	0.66	1.3628	1.4768	1.5144
0.39	1.3758	1.4999	1.5432	0.68	1.3625	1.4760	1.5135
0.40	1.3748	1.4981	1.5405	0.70	1.3623	1.4751	1.5126
0.42	1.3729	1.4955	1.5367	0.80	1.3614	1.4729	1.5099
0.44	1.3714	1.4928	1.5332	0.90	1.3604	1.4711	1.5077
0.46	1.3705	1.4902	1.5301	1.00	1.3595	1.4694	1.5055
0.48	1.3696	1.4880	1.5279	1.10	1.3590	1.4675	1.5032
0.50	1.3686	1.4862	1.5257	1.20	1.3587	1.4658	1.5011
0.52	1.3677	1.4845	1.5236	1.30	1.3585	1.4644	1.4987
0.54	1.3666	1.4827	1.5219	1.40	1.3583	1.4630	1.4970
0.56	1.3657	1.4813	1.5200	1.50	1.3581	1.4617	
0.58	1.3647	1.4804	1.5187				

Sellmeier equations (λ in μm, $T = 20$ °C) [3.535]:

$$n_X^2 = 1.4376 + \frac{0.4045\ \lambda^2}{\lambda^2 - 0.01692601} - 0.0005\ \lambda^2 \, ,$$

$$n_Y^2 = 1.6586 + \frac{0.5006\ \lambda^2}{\lambda^2 - 0.023409} - 0.0127\ \lambda^2 \, ,$$

$$n_Z^2 = 1.6714 + \frac{0.5928\ \lambda^2}{\lambda^2 - 0.02534464} - 0.0153\ \lambda^2 \, .$$

Comparison between experimental and theoretical values of phase-matching angle:

XY plane, $\theta = 90°$

Interacting wavelengths [μm]	$\phi_{\exp}$ [deg]	ϕ_{theor} [deg] [3.535]
SFG, e + o $\Rightarrow$ e		
1.0642 + 0.5321 $\Rightarrow$ 0.3547	8.2 [3.534]	9.5

XZ plane, $\phi = 0°$

Interacting wavelengths [μm]	$\theta_{\exp}$ [deg]	θ_{theor} [deg] [3.535]
SHG, o + o $\Rightarrow$ e		
0.486 $\Rightarrow$ 0.243	38.5 [3.536]	36.8
1.0642 $\Rightarrow$ 0.5321	55.1 [3.532]	56.0
SHG, o + e $\Rightarrow$ o		
1.0642 $\Rightarrow$ 0.5321	82.0 [3.532]	80.4

Calculated values of phase-matching and "walk-off" angles:

XY plane, $\theta = 90°$

Interacting wavelengths [μm]	ϕ_{pm}[deg]	ρ_1[deg]	ρ_2[deg]	ρ_3[deg]
SHG, e + o $\Rightarrow$ e				
0.5105 $\Rightarrow$ 0.25525	47.94	4.639		5.783
0.5321 $\Rightarrow$ 0.26605	44.15	4.689		5.712
0.5782 $\Rightarrow$ 0.2891	37.38	4.574		5.368
0.6943 $\Rightarrow$ 0.34715	24.96	3.683		4.103
SFG, eo $\Rightarrow$ e				
0.5782 + 0.5105 $\Rightarrow$ 0.27112	40.23	4.637		5.641
1.0642 + 0.5321 $\Rightarrow$ 0.35473	9.49	1.545		1.786
SFG, o + e $\Rightarrow$ e				
0.5782 + 0.5105 $\Rightarrow$ 0.27112	44.78		4.705	5.631
1.0642 + 0.5321 $\Rightarrow$ 0.35473	33.34		4.442	4.780

XZ plane, $\phi = 0°$, $\theta < V_Z$

Interacting wavelengths [μm]	θ_{pm} [deg]	ρ_3 [deg]
SHG, o + o $\Rightarrow$ e		
0.5105 $\Rightarrow$ 0.25525	39.44	7.722
0.5321 $\Rightarrow$ 0.26605	41.38	7.603
0.5782 $\Rightarrow$ 0.2891	44.69	7.341
0.6943 $\Rightarrow$ 0.34715	50.00	6.784

$1.0642 \Rightarrow 0.5321$	55.98	5.937
$1.3188 \Rightarrow 0.6594$	56.86	5.731
SFG, $o + o \Rightarrow e$		
$0.5105 + 0.5782 \Rightarrow 0.27112$	46.42	7.721
$1.0642 + 0.5321 \Rightarrow 0.35473$	51.41	6.705
$1.3188 + 0.6594 \Rightarrow 0.4396$	54.66	6.209

XZ plane, $\phi = 0°, \theta > V_Z$

Interacting wavelengths [µm]	θ_{pm} [deg]	ρ_2 [deg]
SHG, $o + e \Rightarrow o$		
$1.0642 \Rightarrow 0.5321$	80.42	2.087
$1.3188 \Rightarrow 0.6594$	76.68	2.759

Experimental value of internal angular bandwidth [3.534]:
XY plane, $\theta = 90°$

Interacting wavelengths [µm]	θ_{pm} [deg]	$\Delta\phi^{int}$ [deg]
SFG, $e + o \Rightarrow e$		
$1.0642 + 0.5321 \Rightarrow 0.3547$	8.2	0.04

Effective nonlinearity expressions in the phase-matching direction for three-wave interactions in the principal planes of LFM crystal [3.35, 36]:

XY plane

$$d_{eoe} = d_{oee} = d_{31} \sin^2 \phi + d_{32} \cos^2 \phi\,;$$

YZ plane

$$d_{eoe} = d_{oee} = d_{31} \sin \theta\,;$$

XZ plane, $\theta < V_Z$

$$d_{ooe} = d_{32} \sin \theta\,;$$

XZ plane, $\theta > V_Z$

$$d_{oeo} = d_{eoo} = d_{32} \sin \theta\,.$$

Effective nonlinearity expressions for three-wave interactions in the arbitrary direction of LFM crystal are given in [3.36]
Nonlinear coefficients [3.37]:

$$d_{31}(1.0642\ \mu m) = 0.13\ pm/V\,,$$

$$d_{32}(1.0642\ \mu m) = -0.60\ pm/V\,,$$

$$d_{33}(1.0642\ \mu m) = 0.94\ pm/V\,.$$

Laser-induced surface-damage threshold:

λ [µm]	τ_p [ns]	$I_{thr} \times 10^{-12}$ [W/m^2]	Ref.
0.475	330	1.5	3.101
0.488	cw	> 0.00001	3.532
0.490	330	1.5	3.101

3.3.15 NaCOOH, Sodium Formate

Negative biaxial crystal: $2V_Z = 92.5°$ at $\lambda = 0.54$ µm [3.533] ;
Point group: mm2;
Assignment of dielectric and crystallographic axes:
$X, Y, Z \Rightarrow a, b, c$;
Transparency range at "0" transmittance level: 0.23–2.2 µm [3.533] ;
Linear absorption coefficient α (along X axis) [3.537]:

λ [µm]	α [cm^{-1}]
0.3547	0.013
0.5321	0.003
1.0642	0.010

The graph of n_X, n_Y, n_Z dependences versus wavelength is given in [3.533] ($n_X < n_Y < n_Z$).
Sellmeier equations (λ in µm, $T = 20$ °C) [3.533]:

$$n_X^2 = 1.2646 + \frac{0.6381 \, \lambda^2}{\lambda^2 - 0.01212201} - 0.0011 \, \lambda^2 ,$$

$$n_Y^2 = 1.2589 + \frac{0.8423 \, \lambda^2}{\lambda^2 - 0.01447209} - 0.0005 \, \lambda^2 ,$$

$$n_Z^2 = 1.2515 + \frac{1.0729 \, \lambda^2}{\lambda^2 - 0.01726596} - 0.0013 \, \lambda^2 .$$

Experimental and theoretical values of phase-matching angle and calculated values of "walk-off" angle:
XY plane, $\theta = 90°$

Interacting wavelengths [µm]	ϕ_{exp} [deg]	ϕ_{theor} [deg] [3.533]	ρ_2 [deg]	ρ_3 [deg]
SFG, o + e $\Rightarrow$ e				
1.0642 + 0.5321 $\Rightarrow$ 0.3547	2.2 [3.537]	4.61	0.512	0.559

Experimental values of internal angular bandwidth [3.537]:
XY plane, $\theta = 90°$

Interacting wavelengths [µm]	θ_{pm} [deg]	$\Delta\phi^{int}$ [deg]	$\Delta\theta^{int}$ [deg]
SFG, $o + e \Rightarrow e$			
$1.0642 + 0.5321 \Rightarrow 0.3547$	2.2	0.75	1.8

Effective nonlinearity expressions in the phase-matching direction for three-wave interactions in the principal planes of NaCOOH crystal [3.35, 36]:
XY plane

$$d_{eoe} = d_{oee} = d_{31} \sin^2 \phi + d_{32} \cos^2 \phi \,,$$

YZ plane

$$d_{oeo} = d_{eoo} = d_{31} \sin \theta \,,$$

XZ plane, $\theta < V_Z$

$$d_{ooe} = d_{32} \sin \theta \,,$$

XZ plane, $\theta > V_Z$

$$d_{oeo} = d_{eoo} = d_{32} \sin \theta \,.$$

Effective nonlinearity expressions for three-wave interactions in the arbitrary direction of NaCOOH crystal are given in [3.36]
Nonlinear coefficients:

$$|d_{32}(1.0642 \ \mu m + 0.5321 \ \mu m \Rightarrow 0.3547 \ \mu m)|$$
$$= 1.2 \times d_{36}(\text{KDP}) \pm 20\% = 0.47 \pm 0.09 \ \text{pm/V} \ [3.537, 515, 37] \,;$$

$$d_{31}(1.0642 \ \mu m) \approx 0.047 \ \text{pm/V} \ [3.533, 537, 515, 198, 37] \,;$$

$$d_{32}(1.0642 \ \mu m) = -0.47 \pm 0.09 \ \text{pm/V} \ [3.537, 515, 198, 37] \,;$$

$$d_{33}(1.0642 \ \mu m) \approx 0.70 \ \text{pm/V} \ [3.533, 537, 515, 198, 37] \,.$$

Laser-induced surface-damage threshold [3.537]:

λ [µm]	τ_p [ns]	$I_{thr} \times 10^{-12}$ [W/m^2]	Note
0.3547	8	> 1.2	10 Hz
0.5321	10	> 1.4	10 Hz
1.0642	12	> 1.2	10 Hz

3.3.16 Ba(COOH)$_2$, Barium Formate

Positive biaxial crystal: $2V_Z = 101.3°$ at $\lambda = 0.5321$ μm [3.512];
Point group: 222;
Assignment of dielectric and crystallographic axes:
$X, Y, Z \Rightarrow a, b, c$;
Transparency range: $0.245 - 2.2$ μm [3.512];
Experimental values of refractive indices [3.512]:

λ [μm]	n_X	n_Y	n_Z
0.5321	1.6407	1.6019	1.5773
1.0642	1.6214	1.5819	1.5585

Sellmeier equations (λ in μm, $T = 20$ °C) [3.512]:

$$n_X^2 = 2.619 + \frac{0.0177}{\lambda^2 - 0.039} \; ;$$

$$n_Y^2 = 2.491 + \frac{0.0184}{\lambda^2 - 0.035} \; ;$$

$$n_Z^2 = 2.421 + \frac{0.0160}{\lambda^2 - 0.042} \; .$$

Experimental and theoretical values of phase-matching angle and calculated values of "walk-off" angle:
XZ plane, $\phi = 0°, \theta < V_Z$

Interacting wavelengths [μm]	$\theta_{\exp}$ [deg]	θ_{theor} [deg] [3.512]	ρ_1 [deg]
SHG, e + e $\Rightarrow$ o			
$1.0642 \Rightarrow 0.5321$	33.3 [3.512]	35.26	2.153

Effective nonlinearity expressions in the phase-matching direction for three-wave interactions in the principal planes of Ba(COOH)$_2$ crystal [3.35]:
XY plane

$$d_{\text{eee}} = d_{14} \sin 2\phi \; ;$$

YZ plane

$$d_{\text{eoe}} = d_{\text{oee}} = d_{14} \sin 2\theta \; ;$$

XZ plane, $\theta < V_Z$

$$d_{\text{eeo}} = d_{14} \sin 2\theta \; ;$$

XZ plane, $\theta > V_Z$

$$d_{\text{eoe}} = d_{\text{oee}} = d_{14} \sin 2\theta \; .$$

Nonlinear coefficient:

$$d_{14}(1.064 \text{ μm}) = 0.27 \times d_{36}(\text{KDP}) \pm 15\% = 0.105 \pm 0.016 \text{ pm/V} \text{ [3.512, 37]}$$

3.3.17 Sr(COOH)$_2$, Strontium Formate

Positive biaxial crystal: $2V_Z = 78.8°$ at $\lambda = 0.532$ μm [3.94];
Point group: 222;
Assignment of dielectric and crystallographic axes:
$X, Y, Z \Rightarrow c, a, b$;
Mass density: 2.69 g/cm^3;
Transparency range at "0" transmittance level: 0.25 – 1.7 μm [3.94].
Linear absorption coefficient α [3.94]:

λ [μm]	α [cm^{-1}]
0.235	> 15
0.250	2

Experimental values of refractive indices [3.94]:

λ [μm]	n_X	n_Y	n_Z
0.266	1.613	1.635	1.675
0.3547	1.569	1.587	1.612
0.532	1.545	1.560	1.583
1.064	1.528	1.543	1.563

Experimental and theoretical values of phase-matching angle and calculated values of "walk-off" angle:

YZ plane, $\phi = 90°$

Interacting wavelengths [μm]	θ_{exp} [deg]	θ_{theor} [deg]	ρ_1 [deg]
SHG, e + e $\Rightarrow$ o			
1.064 $\Rightarrow$ 0.532	26 [3.94]	18.60[#]	0.442

XZ plane, $\phi = 0°, \theta > V_Z$

Interacting wavelengths [μm]	θ_{exp} [deg]	θ_{theor} [deg]	ρ_1 [deg]
SHG, e + e $\Rightarrow$ o			
1.064 $\Rightarrow$ 0.532	72.5 [3.94]	73.25[#]	0.730

[#]derived from experimental data on refractive indices [3.94].

Experimental value of internal angular bandwidth [3.94]:
YZ plane, $\phi = 90°$

Interacting wavelengths [µm]	θ_{pm} [deg]	$\Delta\theta^{int}$ [deg]
SHG, $e + e \Rightarrow o$		
$1.064 \Rightarrow 0.532$	26	0.204

Effective nonlinearity expressions in the phase-matching direction for three-wave interactions in the principal planes on $Sr(COOH)_2$ crystal [3.35]:
XY plane

$$d_{eoe} = d_{oee} = d_{14}\sin 2\phi\,;$$

YZ plane

$$d_{eeo} = d_{14}\sin 2\theta\,;$$

XZ plane, $\theta < V_Z$

$$d_{eoe} = d_{oee} = d_{14}\sin 2\theta\,;$$

XZ plane, $\theta > V_Z$

$$d_{eeo} = d_{14}\sin 2\theta\,.$$

Nonlinear coefficient:

$$d_{14}(1.064\ \mu m) = 1.25 \times d_{36}\ (KDP) \pm 16\% = 0.49 \pm 0.08\ pm/V\ [3.94, 37]\,.$$

Laser-induced damage threshold [3.94]:

λ [µm]	τ_p [ns]	$I_{thr} \times 10^{-12}$ [W/m^2]
1.064	≈ 20	> 1.5

3.3.18 $Sr(COOH)_2 \cdot 2H_2O$, Strontium Formate Dihydrate

Negative biaxial crystal: $2V_Z = 64.6°$ at $\lambda = 0.532$ µm [3.94];
Point group: 222;
Assignment of dielectric and crystallographic axes:
$X, Y, Z \Rightarrow a, b, c\,;$
Mass density: 2.25 g/cm^3 [3.94];
Transparency range at "0" transmittance level: $0.25 - 1.4$ µm [3.94];
Linear absorption coefficient α [3.94]:

λ [µm]	α [cm^{-1}]
0.235	> 15
0.250	2

Experimental values of refractive indices [3.94]:

λ [µm]	n_X	n_Y	n_Z
0.266	1.621	1.598	1.543
0.3547	1.570	1.553	1.509
0.532	1.542	1.526	1.488
1.064	1.525	1.509	1.477

Experimental and theoretical values of phase-matching angle and calculated values of "walk-off" angle:

YZ plane, $\phi = 90°$

Interacting wavelengths [µm]	θ_{exp} [deg]	θ_{theor} [deg]	ρ_1 [deg]	ρ_3 [deg]
SHG, e + o $\Rightarrow$ e				
1.064 $\Rightarrow$ 0.532	46 [3.94]	38.56[#]	1.203[#]	1.405[#]
SFG, e + o $\Rightarrow$ e				
1.064 + 0.532 $\Rightarrow$				
$\Rightarrow$ 0.35467	58.5 [3.94]	53.60[#]	1.165[#]	1.559[#]

XZ plane, $\phi = 0°, \theta > V_Z$

Interacting wavelengths [µm]	θ_{exp} [deg]	θ_{theor} [deg]	ρ_1 [deg]	ρ_3 [deg]
SHG, e + o $\Rightarrow$ e				
1.064 $\Rightarrow$ 0.532	71 [3.94]	65.07[#]	1.372[#]	1.525[#]

[#]derived from experimental data on refractive indices [3.94]:

Experimental value of internal angular bandwidth [3.94]:

YZ plane, $\phi = 90°$

Interacting wavelengths [µm]	θ_{pm} [deg]	$\Delta\theta^{int}$ [deg]
SHG, e + o $\Rightarrow$ e		
1.064 $\Rightarrow$ 0.532	46	0.142

Effective nonlinearity in the phase-matching direction for three-wave interactions in the principal planes of $Sr(COOH)_2 \cdot 2H_2O$ crystal [3.35]:

XY plane

$$d_{eeo} = d_{14} \sin 2\phi \, ;$$

YZ plane

$$d_{eoe} = d_{oee} = d_{14} \sin 2\theta \, ;$$

XZ plane, $\theta < V_Z$

$$d_{\mathrm{eeo}} = d_{14} \sin 2\theta \, ;$$

XZ plane, $\theta > V_Z$

$$d_{\mathrm{eoe}} = d_{\mathrm{oee}} = d_{14} \sin(2\theta) \, .$$

Nonlinear coefficient:

$$d_{14}(1.064 \ \mu\mathrm{m}) = 0.8 \times d_{36} \ (\mathbf{KDP}) \pm 25\% = 0.31 \pm 0.08 \ \mathrm{pm/V} \ [3.94, \ 37] \, .$$

Laser-induced damage threshold [3.94]:

λ [μm]	τ_{p} [ns]	$I_{\mathrm{thr}} \times 10^{-12}$ [W/m^2]
1.064	≈ 20	> 1.5

3.3.19 LiGaO$_2$, Lithium Gallium Oxide

Negative biaxial crystal: $2V_Z = 74.5°$ at $\lambda = 0.5 \ \mu$m [3.538];
Point group: mm2;
Assignment of dielectric and crystallographic axes:
$X, Y, Z \Rightarrow b, c, a$ [3.538];
Mass density: 4.187 g/cm^3 [3.64];
Mohs hardness: 7.5 [3.64]
Transparency range: $0.3 - 5 \ \mu$m [3.539]
Experimental values of refractive indices:

λ [μm]	n_X	n_Y	n_Z	Ref.	λ [μm]	n_X	n_Y	n_Z	Ref.
0.41	1.7702			3.539	1.00	1.7160			3.539
0.47	1.7534	1.7835	1.7852	3.538	1.20	1.7122			3.539
0.50	1.7477	1.7768	1.7791	3.538	1.40	1.7095			3.539
0.54	1.7407	1.7683	1.7708	3.538	1.60	1.7070			3.539
0.58	1.7351	1.7626	1.7653	3.538	1.80	1.7045			3.539
0.62	1.7311	1.7589	1.7617	3.538	2.00	1.7025			3.539
0.66	1.7289	1.7578	1.7604	3.538	2.20	1.7005			3.539
0.70	1.7268			3.539	2.40	1.6978			3.539
0.80	1.7218			3.539	2.60	1.6955			3.539
0.90	1.7185			3.539					

Effective nonlinearity expressions in the phase-matching direction for three-wave interactions in the principal planes of LiGaO$_2$ crystal [3.35, 36]:

XY plane

$$d_{\mathrm{ooe}} = d_{31} \cos \phi \, ;$$

YZ plane

$$d_{oeo} = d_{eoo} = d_{32} \cos \theta \, ;$$

XZ plane, $\theta < V_Z$

$$d_{oee} = d_{eoe} = d_{31} \sin^2 \theta + d_{32} \cos^2 \theta \, ;$$

XZ plane, $\theta > V_Z$

$$d_{eeo} = d_{31} \sin^2 \theta + d_{32} \cos^2 \theta \, .$$

Effective nonlinearity expressions for three-wave interactions in the arbitrary direction of $LiGaO_2$ crystal are given in [3.36].

Nonlinear coefficients:

$$d_{31}(1.0642 \ \mu m) = \pm 0.17 \times d_{36} \ (\text{KDP}) \pm 10\%$$
$$= \pm 0.066 \pm 0.007 \ \text{pm/V} \ [3.539, 37] \, ,$$

$$d_{32}(1.0642 \ \mu m) = \mp 0.37 \times d_{36} \ (\text{KDP}) \pm 10\%$$
$$= \mp 0.144 \pm 0.014 \ \text{pm/V} \ [3.539, 37] \, ,$$

$$d_{33}(1.0642 \ \mu m) = \pm 1.45 \times d_{36} \ (\text{KDP}) \pm 10\%$$
$$= \pm 0.566 \pm 0.057 \ \text{pm/V} \ [3.539, 37] \, .$$

3.3.20 α-HIO$_3$, α-Iodic Acid

Negative biaxial crystal: $2V_Z = 47°$ [3.540];
Point group: 222;
Assignment of dielectric and crystallographic axes:
$X, Y, Z \Rightarrow b, c, a \, ;$
Mass density: 4.63 g/cm^3 [3.540];
Transparency range at "0" transmittance level: $0.32 - 1.7 \ \mu m \ (\parallel c)$,
$0.32 - 2.3 \ \mu m \ (\perp c)$ [3.540];
Linear absorption coefficient $\alpha : < 0.3 \ \text{cm}^{-1}$ in the range $0.35 - 1.3 \ \mu m$ [3.541];
Experimental values of refractive indices at $T = 293$ K [3.542]:

λ [μm]	n_X	n_Y	n_Z	λ [μm]	n_X	n_Y	n_Z
0.35	2.1485	2.1265	1.9612	0.42	2.0637	2.0394	1.8952
0.36	2.1330	2.1077	1.9474	0.44	2.0494	2.0246	1.8847
0.37	2.1171	2.0917	1.9360	0.46	2.0378	2.0119	1.8753
0.38	2.1053	2.0782	1.9257	0.48	2.0292	2.0026	1.8685
0.39	2.0929	2.0662	1.9154	0.50	2.0194	1.9926	1.8624
0.40	2.0808	2.0545	1.9086	0.52	2.0126	1.9883	1.8562
0.41	2.0715	2.0465	1.9020	0.54	2.0065	1.9829	1.8522

λ [μm]	n_X	n_Y	n_Z	λ [μm]	n_X	n_Y	n_Z
0.56	2.0010	1.9763	1.8476	0.90	1.9602	1.9346	1.8202
0.58	1.9960	1.9712	1.8436	0.95	1.9569	1.9314	1.8184
0.60	1.9918	1.9665	1.8405	1.00	1.9541	1.9286	1.8150
0.62	1.9884	1.9632	1.8388	1.10	1.9486	1.9260	1.8114
0.64	1.9854	1.9589	1.8368	1.20	1.9436	1.9229	1.8088
0.66	1.9821	1.9560	1.8348	1.30	1.9390	1.9206	1.8063
0.68	1.9791	1.9529	1.8328	1.40	1.9348	1.9180	1.8038
0.70	1.9763	1.9506	1.8311	1.50	1.9310	1.9157	1.8018
0.80	1.9668	1.9409	1.8248	1.60		1.9132	1.7998
0.85	1.9634	1.9377	1.8222				

Optical activity at $T = 300$ K [3.540]:

λ [μm]	ρ [deg/mm]
0.4360	74.5
0.5461	58.7

Experimental values of phase-matching angle ($T = 293$ K) and comparison
between different sets of dispersion relations:
YZ plane, $\phi = 90°$

Interacting wavelengths [μm]	θ_{exp} [deg]	θ_{theor} [deg]		
		[3.458]	[3.542]	[3.543]
SHG, e + o $\Rightarrow$ e				
0.976 $\Rightarrow$ 0.488	57.9 [3.544]	56.9	57.5	58.1
1.029 $\Rightarrow$ 0.5145	52.7 [3.544]	51.9	52.5	52.9
1.0642 $\Rightarrow$ 0.5321	50.4 [3.545]	49.3	49.8	49.9
1.065 $\Rightarrow$ 0.5325	52 [3.540]	49.3	49.8	49.9

XZ plane, $\phi = 0°, \theta > V_Z$

Interacting wavelengths [μm]	θ_{exp} [deg]	θ_{theor} [deg]		
		[3.458]	[3.542]	[3.543]
SHG, e + o $\Rightarrow$ e				
0.976 $\Rightarrow$ 0.488	72.2 [3.544]	71.2	71.4	72.4
1.029 $\Rightarrow$ 0.5145	66.1 [3.544]	65.0	65.4	66.3
1.06 $\Rightarrow$ 0.53	64.9 [3.199]	62.4	62.9	63.6
1.065 $\Rightarrow$ 0.5325	66 [3.540]	62.1	62.5	63.2

Best set of Sellmeier equations (λ in µm, $T = 293$ K) [3.543]:

$$n_X^2 = 3.739 + \frac{0.07128}{\lambda^2 - 0.05132} \, ,$$

$$n_Y^2 = 3.654 + \frac{0.06721}{\lambda^2 - 0.04234} \, ,$$

$$n_Z^2 = 3.239 + \frac{0.05353}{\lambda^2 - 0.017226} \, .$$

Calculated values of phase-matching and "walk-off" angles:
YZ plane, $\phi = 90°$

Interacting wavelengths [µm]	θ_{pm} [deg]	ρ_1 [deg]	ρ_3 [deg]
SHG, e + o $\Rightarrow$ e			
1.0642 $\Rightarrow$ 0.5321	49.92	3.416	3.725
1.3188 $\Rightarrow$ 0.6594	34.55	3.324	3.484

XZ plane, $\phi = 0°, \theta > V_Z$

Interacting wavelengths [µm]	θ_{pm} [deg]	ρ_1 [deg]	ρ_3 [deg]
SHG, e + o $\Rightarrow$ e			
1.0642 $\Rightarrow$ 0.5321	63.21	3.224	3.557
1.3188 $\Rightarrow$ 0.6594	49.22	4.058	4.278

Experimental values of internal angular and spectral bandwidths [3.96]:
XZ plane, $\phi = 0°, \theta > V_Z$

Interacting wavelengths [µm]	θ_{pm} [deg]	$\Delta\theta^{\text{int}}$ [deg]	$\Delta\nu$ [cm^{-1}]
SHG, e + o $\Rightarrow$ e			
1.06 $\Rightarrow$ 0.53	66	0.035	3.38

Temperature tuning of critical SFG process [3.544]:
XZ plane, $\phi = 0°$

Interacting wavelength [µm]	θ_{pm} [deg]	$d\lambda_2/dT$ [nm/K]
SHG, e + o $\Rightarrow$ e		
1.9226 + 0.654 $\Rightarrow$ 0.488	50	0.055

Effective nonlinearity expressions in the phase-matching direction for three-wave interactions in the principal planes of α-HIO$_3$ crystal [3.35]:

XY plane

$$d_{\text{eeo}} = d_{14} \sin 2\phi \; ;$$

YZ plane

$$d_{\text{eoe}} = d_{\text{oee}} = d_{14} \sin 2\theta \; ;$$

XZ plane, $\theta < V_Z$

$$d_{\text{eeo}} = d_{14} \sin 2\theta \; ;$$

XZ plane, $\theta > V_Z$

$$d_{\text{eoe}} = d_{\text{oee}} = d_{14} \sin 2\theta \; .$$

Nonlinear coefficient:

$$d_{14}(1.064 \; \mu\text{m}) = 20 \times d_{11}(\text{SiO}_2) \pm 25\%$$
$$= 6.0 \pm 1.5 \; \text{pm/V} \, [3.540], [3.37] \; ,$$

$$d_{14}(1.1523 \; \mu\text{m}) = 10.9 \times d_{36}(\text{ADP}) \pm 14\%$$
$$= 5.1 \pm 0.7 \; \text{pm/V} \, [3.546, 37] \; .$$

Laser-induced surface-damage threshold:

λ [μm]	τ_{p} [ns]	$I_{\text{thr}} \times 10^{-12}$ [W/m^2]	Ref.	Note
0.488	cw	>0.0025	3.540	
0.528	0.007	>70	3.68	2Hz
0.53	15	0.55	3.199	
	0.006	>8	3.547	
0.532	0.03	>8	3.548	25 Hz
	0.03	>55	3.549	
	0.035	80–100	3.222	1 Hz
	0.035	40–50	3.222	12.5 Hz

3.3.21 K$_2$La(NO$_3$)$_5 \cdot$ 2H$_2$O, Potassium Lanthanum Nitrate Dihydrate (KLN)

Negative biaxial crystal: $2V_Z = 111°$ at $\lambda = 0.5461 \; \mu\text{m}$ [3.550]
Point group: mm2;
Assignment of dielectric and crystallographic axes:
$X, Y, Z \Rightarrow b, c, a$;
Transparency range at "0" transmittance level: $0.335 - > 1.1 \; \mu\text{m}$ [3.550];
Linear absorption coefficient:

$$\alpha < 0.03 \, \text{cm}^{-1} \; \text{at} \; \lambda = 1.064 \, \mu\text{m} \, [3.550];$$

Experimental values of refractive indices [3.550]:

λ [µm]	n_X	n_Y	n_Z
0.3650	1.5297	1.5820	1.6063
0.4005	1.5201	1.5702	1.5936
0.4872	1.5062	1.5530	1.5760
0.5461	1.5008	1.5456	1.5682
0.6476	1.4950	1.5387	1.5601
0.7500	1.4915	1.5341	1.5556
0.8500	1.4891	1.5306	1.5518
0.9500	1.4872	1.5285	1.5496
1.0500	1.4857	1.5269	1.5475

Sellmeier equations (λ in µm, $T = 20\ °C$) [3.550]:

$$n_X^2 = 2.20094 + \frac{0.0142619}{\lambda^2 - 0.0313420} - 0.00617543\,\lambda^2\ ,$$

$$n_Y^2 = 2.31901 + \frac{0.0200108}{\lambda^2 - 0.0247406} - 0.00586460\,\lambda^2\ ,$$

$$n_Z^2 = 2.38504 + \frac{0.0208525}{\lambda^2 - 0.0269388} - 0.00873084\,\lambda^2\ .$$

Experimental and theoretical values of phase-matching angle and calculated values of "walk-off" angle:

XY plane, $\theta = 90°$

Interacting wavelengths [µm]	ϕ_{exp} [deg]	ϕ_{theor} [deg] [3.550]	ρ_3 [deg]
SHG, o + o $\Rightarrow$ e			
1.0642 $\Rightarrow$ 0.5321	0.8 [3.550]	4.17	0.26
SFG, o + o $\Rightarrow$ e			
1.0642 + 0.5321 $\Rightarrow$ 0.35473	42.6 [3.550]	41.64	1.94

YZ plane, $\phi = 90°$

Interacting wavelengths [µm]	θ_{exp} [deg]	θ_{theor} [deg] [3.550]	ρ_2 [deg]
SHG, o + e $\Rightarrow$ o			
1.0642 + 0.5321 $\Rightarrow$ 0.35473	42.1 [3.550]	41.69	0.81

XZ plane, $\phi = 0°, \theta < V_Z$

Interacting wavelengths [μm]	θ_{exp} [deg]	θ_{theor} [deg] [3.550]	ρ_1 [deg]	ρ_3 [deg]
SHG, e + e $\Rightarrow$ o 1.0642 $\Rightarrow$ 0.5321	19.8 [3.550]	20.42	1.48	1.60

Experimental values of internal angular bandwidth [3.550]:
XY plane, $\theta = 90°$

Interacting wavelengths [μm]	ϕ_{pm} [deg]	$\Delta\phi^{\mathrm{int}}$ [deg]
SHG, o + o $\Rightarrow$ e 1.0642 $\Rightarrow$ 0.5321	0.8	1.107

XZ plane, $\phi = 0°, \theta < V_Z$

Interacting wavelengths [μm]	θ_{pm} [deg]	$\Delta\theta^{\mathrm{int}}$ [deg]
SHG, e + o $\Rightarrow$ e 1.0642 $\Rightarrow$ 0.5321	19.8	0.123

Effective nonlinear expressions in the phase-matching direction for three-wave interactions in the principal planes of KLN crystal [3.35,36]:

XY plane

$$d_{\mathrm{ooe}} = d_{31} \cos \phi \; ;$$

YZ plane

$$d_{\mathrm{oeo}} = d_{\mathrm{eoo}} = d_{32} \cos \theta \; ;$$

XZ plane, $\theta < V_Z$

$$d_{\mathrm{oee}} = d_{\mathrm{eoe}} = d_{31} \sin^2 \theta + d_{32} \cos^2 \theta \; ;$$

XZ plane, $\theta > V_Z$

$$d_{\mathrm{eeo}} = d_{31} \sin^2 \theta + d_{32} \cos^2 \theta \; .$$

Effective nonlinearity expressions for three-wave interactions in the arbitrary direction of KLN crystal are given in [3.36]

Nonlinear coefficients [3.550]:

$$d_{31}(1.0642\,\mu\mathrm{m}) = \mp 1.13 \pm 0.15\,\mathrm{pm/V} \; ,$$

$$d_{32}(1.0642\,\mu m) = \pm 1.10 \pm 0.10\,\text{pm/V},$$

$$|d_{33}(1.0642\,\mu m)| = 0.13 \pm 0.10\,\text{pm/V}\,.$$

3.3.22 CsTiOAsO$_4$, Cesium Titanyl Arsenate (CTA)

Positive biaxial crystal: $2V_Z = 52.9°$ at $\lambda = 0.5321\,\mu m$ [3.551];
Point group: mm2;
Assignment of dielectric and crystallographic axes:
$X, Y, Z \Rightarrow a, b, c$;
Transparency range at "0" transmittance level: $0.35 - 5.3\,\mu m$[3.551];
Sellmeier equations (λ in μm, $T = 20\,°C$) [474];

$$n_X^2 = 2.34498 + \frac{1.04863\,\lambda^2}{\lambda^2 - (0.22044)^2} - 0.01483\,\lambda^2\,,$$

$$n_Y^2 = 2.74440 + \frac{0.70733\,\lambda^2}{\lambda^2 - (0.26033)^2} - 0.01526\,\lambda^2\,,$$

$$n_Z^2 = 2.53666 + \frac{1.10600\,\lambda^2}{\lambda^2 - (0.24988)^2} - 0.01711\,\lambda^2\,.$$

Experimental and theoretical values of phase-matching angle and calculated values of "walk off" angle:
XY plane, $\theta = 90°$

Interacting wavelengths [μm]	ϕ_{exp} [deg]	ϕ_{theor} [deg] [3.551]	ρ_1 [deg]	ρ_3 [deg]
SHG, e + o $\Rightarrow$ e 1.3188 $\Rightarrow$ 0.6594	64.5 [3.551]	62.85	0.378	0.369

Experimental value of internal angular bandwidth [3.551]:
XY plane, $\theta = 90°$

Interacting wavelengths [μm]	ϕ_{pm} [deg]	$\Delta\phi^{int}$ [deg]
SHG, e + o $\Rightarrow$ e 1.3188 $\Rightarrow$ 0.6594	64.5	0.5

Effective nonlinearity expressions in the phase-matching direction for three-wave interactions in the principal planes of CTA crystal [3.35,36]:

XY plane

$$d_{\text{eoe}} = d_{\text{oee}} = d_{31} \sin^2 \phi + d_{32} \cos^2 \phi \; ;$$

YZ plane

$$d_{\text{oeo}} = d_{\text{eoo}} = d_{31} \sin \theta \; ;$$

XZ plane, $\theta < V_Z$

$$d_{\text{ooe}} = d_{32} \sin \theta \; ;$$

XZ plane, $\theta > V_Z$

$$d_{\text{oeo}} = d_{\text{eoo}} = d_{32} \sin \theta \; .$$

Effective nonlinearity expressions for three-wave interactions in the arbitrary direction of CTA crystal are given in [3.36]

Nonlinear coefficients [3.551]:

$$d_{31}(1.0642 \,\mu\text{m}) = 2.1 \pm 0.4 \,\text{pm/V} \; ,$$

$$d_{32}(1.0642 \,\mu\text{m}) = 3.4 \pm 0.7 \,\text{pm/V} \; ,$$

$$d_{33}(1.0642 \,\mu\text{m}) = 18.1 \pm 1.8 \,\text{pm/V} \; .$$

3.3.23 NaNO$_2$, Sodium Nitrite

Positive biaxial crystal: $2V_Z = 62.5°$ at $\lambda = 0.5325 \,\mu\text{m}$ [3.552];
Point group: mm2;
Assignment of dielectric and crystallographic axes:
$X, Y, Z \Rightarrow a, c, b$;
Mass density: 2.168 g/cm^3;
Transparency range: $0.35 - 3.4 \,\mu\text{m}$ with the window in $5 - 8 \,\mu\text{m}$ range [3.553,554];
Experimental values of refractive indices:

λ [μm]	n_X	n_Y	n_Z	Ref.
0.5325	1.3475	1.4147	1.6643	3.552
0.5762	1.3455	1.4125	1.6547	3.553
1.0650	1.3395	1.4036	1.6365	3.552
1.1523	1.3353	1.4029	1.6319	3.553
1.3673		1.4018	1.6214	3.554
1.5295		1.4010	1.6160	3.554
1.7109		1.4010	1.6136	3.554

λ [μm]	n_X	n_Y	n_Z	Ref.
2.2500		1.3997	1.6102	3.554
3.4000		1.3980	1.5933	3.554
4.4000		1.3950	1.5400	3.554
5.4000		1.3907	1.4950	3.554
6.0000		1.3880	1.4626	3.554

Sellmeier equations (λ in μm, $T = 293$ K) [3.553]:

$$n_X^2 = 1 + \frac{0.727454\,\lambda^2}{\lambda^2 - (0.108759)^2} \; ,$$

$$n_Y^2 = 1 + \frac{0.978108\,\lambda^2}{\lambda^2 - (0.105970)^2} \; ,$$

$$n_Z^2 = 1 + \frac{1.616683\,\lambda^2}{\lambda^2 - (0.149021)^2} \; .$$

Experimental and theoretical values of phase-matching angle and calculated values of "walk-off" angle:
XZ plane, $\phi = 0°, \theta < V_Z$

Interacting wavelengths [μm]	θ_{exp} [deg]	θ_{theor} [deg] [3.553]	ρ_1 [deg]	ρ_3 [deg]
SHG, e + o $\Rightarrow$ e				
1.1523 $\Rightarrow$ 0.57615	27.1 [3.553]	27.60[#] 34.35[*]	8.309[#]	8.531[#]

XZ plane, $\phi = 0°, \theta < V_z$

Interacting wavelengths [μm]	θ_{exp} [deg]	θ_{theor} [deg] [3.553]	ρ_1 [deg]
SHG, e + e $\Rightarrow$ o			
1.1523 $\Rightarrow$ 0.57615	34.6 [3.553]	34.56[#] 39.34[*]	9.801[#]

[#]derived from experimental data on refractive indices;
[*]derived from Sellmeier equations.

Experimental values of internal angular bandwidth [3.553]:
XZ plane, $\phi = 0°, \theta < V_Z$

Interacting wavelengths [μm]	θ_{pm} [deg]	$\Delta\theta^{int}$ [deg]
SHG, e + o $\Rightarrow$ e		
$1.1523 \Rightarrow 0.57615$	27.08	0.407

XZ plane, $\phi = 0°, \theta > V_Z$

Interacting wavelengths [μm]	θ_{pm} [deg]	$\Delta\theta^{int}$ [deg]
SHG, e + e $\Rightarrow$ o		
$1.1523 \Rightarrow 0.57615$	34.60	0.22

Effective nonlinearity expressions in the phase-matching direction for three-wave interactions in the principal planes of $NaNo_2$ crystal [3.35,36]:

XY plane

$$d_{ooe} = d_{32} \cos\phi \ ;$$

YZ plane

$$d_{oeo} = d_{eoo} = d_{31} \cos\theta \ ;$$

XZ plane, $\theta < V_Z$

$$d_{eoe} = d_{oee} = d_{32} \sin^2\theta + d_{31} \cos^2\theta \ ;$$

XZ plane, $\theta > V_Z$

$$d_{eeo} = d_{32} \sin^2\theta + d_{31} \cos^2\theta \ .$$

Effective nonlinearity expressions for three-wave interactions in the arbitrary direction of $NaNO_2$ crystal are given in [3.36]

Nonlinear coefficients:

$$d_{31}(1.1523\,\mu m) = 0.174 \times d_{36}\,(KDP) \pm 28\%$$
$$= 0.068 \pm 0.019\,pm/V\ [3.553, 37] \ ,$$

$$d_{32}(1.1523\,\mu m) = -3.367 \times d_{36}\,(KDP) \pm 0.5\%$$
$$= -1.313 \pm 0.004\,pm/V\ [3.553, 37] \ ,$$

$$|d_{33}(1.06\,\mu m)| = 0.24 \times d_{36}\,(KDP) \pm 25\%$$
$$= 0.094 \pm 0.023\,pm/V\ [3.553, 37] \ .$$

3.3.24 $Ba_2NaNb_5O_{15}$, Barium Sodium Niobate ("Banana")

Negative biaxial crystal: $2V_Z = 13°$ [3.555];
Point group: mm2;
Assignment of dielectric and crystallographic axes:
$X, Y, Z \Rightarrow a, b, c$;
Mass density: $5.4076\,\text{g/cm}^3$ [3.555], $5.42\,\text{g/cm}^3$ [3.556];
Transparency range at "0" transmittance level: $0.37 - 5$ µm [3.555, 557];

Linear absorption coefficient α:

λ [µm]	α [cm^{-1}]	Ref.	Note
0.5321	0.04	3.556	NCSHG direction
	0.051–0.067	3.558	along a axis
1.0642	<0.002	3.556	NCSHG direction
	0.003	3.558	along a axis
	0.002	3.316	along b axis

Experimental values of refractive indices [3.555]:

λ [µm]	n_X	n_Y	n_Z
0.4579	2.4284	2.4266	2.2931
0.4765	2.4094	2.4076	2.2799
0.4880	2.3991	2.3974	2.2727
0.4965	2.3920	2.3903	2.2678
0.5017	2.3879	2.3862	2.2649
0.5145	2.3786	2.3767	2.2583
0.5321	2.3672	2.3655	2.2502
0.6328	2.3222	2.3205	2.2177
1.0642	2.2580	2.2567	2.1700

Temperature derivative of n_X and n_Z at $\lambda = 1.064$ µm
(n_Y depends on T only slightly) [3.555]:

$$\mathrm{d}n_X/\mathrm{d}T = -2.5 \times 10^{-5}\,\text{K}^{-1}\,,$$

$$\mathrm{d}n_Z/\mathrm{d}T = +8.0 \times 10^{-5}\,\text{K}^{-1}\,.$$

Experimental values of phase-matching angle ($T = 293$ K) and comparison between different sets of dispersion relations:
YZ plane, $\phi = 90°$

Interacting wavelengths [µm]	θ_{exp} [deg]	θ_{theor} [deg]	
		[3.458]	[3.555]
SHG, o + o $\Rightarrow$ e			
1.0642 $\Rightarrow$ 0.5321	73.8 [3.555]	74.1	75.0

XZ plane, $\phi = 0°, \theta > V_Z$

Interacting wavelengths [µm]	θ_{exp} [deg]	θ_{theor} [deg]	
		[3.458]	[3.555]
SHG, o + o $\Rightarrow$ e			
1.0642 $\Rightarrow$ 0.5321	75.4 [3.555]	74.6	75.3

Note: The PM angle values are strongly dependent on melt stoichiometry

Experimental values of NCPM temperature and temperature bandwidth:
along a axis

Interacting wavelengths [µm]	T [°C]	ΔT [°C]	Ref.
SHG, o + o $\Rightarrow$ e			
1.0642 $\Rightarrow$ 0.5321	85	0.45–0.47	3.558
	85		3.559
	86–87	0.45	3.300
	89	0.5	3.555
1.08 $\Rightarrow$ 0.54		0.42	3.560

along b axis

Interacting wavelengths [µm]	T [°C]	ΔT [°C]	Ref.
SHG, o + o $\Rightarrow$ e			
1.0642 $\Rightarrow$ 0.5321	97		3.561
	101	0.5	3.555

Note: The NCPM temperature values are strongly dependent on melt stoichiometry

Best set of Sellmeier equations (λ in μm, $T = 293$ K) [3.555]:

$$n_X^2 = 1 + \frac{3.9495\,\lambda^2}{\lambda^2 - 0.04038894} \; ,$$

$$n_Y^2 = 1 + \frac{3.9495\,\lambda^2}{\lambda^2 - 0.04014012} \; ,$$

$$n_Z^2 = 1 + \frac{3.6008\,\lambda^2}{\lambda^2 - 0.03219871} \; .$$

Calculated values of phase-matching and "walk-off" angles:
YZ plane, $\phi = 90°$

Interacting wavelengths [μm]	θ_{pm} [deg]	ρ_3 [deg]
SHG, o + o $\Rightarrow$ e		
1.0642 $\Rightarrow$ 0.5321	75.03	1.384
1.3188 $\Rightarrow$ 0.6594	53.44	2.442

XZ plane, $\phi = 0°, \theta > V_Z$

Interacting wavelengths [μm]	θ_{pm} [deg]	ρ_3 [deg]
SHG, o + o $\Rightarrow$ e		
1.0642 $\Rightarrow$ 0.5321	75.31	1.372
1.3188 $\Rightarrow$ 0.6594	53.63	2.450

Temperature variation of birefringence for noncritical SHG process [3.555]:
along b axis (1.0642 μm $\Rightarrow$ 0.5321 μm)

$$\mathrm{d}[n_Z(2\omega) - n_X(\omega)]/\mathrm{d}T = 1.05 \times 10^{-4} \; \mathrm{K}^{-1} \; .$$

Effective nonlinearity expressions in the phase-matching direction for three-wave interactions in the principal planes of $\mathrm{Ba_2NaNb_5O_{15}}$ crystal [3.35,36]:
XY plane

$$d_{\mathrm{eeo}} = d_{31} \sin^2 \phi + d_{32} \cos^2 \phi \; ;$$

YZ plane

$$d_{\mathrm{ooe}} = d_{31} \sin \theta \; ;$$

XZ plane, $\theta < V_Z$

$$d_{\mathrm{oeo}} = d_{\mathrm{eoo}} = d_{32} \sin \theta \; ;$$

XZ plane, $\theta > V_Z$

$$d_{\mathrm{ooe}} = d_{32} \sin \theta \; .$$

Effective nonlinearity expressions for three-wave interactions in the arbitrary direction of $Ba_2NaNb_5O_{15}$ crystal are given in [3.36].

Nonlinear coefficients:

$$d_{31}(1.0642\,\mu m) = 40 \times d_{11}(SiO_2) \pm 5\%$$
$$= 12 \pm 0.6\,pm/V\,[3.555, 37]$$

$$d_{32}(1.0642\,\mu m) = 40 \times d_{11}(SiO_2) \pm 10\%$$
$$= 12 \pm 1.2\,pm/V\,[3.555, 37]\ ,$$

$$d_{33}(1.0642\,\mu m) = 55 \times d_{11}(SiO_2) \pm 7\%$$
$$= 16.5 \pm 1.2\,pm/V\,[3.555, 37]\ .$$

Laser-induced damage threshold:

λ [μm]	τ_p [ns]	$I_{thr} \times 10^{-12}$ [W/m^2]	Ref.	Note
0.5321	cw	>0.0005	3.561	
	450	0.002	3.562	2 kHz
	0.05	0.72	3.563	1 kHz
1.0642	450	0.04	3.562	2 kHz
	0.08	>0.025	3.558	500 MHz

Thermal conductivity coefficient [3.556]:

$$\kappa = 3.5\,W/mK\ .$$

3.3.25 $K_2Ce\,(NO_3)_5 \cdot 2H_2O$, Potassium Cerium Nitrate Dihydrate (KCN)

Negative biaxial crystal: $2V_Z = 115.2°$ at $\lambda = 0.5461\,\mu m$ [3.550];
Point group: mm2;
Assignment of dielectric and crystallographic axes:
$X, Y, Z \Rightarrow b, c, a$;
Transparency range at "0" transmittance level: $0.39 -> 1.1\,\mu m$ [3.550];
Linear absorption coefficient:

$$\alpha < 0.03\ cm^{-1}\ at\ \lambda = 1.064\,\mu m\ [3.550];$$

Experimental values of refractive indices [3.550]:

λ [μm]	n_X	n_Y	n_Z
0.3650	1.5340	1.5912	1.6142
0.4005	1.5238	1.5775	1.5999
0.4872	1.5099	1.5597	1.5811
0.5461	1.5041	1.5524	1.5732

λ [µm]	n_X	n_Y	n_Z
0.6476	1.4983	1.5443	1.5653
0.7500	1.4947	1.5398	1.5603
0.8500	1.4924	1.5365	1.5567
0.9500	1.4905	1.5343	1.5542
1.0500	1.4890	1.5324	1.5519

Sellmeier equations (λ in µm, $T = 20\ °C$)[3.550]:

$$n_X^2 = 2.21109 + \frac{0.0140950}{\lambda^2 - 0.0345830} - 0.0063894\,\lambda^2\ ,$$

$$n_Y^2 = 2.33882 + \frac{0.0193380}{\lambda^2 - 0.0333504} - 0.0079345\,\lambda^2\ ,$$

$$n_Z^2 = 2.40514 + \frac{0.0194084}{\lambda^2 - 0.0371520} - 0.0135716\,\lambda^2\ .$$

Experimental and theoretical values of phase-matching angle and calculated values of "walk-off" angle:

XY plane, $\theta = 90°$

Interacting wavelengths [µm]	ϕ_{exp} [deg]	ϕ_{theor} [deg] [3.550]	ρ_3 [deg]
SHG, o + o $\Rightarrow$ e			
1.0642 $\Rightarrow$ 0.5321	10.2 [3.550]	11.74	0.74

XZ plane, $\phi = 0°, \theta < V_Z$

Interacting wavelengths [µm]	θ_{exp} [deg]	θ_{theor} [deg] [3.550]	ρ_1 [deg]	ρ_3 [deg]
SHG, e + o $\Rightarrow$ e				
1.0642 $\Rightarrow$ 0.5321	21.5 [3.550]	22.58	1.63	1.78

Experimental value of internal angular bandwidth [3.550]:

XY plane, $\theta = 90°$

Interacting wavelengths [µm]	ϕ_{pm} [deg]	$\Delta\phi^{\text{int}}$ [deg]
SHG, o + o $\Rightarrow$ e		
1.0642 $\Rightarrow$ 0.5321	10.2	0.152

Effective nonlinearity expressions in the phase-matching direction for three-wave interactions in the principal planes of KCN crystal [3.35,36]:

XY plane

$$d_{\mathrm{ooe}} = d_{31} \cos \phi \; ;$$

YZ plane

$$d_{\mathrm{oeo}} = d_{\mathrm{eoo}} = d_{32} \cos \theta \; ;$$

XZ plane, $\theta < V_Z$

$$d_{\mathrm{oee}} = d_{\mathrm{eoe}} = d_{31} \sin^2 \theta + d_{32} \cos^2 \theta \; ;$$

XZ plane, $\theta > V_Z$

$$d_{\mathrm{eeo}} = d_{31} \sin^2 \theta + d_{32} \cos^2 \theta \; .$$

Effective nonlinearity expressions for three-wave interactions in the arbitrary direction of KCN crystal are given in [3.36]
Nonlinear coefficients [3.550]:

$$d_{31}(1.0642 \,\mu\mathrm{m}) = \mp 1.13 \pm 0.15 \,\mathrm{pm/V} \; ,$$

$$d_{32}(1.0642 \,\mu\mathrm{m}) = \pm 1.10 \pm 0.10 \,\mathrm{pm/V} \; ,$$

$$|d_{33}(1.0642 \,\mu\mathrm{m})| = 0.13 \pm 0.10 \,\mathrm{pm/V} \; .$$

3.3.26 $K_3Li_2Nb_5O_{15}$, Potassium Lithium Niobate

Negative uniaxial crystal: $n_\mathrm{o} > n_\mathrm{e}$;
Point group: 4mm;
Mass density: 4.3 g/cm^3 [3.273];
Transparency range: $0.35 - 5 \,\mu$m [3.564, 315];
Linear absorption coefficient:

$$\alpha = 0.004 \;\mathrm{cm}^{-1} \;\text{at}\; \lambda = 1.064 \,\mu\mathrm{m} \;[3.315]$$

Experimental values of refractive indices at $T = 303$ K [3.517]:

λ [μm]	n_o	n_e
0.4500	2.4049	2.2512
0.4750	2.3751	2.2315
0.5000	2.3546	2.2144
0.5250	2.3349	2.2010
0.5321	2.3260	2.1975
0.5500	2.3156	2.1900
0.5750	2.3016	2.1801
0.6000	2.2899	2.1720
0.6250	2.2799	2.1645

λ [μm]	n_o	n_e
0.6328	2.2770	2.1630
0.6500	2.2711	2.1586
0.6750	2.2361	2.1529
1.0642	2.2080	2.1120

Sellmeier equations (λ in μm, $T = 303$ K) [3.517]:

$$n_\mathrm{o}^2 = 1 + \frac{3.708\,\lambda^2}{\lambda^2 - 0.04601} \; ,$$

$$n_\mathrm{e}^2 = 1 + \frac{3.349\,\lambda^2}{\lambda^2 - 0.03564} \; .$$

Experimental and theoretical values of phase-matching angle:

Interacting wavelengths [μm]	θ_exp [deg]	θ_theor [deg] [3.517]
SHG, o + o $\Rightarrow$ e		
0.82 $\Rightarrow$ 0.41	90 [3.315]	no pm

Calculated values of phase-matching and "walk-off" angles:

Interacting wavelengths [μm]	θ_pm [deg]	ρ_3 [deg]
SHG, o + o $\Rightarrow$ e		
2.9365 $\Rightarrow$ 1.46825	22.64	1.75
2.098 $\Rightarrow$ 1.049	32.07	2.30
1.3188 $\Rightarrow$ 0.6594	54.03	2.73
1.0642 $\Rightarrow$ 0.5321	75.45	1.54
1.053 $\Rightarrow$ 0.5265	77.46	1.35
1.047 $\Rightarrow$ 0.5235	78.69	1.23

Effective nonlinearity expression in the phase-matching direction [3.100]:

$$d_\mathrm{ooe} = d_{31} \sin \theta.$$

Nonlinear coefficients:

$$d_{31}(0.8\,\mu\mathrm{m}) = 11.8 \; \mathrm{pm/V} \; [3.315] \; ,$$

$$d_{31}(1.0642\,\mu\mathrm{m}) = 19.3 \times d_{11}(\mathrm{SiO_2}) \pm 20\%$$
$$= 5.8 \pm 1.2 \, \mathrm{pm/V} \; [3.565, 37] \; ,$$

$$d_{33}(1.0642\,\mu\mathrm{m}) = 35 \times d_{11}(\mathrm{SiO_2}) \pm 15\%$$
$$= 10.5 \pm 1.5 \, \mathrm{pm/V} \; [3.565, 37] \; .$$

3.3.27 HgGa$_2$S$_4$, Mercury Thiogallate

Negative uniaxial crystal: $n_o > n_e$;
Point group: $\bar{4}$;
Mass density: 4.95 g/cm^3 [3.338];
Mohs hardness: 3 – 3.5;
Transparency range at "0" transmittance level: 0.55 – 13 μm [3.566];
Linear absorption coefficient α:

λ [μm]	α [cm^{-1}]	Ref.	Note
0.53	8	3.567	e – wave, SHG direction
	11	3.566	
0.96	0.25	3.568	e – wave, SFG direction
1.06	0.1	3.567	o – wave, SHG direction
	0.25	3.568	o – wave, SFG direction
10.6	1.2	3.568	o – wave, SFG direction

Experimental values of refractive indices at $T = 293$ K [3.569]:

λ [μm]	n_o	n_e
0.5495	2.6592	2.5979
0.5747	2.6334	2.5748
0.6009	2.6112	2.5549
0.6328	2.5890	2.5349
0.6500	2.5796	2.5264
1.0760	2.477	2.432
1.1500	2.472	2.428
2.6500	2.444	2.403
3.5400	2.439	2.398
7.1500	2.414	2.372
8.7300	2.400	2.358
10.400	2.380	2.337
11.000	2.369	2.329

Sellmeier equations (λ in μm, $T = 20$ °C) [3.569]:

$$n_o^2 = 6.20815221 + \frac{63.70629851}{\lambda^2 - 225} + \frac{0.23698804}{\lambda^2 - 0.09568646} \; ,$$

$$n_e^2 = 6.00902670 + \frac{63.28065920}{\lambda^2 - 225} + \frac{0.21489656}{\lambda^2 - 0.09214633} \; .$$

Calculated values of phase-matching and "walk-off" angles:

Interacting wavelengths [µm]	θ_{pm} [deg]	ρ_1 [deg]	ρ_3 [deg]
SHG, o + o $\Rightarrow$ e			
9.6 $\Rightarrow$ 4.8	68.38		0.66
5.3 $\Rightarrow$ 2.65	31.80		0.89
4.8 $\Rightarrow$ 2.4	31.53		0.88
2.9365 $\Rightarrow$ 1.46825	42.22		1.00
2.1284 $\Rightarrow$ 1.0642	64.40		0.80
SFG, o + o $\Rightarrow$ e			
10.6 + 1.0642 $\Rightarrow$ 0.9671	41.62		1.05
SHG, e + o $\Rightarrow$ e			
5.3 $\Rightarrow$ 2.65	47.95	0.97	0.97
4.8 $\Rightarrow$ 2.4	47.39	0.97	0.98
2.9365 $\Rightarrow$ 1.46825	70.02	0.62	0.64
SFG, e + o $\Rightarrow$ e			
10.6 + 5.3 $\Rightarrow$ 3.533	70.21	0.63	0.61
9.6 + 4.8 $\Rightarrow$ 3.2	54.49	0.94	0.92
10.6 + 1.0642 $\Rightarrow$ 0.9671	43.93	1.01	1.06

Effective nonlinearity expressions in the phase-matching direction [3.100]:

$$d_{ooe} = d_{36} \sin\theta \sin 2\phi + d_{31} \sin\theta \cos 2\phi \ ,$$

$$d_{eoe} = d_{oee} = d_{36} \sin 2\theta \cos 2\phi + d_{31} \sin 2\theta \sin 2\phi \ .$$

Nonlinear coefficients:

$$|d_{36}(1.064\,\mu m)| = 80 \times d_{11}(SiO_2) \pm 30\%$$
$$= 24.0 \pm 7.2 \text{ pm/V } [3.566, 37] \ ,$$

$$|d_{36}(1.064\,\mu m)| = 1.08 \times d_{36}(AgGaS_2) \pm 15\%$$
$$= 20.0 \pm 3.0 \text{ pm/V } [3.567, 344, 37] \ ,$$

$$|d_{31}(1.064\,\mu m)| = 0.33 \times |d_{36}(HgGa_2S_4)|$$
$$= 6.7 \pm 1.0 \text{ pm/V } [3.576, 344, 37] \ .$$

Laser-induced surface-damage threshold [3.568]:

λ [µm]	τ_p [ns]	$I_{thr} \times 10^{-12}$ [W/m^2]
1.064	30	0.6
10.6	cw	> 0.00000016

3.3.28 HgS, Cinnibar

Positive uniaxial crystal: $n_e > n_o$;
Point group: 32;
Mass density: 8.10 g/cm^3 [3.64];
Mohs hardness: 2 – 2.5 [3.64], 3 [3.338];
Transparency range at "0" transmittance level: 0.62 – 13 μm [3.570];
Linear absorption coefficient α [3.571]:

λ [μm]	α [cm^{-1}]	Note
0.6328	1.7	o – wave, DFG direction
0.6729	1.4	e – wave, DFG direction
5.3	0.032	o – wave, SHG direction
10.6	0.073	e – wave, SHG and DFG directions

Experimental values of refractive indices [3.570]:

λ [μm]	n_o	n_e	λ [μm]	n_o	n_e
0.62	2.9028	3.2560	2.80	2.6414	2.9052
0.65	2.8655	3.2064	3.00	2.6401	2.9036
0.68	2.8384	3.1703	3.20	2.6387	2.9017
0.70	2.8224	3.1489	3.40	2.6375	2.9001
0.80	2.7704	3.0743	3.60	2.6358	2.8987
0.90	2.7383	3.0340	3.80	2.6353	2.8971
1.00	2.7120	3.0050	4.00	2.6348	2.8963
1.20	2.6884	2.9680	5.00	2.6267	2.8863
1.40	2.6730	2.9475	6.00	2.6233	2.8799
1.60	2.6633	2.9344	7.00	2.6156	2.8741
1.80	2.6567	2.9258	8.00	2.6112	2.8674
2.00	2.6518	2.9194	9.00	2.6066	2.8608
2.20	2.6483	2.9146	10.00	2.6018	2.8522
2.40	2.6455	2.9108	11.00	2.5914	2.8434
2.60	2.6433	2.9079			

Optical activity [3.194]:

λ [μm]	ρ [deg/mm]	λ [μm]	ρ [deg/mm]
0.6058	447	0.7281	145
0.6131	393.5	0.7789	113.5
0.6278	319	0.8296	92.5
0.6424	270.5	0.8757	74.5
0.6571	237.5	0.9196	65.5
0.6681	218	0.9527	59
0.6770	200	0.9967	51.5

Experimental values of phase-matching angle ($T = 293$ K) and comparison between different sets of dispersion relations:

Interacting wavelengths [μm]	θ_{exp} [deg]	θ_{theor} [deg]		
		[3.458]	[3.362]	[3.543]
SHG, e + e $\Rightarrow$ o				
$10.6 \Rightarrow 5.3$	20.8 [3.571]	21.2	23.0	21.3
	21.5 [3.572]			
SFG, e + e $\Rightarrow$ o				
$10.6 + 0.6729 \Rightarrow 0.6328$	25.3 [3.571]	no pm	25.8	25.7

Best set of dispersion relations (λ in μm, $T = 20$ °C) [3.543]:

$$n_\text{o}^2 = 7.8113 + \frac{0.3944}{\lambda^2 - 0.1172} + \frac{604.5}{\lambda^2 - 682.5} \ ,$$

$$n_\text{e}^2 = 9.3139 + \frac{0.5870}{\lambda^2 - 0.1166} + \frac{542.6}{\lambda^2 - 540.8} \ .$$

Calculated values of phase-matching and "walk-off" angles:

Interacting wavelengths [μm]	θ_{pm} [deg]	ρ_1 [deg]	ρ_2 [deg]
SHG, e + e $\Rightarrow$ o			
$10.6 \Rightarrow 5.3$	21.32	3.19	3.19
$9.6 \Rightarrow 4.8$	19.09	2.93	2.93
$5.3 \Rightarrow 2.65$	14.42	2.32	2.32
$4.8 \Rightarrow 2.4$	14.82	2.38	2.38
$2.65 \Rightarrow 1.325$	23.44	3.61	3.61
$2.4 \Rightarrow 1.2$	26.00	3.93	3.93
SHG, o + e $\Rightarrow$ o			
$10.6 \Rightarrow 5.3$	30.68		4.22
$9.6 \Rightarrow 4.8$	27.36		3.93
$5.3 \Rightarrow 2.65$	20.54		3.19
$4.8 \Rightarrow 2.4$	21.12		3.27
$2.65 \Rightarrow 1.325$	33.85		4.69
$2.4 \Rightarrow 1.2$	37.78		4.98
SFG, o + e $\Rightarrow$ o			
$10.6 + 5.3 \Rightarrow 3.533$	19.75		3.08
$9.6 + 4.8 \Rightarrow 3.2$	18.65		2.94
SFG, e + o $\Rightarrow$ o			
$10.6 + 5.3 \Rightarrow 3.533$	28.96	4.18	
$9.6 + 2.4 \Rightarrow 1.92$	27.16	4.00	

Effective nonlinearity expressions in the phase-matching direction [3.100]:

$$d_{\text{eeo}} = d_{11} \cos^2 \theta \sin 3\phi \ ,$$

$$d_{\text{oeo}} = d_{\text{eoo}} = d_{11} \cos\theta \cos 3\phi \ .$$

Nonlinear coefficient:

$$d_{11}(10.6\,\mu\text{m}) = 50 \pm 16\,\text{pm/V} \ [3.365] :$$

Laser-induced surface-damage threshold [3.365]:

λ [µm]	τ_{p} [ns]	$I_{\text{thr}} \times 10^{-12}$ [W/m^2]
1.06	17	0.4

3.3.29 Ag$_3$SbS$_3$, Pyrargyrite

Negative uniaxial crystal: $n_{\text{o}} > n_{\text{e}}$;
Point group: 3m;
Mass density: 5.83 g/cm^3 [3.64];
Mohs hardness: 2 – 2.5 [3.64];
Transparency range at "0" transmittance level: 0.7–14 µm [3.573];
Linear absorption coefficient α:

λ [µm]	α [cm^{-1}]	Ref.	Note
0.967	≈ 0.7	3.574	e – wave, SFG direction
1.064	≈ 0.7	3.574	o – wave, SFG direction
10.6	≈ 0.7	3.574	o – wave, SFG direction
	0.5	3.455	o – wave, SHG direction
	0.34	3.575	o – wave, $\parallel c$
	0.08	3.575	e – wave, $\perp c$
13.5	<1	3.573	

The graph of n_{o} and n_{e} dependences versus wavelength is given in [3.573].
Sellmeier equations (λ in µm, $T = 20$ °C) [3.573]:

$$n_{\text{o}}^2 = 1 + \frac{6.585\,\lambda^2}{\lambda^2 - 0.16} + \frac{0.1133\,\lambda^2}{\lambda^2 - 225} \ ,$$

$$n_{\text{e}}^2 = 1 + \frac{5.845\,\lambda^2}{\lambda^2 - 0.16} + \frac{0.0202\,\lambda^2}{\lambda^2 - 225} \ .$$

Experimental and theoretical values of phase-matching angle and calculated
values of "walk-off" angle:

Interacting wavelengths [µm]	θ_{exp} [deg]	θ_{theor} [deg] [3.573]	ρ_3 [deg] [3.573]
SHG, o + o $\Rightarrow$ e			
10.59 $\Rightarrow$ 5.295	30 [3.455]	23.34	2.18
10.6 $\Rightarrow$ 5.3	29 [3.575]	23.37	2.18
	27.6 [3.574]		

SFG, $o + o \Rightarrow e$

$10.6 + 1.064 \Rightarrow 0.967$	27.3 [3.574]	39.65	2.98

Effective nonlinearity expressions in the phase-matching direction [3.100]:

$$d_{\text{ooe}} = d_{31} \sin\theta - d_{22} \cos\theta \sin 3\phi \ ,$$
$$d_{\text{eoe}} = d_{\text{oee}} = d_{22} \cos^2\theta \cos 3\phi \ .$$

Nonlinear coefficients:

$$d_+(10.6\,\mu\text{m}) = (7.5 \pm 0.3)^{-1} \times |d_{36}(\text{GaAs})|$$
$$= 11.1 \pm 0.4 \ \text{pm/V} \ [3.576, 37].$$

Using the ratio

$$|d_{22}(\text{Ag}_3\text{SbS}_3)| \, / \, |d_{31}(\text{Ag}_3\text{SbS}_3)| = 1.05 \pm 0.04 \ [3.575]$$

and the values

$$\theta_{\text{pm}} \ (10.6 \Rightarrow 5.3\,\mu\text{m}, \ o + o \Rightarrow e) = 29°, \ \rho = 2.4° \ [3.575]$$

from the equation

$$d_+ = |d_{31}| \sin(\theta_{\text{pm}} + \rho) + |d_{22}| \cos(\theta_{\text{pm}} + \rho)$$

we deduce

$$|d_{22}(10.6\,\mu\text{m})| = 8.2 \pm 0.8 \ \text{pm/V} \ ,$$

$$|d_{31}(10.6\,\mu\text{m})| = 7.8 \pm 0.5 \ \text{pm/V}.$$

Laser-induced surface-damage threshold [3.365]:

λ [μm]	τ_p [ns]	$I_{\text{thr}} \times 10^{-12}$ [W/m^2]
1.06	17	>0.09
10.6	200	>0.46

3.3.30 Se, Selenium

Positive uniaxial crystal: $n_e > n_o$;
Point group: 32;
Mass density: 4.79 g/cm^3 [3.59];
Mohs hardness: 2 [3.59];
Transparency range at "0" transmittance level: $0.7 - 21$ μm [3.577, 578];
Linear absorption coefficient α:

λ [μm]	α [cm^{-1}]	Ref.	Note
5.3	1.40 ± 0.05	3.579	$\parallel c$
10.6	1.09 ± 0.02	3.579	$\parallel c$

λ [μm]	α [cm^{-1}]	Ref.	Note
14	2.8 ± 0.5	3.580	o − wave, $\perp c$
28	50 ± 5	3.580	o − wave, $\perp c$

Experimental values of refractive indices at 296 K [3.581]:

λ [μm]	n_o	n_e
1.064	1.790 ± 0.008	3.608 ± 0.008
1.1523	2.737 ± 0.008	3.573 ± 0.008
3.3913	2.650 ± 0.01	3.460 ± 0.01
10.6	2.640 ± 0.01	3.410 ± 0.01

Optical activity:

λ [μm]	ρ [deg/mm]	Ref.
0.70	440 ± 20	3.582
0.79	300 ± 15	3.582
0.91	200 ± 15	3.582
1.00	150 ± 10	3.582
1.14	100 ± 10	3.582
3.39	4.8 ± 0.5	3.579
10.6	2.5 ± 0.5	3.579

Experimental values of phase-matching angle:

Interacting wavelengths [μm]	θ_{pm} [deg]
SHG, e + e ⇒ o	
$10.6 \Rightarrow 5.3$	5.5 ± 0.3 [3.579]
	6.5 [3.577]
	≈ 10 [3.583]

Effective nonlinearity expressions in phase-matching direction [3.100]:

$$d_{\mathrm{eeo}} = d_{11} \cos^2 \theta \sin 3\phi$$

$$d_{\mathrm{oeo}} = d_{\mathrm{eoo}} = d_{11} \cos \theta \cos 3\phi$$

Nonlinear coefficient:

$$d_{11}(10.6 \ \mu\mathrm{m}) = 97 \pm 25 \ \mathrm{pm/V} \ [3.579]$$

Thermal conductivity coefficient [3.584]:

T [K]	κ [W/mK], $\parallel$ c	κ [W/mK], $\perp$ c
273	4.81	1.37
298	4.52	1.31

3.3.31 Tl$_3$AsS$_3$, Thallium Arsenic Selenide (TAS)

Negative uniaxial crystal: $n_\mathrm{o} > n_\mathrm{e}$;
Point group: 3m;
Mass density: 7.83 [3.585];
Mohs hardness: $2 - 3$ [3.586];
Transparency range at 0.5 transmittance level for a 6 mm long crystal: $1.28 - 17$ μm [3.586];
Linear absorption coefficient α:

λ [μm]	α [cm^{-1}]	Ref.	Note
2–12	< 0.02	3.585	
10.6	0.082	3.454	SHG direction
	0.038	3.586	

Experimental values of refractive indices at 300 K [3.587]:

λ [μm]	n_o	n_e
2.056	3.419	3.227
3.059	3.380	3.190
4.060	3.364	3.177
5.035	3.357	3.171
5.856	3.354	3.168
6.945	3.349	3.164
7.854	3.345	3.162
9.016	3.340	3.158
9.917	3.336	3.155
10.961	3.331	3.152
12.028	3.327	3.147

Temperature derivative of refractive indices at

$\lambda = 2 - 10.6$ μm $(T = 80 - 300$ K) [3.587] :

$$\frac{\mathrm{d}n_\mathrm{o}}{\mathrm{d}T} = -4.52 \times 10^{-5}\,\mathrm{K}^{-1} \; ;$$

$$\frac{\mathrm{d}n_\mathrm{e}}{\mathrm{d}T} = +3.55 \times 10^{-5}\,\mathrm{K}^{-1} \; .$$

Sellmeier equations (λ in μm, $T = 27\,°$C) [3.587]:

$$n_\mathrm{o}^2 = 1 + \frac{10.210\,\lambda^2}{\lambda^2 - 0.197136} + \frac{0.522\,\lambda^2}{\lambda^2 - 625} \; ,$$

$$n_\mathrm{e}^2 = 1 + \frac{8.993\,\lambda^2}{\lambda^2 - 0.197136} + \frac{0.308\,\lambda^2}{\lambda^2 - 625} \; .$$

Calculated values of phase-matching and "walk-off" angles:

Interacting wavelengths [μm]	θ_{pm} [deg]	ρ_1 [deg]	ρ_2 [deg]	ρ_3 [deg]
SHG, $o + o \Rightarrow e$				
$10.6 \Rightarrow 5.3$	19.10			2.12
$9.6 \Rightarrow 4.8$	18.54			2.07
$5.3 \Rightarrow 2.65$	24.79			2.60
$4.8 \Rightarrow 2.4$	27.26			2.77
$2.9365 \Rightarrow 1.46825$	48.74			3.26
SFG, $o + o \Rightarrow e$				
$10.6 + 2.65 \Rightarrow 2.12$	25.21			2.64
$9.6 + 2.4 \Rightarrow 1.92$	27.65			2.81
SHG, $e + o \Rightarrow e$				
$10.6 \Rightarrow 5.3$	26.79	2.65		2.72
$9.6 \Rightarrow 4.8$	26.03	2.62		2.67
$5.3 \Rightarrow 2.65$	35.77	3.16		3.18
$4.8 \Rightarrow 2.4$	39.78	3.25		3.27
SFG, $e + o \Rightarrow e$				
$10.6 + 5.3 \Rightarrow 3.533$	23.81	2.45		2.52
$9.6 + 4.8 \Rightarrow 3.2$	25.06	2.55		2.62
SFG, $o + e \Rightarrow e$				
$10.6 + 5.3 \Rightarrow 3.533$	34.84		3.13	3.14
$9.6 + 4.8 \Rightarrow 3.2$	36.84		3.19	3.20

Experimental values of internal angular bandwidth:

Interacting wavelengths [μm]	$\Delta\theta^{int}$ [deg]	Ref.
SHG, $o + o \Rightarrow e$		
$9.6 \Rightarrow 4.8$	0.27	3.588
$10.6 \Rightarrow 5.3$	0.30	3.589

Effective nonlinearity expressions in the phase-matching direction [3.100]:

$$d_{ooe} = d_{31} \sin \theta - d_{22} \cos \theta \sin 3\phi \,,$$

$$d_{eoe} = d_{oee} = d_{22} \cos^2 \theta \cos 3\phi \,.$$

Nonlinear coefficient:

$$d_+(10.6 \,\mu m) = (3.47 \pm 1.04) \times d_+(Ag_3AsS_3)$$
$$= 67.5 \pm 31.3 \text{ pm/V} \,[3.586, 455, 37] \,,$$

$$d_+(10.6 \,\mu m) = (3.3 \pm 1.0) \times d_+(Ag_3SbS_3)$$
$$= 36.5 \pm 12.5 \text{ pm/V} \,[3.586, 576, 37] \,.$$

Laser-induced surface-damage threshold:

λ [μm]	τ_p [ns]	$I_{thr} \times 10^{-12}$ [W/m^2]	Ref.
9.6	70	> 0.054	3.588
10.6	150	0.1-0.17	3.368
10.6	200	0.16	3.586

3.3.32 Te, Tellurium

Positive uniaxial crystal: $n_e > n_o$;
Point group: 32;
Mass density: 6.25 g/cm^3 [3.59];
Mohs hardness: 2 − 2.5 [3.59];
Transparency range at "0" transmittance level: 3.5 − 36 μm [3.590, 578, 591];
Linear absorption coefficient α:

λ [μm]	α [cm^{-1}]	Ref.	Note
5.3	1.32	3.451	o − wave, SHG direction
10.6	0.96	3.451	e − wave, SHG direction
	0.5–1.0	3.576	e − wave, SHG direction
	0.2–0.6	3.592	e − wave, SHG direction
14	1.1 ± 0.4	3.580	o − wave, $\perp c$
28	4.4 ± 0.04	3.580	o − wave, $\perp c$

Two-photon absorption coefficient β [3.593]:

Wavelengths of absorbed photons [μm]	$\beta \times 10^9$ [m/W]
5.3 + 5.3	8
5.3 + 10.6	2

Experimental values of refractive indices:

λ [μm]	n_o	n_e	Ref.	λ [μm]	n_o	n_e	Ref.
4.0	4.929	6.372	3.578	10.6	4.792	6.247	3.590
5.0	4.864	6.316	3.578	10.8	4.791	6.246	3.590
6.0	4.838	6.286	3.578	11.4	4.789	6.243	3.590
7.0	4.821	6.257	3.578	12.0	4.785	6.240	3.590
8.0	4.809	6.253	3.578	12.8	4.781	6.235	3.590
8.5	4.801	6.260	3.590	13.7	4.776	6.231	3.590
8.8	4.799	6.258	3.590	14.0	4.775	6.230	3.590
9.3	4.798	6.255	3.590	14.7	4.772	6.227	3.590
9.7	4.795	6.252	3.590	15.9	4.767	6.222	3.590
10.2	4.793	6.249	3.590	17.2	4.761	6.216	3.590

λ [μm]	n_o	n_e	Ref.	λ [μm]	n_o	n_e	Ref.
18.9	4.753	6.210	3.590	26.3	4.722	6.188	3.590
20.8	4.744	6.203	3.590	28.0	4.716	6.183	3.590
23.4	4.734	6.196	3.590	30.3	4.706	6.180	3.590

Optical activity [3.594]:

λ [μm]	ρ [deg/mm]
3.94	140
4.34	93.3
5.00	55.6
5.76	37.1
7.02	23.4

Experimental values of phase-matching angle ($T = 293$ K) and comparison between different sets of dispersion relations:

Interacting wavelengths [μm]	θ_{exp} [deg]	θ_{theor} [deg]		
		[3.543]	[3.362]$^{\#}$	[3.362]*
SHG, e + e $\Rightarrow$ o				
10.6 $\Rightarrow$ 5.3	14.17 [3.595]	14.28	14.22	15.28
	14.83 [3.596]			
	14.07 [3.590]			
	14.75 [3.597]			
23.4 $\Rightarrow$ 11.7	12.19 [3.362]	6.12	9.14	12.49
26.6 $\Rightarrow$ 13.3	13.33 [3.362]	5.36	9.56	13.53
28.0 $\Rightarrow$ 14.0	14.07 [3.362]	5.09	9.82	14.07
SHG, o + e $\Rightarrow$ e				
10.6 $\Rightarrow$ 5.3	20.42 [3.598]	20.22	20.13	21.64

Note: [3.362]$^{\#}$ - a set for 4.0–14.0 μm spectral range;
      [3.362]* - a set for 8.5–30.3 μm spectral range.

Best set of dispersion relations for 4.0–14.0 μm spectral range (λ in μm, $T = 293$ K) [3.362]:

$$n_o^2 = 18.5346 + \frac{4.3289\,\lambda^2}{\lambda^2 - 3.9810} + \frac{3.780\,\lambda^2}{\lambda^2 - 11813} \ ,$$

$$n_e^2 = 29.5222 + \frac{9.3068\,\lambda^2}{\lambda^2 - 2.5766} + \frac{9.235\,\lambda^2}{\lambda^2 - 13521} \ .$$

Calculated values of phase-matching and "walk-off" angles:

Interacting wavelengths [μm]	θ_{pm} [deg]	ρ_1 [deg]	ρ_2 [deg]
SHG, e + e ⇒ o			
28 ⇒ 14	9.82	3.98	3.98
14 ⇒ 7	10.90	4.42	4.42
10.6 ⇒ 5.3	14.22	5.72	5.72
9.6 ⇒ 4.8	15.90	6.36	6.36
SHG, o + e ⇒ o			
28 ⇒ 14	13.89		5.58
14 ⇒ 7	15.43		6.19
10.6 ⇒ 5.3	20.13		7.93
9.6 ⇒ 4.8	22.52		8.77
SFG, o + e ⇒ o			
28 + 14 ⇒ 9.333	11.32		4.59
SFG, e + o ⇒ o			
28 + 14 ⇒ 9.333	16.09	6.44	

Experimental values of internal angular bandwidth:

Interacting wavelengths [μm]	θ_{pm} [deg]	$\Delta\theta^{int}$ [deg]	Ref.
SHG, e + e ⇒ o			
10.6 ⇒ 5.3	14.17	0.19	3.595
	≈ 14.5	0.20	3.451

Effective nonlinearity expressions in the phase-matching direction [3.100]:

$$d_{eeo} = d_{11} \cos^2 \theta \sin 3\phi,$$
$$d_{oeo} = d_{eoo} = d_{11} \cos \theta \cos 3\phi.$$

Nonlinear coefficient:

$$d_{11}(10.6\,\mu m) = 7.2 \times d_{36}(GaAs) \pm 4\% = 598 \pm 25 \text{ pm/V [3.576, 37]},$$

$$d_{11}(10.6\,\mu m) = 670 \pm 209 \text{ pm/V [3.599]},$$

$$d_{11}(28\,\mu m) = 570 \pm 190 \text{ pm/V [3.590]}.$$

Laser-induced surface-damage threshold:

λ [μm]	τ_p [ns]	$I_{thr} \times 10^{-12}$ [W/m²]	Ref.
10.6	cw	0.0000015	3.599
	190	0.1 – 0.6	3.365
	150	0.02 – 0.04	3.599

Thermal conductivity coefficient [3.584]:

T [K]	κ [W/mK], $\parallel c$	κ [W/mK], $\perp c$
273	3.60	2.08
298	3.38	1.97

3.4 Other Organic Nonlinear Optical Crystals

3.4.1 $C_{12}H_{22}O_{11}$, Sucrose (Saccharose)

Negative biaxial crystal: $2V_Z = 132.3°$ at $\lambda = 0.5321$ μm [3.600];
Point group: 2;
Assignment of dielectric and crystallographic axes:
$Y \parallel b$, the axes a and c lie in XZ plane, the angle between them is $\beta = 103.5°$, the
angle between the axes Z and c is $\alpha = 23.5°$ (Fig. 3.5) [3.600];
Mohs hardness: > 2.5 [3.600];
Transparency range at "0" transmittance level: $0.19 - 1.42$ μm [3.600];
Experimental values of refractive indices [3.600]:

λ [μm]	n_X	n_Y	n_Z
0.5321	1.5404	1.5681	1.5737
1.0642	1.5278	1.5552	1.5592

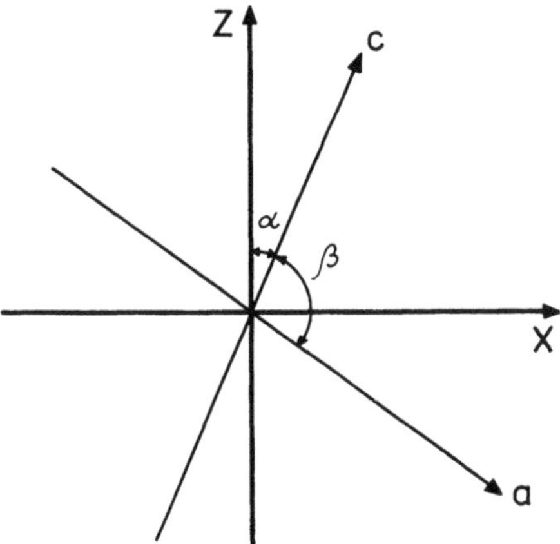

Fig. 3.5. Dielectric (X, Y, Z) and crystallographic (a, b, c) axes of sucrose crystal. The Y axis is parallel to the b axis and normal to the plane of the figure

Sellmeier equations (λ in μm, $T = 20°$ C) [3.600]:

$$n_X^2 = 1.8719 + \frac{0.4660\,\lambda^2}{\lambda^2 - 0.0214} - 0.0113\,\lambda^2 ,$$

$$n_Y^2 = 1.9703 + \frac{0.4502\,\lambda^2}{\lambda^2 - 0.0238} - 0.0101\,\lambda^2 \;,$$

$$n_Z^2 = 2.0526 + \frac{0.3909\,\lambda^2}{\lambda^2 - 0.0252} - 0.0187\,\lambda^2 \;,$$

Experimental and theoretical values of phase-matching angle and calculated values of "walk-off" angle:

XY plane, $\theta = 90°$

Interacting wavelengths [μm]	$\phi_{\exp}$ [deg]	ϕ_{theor} [deg] [3.600]	ρ_1 [deg]	ρ_3 [deg]
SHG, $e + o \Rightarrow e$ $1.0642 \Rightarrow 0.5321$	60.5 [3.600]	61.38	0.850	0.851

XZ plane, $\phi = 0°, \theta < V_Z$

Interacting wavelengths [μm]	$\theta_{\exp}$ [deg]	θ_{theor} [deg] [3.600]	ρ_1 [deg]	ρ_3 [deg]
SHG, $e + o \Rightarrow e$ $1.0642 \Rightarrow 0.5321$	16.0 [3.600]	14.88	0.570	0.597

Effective nonlinearity expressions in the phase-matching direction [3.35]:

XY plane

$$d_{ooe} = d_{23} \cos \phi \;,$$
$$d_{eoe} = d_{oee} = d_{25} \sin 2\phi \;;$$

YZ plane

$$d_{eeo} = d_{25} \sin(2\theta)$$
$$d_{oeo} = d_{eoo} = d_{21} \cos \theta \;;$$

XZ plane, $\theta < V_Z$

$$d_{eoe} = d_{oee} = d_{21} \cos^2 \theta + d_{23} \sin^2 \theta - d_{25} \sin 2\theta \;;$$

XZ plane, $\theta > V_Z$

$$d_{eeo} = d_{21} \cos^2 \theta + d_{23} \sin^2 \theta - d_{25} \sin 2\theta \;.$$

Laser-induced surface-damage threshold [3.600]:

λ [μm]	τ_p [ns]	$I_{\mathrm{thr}} \times 10^{-12}$ [W/m^2]
1.06	10	> 5

3.4.2 L-Arginine Phosphate Monohydrate (LAP)

Negative biaxial crystal: $2V_Z = 141.3°$ at $\lambda = 0.5321$ µm [3.112];
Point group: 2;
Assignment of dielectric and crystallographic axes:
$Y \parallel b$, the axes a and c lie in XZ plane, the angle between them is $\beta = 98°$, the angle between the axes Z and c is $\alpha = 35°$ (Fig. 3.6) [3.112];
Transparency range at "0" transmittance level: $0.23 - 1.25$ µm [3.112];
Linear absorption coefficient α:

λ [µm]	α [cm^{-1}]	Ref.	Note
0.230	0.1	3.601	
0.5265	0.01	3.66	
0.5321	< 0.01	3.112	
0.910	0.032	3.112	along X
	0.055	3.112	along Y
	0.051	3.112	along Z
1.040	0.113	3.112	along X
	0.219	3.112	along Y
	0.315	3.112	along Z
1.053	0.09	3.66	
1.0642	0.097	3.112	along X
	0.145	3.112	along Y
	0.184	3.112	along Z

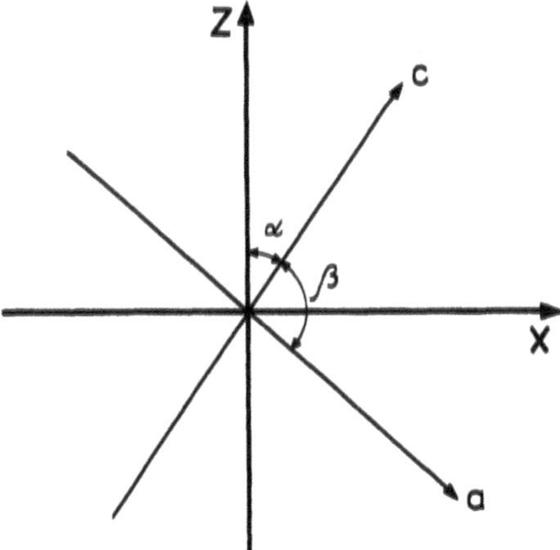

Fig. 3.6. Dielectric (X, Y, Z) and crystallographic (a, b, c) axes of LAP and DLAP crystals. The Y axis is parallel to the b axis and normal to the plane of the figure

Sellmeier equations (λ in µm, $T = 25$ °C) [3.112]:

$$n_X^2 = 2.2439 + \frac{0.0117}{\lambda^2 - 0.0179} - 0.0111 \, \lambda^2 ,$$

$$n_Y^2 = 2.4400 + \frac{0.0158}{\lambda^2 - 0.0191} - 0.0212\,\lambda^2\,,$$

$$n_Z^2 = 2.4590 + \frac{0.0177}{\lambda^2 - 0.0226} - 0.0162\,\lambda^2\,,$$

Experimental and theoretical values of phase-matching angle and calculated values of "walk-off" angle:

XY plane, $\theta = 90°$

Interacting wavelengths [μm]	ϕ_{exp} [deg]	ϕ_{theor} [deg] [3.112]	ρ_1 [deg]	ρ_3 [deg]
SHG, o + o ⇒ e				
0.5321 ⇒ 0.26605	60 [3.112]	61.65		2.498
1.0642 ⇒ 0.5321	25.5 [3.112]	24.02		1.919
SFG, o + o ⇒ e				
1.0642 + 0.5321 ⇒				
⇒ 0.35473	35.4 [3.112]	35.73		2.618
SHG, e + o ⇒ e				
1.0642 ⇒ 0.5321	40.8 [3.112]	40.00	2.290	2.485
SFG, e + o ⇒ e				
1.0642 + 0.5321 ⇒				
⇒ 0.35473	43.2 [3.112]	46.28	2.302	2.711

XZ plane, $\phi = 0°, \theta < V_Z$

Interacting wavelengths [μm]	θ_{exp} [deg]	θ_{theor} [deg] [3.112]	ρ_1 [deg]	ρ_3 [deg]
SHG, e + o ⇒ e				
1.0642 ⇒ 0.5321	40 [3.112]	40.59	2.568	2.774
SFG, e + o ⇒ e				
1.0642 + 0.5321 ⇒				
⇒ 0.35473	34.8 [3.112]	33.86	2.381	2.891

Effective nonlinearity expressions in the phase-matching direction [3.35]:

XY plane

$$d_{\mathrm{ooe}} = d_{23} \cos\phi\,,$$

$$d_{\mathrm{eoe}} = d_{\mathrm{oee}} = d_{25} \sin(2\phi)\,;$$

YZ plane

$$d_{\mathrm{eeo}} = d_{25} \sin(2\theta)$$

$$d_{\mathrm{oeo}} = d_{\mathrm{eoo}} = d_{21} \cos\theta\,;$$

XZ plane, $\theta < V_Z$

$$d_{\text{eoe}} = d_{\text{oee}} = d_{21} \cos^2 \theta + d_{23} \sin^2 \theta - d_{25} \sin 2\theta \; ;$$

XZ plane, $\theta > V_Z$

$$d_{\text{eeo}} = d_{21} \cos^2 \theta + d_{23} \sin^2 \theta - d_{25} \sin 2\theta \; .$$

Nonlinear coefficients [3.112, .37]:

$$d_{21}(1.0642 \; \mu\text{m}) = 0.40 \; \text{pm/V} \; ,$$
$$d_{22}(1.0642 \; \mu\text{m}) = 0.37 \; \text{pm/V} \; ,$$
$$d_{23}(1.0642 \; \mu\text{m}) = -0.84 \; \text{pm/V} \; ,$$
$$d_{25}(1.0642 \; \mu\text{m}) = -0.58 \; \text{pm/V} \; .$$

Laser-induced damage threshold:

λ [μm]	τ_p [ns]	$I_{\text{thr}} \times 10^{-12}$ [W/m^2]	Ref.
0.5265	20	300	3.66
	0.6	600	3.66
1.053	25	130	3.66
	1	630	3.66
1.0642	1	100–130	3.112

Thermal conductivity coefficient [3.602]:
$$\kappa = 0.59 \; \text{W/mK} \; .$$

3.4.3 Deuterated L-Arginine Phosphate Monohydrate (DLAP)

Negative biaxial crystal: $2V_Z = 142.6°$ at $\lambda = 0.5321 \; \mu\text{m}$ [3.112];
Point group: 2;
Assignment of dielectric and crystallographic axes:
$Y \parallel b$, the axes a and c lie in XZ plane, the angle between them is $\beta = 98°$, the angle between the axes Z and c is $\alpha = 35°$ (Fig. 3.6) [3.112];
Mass density: $\approx 1.5 \; \text{g/cm}^3$ [3.603];
Transparency range at "0" transmittance level: $0.22 - 1.30 \; \mu\text{m}$ [3.112];
Linear absorption coefficient α:

λ [μm]	α [cm^{-1}]	Ref.	Note
0.266	0.074	3.112	along X
	0.131	3.112	along Y
	0.184	3.112	along Z
0.3547	0.025	3.112	along X
	0.053	3.112	along Y
	0.039	3.112	along Z
0.5265	0.01	3.66	
0.5321	< 0.01	3.112	
0.910	0.028	3.112	along X
	0.037	3.112	along Y

λ [μm]	α [cm^{-1}]	Ref.	Note
	0.044	3.112	along Z
1.040	0.012	3.112	along X
	0.014	3.112	along Y
	0.009	3.112	along Z
1.053	0.02	3.66	
1.064	0.012	3.112	along X
	0.014	3.112	along Y
	0.009	3.112	along Z
1.180	0.385	3.112	along X
	0.394	3.112	along Y
	0.557	3.112	along Z

Temperature derivative of refractive indices [3.604]:

λ [μm]	$\mathrm{d}n_X/\mathrm{d}T \times 10^5$ [K^{-1}]	$\mathrm{d}n_Y/\mathrm{d}T \times 10^5$ [K^{-1}]	$\mathrm{d}n_Z/\mathrm{d}T \times 10^5$ [K^{-1}]
0.5321	−3.64	−5.34	−6.69
1.0642	−3.73	−5.30	−6.30

Sellmeier equations (λ in μm, $T = 25\ °C$) [3.112]:

$$n_X^2 = 2.2352 + \frac{0.0118}{\lambda^2 - 0.0146} - 0.00683\,\lambda^2 \ ,$$

$$n_Y^2 = 2.4313 + \frac{0.0151}{\lambda^2 - 0.0214} - 0.0143\,\lambda^2 \ ,$$

$$n_Z^2 = 2.4484 + \frac{0.0172}{\lambda^2 - 0.0229} - 0.0115\,\lambda^2 \ .$$

Experimental and theoretical values of phase-matching angle and calculated values of "walk-off" angle:

XY plane, $\theta = 90°$

Interacting wavelengths [μm]	$\phi_{\exp}$ [deg]	ϕ_{theor} [deg] [3.112]	ρ_1 [deg]	ρ_3 [deg]
SHG, $o + o \Rightarrow e$				
$1.0642 \Rightarrow 0.5321$	22.2 [3.604]	22.98		1.852
SHG, $e + o \Rightarrow e$				
$1.0642 \Rightarrow 0.5321$	37.5 [3.604]	37.81	2.290	2.446

XZ plane, $\phi = 0°, \theta < V_Z$

Interacting wavelengths [μm]	$\theta_{\exp}$ [deg]	θ_{theor} [deg] [3.112]	ρ_1 [deg]	ρ_3 [deg]
SHG, $e + o \Rightarrow e$				
$1.0642 \Rightarrow 0.5321$	42.8 [3.604]	43.34	2.588	2.785

Experimental values of internal angular, temperature and spectral bandwidths
($T = 20$ °C) [3.603]:
XY plane, $\theta = 90°$

Interacting wavelengths [μm]	ϕ_{pm} [deg]	$\Delta\phi^{int}$ [deg]	ΔT [° C]	Δv [cm^{-1}]
SHG, o + o $\Rightarrow$ e				
1.0642 $\Rightarrow$ 0.5321	22.2	0.036	5.4	20.2
SHG, e + o $\Rightarrow$ e				
1.0642 $\Rightarrow$ 0.5321	37.5	0.072	14.6	20.1

Effective nonlinearity expressions in the phase-matching direction [3.35]:
XY plane

$$d_{ooe} = d_{23} \cos\phi \ ,$$
$$d_{eoe} = d_{oee} = d_{25} \sin 2\phi \ ;$$

YZ plane

$$d_{eeo} = d_{25} \sin 2\theta \ ,$$
$$d_{oeo} = d_{eoo} = d_{21} \cos\theta \ ;$$

XZ plane, $\theta < V_Z$

$$d_{eoe} = d_{oee} = d_{21} \cos^2\theta + d_{23} \sin^2\theta - d_{25} \sin 2\theta \ ;$$

XZ plane, $\theta > V_Z$

$$d_{eeo} = d_{21} \cos^2\theta + d_{23} \sin^2\theta - d_{25} \sin 2\theta \ .$$

Nonlinear coefficients [3.112, .37]:

$$d_{21}(1.0642 \, \mu m) = 0.40 \ pm/V \ ,$$
$$d_{22}(1.0642 \, \mu m) = 0.37 \ pm/V \ ,$$
$$d_{23}(1.0642 \, \mu m) = -0.84 \ pm/V \ ,$$
$$d_{25}(1.0642 \, \mu m) = -0.58 \ pm/V \ .$$

Laser-induced damage threshold:

λ [μm]	τ_p [ns]	$I_{thr} \times 10^{-12}$ [W/m^2]	Ref.
0.308	17	0.3	3.605
0.5265	20	380	3.66
	0.6	670	3.66
1.053	25	330	3.66
	1	870	3.66
1.0642	1	90–130	3.112

3.4.4 L-Pyrrolidone-2-carboxylic Acid (L-PCA)

Positive biaxial crystal: $2V_Z = 51.7°$ at $\lambda = 0.5321\mu m$ [3.606];
Point group: 222;
Assignment of dielectric and crystallographic axes:
$X, Y, Z \Rightarrow b, a, c$ [3.606]
Calculated mass density: 1.44 g/cm^3;
Vickers hardness: 33 [3.606];
Transparency range: $0.26 - 1.064$ μm [3.606].
Sellmeier equations (λ in μm, $T = 20\,°C$) [3.606]:

$$n_X^2 = 2.1907 + \frac{0.0142}{\lambda^2 - 0.0124} \ ,$$

$$n_Y^2 = 2.2629 + \frac{0.0139}{\lambda^2 - 0.0148} \ ,$$

$$n_Z^2 = 2.5858 + \frac{0.0242}{\lambda^2 - 0.0217} \ .$$

Experimental and theoretical values of phase-matching angle and calculated values of "walk-off" angle:
XY plane, $\theta = 90°$

Interacting wavelengths [μm]	ϕ_{exp} [deg]	ϕ_{theor} [deg] [3.606]	ρ_1 [deg]	ρ_3 [deg]
SHG, e + o $\Rightarrow$ e				
$0.5321 \Rightarrow 0.26605$	42 [3.606]	41.44	0.89	0.89

YZ, plane, $\phi = 90°$

Interacting wavelengths [μm]	θ_{exp} [deg]	θ_{theor} [deg] [3.606]	ρ_1 [deg]
SHG, e + e $\Rightarrow$ o			
$0.5321 \Rightarrow 0.26605$	37 [3.606]	36.92	3.92

XZ plane, $\phi = 0°, \theta > V_Z$

Interacting wavelengths [μm]	θ_{exp} [deg]	θ_{theor} [deg] [3.600]	ρ_1 [deg]
SHG, e + e $\Rightarrow$ o			
$0.5321 \Rightarrow 0.26605$	54 [3.606]	54.14	4.94
$1.0642 \Rightarrow 0.5321$	33.5 [3.606]	33.85	4.31

Experimental value of internal angular bandwidth [3.606]:
XY plane, $\theta = 90°$

Interacting wavelengths [μm]	ϕ_{pm} [deg]	$\Delta\phi^{int}$ [deg]
SHG, $e + o \Rightarrow e$		
$0.5321 \Rightarrow 0.26605$	42	0.123

Effective nonlinearity expressions in phase-matching direction for three-wave interactions in the principal planes of L-PCA crystal [3.35]:

XY plane

$$d_{eoe} = d_{oee} = d_{14} \sin 2\phi \; ;$$

YZ plane

$$d_{eeo} = d_{14} \sin 2\theta \; ;$$

XZ plane, $\theta < V_Z$

$$d_{eoe} = d_{oee} = d_{14} \sin 2\theta \; ;$$

XZ plane, $\theta > V_Z$

$$d_{eeo} = d_{14} \sin 2\theta \; ;$$

Nonlinear coefficient

$$d_{14}(0.5321) \; \mu m = 0.29 \; pm/V \; [3.606, .37]$$

3.4.5 $CaC_4H_4O_6 \cdot 4H_2O$, Calcium Tartrate Tetrahydrate (L-CTT)

Positive biaxial crystal: $2V_Z = 65.8°$ at $\lambda = 0.5321$ μm [3.607];
Point group: mm2;
Assignment of dielectric and crystallographic axes:
$X, Y, Z \Rightarrow a, c, b$
Transparency range at "0" transmittance level: $0.28 - 1.4$ μm [3.607];
Experimental values of refractive indices [3.607]:

λ [μm]	n_X	n_Y	n_Z
0.4880	1.5306	1.5428	1.5649
0.5145	1.5270	1.5388	1.5613
0.5321	1.5264	1.5364	1.5611
1.0642	1.5125	1.5220	1.5477

Sellmeier equations (λ in µm, $T = 293$ K) [3.607]:

$$n_X^2 = 1 + \frac{1.26\,\lambda^2}{\lambda^2 - 0.0127273} \; ,$$

$$n_Y^2 = 1 + \frac{1.30\,\lambda^2}{\lambda^2 - 0.0121495} \; ,$$

$$n_Z^2 = 1 + \frac{1.38\,\lambda^2}{\lambda^2 - 0.0094521} \; .$$

Calculated values of phase-matching and "walk-off" angles:
XZ plane, $\phi = 0°, \theta > V_Z$

Interacting wavelengths [µm]	θ_{pm} [deg]	ρ_1 [deg]
SHG, e + e $\Rightarrow$ o		
$1.0642 \Rightarrow 0.5321$	58.43	1.299
$1.3188 \Rightarrow 0.6594$	50.07	1.438

Effective nonlinearity expressions in the phase-matching direction for three-wave interactions in the principal planes of LCTT crystal (Kleinman symmetry relations are not valid) [3.35, .36]:

XY plane

$$d_{ooe} = d_{32} \cos \theta \; ;$$

YZ plane

$$d_{oeo} = d_{eoo} = d_{15} \cos \theta \; ;$$

XZ plane, $\theta < V_Z$

$$d_{eoe} = d_{oee} = d_{24} \sin^2 \phi + d_{15} \cos^2 \phi \; ;$$

XZ plane, $\theta > V_Z$

$$d_{eeo} = d_{32} \sin^2 \phi + d_{31} \cos^2 \phi \; .$$

Effective nonlinearity expressions for three-wave interactions in the arbitrary direction of LCTT crystal are given in [3.36]
Nonlinear coefficients [3.607, .37]:

$$d_{15}(1.0642 \text{ µm}) = 1.73 \pm 0.03 \text{ pm/V} \; ,$$

$$d_{24}(1.0642 \text{ µm}) = 0.90 \pm 0.03 \text{ pm/V} \; ,$$

$$d_{31}(1.0642 \text{ µm}) < 0.015 \text{ pm/V} \; ,$$

$$d_{32}(1.0642 \text{ µm}) = 0.20 \pm 0.02 \text{ pm/V} \; ,$$

$$d_{33}(1.0642 \text{ µm}) = 0.14 \pm 0.02 \text{ pm/V} \; .$$

3.4.6 $(NH_4)_2C_2O_4 \cdot H_2O$, Ammonium Oxalate (AO)

Negative biaxial crystal: $2V_Z = 64.17°$ at $\lambda = 0.5461$ μm [3.608];
Point group: 222;
Transparency range: $0.3 - 1.1$ μm [3.609];
Experimental values of refractive indices [3.608]:

λ [μm]	n_X	n_Y	n_Z
0.4471	1.6119	1.5599	1.4460
0.4713	1.6084	1.5561	1.4447
0.4922	1.6050	1.5544	1.4435
0.5016	1.6037	1.5536	1.4426
0.5461	1.5993	1.5493	1.4406
0.5780	1.5965	1.5470	1.4391
0.5876	1.5952	1.5469	1.4388
0.6678	1.5892	1.5426	1.4362
0.7016	1.5874	1.5408	1.4352
1.014	1.5763	1.5312	1.4295
1.129	1.5728	1.5284	1.4276
1.367	1.5652	1.5222	1.4235

Experimental values of phase-matching angle:
XZ plane, $\phi = 0°, \theta < V_Z$

Interacting wavelengths [μm]	θ_{pm} [deg]
SHG, e + e ⇒ o	
1.06 ⇒ 0.53	23.17 [3.610]
1.1523 ⇒ 0.57615	23.6 [3.608]

XZ plane, $\phi = 0°, \theta < V_Z$

Interacting wavelengths [μm]	θ_{pm} [deg]
SHG, e + o ⇒ e	
1.06 ⇒ 0.53	45.75 [3.610]
1.1523 ⇒ 0.57615	46.5 [3.608]

Effective nonlinearity expressions in the phase-matching direction in the principal planes of AO crystal [3.35]:
XY plane

$$d_{eeo} = d_{14} \sin 2\phi \; ;$$

YZ plane

$$d_{eoe} = d_{oee} = d_{14} \sin 2\theta \; ;$$

XZ plane, $\theta < V_Z$

$$d_{eeo} = d_{14} \sin 2\theta \; ;$$

XZ plane, $\theta > V_Z$

$$d_{eoe} = d_{oee} = d_{14} \sin 2\theta \; ;$$

Nonlinear coefficient:

$$d_{14}(1.06 \; \mu m) = 0.9 \times d_{36}(\text{KDP}) = 0.31 \; \text{pm/V} \; [3.609, 37]$$

Laser-induced damage threshold [3.609]:

λ [μm]	τ_p [ns]	$I_{thr} \times 10^{-12}$ [W/m^2]
1.06	≈ 20	8

3.4.7 m-Bis(aminomethyl)benzene (BAMB)

Negative biaxial crystal: $2V_Z = 57.25°$ at $\lambda = 0.5321 \; \mu m$ [3.611];
Point group: mm2;
Assignment of dielectric and crystallographic axes:
$X, Y, Z \Rightarrow a, b, c$;
Mass density: 1.26 g/cm^3;
Transparency range at "0" transmittance level: $0.33 - 1.42 \; \mu m$ [3.611];
Experimental values of refractive indices [3.611]:

λ [μm]	n_X	n_Y	n_Z
0.436	1.8632	1.8019	1.6433
0.492	1.8320	1.7778	1.6296
0.532	1.8189	1.7676	1.6226
0.546	1.8150	1.7644	1.6205
0.577	1.8071	1.7583	1.6163
0.579	1.8069	1.7579	1.6161
0.589	1.8047	1.7564	1.6150
0.633	1.7967	1.7499	1.6108
1.064	1.7644	1.7240	1.5930
1.153	1.7618	1.7220	1.5916

Experimental values of phase-matching angle:
YZ plane, $\phi = 90°$

Interacting wavelengths [μm]	θ_{pm} [deg]
SHG, o + o $\Rightarrow$ e	
1.0642 $\Rightarrow$ 0.5321	8 [3.611]

XZ plane, $\phi = 0°, \theta < V_Z$

Interacting wavelengths [µm]	θ_{pm} [deg]
SHG, o + o ⇒ e 1.0642 ⇒ 0.5321	42 [3.611]

Experimental value of internal angular bandwidth [3.611]:
YZ plane, $\phi = 90°$

Interacting wavelengths [µm]	θ_{pm} [deg]	$\Delta\theta^{int}$ [deg]
SHG, o + o ⇒ e 1.0642 ⇒ 0.5321	8	0.098

Effective nonlinearity expressions in the phase-matching direction in the principal planes of **BAMB** crystal [3.35, .36]:

XY plane

$$d_{eeo} = d_{31} \sin^2 \phi + d_{32} \cos^2 \phi \; ;$$

YZ plane

$$d_{ooe} = d_{31} \sin \theta \; ;$$

XZ plane, $\theta < V_Z$

$$d_{oeo} = d_{eoo} = d_{32} \sin \theta \; ;$$

XZ plane, $\theta > V_Z$

$$d_{ooe} = d_{32} \sin \theta \; ;$$

Effective nonlinearity expressions for three-wave interactions in the arbitrary direction of **BAMB** crystal are given in [3.36]

Nonlinear coefficients [3.611, .37]:

$$d_{31}(1.0642 \; \mu m) = 0.95 \times d_{36}(KDP) \pm 20\% = 0.37 \pm 0.07 \; pm/V \; ,$$

$$d_{32}(1.0642 \; \mu m) = 2.45 \times d_{36}(KDP) \pm 20\% = 0.96 \pm 0.19 \; pm/V \; ,$$

$$d_{33}(1.0642 \; \mu m) = 1.8 \times d_{36}(KDP) \pm 20\% = 0.70 \pm 0.14 \; pm/V \; .$$

Laser-induced surface-damage threshold [3.611]:

λ [µm]	τ_p [ns]	$I_{thr} \times 10^{-12}$ [W/m²]
1.06	40	2

3.4.8 3-Methoxy-4-hydroxy-benzaldehyde (MHBA)

Positive biaxial crystal: $2V_Z = 89.5°$ at $\lambda = 0.5461$ μm [3.612];
Point group: 2;
Assignment of dielectric and crystallographic axes:
$X, Y, Z \Rightarrow a, b, c$;
Calculated mass density: 1.34 g/cm^3 [3.613];
Mohs hardness: 1.67 [3.613];
Transparency range at "0" transmittance level: $0.37 - 2.2$ μm [3.612];
Linear absorption coefficient α [3.612]:

λ [μm]	α [cm^{-1}]
0.415	1.42
0.532	0.95
0.830	0.53
1.064	0.53

Experimental values of refractive indices [3.612]:

λ [μm]	n_X	n_Y	n_Z
0.4047	1.63352		
0.4358	1.60345		
0.4471	1.59644		1.89349
0.5461	1.55840	1.70018	
0.5875	1.55143	1.69045	1.80896
0.5893	1.55127	1.69039	1.79235
0.6563		1.68352	
0.6678	1.53996	1.67963	1.77105
0.7057	1.53673	1.67668	1.76812

The Sellmeier equations given in [3.612] are incorrect.

Experimental values of the phase-matching angle:
XY plane, $\theta = 90°$

Interacting wavelengths [μm]	ϕ_{pm} [deg]
SHG, o + o $\Rightarrow$ e	
0.83 $\Rightarrow$ 0.415	16 [3.612]
1.0642 $\Rightarrow$ 0.5321	11 [3.612]
SHG, e + o $\Rightarrow$ e	
0.83 $\Rightarrow$ 0.415	58 [3.612]

YZ plane, $\phi = 90°$

Interacting wavelengths [μm]	θ_{pm} [deg]
SHG, o + o $\Rightarrow$ e	
$0.83 \Rightarrow 0.415$	49 [3.612]

XZ plane, $\phi = 0°, \theta < V_Z$

Interacting wavelengths [μm]	θ_{pm} [deg]
SHG, e + o $\Rightarrow$ e	
$1.0642 \Rightarrow 0.5321$	28 [3.612]

XZ plane, $\phi = 0°, \theta > V_Z$

Interacting wavelengths [μm]	θ_{pm} [deg]
SHG, e + e $\Rightarrow$ o	
$1.0642 \Rightarrow 0.5321$	68 [3.612]

Experimental value of internal angular bandwidth [3.612]:
XZ plane, $\phi = 0°$

Interacting wavelengths [μm]	θ_{pm} [deg]	$\Delta\theta^{int}$ [deg]
SHG, e + e $\Rightarrow$ o		
$1.0642 \Rightarrow 0.5321$	68	0.052

Effective nonlinearity expressions in the phase-matching direction [3.35]:
XY plane

$$d_{ooe} = d_{23} \cos \phi \;;$$
$$d_{eoe} = d_{oee} = d_{25} \sin 2\phi \;;$$

YZ plane

$$d_{eeo} = d_{25} \sin 2\theta \;;$$
$$d_{oeo} = d_{eoo} = d_{21} \cos \theta \;;$$

XZ plane, $\theta < V_Z$

$$d_{eoe} = d_{oee} = d_{21} \cos^2 \theta + d_{23} \sin^2 \theta - d_{25} \sin 2\theta \;;$$

XZ plane, $\theta > V_Z$

$$d_{eeo} = d_{21} \cos^2 \theta + d_{23} \sin^2 \theta - d_{25} \sin 2\theta \;.$$

Nonlinear coefficients [3.612, 37]:

$$d_{21}(1.0642 \ \mu\text{m}) = 3.9 \pm 0.8 \ \text{pm/V} \ ,$$
$$d_{22}(1.0642 \ \mu\text{m}) = 9.8 \pm 1.0 \ \text{pm/V} \ ,$$
$$d_{23}(1.0642 \ \mu\text{m}) = 13.0 \pm 1.3 \ \text{pm/V} \ ,$$
$$d_{25}(1.0642 \ \mu\text{m}) = 3.2 \pm 0.6 \ \text{pm/V} \ .$$

Laser-induced damage threshold [3.612]:

λ [μm]	τ_p [ns]	$I_\text{thr} \times 10^{-12}$ [W/m^2]
1.064	10	20

3.4.9 2-Furyl Methacrylic Anhydride (FMA)

Positive uniaxial crystal: $n_\text{e} > n_\text{o}$;
Point group: 4mm;
Transparency range at "0" transmittance level: $0.38 - 1.1 \ \mu\text{m}$ [3.614];
Experimental values of refractive indices [3.614]:

λ [μm]	n_o	n_e
0.4305	1.751	2.137
0.4535	1.721	2.064
0.4880	1.691	2.007
0.5145	1.685	1.983
0.5321	1.671	1.958
0.6328	1.641	1.887
0.8330	1.619	1.841
1.0642	1.612	1.821
1.1523	1.617	1.811

Sellmeier equations (λ in μm, $T = 20\,°\text{C}$) [3.614]:

$$n_\text{o}^2 = 1.804 + \frac{0.6884 \ \lambda^2}{\lambda^2 - 0.08301} + 0.0527 \ \lambda^2 \ ,$$
$$n_\text{e}^2 = 2.097 + \frac{1.1090 \ \lambda^2}{\lambda^2 - 0.10172} - 0.008748 \ \lambda^2 \ .$$

Experimental and theoretical values of phase-matching angle and calculated values of "walk-off" angle:

Interacting wavelengths [μm]	θ_exp [deg]	θ_theor [deg] [3.614]	ρ_1 [deg]
SHG, e + o $\Rightarrow$ o			
1.0642 $\Rightarrow$ 0.5321	51.2 [3.614]	50.80	6.766

Experimental values of NCPM temperature [3.614]:

Interacting wavelengths [μm]	T [°C]
SHG, e + o $\Rightarrow$ o	
$0.9038 \Rightarrow 0.4519$	0
$0.9076 \Rightarrow 0.4538$	19
$0.9108 \Rightarrow 0.4554$	38

Experimental value of internal angular bandwidth [3.614]:

Interacting wavelengths [μm]	θ_{pm} [deg]	$\Delta\theta^{\mathrm{int}}$ [deg]
SHG, e + o $\Rightarrow$ o		
$1.0642 \Rightarrow 0.5321$	51.2	0.031

Temperature tuning of noncritical SHG [3.614]:

Interacting wavelengths [μm]	$\mathrm{d}\lambda_1/\mathrm{d}T$ [nm/K]
SHG, e + o $\Rightarrow$ e	
$0.0976 \Rightarrow 0.4538$	0.18

Effective nonlinearity expression in the phase-matching direction [3.100]:

$$d_{\mathrm{oeo}} = d_{\mathrm{eoo}} = d_{31} \sin\theta .$$

Nonlinear coefficients [3.614, 37]:

$$d_{31}(1.0642\,\mu\mathrm{m}) = 12 \ \mathrm{pm/V} ,$$
$$d_{33}(1.0642\,\mu\mathrm{m}) = 18 \ \mathrm{pm/V} .$$

3.4.10 3-Methyl-4-nitropyridine-1-oxide (POM)

Positive biaxial crystal: $2V_Z = 68.87°$ at $\lambda = 0.5461$ μm [3.615];
Point group: 222;
Assignment of dielectric and crystallographic axes:
$X, Y, Z \Rightarrow c, a, b$;
Transparency range: $0.4 - 2.3$ μm [3.615];
Linear absorption coefficient α:

λ [μm]	α [cm^{-1}]	Ref.
0.5321	1.88	3.615
	1.2	3.616
1.0642	0.77	3.615

Experimental values of refractive indices [3.615]:

λ [μm]	n_X	n_Y	n_Z
0.435	1.717		
0.468	1.690	1.809	2.114
0.480	1.682	1.793	2.082
0.509	1.668	1.766	2.028
0.532	1.660	1.750	1.997
0.546	1.656	1.742	1.981
0.579	1.648	1.728	1.953
0.644	1.637	1.709	1.915
1.064	1.625	1.668	1.829

Sellmeier equations (λ in μm, $T = 20\ °C$) [3.615]:

$$n_X^2 = 2.4529 + \frac{0.1641\,\lambda^2}{\lambda^2 - 0.1280}\ ,$$

$$n_Y^2 = 2.4315 + \frac{0.3556\,\lambda^2}{\lambda^2 - 0.1276} - 0.0579\,\lambda^2\ ,$$

$$n_Z^2 = 2.5521 + \frac{0.7962\,\lambda^2}{\lambda^2 - 0.1289} - 0.0941\,\lambda^2\ .$$

Experimental and theoretical values of phase-matching angle and calculated values of "walk-off" angle:
XZ plane, $\phi = 0°, \theta > V_Z$

Interacting wavelengths [μm]	θ_{exp} [deg]	θ_{theor} [deg] [3.615]	ρ_1 [deg]
SHG, e + e $\Rightarrow$ o			
1.0642 $\Rightarrow$ 0.5321	54.3 [3.615]	54.12	6.640
1.3188 $\Rightarrow$ 0.6594	44.2 [3.617]	45.28	6.010
1.34 $\Rightarrow$ 0.67	43.8 [3.617]	44.90	5.943

Experimental values of internal angular bandwidth:
XZ plane, $\phi = 0°$

Interacting wavelengths [μm]	θ_{pm} [deg]	$\Delta\theta^{\mathrm{int}}$ [deg]	Ref.
SHG, e + e $\Rightarrow$ o			
1.0642 $\Rightarrow$ 0.5321	54.3	0.025	3.615
1.3188 $\Rightarrow$ 0.6594	44.2	0.021	3.617
1.34 $\Rightarrow$ 0.67	43.8	0.020	3.617

Effective nonlinearity expressions in the phase-matching direction for three-wave interactions in the principal planes of POM crystal [3.35]:

XY plane

$$d_{\mathrm{eoe}} = d_{\mathrm{oee}} = d_{14} \sin 2\phi \; ;$$

YZ plane

$$d_{\mathrm{eeo}} = d_{14} \sin 2\theta \; ;$$

XZ plane, $\theta < V_Z$

$$d_{\mathrm{eoe}} = d_{\mathrm{oee}} = d_{14} \sin 2\theta \; ;$$

XZ plane, $\theta > V_Z$

$$d_{\mathrm{eeo}} = d_{14} \sin 2\theta \; ;$$

Nonlinear coefficients [3.615, 37]:

$$d_{14}(1.064 \; \mu\mathrm{m}) = 20 \times d_{11}(\mathrm{SiO_2}) \pm 15\% = 6 \pm 0.9 \; \mathrm{pm/V} \; ,$$

$$d_{14}(1.064 \; \mu\mathrm{m}) = 13.5 \times d_{36}(\mathrm{KDP}) \pm 10\% = 5.3 \pm 0.5 \; \mathrm{pm/V} \; ,$$

Laser-induced damage threshold:

λ [μm]	τ_{p} [ns]	$I_{\mathrm{thr}} \times 10^{-12}$ [W/m^2]	Ref.
0.5321	15	0.5	3.616
	0.02	> 1.5	3.615
	0.025	> 2.7	3.616
0.5927	1	1	3.618
0.62	0.0001	10000 (?)	3.619
1.0642	0.02	> 20	3.615

3.4.11 Thienylchalcone (T-17)

Positive biaxial crystal: $2V_Z = 82.6°$ at $\lambda = 0.5321 \; \mu\mathrm{m}$ [3.230];
Point group: 2;
Assignment of dielectric and crystallographic axes:
$Y \parallel b$, the axes a and c lie in XZ plane, the angle between them is $\beta = 109.9°$, $Z \parallel a$ (Fig. 3.7) [3.230];
Mass density: 1.27 g/cm^3 [3.230];
Vickers hardness: 17 [3.230];
Transparency range at "0" transmittance level: $\approx 0.4 - 1.06 \; \mu\mathrm{m}$ [3.230];

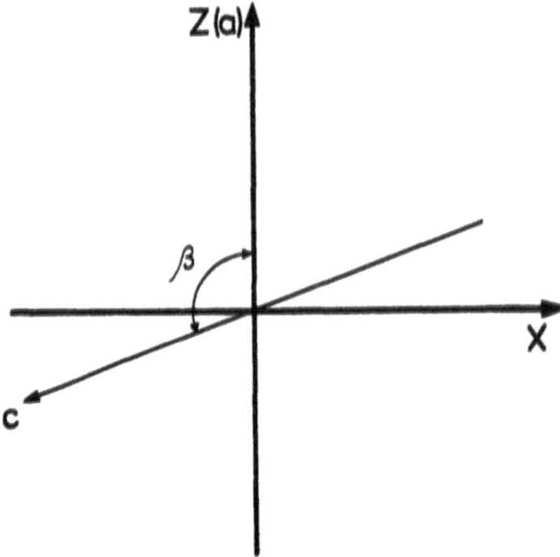

Fig. 3.7. Dielectric (X, Y, Z) and crystallographic (a, b, c) axes of T-17 crystal. The Y axis is parallel to the b axis and normal to the plane of the figure

Sellmeier equations (λ in μm, $T = 20°$ C) [3.230]:

$$n_X^2 = 2.6311 + \frac{0.059014}{\lambda^2 - 0.121160} + 0.25553 \times 10^{-5}\,\lambda^2 \ ,$$

$$n_Y^2 = 2.8265 + \frac{0.037232}{\lambda^2 - 0.098256} - 3.02020 \times 10^{-5}\,\lambda^2 \ ,$$

$$n_Z^2 = 3.0468 + \frac{0.078174}{\lambda^2 - 0.098845} - 0.61590 \times 10^{-5}\,\lambda^2 \ .$$

Experimental and theoretical values of phase-matching angle and calculated values of "walk-off" angle:
XZ plane, $\phi = 0°, \theta > V_Z$

Interacting wavelengths [μm]	$\theta_{\exp}$ [deg]	θ_{theor} [deg] [3.230]	ρ_1 [deg]
SHG, e + e ⇒ o 1.6042 ⇒ 0.5321	61.6 [3.230]	63.87	3.532

Experimental values of internal angular and temperature bandwidths [3.230]:
XZ plane, $\phi = 0°, \theta > V_Z$

Interacting wavelengths [μm]	θ_{pm} [deg]	$\Delta\theta^{\mathrm{int}}$ [deg]	$\Delta\theta^{\mathrm{int}}$ [deg]	ΔT [deg]
SHG, e + e ⇒ o 1.0642 ⇒ 0.5321	61.6	0.030	0.690	2.2

Effective nonlinearity expressions in the phase-matching direction [3.35]:

XY plane

$$d_{ooe} = d_{23}\cos\phi \ ,$$
$$d_{eoe} = d_{oee} = d_{25}\sin 2\phi \ ,$$

YZ plane

$$d_{eeo} = d_{25}\sin 2\theta \ ,$$
$$d_{oeo} = d_{eoo} = d_{21}\cos\theta \ ,$$

XZ plane, $\theta > V_Z$

$$d_{eoe} = d_{oee} = d_{21}\cos^2\theta + d_{23}\sin^2\theta - d_{25}\sin 2\theta \ ;$$

XZ plane, $\theta > V_Z$

$$d_{eeo} = d_{21}\cos^2\theta + d_{23}\sin^2\theta - d_{25}\sin 2\theta \ .$$

Nonlinear coefficients [3.230, 37]:

XZ plane, $\theta > V_Z$

$$d_{eeo}(1.0642 \Rightarrow 0.5321\,\mu m) = 0.226 \times d_{21} + 0.774 \times d_{23} - 0.837 \times d_{25}$$
$$= 6.3\,\mathrm{pm/V} \ .$$

3.4.12 5-Nitrouracil (5NU)

Positive biaxial crystal: $2V_Z = 92.9°$ at $\lambda = 0.546\,\mu m$ [3.620];
Point group: 222;
Assignment of dielectric and crystallographic axes:
$X, Y, Z \Rightarrow b, c, a$;
Transparency range: $0.41 - 2.4\,\mu m$ [3.620];
Experimental values of refractive indices [3.620]:

$\lambda[\mu m]$	n_X	n_Y	n_Z
0.435	2.0051	1.7797	1.6351
0.468	1.9737	1.7566	1.6113
0.480	1.9668	1.7500	1.6065
0.509	1.9537	1.7441	1.5958
0.518	1.9411	1.7375	1.5894
0.546	1.9315	1.7242	1.5850
0.579	1.9190	1.7176	1.5787
0.589	1.9135	1.7156	1.5758
0.636	1.9014	1.7070	1.5694
0.644	1.9010	1.7050	1.5670
1.0642	1.8517	1.6799	1.5341
1.3188	1.8362	1.6719	1.5248

Sellmeier equations (λ in μm, $T = 20\,°C$) [3.620]:

$$n_X^2 = 2.390 + \frac{1.033\,\lambda^2}{\lambda^2 - 0.0700} - 0.0549\,\lambda^2 \;,$$

$$n_Y^2 = 1.892 + \frac{0.870\,\lambda^2}{\lambda^2 - 0.0599} \;,$$

$$n_Z^2 = 2.098 + \frac{0.290\,\lambda^2}{\lambda^2 - 0.0947} - 0.0485\,\lambda^2 \;.$$

Experimental and theoretical values of phase-matching angle and calculated values of "walk-off" angle:

XZ plane, $\phi = 0°, \theta < V_Z$

Interacting wavelengths [μm]	$\theta_{\exp}$ [deg]	θ_{theor} [deg] [3.620]	ρ_1 [deg]
SHG, e + e $\Rightarrow$ o			
1.0642 $\Rightarrow$ 0.5321	37.2 [3.620]	34.41	10.46
1.338 $\Rightarrow$ 0.669	40.2 [3.620]	36.79	10.58

XZ plane, $\phi = 0°, \theta > V_Z$

Interacting wavelengths [μm]	$\theta_{\exp}$ [deg]	θ_{theor} [deg] [3.620]	ρ_1 [deg]	ρ_3 [deg]
SHG, o + e $\Rightarrow$ e				
1.0642 $\Rightarrow$ 0.5321	67.7 [3.620]	67.60	6.56	6.91
1.338 $\Rightarrow$ 0.669	60.0 [3.620]	59.05	8.52	8.74
1.907 $\Rightarrow$ 0.9535	61.2 [3.620]	56.57	9.12	9.02

Effective nonlinearity expressions in the phase-matching direction for three-wave interactions in the principal planes of 5-NU crystal [3.35]:

XY plane

$$d_{\text{eeo}} = d_{14} \sin 2\phi \;;$$

YZ plane

$$d_{\text{eoe}} = d_{\text{oee}} = d_{14} \sin 2\theta \;;$$

XZ plane, $\theta < V_Z$

$$d_{\text{eeo}} = d_{14} \sin 2\theta \;;$$

XZ plane, $\theta > V_Z$

$$d_{\text{eoe}} = d_{\text{oee}} = d_{14} \sin 2\theta \;.$$

Nonlinear coefficient:

$$d_{14}(1.064\,\mu m) = 8.4 \pm 1.3\,\text{pm/V}\ [3.620]\ .$$

Laser-induced damage threshold [3.620]:

λ [μm]	τ_p [ns]	$I_{thr} \times 10^{-12}$ [W/m^2]
0.532	6	10
0.593	9	10
1.0642	10	30
1.338	0.16	68

3.4.13 2-(N-Prolinol)-5-nitropyridine (PNP)

Negative biaxial cyrstal: $2V_Z = 64.6°$ at $\lambda = 0.58\,\mu m$ [3.621];
Point group: 2;
Assignment of dielectric and crystallographic axes of PNP is given in [3.622];
Transparency range at "0" transmittance level [3.621]:
0.49 – 2.08 μm along X, Y axes; 0.466 – 2.3 μm along Z axis;
Experimental values of refractive indices [3.621]:

λ [μm]	n_X	n_Y	n_Z
0.4880	2.239	1.929	1.477
0.5145	2.164	1.873	1.474
0.580	2.040	1.813	1.468
0.600		1.801	1.468
0.6328	1.990	1.788	1.467
1.0642	1.880	1.732	1.456

Sellmeier equations (λ in μm, $T = 20\,°\text{C}$) [3.621]:

$$n_X^2 = 2.3454 + \frac{1.029757\,\lambda^2}{\lambda^2 - (0.3830)^2}\ ,$$

$$n_Y^2 = 2.5658 + \frac{0.375380\,\lambda^2}{\lambda^2 - (0.4006)^2}\ ,$$

$$n_Z^2 = 2.0961 + \frac{0.029386\,\lambda^2}{\lambda^2 - (0.4016)^2}\ .$$

Experimental and theoretical values of phase-matching angle and calculated value of "walk-off" angle:
XZ plane, $\phi = 0°, \theta < V_Z$

Interacting wavelengths [μm]	$\theta_{\exp}$ [deg]	θ_{theor} [deg] [3.621]	ρ_1 [deg]
SHG, e + e $\Rightarrow$ o			
1.0642 $\Rightarrow$ 0.5321	21 [3.621]	11.92	7.349

Effective nonlinearity expressions in the phase-matching direction [3.35]:
XY plane

$$d_{\text{eeo}} = d_{25} \sin 2\phi \ ,$$
$$d_{\text{oeo}} = d_{\text{eoo}} = d_{23} \cos \phi \ ;$$

YZ plane

$$d_{\text{ooe}} = d_{21} \cos \theta \ ,$$
$$d_{\text{eoe}} = d_{\text{oee}} = d_{25} \sin 2\theta \ ;$$

XZ plane, $\theta < V_Z$

$$d_{\text{eeo}} = d_{21} \cos^2 \theta + d_{23} \sin^2 \theta - d_{25} \sin 2\theta \ ;$$

XZ plane, $\theta > V_Z$

$$d_{\text{eoe}} = d_{\text{oee}} = d_{21} \cos^2 \theta + d_{23} \sin^2 \theta - d_{25} \sin 2\theta \ .$$

Nonlinear coefficients [3.622]:

$$d_{21}(1.064\,\mu\text{m}) = 48 \pm 11 \ \text{pm/V} \ ,$$
$$d_{22}(1.064\,\mu\text{m}) = 17 \pm 4 \ \text{pm/V} \ .$$

3.4.14 2-Cyclooctylamino-5-nitropyridine (COANP)

Positive biaxial crystal: $2V_Z = 36.13°$ at $\lambda = 0.547\,\mu\text{m}$ (at $\lambda = 0.497\,\mu\text{m}$ COANP becomes uniaxial) [3.623];
Point group: mm2;
Assignment of dielectric and crystallographic axes:
$X, Y, Z \Rightarrow c, a, b$;
Mass density: $1.24\,\text{g/cm}^3$ [3.624];
Transparency range at 0.5 transmittance level for 0.9 mm long crystal:
$0.47 - 1.5\,\mu\text{m}$ (along a axis) [3.624];
Linear absorption coefficient α [3.624]:

λ [μm]	α [cm^{-1}]
0.532	3
1.064	0.8
1.35	< 1

Experimental values of refractive indices [3.623]:

λ [μm]	n_X	n_Y	n_Z
0.480	1.776	1.766	2.505
0.547	1.687	1.700	1.839
0.577	1.663	1.690	1.824
0.650	1.643	1.668	1.772
1.064	1.604	1.636	1.715

Experimental value of phase-matching angle:
YZ plane, $\phi = 90°$

Interacting wavelengths [μm]	θ_{pm} [deg]
SHG, $e + e \Rightarrow o$	
$1.0642 \Rightarrow 0.5321$	63.6 [3.624]

Effective nonlinearity expressions in the phase-matching direction for three-wave interactions in the principal planes of COANP crystal [3.35, 36]:
XY plane

$$d_{\text{ooe}} = d_{32} \sin \phi \; ;$$

YZ plane

$$d_{\text{eeo}} = d_{32} \sin^2 \theta + d_{31} \cos^2 \theta \; ;$$

XZ plane, $\theta < V_z$

$$d_{\text{ooe}} = d_{31} \cos \theta \; ;$$

XZ plane, $\theta > V_z$

$$d_{\text{oeo}} = d_{\text{eoo}} = d_{31} \cos \theta \; .$$

Effective nonlinearity expressions for three-wave interactions in the aribitrary direction of COANP crystal are given in [3.36]
Nonlinear coefficients [3.624, 623, 37]:

$$d_{31}(1.0642\,\mu\text{m}) = 11.3 \pm 1.5\,\text{pm/V} \; ,$$

$$d_{32}(1.0642\,\mu\text{m}) = 24 \pm 12\,\text{pm/V} \; ,$$

$$d_{33}(1.0642\,\mu\text{m}) = 10.8 \pm 1.5\,\text{pm/V} \; .$$

Laser-induced damage threshold [3.624]:

λ [μm]	τ_p [ns]	$I_{\text{thr}} \times 10^{-12}$ [W/m^2]
1.064	250	> 0.015

3.4.15 L-N-(5-Nitro-2-pyridyl)leucinol (NPLO)

Positive biaxial crystal: $2V_Z = 43°$ at $\lambda = 0.514\,\mu m$ [3.625];
Point group: 2;
Assignment of dielectric and crystallographic axes:
$Y \parallel b$, the axes a and c lie in XZ plane, the angle between them is $\beta = 110.4°$, the angle between the axes Z and c is $\alpha = 56°$ (Fig. 3.8) [3.625];
Mass density: 1.24 g/cm^3 [3.625];
Vickers hardness: 18 [3.625];
Transparency range at "0" transmittance level: $0.47 - > 1.06$ µm;
Experimental values of refractive indices [3.625]:

λ [µm]	n_X	n_Y	n_Z
0.4880	1.470	1.712	2.218
0.5145	1.463	1.681	2.116
0.6328	1.457	1.631	1.933
1.0642	1.451	1.598	1.812

Sellmeier equations (λ in µm, $T = 20\,°C$) [3.625]:

$$n_X^2 = 2.1240 + \frac{0.0011}{\lambda^2 - 0.2108} - 0.0174\,\lambda^2 \ ,$$

$$n_Y^2 = 2.5607 + \frac{0.0257}{\lambda^2 - 0.1700} - 0.0299\,\lambda^2 \ ,$$

$$n_Z^2 = 3.2123 + \frac{0.1302}{\lambda^2 - 0.1625} - 0.0559\,\lambda^2 \ .$$

Experimental and theoretical values of phase-matching angle and calculated values of "walk-off" angle:

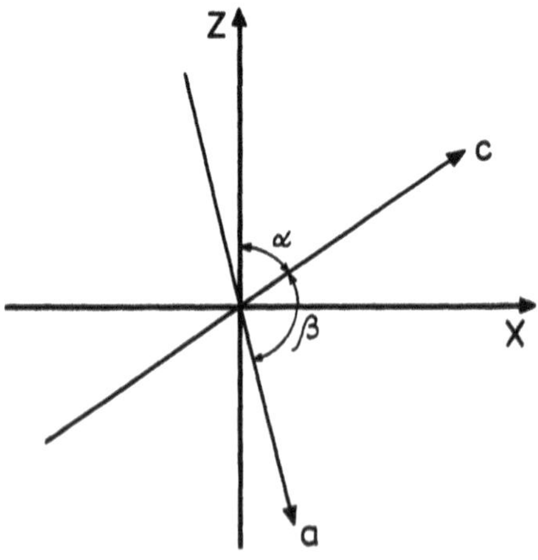

Fig. 3.8. Dielectric (X, Y, Z) and crystallographic (a, b, c) axes of NPLO crystal. The Y axis is parallel to the b axis and normal to the plane of the figure

XZ plane, $\phi = 0°, \theta < V_Z$

Interacting wavelengths [μm]	θ_{exp} [deg]	θ_{theor} [deg] [3.625]	ρ_1 [deg]	ρ_3 [deg]
SHG, $e + e \Rightarrow 0$				
$1.0642 \Rightarrow 0.5321$	33 [3.625]	30.52	9.811	14.123

XZ plane, $\phi = 0°, \theta > V_Z$

Interacting wavelengths [μm]	θ_{exp} [deg]	θ_{theor} [deg] [3.625]	ρ_1 [deg]
SHG, $e + e \Rightarrow o$			
$1.0642 \Rightarrow 0.5321$	51.7 [3.625]	55.30	12.496

Experimental values of the internal angular bandwidth [3.625]:
XZ plane, $\phi = 0°, \theta < V_Z$

Interacting wavelengths [μm]	θ_{pm} [deg]	$\Delta\theta^{\text{int}}$ [deg]
SHG, $e + o \Rightarrow e$		
$1.0642 \Rightarrow 0.5321$	33	0.12

XZ plane, $\phi = 0°, \theta > V_Z$

Interacting wavelengths [μm]	θ_{pm} [deg]	$\Delta\theta^{\text{int}}$ [deg]
SHG, $e + e \Rightarrow o$		
$1.0642 \Rightarrow 0.5321$	51.7	0.11

Effective nonlinearity expressions in the phase-matching direction [3.35]:
XY plane

$$d_{\text{ooe}} = d_{23} \cos \phi ,$$

$$d_{\text{eoe}} = d_{\text{oee}} = d_{25} \sin 2\phi ;$$

YZ plane

$$d_{\text{eeo}} = d_{25} \sin 2\theta ,$$

$$d_{\text{oeo}} = d_{\text{eoo}} = d_{21} \cos \theta ;$$

XZ plane, $\theta < V_Z$

$$d_{\text{eoe}} = d_{\text{oee}} = d_{21} \cos^2 \theta + d_{23} \sin^2 \theta - d_{25} \sin 2\theta ;$$

XZ plane, $\theta > V_Z$

$$d_{\text{eeo}} = d_{21} \cos^2 \theta + d_{23} \sin^2 \theta - d_{25} \sin 2\theta .$$

Nonlinear coefficients [3.625, 37]:

XZ plane, $\theta < V_Z$

$$d_{\text{eoe}}(1.0642 \Rightarrow 0.5321\ \mu\text{m}) = d_{\text{oee}}(1.0642 \Rightarrow 0.5321\ \mu\text{m})$$
$$= 0.703 \times d_{21} + 0.297 \times d_{23} - 0.914 \times d_{25}$$
$$= 2.7\ \text{pm/V}\ ;$$

XZ plane, $\theta > V_Z$

$$d_{\text{eeo}}(1.0642 \Rightarrow 0.5321\mu\text{m}) = 0.322 \times d_{21} + 0.678 \times d_{23} - 0.935 \times d_{25}$$
$$= 33.2\ \text{pm/V}\ .$$

Laser-induced surface-damage threshold [3.625]:

λ [µm]	τ_p [ns]	$I_{\text{thr}} \times 10^{-12}$ [W/m^2]
1.064	8	60

3.4.16 C$_6$H$_4$(NO$_2$)$_2$, m-Dinitrobenzene (MDNB)

Negative biaxial crystal: $2V_Z = 51.15°$ at $\lambda = 0.5321\mu$m [3.611];
Point group: mm2;
Assignment of dielectric and crystallographic axes:
$X, Y, Z \Rightarrow a, b, c$;
Mass density: 1.57 g/cm^3;
Transparency range at "0" transmittance level: $0.48 - 1.57$ µm [3.611];

Experimental values of refractive indices [3.611]:

λ [µm]	n_X	n_Y	n_Z
0.436	1.8025	1.7361	1.5072
0.492	1.7731	1.7104	1.4964
0.532	1.7592	1.6983	1.4912
0.546	1.7553	1.6950	1.4896
0.577	1.7480	1.6886	1.4869
0.579	1.7476	1.6882	1.4865
0.589	1.7456	1.6865	1.4859
0.633	1.7381	1.6798	1.4827
1.064	1.7093	1.6539	1.4707
1.153	1.7072	1.6520	1.4698

Experimental values of phase-matching angle:
XZ plane, $\phi = 0°, \theta > V_Z$

Interacting wavelengths [µm]	θ_{pm} [deg]
SHG, $o + o \Rightarrow e$	
$1.0642 \Rightarrow 0.5321$	35 [3.611]
$1.1523 \Rightarrow 0.57615$	34.75 [3.626]

Experimental value of internal angular bandwidth [3.611]:
XZ plane, $\phi = 0°$

Interacting wavelengths [µm]	θ_{pm} [deg]	$\Delta\theta^{int}$ [deg]
SHG, $o + o \Rightarrow e$		
$1.0642 \Rightarrow 0.5321$	35	0.029

Effective nonlinearity expressions in the phase-matching direction for three-wave interactions in the principal planes of MDNB crystal [3.35, 36]:
XY plane

$$d_{eeo} = d_{31} \sin^2 \phi + d_{32} \cos^2 \phi \; ;$$

YZ plane

$$d_{ooe} = d_{31} \sin \theta \; ;$$

XZ plane, $\theta < V_Z$

$$d_{oeo} = d_{eoo} = d_{32} \sin \theta \; ;$$

XZ plane, $\theta > V_Z$

$$d_{ooe} = d_{32} \sin \theta \; .$$

Effective nonlinearity expressions for three–wave interactions in the arbitrary direction of MDNB crystal are given in [3.36]
Nonlinear coefficients [3.611, 37]:

$$d_{31}(1.0642\,\mu m) = 2.75 \times d_{36}\,(KDP) \pm 20\% = 1.1 \pm 0.2\,pm/V \; ,$$

$$d_{32}(1.0642\,\mu m) = 5.5 \times d_{36}\,(KDP) \pm 20\% = 2.1 \pm 0.4\,pm/V \; ,$$

$$d_{33}(1.0642\,\mu m) = 1.7 \times d_{36}\,(KDP) \pm 25\% = 0.7 \pm 0.2\,pm/V \; .$$

Laser-induced surface-damage threshold [3.611]:

λ [µm]	τ_p [ns]	$I_{thr} \times 10^{-12}$ [W/m^2]
1.06	40	2

3.4.17 4-(N,N-Dimethylamino)-3-acetamidonitrobenzene (DAN)

Positive biaxial crystal: $2V_Z = 81.7°$ at $\lambda = 0.5321$ μm [3.627];
Point group: 2;
Assignment of dielectric and crystallographic axes:
$Y \parallel b$, the axes a and c lie in XZ plane, the angle between them is $\beta = 94.4°$, the
angle between the axes X and c is $\alpha = 50.6°$ (Fig. 3.9) [3.628, 629];
Transparency range at "0" transmittance level: 0.485–2.27 μm [3.629];
Linear absorption coefficient α

λ [μm]	α [cm^{-1}]	Ref
0.5–2.0	< 1	3.628
1.0	1.5	3.627
	1.2	3.629

Experimental values of refractive indices [3.627]:

λ [μm]	n_X	n_Y	n_Z
0.4965	1.574	1.779	2.243
0.5145	1.557	1.748	2.165
0.5321	1.554	1.732	2.107
0.5850	1.545	1.701	2.005
0.6328	1.539	1.682	1.949
1.0642	1.517	1.636	1.843

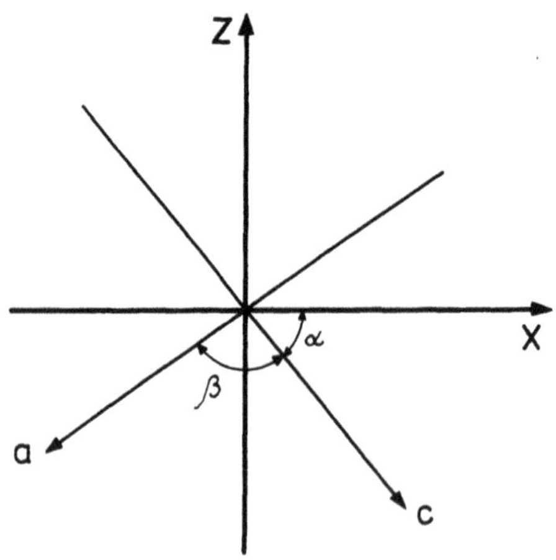

Fig. 3.9. Dielectric (X, Y, Z) and crystallographic (a, b, c) axes of DAN crystal. The Y axis is parallel to the b axis and normal to the plane of the figure

Sellmeier equations (λ in μm, $T = 20$ °C) [3.629];

$$n_X^2 = 2.1390 + \frac{0.147408\,\lambda^2}{\lambda^2 - (0.3681)^2} \, ,$$

$$n_Y^2 = 2.3290 + \frac{0.307173\,\lambda^2}{\lambda^2 - (0.3933)^2} \ ,$$

$$n_Z^2 = 2.5379 + \frac{0.719557\,\lambda^2}{\lambda^2 - (0.4194)^2} \ .$$

Experimental and theoretical values of the phase-matching angle and calculated values of "walk-off" angle:
XZ plane, $\phi = 0°, \theta < V_Z$

Interacting wavelengths [µm]	$\theta_{\exp}$ [deg]	θ_{theor} [deg] [3.629]	ρ_1 [deg]	ρ_3 [deg]
SHG, e + o $\Rightarrow$ e				
1.0642 $\Rightarrow$ 0.5321	20.9 [3.629]	18.42	5.653	8.120
1.3188 $\Rightarrow$ 0.6594	27.6 [3.629]	29.55	8.224	9.949

XZ plane, $\phi = 0°, \theta > V_Z$

Interacting wavelengths [µm]	$\theta_{\exp}$ [deg]	θ_{theor} [deg] [3.629]	ρ_1 [deg]
SHG, e + e $\Rightarrow$ o			
1.0642 $\Rightarrow$ 0.5321	57.3 [3.629]	58.58	10.498
1.3188 $\Rightarrow$ 0.6594	49.4 [3.629]	49.62	10.623

Experimental values of the internal angular bandwidth:
XZ plane, $\phi = 0°, \theta > V_Z$

Interacting wavelengths [µm]	θ_{pm} [deg]	$\Delta\theta^{\text{int}}$ [deg]
SHG, e + e $\Rightarrow$ o		
1.0642 $\Rightarrow$ 0.5321	57.3	0.007 [3.629]
		0.011 [3.628]

Effective nonlinearity expressions in the phase-matching direction [3.35]:
XY plane

$$d_{\text{ooe}} = d_{23} \cos\phi \ ,$$

$$d_{\text{eoe}} = d_{\text{oee}} = d_{25} \sin 2\phi \ ;$$

YZ plane

$$d_{\text{eeo}} = d_{25} \sin 2\theta \ ,$$

$$d_{\text{oeo}} = d_{\text{eoo}} = d_{21} \cos\theta \ ;$$

XZ plane, $\theta < V_Z$

$$d_{\text{eoe}} = d_{\text{oee}} = d_{21}\cos^2\theta + d_{23}\sin^2\theta - d_{25}\sin 2\theta \ ;$$

XZ plane, $\theta > V_Z$

$$d_{\text{eeo}} = d_{21}\cos^2\theta + d_{23}\sin^2\theta - d_{25}\sin 2\theta \ .$$

Nonlinear coefficients [3.629, 323, 37]:

$$d_{21}(1.0642\,\mu\text{m}) = 1.1 \pm 1.5\,\text{pm/V} \ ,$$

$$d_{22}(1.0642\,\mu\text{m}) = 3.9 \pm 0.8\,\text{pm/V} \ ,$$

$$d_{23}(1.0642\,\mu\text{m}) = 37.5 \pm 11.3\,\text{pm/V} \ ,$$

$$d_{25}(1.0642\,\mu\text{m}) = 1.1 \pm 1.5\,\text{pm/V} \ .$$

Laser-induced damage threshold [3.629]:

λ [μm]	τ_p[ns]	$I_{\text{thr}} \times 10^{-12}$[W/m^2]	Note	
1.064	15	0.8	30	Hz
	0.1	50		

3.4.18 Methyl-(2,4-dinitrophenyl)-aminopropanoate (MAP)

Positive biaxial crystal: $2V_Z = 79.9°$ at $\lambda = 0.5321$ μm [3.630];
Point group: 2;
Assignment of dielectric and crystallographic axes:
$Y \parallel b$, the axes a and c lie in XZ plane, the angle between them is $\beta = 95.6°$, the angle between the axes Z and a is $\alpha = 37°$ (Fig. 3.10) [3.630];
Transparency range at "0" transmittance level: $0.5 - 2.2$ μm [3.630];

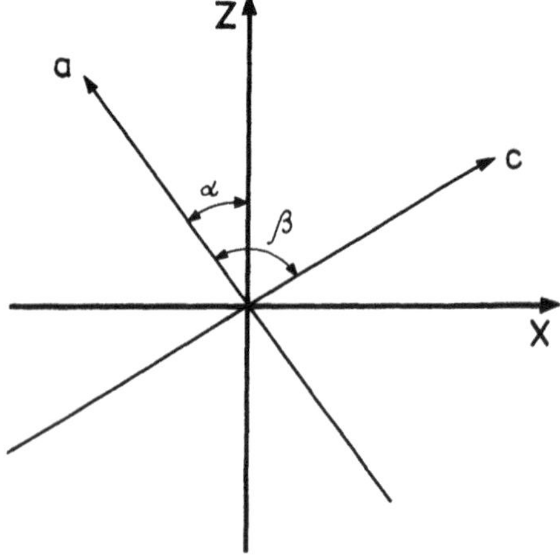

Fig. 3.10. Dielectric $(X,\ Y,\ Z)$ and crystallographic $(a,\ b,\ c)$ axes of MAP crystal. The Y axis is parallel to the b axis and normal to the plane of the figure

Linear absorption coefficient α:

$\alpha = 3.7$ cm^{-1} at $\lambda = 0.5321\ \mu$m [3.630] ;

Experimental values of refractive indices:

λ [μm]	n_X	n_Y	n_Z
0.5321	1.5568	1.7100	2.0353
1.0642	1.5078	1.5991	1.8439

Sellmeier equations (λ in μm, $T = 20\ ^{\circ}$C) [3.630]:

$$n_X^2 = 2.1713 + \frac{0.10305\,\lambda^2}{\lambda^2 - 0.16951} - 0.01667\,\lambda^2 \ ,$$

$$n_Y^2 = 2.3100 + \frac{0.22580\,\lambda^2}{\lambda^2 - 0.17988} - 0.01886\,\lambda^2 \ ,$$

$$n_Z^2 = 2.7523 + \frac{0.60790\,\lambda^2}{\lambda^2 - 0.16060} - 0.05361\,\lambda^2 \ .$$

Experimental and theoretical values of phase-matching angle and calculated values of "walk-off" angle:
YZ plane, $\phi = 90°$

Interacting wavelengths [μm]	θ_{exp} [deg]	θ_{theor} [deg] [3.630]	ρ_2 [deg]
SHG, o + e $\Rightarrow$ o			
1.0642 $\Rightarrow$ 0.5321	11 [3.630]	10.40	2.541

XZ plane, $\phi = 0°, \theta > V_Z$

Interacting wavelengths [μm]	θ_{exp} [deg]	θ_{theor} [deg] [3.630]	ρ_1 [deg]
SHG, e + e $\Rightarrow$ o			
1.0642 $\Rightarrow$ 0.5321	56 [3.630]	55.03	11.316

Effective nonlinearity expressions in the phase-matching direction [3.35]:
XY plane

$d_{\mathrm{ooe}} = d_{23}\cos\phi \ ,$

$d_{\mathrm{eoe}} = d_{\mathrm{oee}} = d_{25}\sin 2\phi \ ;$

YZ plane

$d_{\mathrm{eeo}} = d_{25}\sin 2\theta \ ,$

$d_{\mathrm{oeo}} = d_{\mathrm{eoo}} = d_{21}\cos\theta \ ;$

XZ plane, $\theta < V_Z$

$$d_{\text{eoe}} = d_{\text{oee}} = d_{21} \cos^2 \theta + d_{23} \sin^2 \theta - d_{25} \sin 2\theta \ ;$$

XZ plane, $\theta > V_Z$

$$d_{\text{eeo}} = d_{21} \cos^2 \theta + d_{23} \sin^2 \theta - d_{25} \sin 2\theta \ .$$

Nonlinear coefficients (in crystallographic reference frame a, b, c) [3.630]:

$$d_{21}(1.0642\,\mu\text{m}) = \pm(23.9 \pm 3.0)\,\text{pm/V} \ ,$$

$$d_{22}(1.0642\,\mu\text{m}) = \pm(26.3 \pm 3.0)\,\text{pm/V} \ ,$$

$$d_{23}(1.0642\,\mu\text{m}) = \pm(5.3 \pm 1.2)\,\text{pm/V} \ ,$$

$$d_{25}(1.0642\,\mu\text{m}) = \mp(0.8 \pm 0.6)\,\text{pm/V} \ .$$

The transformation of **d**-tensor coefficients to dielectric reference frame (X, Y, Z) is performed in [3.630]

Laser-induced damage threshold [3.630]:

λ [μm]	τ_p [ns]	$I_{\text{thr}} \times 10^{-12}$ [W/m^2]
0.5321	7	> 1.5
1.0642	10	30

3.4.19 m-Nitroaniline (MNA)

Negative biaxial crystal: $2V_Z = 104°$ at $\lambda = 0.5321\,\mu$m [3.631];
Point group: mm2;
Assignment of dielectric and crystallographic axes:
$X, Y, Z \Rightarrow c, b, a$;
Transparency range at "0" transmittance level: $0.5 - 2$ μm [3.632];
Linear absorption coefficient α [3.632]:

λ [μm]	α [cm^{-1}]	Note
0.5315	4	along b, $\mathbf{E} \parallel c$
	6	along b, $\mathbf{E} \parallel a$

Experimental values of refractive indices [3.631]:

λ [μm]	n_X	n_Y	n_Z
0.5321	1.6982	1.7533	1.7887
1.0642	1.6283	1.6815	1.7168

Experimental values of phase-matching angle:
XY plane, $\theta = 90°$

Interacting wavelengths [μm]	ϕ_{pm} [deg]
SHG, o + o $\Rightarrow$ e	
1.063 $\Rightarrow$ 0.5315	55 [3.632]
1.0462 $\Rightarrow$ 0.5321	55 [3.633]

YZ plane, $\phi = 90°$

Interacting wavelengths [μm]	θ_{pm} [deg]
SHG, e + e $\Rightarrow$ o	
1.063 $\Rightarrow$ 0.5315	44 [3.632]

Experimental values of internal angular bandwidth:
XY plane, $\theta = 90°$

Interacting wavelengths [μm]	ϕ_{pm} [deg]	$\Delta\theta^{int}$ [deg]	$\Delta\phi^{int}$ [deg]	Ref.
SHG, o + o $\Rightarrow$ e				
1.063 $\Rightarrow$ 0.5315	55	≈ 2	≈ 0.17	3.632

Temperature variation of the phase-matching angle [3.632]:
XY plane, $\theta = 90°$

Interacting wavelengths [μm]	T [°C]	ϕ_{pm} [deg]	$d\phi_{pm}/dT$ [deg/K]
SHG, o + o $\Rightarrow$ e			
1.063 $\Rightarrow$ 0.5315	20	55	−0.006

Effective nonlinearity expressions in the phase-matching direction for three-wave interactions in the principal planes of MNA crystal [3.35], [3.36]:

XY plane

$$d_{ooe} = d_{31} \sin \phi \; ;$$

YZ plane

$$d_{eeo} = d_{31} \sin^2 \theta + d_{32} \cos^2 \theta \; ;$$

XZ plane, $\theta < V_Z$

$$d_{ooe} = d_{32} \cos \theta \; ;$$

XZ plane, $\theta > V_Z$

$$d_{oeo} = d_{eoo} = d_{32} \cos \theta \; .$$

Effective nonlinearity expressions for three-wave interactions in the arbitrary direction of MNA crystal are given in [3.36]
Nonlinear coefficients [3.633, 37]:

$$d_{31}(1.0642\,\mu m) = 90 \times d_{36}\,(KDP) = 35.1\,pm/V\ ,$$

$$d_{32}(1.0642\,\mu m) = 0.8 \times d_{36}\,(KDP) = 0.3\,pm/V\ ,$$

$$d_{33}(1.0642\,\mu m) = 90 \times d_{36}\,(KDP) = 35.1\,pm/V\ .$$

Laser-induced damage threshold [3.632]:

λ [μm]	τ_p [ns]	$I_{thr} \times 10^{-12}$ [W/m^2]
1.063	25	> 2

3.4.20 N-(4-Nitrophenyl)-N-methylaminoacetonitrile (NPAN)

Biaxial crystal;
Point group: mm2;
Assignment of dielectric and crystallographic axes:
$X, Y, Z \Rightarrow a, b, c$ [3.634] ;
Calculated molecular mass: 1.34 g/cm^3 [3.634];
Transparency range at "0" transmittance level: $\approx 0.5 - \approx 2.0\,\mu m$ [3.634];
Linear absoprtion coefficient α [3.635]:

λ [μm]	α [cm^{-1}]
0.5321	3.8–5.0
1.0642	1.8–2.3

Experimental values of phase-matching angle:
XY plane, $\theta = 90°$

Interacting wavelengths [μm]	ϕ_{pm} [deg]
SHG, e + e $\Rightarrow$ o	
1.0642 $\Rightarrow$ 0.5321	4.9 [3.635]
1.21 $\Rightarrow$ 0.605	20.6 [3.635]
1.30 $\Rightarrow$ 0.65	26.5 [3.635]

Experimental values of internal angular and temperature bandwidths [3.635]
XY plane, $\theta = 90°$

Interacting wavelengths [μm]	ϕ_{pm} [deg]	$\Delta\phi^{int}$ [deg]	ΔT [°C]
SHG, $e + e \Rightarrow o$			
$1.0642 \Rightarrow 0.5321$	4.9	0.118	3.7
$1.21 \Rightarrow 0.605$	20.6	0.105	4.8
$1.30 \Rightarrow 0.65$	26.5	0.185	3.8

Temperature variation of phase-matching angle [3.635]:

Interacting wavelengths [μm]	T [°C]	ϕ_{pm} [deg]	$d\phi_{pm}/dT$ [deg/K]
SHG, $e + e \Rightarrow o$			
$1.0642 \Rightarrow 0.5321$	20	4.9	-0.040
$1.21 \Rightarrow 0.605$	20	20.6	-0.035
$1.30 \Rightarrow 0.65$	20	26.5	-0.038

Effective nonlinearity expressions in the phase-matching direction for three-wave interactions in the principal planes of NPAN crystal [3.35, 36]:

XY plane

$$d_{eeo} = d_{31} \sin^2 \phi + d_{32} \cos^2 \phi \; ;$$

YZ plane

$$d_{ooe} = d_{31} \sin \theta \; ;$$

XZ plane, $\theta < V_Z$

$$d_{oeo} = d_{eoo} = d_{32} \sin \theta \; ;$$

XZ plane, $\theta > V_Z$

$$d_{ooe} = d_{32} \sin \theta \; .$$

Effective nonlinearity expressions for three-wave interactions in the arbitrary direction of NPAN crystal are given in [3.36]
Nonlinear coefficients [3.635]:

$$d_{31}(1.0642 \, \mu m) \approx 6 \, pm/V \; ,$$

$$d_{32}(1.0642 \, \mu m) = 57 \, pm/V \; ,$$

$$d_{33}(1.0642 \, \mu m) = 27 \, pm/V \; .$$

3.4.21 N-(4-Nitrophenyl)-L-prolinol (NPP)

Negative biaxial crystal: $2V_Z = 55.3°$ at $\lambda = 0.5461$ μm [3.636];
Point group: 2;
Assignment of dielectric and crystallographic axes of NPP is given in [3.636];
Transparency range at "0" transmittance level: $0.51 - 2.0$ μm [3.636];
Linear absorption coefficient α [3.618]:

λ [μm]	α [cm^{-1}]
0.5927	3.4
1.0	1.5
1.455	1.5

Experimental values of refractive indices [3.636]:

λ [μm]	n_X	n_Y	n_Z
0.436	2.630		
0.509	2.355	2.116	1.497
0.5321	2.277	2.024	
0.5461	2.231	1.982	1.491
0.577	2.153	1.927	1.495
0.589	2.128	1.911	1.484
0.6328	2.066	1.876	1.478
0.644	2.055	1.857	1.474
0.690	2.051	1.857	1.474
1.0642	1.926	1.774	1.457
1.338	1.917	1.757	1.440

Sellmeier equations (λ in μm, $T = 20\,°C$) [3.636]:

$$n_X^2 = 2.3532 + \frac{1.1299\,\lambda^2}{\lambda^2 - 0.1678} + 0.0392\,\lambda^2\;,$$

$$n_Y^2 = 2.8137 + \frac{0.3655\,\lambda^2}{\lambda^2 - 0.2030} - 0.0816\,\lambda^2\;,$$

$$n_Z^2 = 2.1268 + \frac{0.0527\,\lambda^2}{\lambda^2 - 0.1550} - 0.0608\,\lambda^2\;.$$

Experimental and theoretical values of phase-matching angle and calculated values of "walk-off" angle:

XZ plane, $\phi = 0°, \theta < V_Z$

Interacting wavelengths [μm]	θ_{exp} [deg]	θ_{theor} [deg] [3.636]	ρ_1 [deg]
SHG, e + e $\Rightarrow$ o			
1.1854 $\Rightarrow$ 0.5927	9.4 [3.637]	6.95	5.144
1.24 $\Rightarrow$ 0.62	14.8 [3.638]	12.41	8.763

Effective nonlinearity expressions in the phase-matching direction [3.35]:

XY plane

$$d_{\text{eeo}} = d_{25} \sin 2\phi \ ,$$

$$d_{\text{oeo}} = d_{\text{eoo}} = d_{23} \cos \phi \ ;$$

YZ plane

$$d_{\text{ooe}} = d_{21} \cos \theta \ ,$$

$$d_{\text{eoe}} = d_{\text{oee}} = d_{25} \sin 2\theta \ ;$$

XZ plane, $\theta < V_Z$

$$d_{\text{eeo}} = d_{21} \cos^2 \theta + d_{23} \sin^2 \theta - d_{25} \sin 2\theta \ ;$$

XZ plane, $\theta > V_Z$

$$d_{\text{eoe}} = d_{\text{oee}} = d_{21} \cos^2 \theta + d_{23} \sin^2 \theta - d_{25} \sin 2\theta \ .$$

Nonlinear coefficients [3.636]:

$$d_{21}(1.34 \, \mu\text{m}) = 56.5 \pm 5 \, \text{pm/V} \ ,$$

$$d_{22}(1.34 \, \mu\text{m}) = 18.7 \pm 2 \, \text{pm/V} \ .$$

Laser-induced damage threshold [3.619]:

λ [μm]	τ_p [ns]	$I_{\text{thr}} \times 10^{-12}$ [W/m²]
0.62	0.0001	100

3.4.22 3-Methyl-4-methoxy-4′-nitrostilbene (MMONS)

Positive biaxial crystal: $2V_Z = 70.2°$ at $\lambda = 0.543 \, \mu\text{m}$ [3.639];
Point group: mm2;
Assignment of dielectric and crystallographic axes:
$X, Y, Z \Rightarrow a, b, c$ [3.639] ;
Calculated mass density: $1.282 \, \text{g/cm}^3$ [3.639];
Transparency range: $0.51 - 2.1 \, \mu\text{m}$ [3.639];

Experimental values of refractive indices [3.639]:

λ [μm]	n_X	n_Y	n_Z
0.543	1.597	1.756	2.312
0.6328	1.569	1.693	2.129
1.0642	1.530	1.630	1.961
1.3188	1.525	1.622	1.940

Sellmeier equations (λ in μm, $T = 20\,°\mathrm{C}$) [3.639]:

$$n_X^2 = 1.987 + \frac{0.314\,\lambda^2}{\lambda^2 - (0.363)^2}\ ,$$

$$n_Y^2 = 2.184 + \frac{0.405\,\lambda^2}{\lambda^2 - (0.403)^2}\ ,$$

$$n_Z^2 = 2.507 + \frac{1.130\,\lambda^2}{\lambda^2 - (0.421)^2}\ .$$

Experimental and theoretical values of phase-matching angle and calculated values of "walk-off" angle:
XZ plane, $\phi = 0°, \theta > V_Z$

Interacting wavelengths [μm]	$\theta_{\exp}$ [deg]	θ_{theor} [deg] [3.639]	ρ_1 [deg]
SHG, e + o ⇒ o			
1.047 ⇒ 0.5325	77.6 [3.639]	77.57	7.50
1.0642 ⇒ 0.5321	73.2 [3.639]	73.18	9.58

Experimental values of internal angular and temperature bandwidths [3.639]:
XZ plane, $\phi = 0°$

Interacting wavelengths [μm]	θ_{pm} [deg]	$\Delta\phi^{\mathrm{int}}$ [deg]	ΔT [°C]
SHG, e + o ⇒ o			
1.047 ⇒ 0.5325	77.6	0.047	
1.0642 ⇒ 0.5321	73.2	0.035	0.17

Effective nonlinearity expressions in the phase-matching direction for three-wave interactions in the principal planes of MMONS crystal [3.35], [3.36]:
XY plane

$$d_{\mathrm{eoe}} = d_{\mathrm{oee}} = d_{31} \sin^2 \phi + d_{32} \cos^2 \phi\ ;$$

YZ plane

$$d_{\text{eoe}} = d_{\text{oee}} = d_{31} \sin \theta \; ;$$

XZ plane, $\theta < V_Z$

$$d_{\text{ooe}} = d_{32} \sin \theta \; ;$$

XZ plane, $\theta > V_Z$

$$d_{\text{oeo}} = d_{\text{eoo}} = d_{32} \sin \theta \; .$$

Effective nonlinearity expression for three-waves interactions in the arbitrary direction of MMONS crystal are given in [3.36]
Nonlinear coefficients [3.639, 37]:

$$d_{32}(1.0642 \, \mu\text{m}) = 25 \pm 5 \, \text{pm/V} \; ,$$

$$d_{33}(1.0642 \, \mu\text{m}) = 111 \pm 22 \, \text{pm/V} \; .$$

3.5 Properties of Crystalline Quartz (α-SiO$_2$)

Positive uniaxial crystal: $n_e > n_o$;
Point group: 32;
Mass denisty: 2.649 g/cm^3 [3.59];
Mohs hardness: 7 [3.64];
Transparency range at 0.5 transmittance level for the 10 mm long crystal (along c axis): $0.193 - 3.6 \, \mu$m [3.640, 641];
Linear absorption coefficient α (along c axis) [3.59]:

λ [μm]	α [cm^{-1}]
2.9	1
3.0	0.5
3.3	0.06
3.5	0.2
3.8	0.87

Two-photon absorption coefficient β (along c axis):

λ [μm]	$\beta \times 10^{13}$ [m/W]	Ref.
0.216	40 ± 7	3.399
0.266	4.5	3.71
0.270	< 1.5	3.399

Experimental values of refractive indices ($T = 291$ K) [3.642]:

λ [μm]	n_O	n_e	λ [μm]	n_O	n_e
0.185467	1.67578	1.68997	0.508582	1.548229	1.557475
0.193583	1.65999	1.67343	0.518362	1.547651	1.556887
0.20006	1.64927	1.66227	0.53385	1.546799	1.555996
0.20255	1.64557	1.65842	0.546072	1.546174	1.555350
0.204448	1.64288	1.65562	0.579066	1.544667	1.553791
0.21107	1.63432	1.64671	0.587563	1.544316	1.553428
0.214439	1.63039	1.64262	0.58929	1.544246	1.553355
0.219462	1.62497	1.63698	0.62782	1.542819	1.551880
0.226503	1.61818	1.62992	0.643847	1.542288	1.551332
0.231288	1.61401	1.62559	0.656278	1.541899	1.550929
0.242796	1.60525	1.61650	0.667815	1.541553	1.550573
0.250329	1.60032	1.61139	0.670786	1.541466	1.550483
0.257304	1.59622	1.60714	0.706520	1.540488	1.549472
0.263155	1.59309	1.60389	0.728135	1.539948	1.548913
0.274867	1.58752	1.59813	0.766494	1.539071	1.548005
0.291358	1.58098	1.59136	0.794763	1.538478	1.547392
0.303412	1.576955	1.58720	0.84467	1.537525	1.54640
0.312279	1.57433	1.584485	1.00000	1.53503	1.54381
0.325253	1.570915	1.58095	1.01406	1.53483	1.54360
0.340365	1.56747	1.577385	1.08303	1.53387	1.54260
0.35868	1.563915	1.573705	1.20000	1.53232	1.54098
0.396848	1.55813	1.56772	1.30000	1.53102	1.53962
0.404656	1.557156	1.56671	1.40000	1.52972	1.53826
0.410174	1.556502	1.566031	1.52961	1.52800	1.53646
0.434047	1.553963	1.563405	1.60000	1.52703	1.53545
0.435834	1.553790	1.563225	1.80000	1.52413	1.53242
0.467815	1.551027	1.560368	2.05820	1.51998	1.52814
0.479991	1.550118	1.559428	2.50000	1.51156	1.51950
0.486133	1.549683	1.558979	3.00000	1.49962	1.50700

Optical activity at $T = 300$ K [3.194]:

λ [μm]	ρ [deg/mm]	λ [μm]	ρ [deg/mm]	λ [μm]	ρ [deg/mm]
0.1800	410.5	0.1925	328.5	0.226909	200.90
0.1825	391.5	0.19303	325.31	0.232749	187.25
0.1850	374.0	0.193518	322.76	0.235923	180.43
0.185398	370.9	0.1950	315.5	0.241331	169.98
0.185735	368.6	0.198979	295.65	0.247482	158.66
0.186209	365.6	0.214702	226.91	0.26283	135.66
0.1875	357.5	0.221003	216.50	0.273955	122.12
0.1900	342.5	0.226334	202.27	0.281329	114.29

λ [µm]	ρ [deg/mm]	λ [µm]	ρ [deg/mm]	λ [µm]	ρ [deg/mm]
0.291216	104.97	0.419144	45.28	0.547155	25.43
0.307573	91.97	0.423362	44.29	0.570025	23.31
0.322579	82.13	0.428241	43.19	0.57696	22.72
0.327100	79.49	0.431509	42.47	0.578216	22.62
0.338399	73.43	0.435274	41.66	0.579066	22.55
0.340365	72.46	0.435834	41.55	0.588997	21.75
0.349058	68.36	0.467816	35.61	0.589593	21.70
0.3694	60.06	0.468014	35.57	0.636235	18.48
0.372762	58.84	0.472216	34.89	0.643847	18.02
0.386553	54.21	0.479991	33.68	0.670785	16.54
0.390648	52.95	0.481054	33.52	0.761	12.59
0.393582	52.07	0.508582	29.73	0.940	8.14
0.397775	50.85	0.510554	29.49	1.1	5.836
0.402187	49.62	0.515325	28.90	1.342	3.89
0.407664	48.14	0.520908	28.25	1.6	2.656
0.411855	47.07	0.535065	26.67	2.1	1.46
0.413469	46.66	0.546074	25.54	2.6	0.922
0.414768	46.34	0.546549	25.49	3.1	0.584

Temperature derivative of refractive indices [3.643]:

λ [µm]	$\mathrm{d}n_o/\mathrm{d}T \times 10^5$ [K^{-1}]	$(\mathrm{d}n_e/\mathrm{d}T) \times 10^5$ [K^{-1}]
0.441	−0.475	−0.593
0.467	−0.485	−0.681
0.480	−0.499	−0.600
0.508	−0.514	−0.616
0.589	−0.529	−0.642
0.643	−0.549	−0.653

Nonlinear coefficient [3.37]:

$$d_{11}(1.064\,\mu\text{m}) = 0.30\,\text{pm/V}$$

Laser-induced breakdown threshold (along c axis) [3.644]:

λ [µm]	τ_p [ns]	$I_{\text{thr}} \times 10^{-12}$ [W/m^2]
1.06	31	4000–6000

Thermal conductivity coefficient:

T [K]	κ [W/mK], $\parallel c$	κ [W/mK], $\perp c$	Ref.
273	11.42	6.82	3.645
293	11.7	6.5	3.58

3.6 New Developments

During the time taken to publish this book a number of new works devoted to the properities of nonlinear crystals has appeared. In order to update the material presented in this chapter the most important achievements are briefly discussed below.

CLBO

First, the new nonlinear crystal from the borate family, namely cesium lithium borate ($CsLiB_6O_{10}$ or CLBO) should be mentioned [3.646, 647]. It is a negative uniaxial crystal of point group $\bar{4}2m$, and is transparent from 0.18 to 2.75 μm. The Sellmeier equations for CLBO at room temperature are as follows (λ in μm) [3.646]:

$$n_O^2 = 2.208964 + \frac{0.010493}{\lambda^2 - 0.012865} - 0.011306\,\lambda^2\ ,$$

$$n_e^2 = 2.058791 + \frac{0.008711}{\lambda^2 - 0.011393} - 0.006069\,\lambda^2\ .$$

The CLBO nonlinear coefficient, measured in [3.646], is equal to: $d_{36}(1.064\ \mu m) = 2.2 \times d_{36}(KDP) = 0.86\,pm/V$. The laser-induced damage threshold at $\lambda = 1.064$ μm is $25\,GW/cm^2$ for 1.1 ns pulses [3.647].

BBO

The "improved" set of dispersion relations recently proposed in [3.650] is much worse than the set presented above [3.145].

LBO

The improved set of LBO dispersion relations have been reported by *Kato* (λ in μm, $T = 293$ K) [3.648]:

$$n_X^2 = 2.4542 + \frac{0.01125}{\lambda^2 - 0.01135} - 0.01388\,\lambda^2\ ,$$

$$n_Y^2 = 2.5390 + \frac{0.01277}{\lambda^2 - 0.01189} - 0.01849\,\lambda^2$$
$$+ 4.3025 \times 10^{-5}\,\lambda^4 - 2.9131 \times 10^{-5}\,\lambda^6\ ,$$

$$n_Z^2 = 2.5865 + \frac{0.01310}{\lambda^2 - 0.01223} - 0.01862\,\lambda^2$$
$$+ 4.5778 \times 10^{-5}\,\lambda^4 - 3.2526 \times 10^{-5}\,\lambda^6\ .$$

New information concerning the temperature derivative of LBO refractive indices is now available, e.g., for the spectral range $0.4 - 1.0$ μm and temperature range $293 - 383$ K (λ in μm) [3.648]:

$$dn_X/dT = -(3.76\,\lambda - 2.3) \times 10^{-6}\,\mathrm{K}^{-1}\ ,$$

$$dn_Y/dT = -(19.40\,\lambda - 6.01\,\lambda) \times 10^{-6}\,\mathrm{K}^{-1}\ ,$$

$$dn_Z/dT = -(9.70 - 1.50\,\lambda) \times 10^{-6}\,\mathrm{K}^{-1}\ ,$$

and for $\lambda = 0.6328\,\mu\mathrm{m}$ and a temperature range of $293 - 473$ K (λ in $\mu\mathrm{m}$, T in K) [3.649]:

$$dn_X/dT = [0.20342 - (1.9697 \times 10^{-2})(T - 273)$$
$$- (1.4415 \times 10^{-5})(T - 273)^2] \times 10^{-6}\ \mathrm{K}^{-1}\ ,$$

$$dn_Y/dT = -[10.748 + (7.1034 \times 10^{-2})(T - 273)$$
$$+ (5.7387 \times 10^{-5})(T - 273)^2] \times 10^{-6}\ \mathrm{K}^{-1}\ ,$$

$$dn_Z/dT = -[0.85998 + (1.5476 \times 10^{-1})(T - 273) - (9.4675 \times 10^{-4})(T - 273)^2$$
$$+ (2.2375 \times 10^{-6})(T - 273)^3] \times 10^{-6}\ \mathrm{K}^{-1}\ .$$

CBO

Improved dispersion relations for CBO have been published by *Kato* (λ in $\mu\mathrm{m}$, $T = 293$ K) [3.651]:

$$n_X^2 = 2.3035 + \frac{0.01378}{\lambda^2 - 0.01498} - 0.00612\,\lambda^2\ ,$$

$$n_Y^2 = 2.3704 + \frac{0.01528}{\lambda^2 - 0.01581} - 0.00939\,\lambda^2\ ,$$

$$n_Z^2 = 2.4753 + \frac{0.01806}{\lambda^2 - 0.01752} - 0.01654\,\lambda^2\ .$$

KTP

New data on the temperature derivative of refractive indices of flux-grown KTP have been reported for $T = 288 - 313$ K [3.652]:

$\lambda[\mu\mathrm{m}]$	$dn_X/dT \times 10^6[\mathrm{K}^{-1}]$	$dn_Y/dT \times 10^6[\mathrm{K}^{-1}]$	$dn_Z/dT \times 10^6[\mathrm{K}^{-1}]$
1.0642	6.1	8.3	14.5

KTA

The "infrared-corrected" Sellmeier equations proposed in [3.653] (λ in $\mu\mathrm{m}$, $T = 293$ K) are:

$$n_X^2 = 1.90713 + \frac{1.23552\,\lambda^2}{\lambda^2 - (0.19692)^2} - 0.01025\,\lambda^2\ ,$$

$$n_Y^2 = 2.15912 + \frac{1.00099\,\lambda^2}{\lambda^2 - (0.21844)^2} - 0.01096\,\lambda^2 \; ,$$

$$n_Z^2 = 2.14786 + \frac{1.29559\,\lambda^2}{\lambda^2 - (0.22719)^2} - 0.01436\,\lambda^2 \; .$$

These indeed show better agreement with experiment in the specific case of $1.0642\,\mu m$ pumped OPO in the XZ and YZ plane, but for SHG and SFG processes with shorter wavelength participation ($\lambda_3 = 0.4 - 0.6$ μm) the set from *Kato* [3.434] is preferrable.

RTA

Another KTP isomorph, rubidium titanyl arsenate ($RbTiOAsO_4$ or RTA), has been extensively developed in the last three years. RTA is a positive biaxial crystal of mm2 point group symmetry, and is transparent from 0.35 to 5.8 μm [3.654, 655]. The dispersion relations for RTA are as follows (λ in μm, $T = 293$ K) [3.656]:

$$n_X^2 = 2.22681 + \frac{0.99616\,\lambda^2}{\lambda^2 - (0.21423)^2} - 0.01369\,\lambda^2 \; ,$$

$$n_Y^2 = 1.97756 + \frac{1.25726\,\lambda^2}{\lambda^2 - (0.20448)^2} - 0.00865\,\lambda^2 \; ,$$

$$n_Z^2 = 2.28779 + \frac{1.20629\,\lambda^2}{\lambda^2 - (0.23484)^2} - 0.01583\,\lambda^2 \; .$$

The reported RTA nonlinear coefficients are:

$$d_{31}(1.0642\,\mu m) = 1.4\,pm/V \; [3.654, 37] \; ,$$

$$d_{32}(1.0642\,\mu m) = 4.6\,pm/V \; [3.654, 37] \; ,$$

$$d_{33}(1.0642\,\mu m) = 12.1\,pm/V \; [3.654, 37] \; .$$

AgGaSe$_2$

An improved set of Sellmeier equations, which gives much better agreement with experiment in the case of type I NCPM OPO, has been proposed by *Kato* [3.657] (λ in μm, $T = 293$ K):

$$n_o^2 = 6.85070 + \frac{0.42970}{\lambda^2 - 0.15840} - 0.00125\,\lambda^2 \; ,$$

$$n_e^2 = 6.67920 + \frac{0.45980}{\lambda^2 - 0.21220} - 0.00126\,\lambda^2 \; .$$

4 Applications of Nonlinear Crystals

This chapter is devoted to applications of nonlinear crystals in nonlinear optical devices. It describes the generation of second and higher (up to sixth) optical harmonics of neodymium laser radiation, generation of optical harmonics of powerful wide-aperture neodymium glass laser radiation, generation of optical harmonics of other lasers (ruby, gas, semiconductor, and so on), sum-frequency generation, (including up-conversion of IR radiation to the visible range), difference-frequency generation, parametric light oscillation as a tool for generating tunable radiation, stimulated Raman scattering, and picosecond continuum generation. The chapter contains abundant tabular material on the parameters of converted laser radiation and many references.

4.1 Generation of Neodymium Laser Harmonics

4.1.1 Second-Harmonic Generation of Neodymium Laser Radiation in Inorganic Crystals

Neodymium lasers are typical representatives of the solid-state laser family. Trivalent neodymium ions implanted into various crystals or glass matrices are the active medium of such lasers. Most neodymium lasers generate in the 1.05–1.08 μm, the neodymium phosphate glass laser emits at $\lambda = 1.054\,\mu m$, the neodymium silicate glass laser at $\lambda = 1.060–1.064\,\mu m$ (depending on the glass type), the neodymium-doped yttrium aluminate laser (Nd^{3+}:YAlO$_3$ or Nd:YAP) at $\lambda = 1.0796\,\mu m$, the Nd^{3+}:LiYF$_4$ (Nd:YLF) laser at $\lambda = 1.053\,\mu m$, and the Nd^{3+} : CaWO$_4$ laser at $\lambda = 1.0584\,\mu m$. Most often the neodymium-doped yttrium-aluminium garnet (Nd^{3+} : Y$_3$Al$_5$O$_{12}$ or Nd:YAG) laser is used, which emits at $\lambda = 1.06415\,\mu m$ (*see* Appendix).

Table 4.1 illustrates the results of studying SHG of Nd:YAG laser radiation in different inorganic crystals; for each crystal the type and the angle of phase matching, the intensity I_0 of radiation of the fundamental frequency, second-harmonic pulse duration, crystal length, and energy- or power- conversion efficiency are given.

For SHG of picosecond (or subnanosecond, $\tau_p = 1$–500 ps) Nd: YAG laser radiation use is mainly made of KDP crystals or sometimes DKDP [4.5] or

Table 4.1. Second-harmonic generation of Nd:YAG laser radiation (1.064 → 0.532 μm)

Crystal	Type of interaction	θ_{pm}[deg]	I_O[Wcm^{-2}]	τ_p [ns]	L [mm]	Conversion efficiency[%]	Refs.	Notes
KDP	ooe	41	10^9	0.15	25	32 (energy)	4.1	
	ooe	41	–	0.05	25	60	4.2	
	ooe	41	8×10^9	0.03	14	82 (energy)	4.3	
	ooe	41	7×10^9	0.03	20	81 (energy)	4.3	
	ooe	41.35	–	0.1 ms	40	0.38 (energy)	4.4	Nd:YAG laser cooled to 253 K, $\lambda = 946$ nm
DKDP	eoe	53.5	10^8	18	30	50 (power)	4.5	
	eoe	53.5	3×10^9	0.25	40	70 (power)	4.5	
	eoe	53.5	8×10^7	20	30	50 (energy)	4.6	$P_{2\omega} = 10$ W
	ooe	36.6	3×10^8	8	20	40 (energy)	4.7	
	eoe	53.7	3×10^8	8	20	50 (energy)	4.7	
CDA	ooe	90	2×10^8	10	17.5	57 (power)	4.8	$T = 48\,°$C
	ooe	90	4×10^9	0.007^a	13	25 (energy)	4.9	
DCDA	ooe	90	8×10^7	20	21	40 (energy)	4.6	$T = 90–100\,°$C
	ooe	90	3×10^8	20	16	40 (energy)	4.6	$P_{2\omega} = 10$ W
	ooe	90	2×10^8	10	13.5	45 (power)	4.8	$T = 112\,°$C
	ooe	90	9×10^7	–	29	50 (power)	4.10	
	ooe	90	–	15	20	57	4.11	$P_{2\omega} = 6$ W
RDA	ooe	50	–	10	–	34 (power)	4.12	$T = 25\,°$C
RDP	ooe	50.8	2×10^8	10	15.3	36 (power)	4.13	
	eoe	83.1	2×10^8	10	15.3	11 (power)	4.13	
LiIO$_3$	ooe	30	7×10^7	–	18	44 (power)	4.14	
	ooe	30	3×10^9	0.04	5	50	4.15	
LiNbO$_3$	ooe	90	2×10^7	10	20	40	4.16	$T = 120\,°$C
LFM	ooe	55.1	3.7×10^7	–	15	36	4.17	Single-pulse regime
	ooe	55.1	6.2×10^3	–	15	0.08	4.17	Free running regime

KTP	eoe	26^b	–	10	–	22	4.18	
	eoe	26^b	–	0.04	5	18	4.19	
	eoe	25.2^b	–	0.07	7.2	52	4.20	Quasi-continuous train
	eoe	25^b	2.5×10^8	15	4	60	4.21	
	eoe	30^b	2×10^7	35	9	40 (energy)	4.22	Multimode regime
	eoe	30^b	9×10^7	35	4	45 (energy)	4.22	Multimode regime
	eoe	30^b	10^8	30	5.1	60 (energy)	4.22	Two-pass regime
	eoe	30^b	10^8	30	8	50 (energy)	4.22	Gaussian beam
	eoe	26^b	–	0.2	5	55	4.23	$E_{2\omega} = 0.19$ J
	eoe	23^b	2.5×10^8	10	3	30	4.24	
	eoe	23^b	3.2×10^8	8.5	4.5	55 (power)	4.25	
	eoe	–	–	8	7	80 (energy)	4.26	$E_{2\omega} = 0.72$ J, $T = 55$ °C
"Banana"	ooe	74	–	–	3	20	4.27	YZ plane, $P_{2\omega} = 0.55$ W
$KNbO_3$	ooe	19	4.7×10^7	11	4.8	40 (energy)	4.28	Diode-pumped Nd:YLF laser (1.047 µm)
BBO	ooe	–	1.9×10^8	14	6	47	4.29	$P_{2\omega} = 4.5$ W
	ooe	–	1.67×10^8	14	6	38	4.29	$P_{2\omega} = 8.5$ W
	ooe	–	2.53×10^8	14	6	37	4.29	$P_{2\omega} = 36$ W
	ooe	21	2×10^9	1	6.8	68 (energy)	4.30	
	ooe	21	2.5×10^8	8	6.8	58 (energy)	4.30	
	ooe	22.8	–	10	6	33 (power)	4.31	Double-pass configuration
	ooe	22.8	1.4×10^8	–	7	32 (power)	4.32	
	ooe	22.8	1.6×10^8	8	7.5	55–60 (energy)	4.7	
LBO	ooe	0^b	10^9	0.035	15	65 (energy)	4.33	$T = 148.5 \pm 0.5$ °C
	ooe	0^b	5×10^8	10	12.5	60 (energy)	4.34	$T = 149$ °C
	ooe	0^b	5.2×10^8	10	11	32	4.35	$T = 151$°C
	ooe	12^b	$(5-8) \times 10^8$	9	14	70 (energy)	4.36	
	ooe	12^b	1.4×10^8	8	17	55–60 (energy)	4.7	
	ooe	0^b	2.54×10^8	–	4	12	4.37	16 layers (sheets) of LBO ($L = 16 \times 242$ µm $= 4$ mm)

[a] Neodymium silicate glass laser ($\lambda = 1.062$ µm).
[b] φpm.

LiIO$_3$ [4.15] crystals. A KDP crystal has a high breakdown threshold; non-linear processes of the two-photon absorption type, SRS, and picosecond continuum generation are virtually absent. To attain an optimum efficiency of SHG of powerful laser radiation, large-aperture laser beams must be used. This calls for large-aperture crystals. At present KDP crystals can be grown which possess a high optical quality and have an aperture of several tens of centimeters.

Matveets et al. [4.3] have studied SHG of Nd:YAG laser radiation ($\tau_\mathrm{p} = 30$ ps) in KDP crystals of different lengths (from 3.5 to 40 mm). Maximum energy conversion efficiency ($\eta = 82\%$) was attained for a crystal 14 mm long at fundamental pulse energy $E = 32$ mJ. With due regard for Fresnel losses and depolarization, this corresponds to 92% energy-conversion efficiency.

In a nanosecond regime with a rather low pulse-repetition frequency (several tens of hertz), use is mainly made of CDA and DCDA crystals. They have a phase-matching angle of 90° and, hence, a great angular bandwidth of SHG and a birefringence angle equal to zero. This permits focusing of the fundamental radiation into the crystal. Power-conversion efficiencies in CDA and DCDA were 57% and 45%, respectively [4.8]. A 50 MW Nd:YAG laser with a pulse duration of 12 ns was used as a source of fundamental radiation; SHG was attained for a collimated pump beam (1.1 mrad divergence) 5 mm in diameter. Second-harmonic power of 20–30 MW was observed for an 10 ns output pulse. Because of a large angular bandwidth at $\theta = 90°$ ($L\,\Delta\theta = 50$ cm mrad), CDA and DCDA crystals are especially suitable for frequency doubling of multimode pulsed neodimium laser radiation. In contrast to CDA and DCDA, LiIO$_3$ crystals do not require thermal stabilization but have, however, a significant drawback, namely, a large birefringence angle and a small angular bandwidth. This results in an aperture effect: a decrease in conversion efficiency because of an extraordinary wave energy "walk-off" (Sect. 2.12). To solve this problem, a special SHG scheme is used with several (two or more) sequentially arranged crystals with an identical cut [4.38]. Even crystals in this scheme are turned with respect to odd ones by 180° around the K vector direction. Thus, an extraordinary wave energy walk-off in odd crystals is compensated for by that in the even ones. The length of each crystal must be less than the aperture length L_a, determined as

$$L_\mathrm{a} = d/\rho \; , \tag{4.1}$$

where d is the input beam aperture and ρ is the walk-off (birefringence) angle.

LiNbO$_3$ crystals are also used for the prior purpose, but they have drawbacks such as photorefractive effect ("optical damage" or a change in refraction indices under the effect of laser radiation). They also need thermal stabilization. To greatly decrease the photorefractive effect on the SHG conversion efficiency, MgO-doped LiNbO$_3$ (MgO concentration $>4.5\%$) [4.39–41] or LiNbO$_3$ crystals grown from congruent melt [4.42] are used, which ensure a conversion efficiency of up to 50%. Table 4.2 shows the data on SHG of Nd:YAG laser radiation ($\lambda = 1.064\,\mu\text{m}$, $E = 100\,\text{mJ}$, $\tau_\text{p} = 14\,\text{ns}$, $f = 20\,\text{Hz}$, $I_0 = 35\,\text{MWcm}^{-2}$) in these crystals and also in LiIO_3, DCDA, DKDP, and KTP. The possibility of suppression of the photorefractive effect by heating the LiNbO_3 crystal over 170° should also be mentioned.

Among the crystals that double the frequency of Nd:YAG laser radiation, potassium titanyl phosphate (KTiOPO_4 or KTP) is of special interest. Possessing a very large nonlinearity ($d_{31} = 6.5 \times 10^{-12}\,\text{m/V}$, $d_{32} = 5 \times 10^{-12}\,\text{m/V}$), this crystal has large angular ($\Delta\theta L = 15\text{–}68$ mrad cm) and temperature ($\Delta TL = 20\text{–}25\,°\text{C}$ cm) bandwidths for SHG of 1.06 µm radiation. These exceed similar parameters for KDP, DKDP, and other crystals by almost an order of magnitude. Besides, it is nonhygroscopic and has a rather high surface-damage threshold.

The direction with $\varphi = 23°$ and $\theta = 90°$ has the highest d_{eff} value and is more advantageous than other directions since its angular bandwidth is maximum and the birefringence angle is minimum. Experimental values determined for a crystal 1 cm in length are $\Delta\varphi = 32' \pm 5'$ and $\Delta T = 20\,°\text{C}$ [4.24]. Table 4.1 illustrates the results of experimental studies of SHG of Nd:YAG laser radiation in KTP. In all cases interaction of the eoe type in the XY plane was used. The experiments of *Moody* et al. [4.23] were carried out with a Nd:YAG laser generating trains of pulses of 175±25 ps duration. A 3×3×5 mm KTP crystal was used, and radiation was focused into the crystal to a spot 390 µm in diameter. Efficiency of conversion to the second harmonic equal to 55% was attained. *Driscoll* et al. [4.22] studied in detail SHG of Nd:YAG laser radiation operating in single and multimode regimes with KTP crystals of different lengths (4–9 mm). In the 9 mm crystal, due to back transformation of the second-harmonic radiation to the fundamental one, a lowered conversion efficiency was observed. Maximum energy-conversion efficiency attained in a two-pass scheme with relatively short crystals ($L = 5.1$ mm) amounted to 60%.

For SHG of 1.064 µm radiation in a "banana" crystal the phase-matching angle was $\theta_{\text{ooe}} = 73°45'$ for the interaction in YZ plane ($\varphi = 90°$, d_{31}) and $\theta_{\text{eeo}} = 75°26'$ when the interaction occured in the XZ plane ($\varphi = 0°$, $T = 25\,°\text{C}$,

Table 4.2. Second-harmonic generation of Nd:YAG laser radiation in various crystals

Nonlinear crystal	L [mm]	θ_{pm} [deg]	E (0.53 µm) [mJ]	P (0.53 µm) [W]	η [%]
LiNbO_3 grown from	9	90	53	1.07	50.9
congruent melt	30	90	52	1.04	49.5
LiIO_3	19	29	29	0.58	27.6
DCDA	37	90	48	0.96	47.6
DKDP	50	53	19	0.39	19.5
KTP	5	24 (φ_{pm})	9.6	0.19	42.6
LiNbO_3:MgO	4	90	23	0.46	23.0
LiNbO_3:MgO	9	90	31	0.63	35.2

d_{32}); at $\theta = 90°$ and $\varphi = 90°$ the phase-matching temperature was $T = 101\,°C$; at $\theta = 90°$ and $\varphi = 0°$, $T = 89\,°C$ [4.43]. Note that the values of θ and T vary for different crystals in the ranges 73–75° and 75–77° for θ and 90–110 °C and 80–100 °C for T, respectively. This crystal is widely used in cw intracavity SHG schemes because of its large nonlinear coefficient.

Crystals of BBO and LBO are very promising for harmonic generation of Nd:YAG lasers due to their large transparency range, high damage threshold, high nonlinearity. For LBO also: large acceptance angle, small walk-off angle, and the possibility of being used under noncritically phase-matched conditions [4.36, 44a,b]. Both crystals are nonhygroscopic and are mechanicaly hard. Conversion efficiencies up to 60–70% to the second harmonic of Q-switched and mode-locked Nd:YAG lasers were attained by use of these crystals (Table 4.1). Noncollinear SHG and THG of the Nd:YAG laser in BBO crystal was studied by *Bhar* et al. [4.45,46].

4.1.2 Second-Harmonic Generation of 1.064 µm Radiation in Organic Crystals

Organic crystals have parameters competitive with widely used crystals of the KDP type, niobates, and formates. Their preparation is cheap, their nonlinear susceptibilities are high, and their birefringence is sufficient for use in frequency converters. Damage thresholds are fairly high; for instance, urea has a breakdown threshold of several GW cm^{-2} at nanosecond pumping, which exceeds that of LiNbO$_3$ and LiIO$_3$. However, organic single crystals have significant drawbacks that limit their application in nonlinear optics: they are hygroscopic and extremely soft so that their surfaces must be protected with coatings.

The efficiency of SHG of 1.064 µm radiation has been studied in poly-crystalline powdery samples [4.47–49]. Optically active amino acids (trypto-phan, asparagine, and others) [4.49], sugars (saccharose, maltose, fructose, galactose, lactose) [4.48], and other organic compounds were investigated. Up to now SHG of Nd:YAG laser radiation has been realized in the following organic single crystals: saccharose (C$_{12}$H$_{22}$O$_{11}$), 3-methyl-4-nitropyridine-1-oxide (POM), methyl-(2,4-dinitrophenyl)-amino-2-propanoate (MAP), meta-nitroaniline (MNA), 2-methyl-4-nitroaniline (MNA*), meta-dinitrobenzene (MDNB), 2-cyclooctylamino-5-nitropyridine (COANP), deuterated L-arginine phosphate (DLAP), 2-(N,N-dimethylamino)-5-nitroacetanilide (DAN), N-(4-nitrophenyl)-N-methylaminoacetonitrile (NPAN), 4-nitrophenol sodium (:Na) salt dihydrate (NPNa), its deuterated analogue (DNPNa), L-N-(5-nitro-2-pyridyl) leucinol (NPLO), and 3-methoxy-4-hydroxy-benzaldehyde (MHBA). In the L-PCA crystal (L-pyrrolidone-2-carboxylic acid) the fourth-harmonic of Nd:YAG laser was obtained by frequency doubling of the second harmonic (Table 4.3). High conversion efficiencies have been attained due to large non-linearities of these crystals. For instance, a conversion efficiency of 30% was attained for a MAP crystal only 1 mm long [4.53]. Conversion efficiencies for

Table 4.3. Second-harmonic generation of Nd:YAG laser radiation in organic crystals

Crystal	Type of interaction	d_{eff}/d_{36}(KDP)	θ_{pm} [deg]	φ_{pm} [deg]	Conversion efficiency [%]	Refs.	Notes
Saccharose	eoe	0.2^a	90	60.5	–	4.50	
	ooe	0.2^a	90	33.7	–	4.50	
	eoe	0.2^a	15.8	0	–	4.50	
POM	eeo	21.8	35.7	90	–	4.51	
	eoe	9.9	12.8	0	–	4.51	
	eeo	13.6	18.1 (1.32 µm)	90	50	4.52	$L=7$ mm, $\tau_{\text{p}} = 160$ ps
	eeo	13.1	17.4 (1.34 µm)	90	–	4.52	
MAP	eoe	38.3	2.2	0	30	4.53	$L=1$ mm
	oeo	37.7	11	90	40	4.53	$L=1.7$ mm
MNA	ooe	37.7	90	55	15	4.54	$L=2.5$ mm, $\Delta\theta = 2.9$ mrad
	eeo	24.1	44	90	–	4.54	
	ooe	11.5	90	14.5	10	4.54	NCSHG in the XY plane, $L=1$ mm
	ooe	6.8	90	8.5	85	4.55	NCSHG in the XY plane, $L=3$ mm
MNA*	–	65.7	–	–	–	4.56	
MDNB	ooe	3.6	35.3	0	0.1–0.5	4.57	$L=2$-4 mm, $I_{\text{o}} = 50$ MW cm^{-2}
COANP	oeo	30.9	26.4	0	3.6	4.58	$L=0.9$ mm, $I_{\text{o}} = 1.3$ MW cm^{-2}, $\tau_{\text{p}} = 250$ ns
DLAP	ooe	0.95	90	22	–	4.59	
	Type II	–	76	42	–	4.59	
DAN	–	–	40	0	20	4.60	$L=2$ mm
NPAN	eeo	129	90	8.5	9	4.61	$L=3$ mm, cw regime
NPNa	Type I	11.5	–	–	5	4.62	$L=1$ mm, diode-pumped Nd:YVO$_4$ laser
DNPNa	–	–	–	–	50	4.62	$L=1.5$ mm, diode-pumped Nd:YVO$_4$ laser
NPLO	eeo	85	51.7	0	–	4.63	
	oee	6.9	33.0	–	–	4.63	
MHBA	–	30	–	–	59	4.64	$L=3$ mm
L-PCA	Type II	0.64	90	42	0.6	4.65	SHG 532 $\rightarrow$ 266 nm, $L=6.9$ mm, $E=14.4$ µJ

^{a} The value of $d_{\text{eff}}/d_{\text{eff}}$(ADP) is given.

the MNA crystal are 15% for collinear SHG and 85% for noncollinear SHG (NCSHG) in the XY plane [4.54,55]. In the second case the effective nonlinearity ($d_{eff} = d_{31}$) and angular bandwidth are maximum. Note that along with urea, which has been successfully used for SHG of dye laser radiation and for optical parametric oscillation, MNA crystals have also had extensive application in nonlinear optics. It is possible to perform SFG and DFG in this crystal within the 0.5–1.5 μm band.

4.1.3 Intracavity SHG

Lasers with cw pumping (including cw lasers) are characterized by a low transmission coefficient (several percent) of the output mirror. As a result, the output radiation power is much lower than the radiation power inside the cavity. Since the efficiency of conversion to the second harmonic depends strongly on the power at the fundamental frequency, it is reasonable to place a nonlinear crystal inside the laser cavity (intracavity second-harmonic generation - ICSHG). Here the output mirror must have high reflectance for the fundamental frequency and high transmittance for the second-harmonic frequency. To enhance the conversion efficiency, the fundamental radiation can be focused into the nonlinear crystal.

To obtain a maximum output power for ICSHG, the optimum ICSHG regime or 100% conversion regime must be realized. Optimum ICSHG regime means the following: a laser with mirrors nontransparent for the fundamental radiation has an output power at the second-harmonic frequency equal to that of the fundamental radiation, under the condition that the intracavity nonlinear crystal is mismatched and the laser output mirror has an optimum transmittance at the fundamental frequency [4.38]. Note that the 100% conversion regime does not mean 100% conversion of the fundamental radiation into the second harmonic. In practice, the ICSHG efficiency does not exceed 20–30% for pulsed lasers and 5–10% for cw lasers [4.38].

For intracavity SHG of the Nd:YAG laser at 1.064 μm, LiIO$_3$ and LiNbO$_3$ crystals are usually used in Q-switched and mode-locked regimes and "banana" crystals in the cw regime. Recently, KTP crystals have also been successfully applied for this purpose. Table 4.4 gives the characteristics of Nd:YAG lasers with ICSHG. In both the cw and Q-switched regimes 100% conversion was attained. The output radiation power achieved 28 W [4.77]. According to *Lavrovskaya* et al. [4.24], when the KTP crystal was placed in an additional cavity inside the laser one, an average power of 7.1 W was attained at $\lambda = 532$ nm. The crystal was strongly heated (up to 100 °C) due to absorption at $\lambda = 532$ nm; however, no damage was observed for several hours of irradiation. With the KTP crystal output powers up to 3 W were obtained for a diode-laser pumped Nd:YAG laser operating in cw and mode-locked regimes [4.79].

For ICSHG of Nd^{3+}:YAlO$_3$ laser radiation ($\lambda = 1.0796$ μm) with acousto-optic Q-switching ($f = 5$ kHz), a KTP crystal 4.4 mm long was used [4.83].

Table 4.4. Intracavity SHG of Nd:YAG laser radiation ($1.064 \rightarrow 0.532$ μm)

Crystal	θ_{pm} [deg]	L [mm]	Mode of Nd:YAG laser operation	$P_{2\omega}$ [W]	η [%]	Refs.
LiIO$_3$	29	–	Q-switched	0.3	100	4.66
	29	20	cw	4	40 (0.12[a])	4.67
	29	–	Continuous pump, mode-locked, $\tau = 800$ ps	5	40 (0.13[a])	4.68
	29	15	$\tau = 180$ μs, $f = 50$ Hz	100 (peak)	0.06[a]	4.69
LiNbO$_3$	90	–	Continuous pump, Q-switched	0.31	100	4.70
	90	1	$\tau = 60$ ns, $f = 400$ Hz	100 (peak)	–	4.71
"Banana"	90	3	cw	1.1	100	4.72
	90	3	cw	0.3	–	4.73
	90	–	Continuous pump, Q-switched	0.016	100	4.74
	90	5	–	0.3–0.5	–	4.75
KTP	26	3.5	Q-switched	5.6	–	4.76
	23	3	Acousto-optic modulation, $f = 9$ kHz	7.1	–	4.24
	–	4.6	Acousto-optic modulation, $f = 4$–25 kHz	28	54 (0.6[a])	4.77
	–	–	Diode laser pumped cw Nd:YAG laser	0.03–0.1	6[a]	4.78
	–	15	Diode laser pumped mode-locked Nd:YAG laser, $\tau = 120$ ps, $f = 160$ MHz	3	56 (1.3[a])	4.79
	–	15	Diode laser pumped cw Nd:YAG laser	2.8	47 (0.94[a])	4.79
KNbO$_3$	70	5	cw	0.15	60	4.80
	90	5	cw	0.366	90	4.80
	60	3.7	Diode laser pumped cw Nd:YAG laser, $\lambda = 946$ nm	0.0031	0.74[a]	4.81
	0	6.2	Diode laser pumped Nd:YAG laser	0.002	1[a]	4.82

[a] Conversion efficiency calculated with respect to the energy of pumping flash lamps or diode lasers.

Here 90° phase matching of type II was realized at $T = 153° \pm 3\,°C$; the crystal faces were antireflection coated at $\lambda = 1.08\,\mu m$ and $\lambda = 0.54\,\mu m$. The average second harmonic power at $\lambda = 0.54\,\mu m$ was 15 W.

Besides the ICSHG with the nonlinear crystal inside the laser cavity, the frequency doubling in an external resonant cavity is also used widely [4.84]. The main advantage of this method is the possibility to optimize independently the laser oscillator and the frequency converter. This allows us, in particular, to generate radiation in a single axial mode and, as a result, to obtain a single-frequency second-harmonic radiation. SHG in external resonant cavities is most suitable for low-power laser-diode-pumped neodymium lasers (Table 4.5) and Ti:sapphire lasers (see below). Conversion efficiencies to the second harmonic were as much as 85% [4.89] and high output powers up to 6.5 W [4.90] were achieved for cw neodymium lasers with KTP and LBO crystals.

4.1.4 Third-Harmonic Generation

Third-harmonic generation (THG) of Nd:YAG laser radiation has been realized in KDP, DKDP, RDA, RDP, LiIO$_3$, BBO, LBO, LiCOOH · H$_2$O (LFM), and NaCOOH crystals by mixing the first and second harmonics (Table 4.6). Since their polarization vectors after frequency doubling are orthogonal, crystals possessing type II phase matching can be easily used for THG. In particular, KDP crystals satisfy this condition and, therefore, they have found wide application for THG. In RDA and RDP crystals, only THG of the ooe type is possible, which calls for additional optical elements to achieve coincidence between polarization vectors of the first- and second-harmonic waves. For instance, a crystalline quartz plate 4.35 mm thick was used for the rotation of the wave polarizations [4.94]. LiIO$_3$ crystals allow phase matching both of type I with $d_{\text{eff}} = d_{31}\sin\theta$ and of type II with $d_{\text{eff}} = (1/2)d_{14}\sin(2\theta)$. However, since d_{14} is negligibly small, the type II interaction in LiIO$_3$ is never used in practice. A LiIO$_3$ crystal was used for THG of Nd:YAG laser radiation of types I and II, the powers of the third harmonic being 22 kW and 0.005 kW, respectively [4.96]. Also, THG of Q-switched Nd:YAG laser radiation was realized in RDA and RDP crystals [4.12,94]. In a DCDA crystal 50 MW radiation at 1.064 μm with pulse duration 12 ns was doubled. Second-harmonic power ($\lambda = 532$ nm) was 18–22 MW at a pulse duration of 10ns. Third-harmonic power ($\lambda = 355$ nm) was 6 MW and 10.5 MW for RDA and RDP crystals, respectively. Highly efficient THG of Q-switched Nd:YAG laser in the LBO crystal with $\eta = 60\%$ has been demonstrated by *Wu* et al. [4.98]. LBO crystals are characterized by high optical quality, small walk-off angle and a three times larger effective nonlinear coefficient than that of KDP.

Direct THG of YAlO$_3$:Nd^{3+} nanosecond laser radiation ($\lambda = 1.079\,\mu m$, $\tau_p = 15$ ns) was realized in LiIO$_3$ [4.101]. The phase-matching angle for the ooe-e type conversion was 82°, and the conversion efficiency η attained was

Table 4.5. Second-harmonic generation of Nd:YAG laser radiation ($1.064 \rightarrow 0.532\ \mu$m) in external resonant cavities

Crystal	θ_{pm} [deg]	T_{pm} [°C]	L [mm]	Mode of laser operation	$P_{2\omega}$ [W]	η [%]	Refs.
LiNbO$_3$:MgO	90	–	12.5	Diode laser pumped, cw	0.03	56	4.85
	90	110	12	Diode laser pumped, cw (monolithic ring frequency doubler)	0.2	65	4.86
	90	107	–	Diode laser pumped, cw (monolithic ring frequency doubler)	0.005	50	4.87
LiNbO$_3^a$	90	233.7	10	Injection-locked Nd:YAG laser	1.6	69	4.88
KTP	90	63	10	cw YAlO$_3$:Nd laser ($\lambda = 1.08\ \mu$m)	0.6	85	4.89
LBO	90 (θ), 0 (φ)	149.5	6	Injection-locked cw Nd:YAG laser	6.5	36	4.90
	90	167	12	Diode laser pumped mode-locked Nd:YLF laser ($\lambda = 1.047\ \mu$m, $\tau = 12$ ps, $f = 225$ MHz)	0.75	54	4.91

^{a}Lithium-rich LiNbO$_3$.

Table 4.6. Third-harmonic generation of Nd:YAG laser radiation (1.064 $\rightarrow$ 0.355 μm)

Crystal	Type of interaction	θ_{pm} [deg]	τ_p [ns]	L [mm]	Conversion efficiency [%]	Refs.	Notes
KDP	eoe	58	0.15	12	32 (energy)	4.1	$I_O = 1$ GW cm^{-2}
	eoe[a]	58	25	–	6 (energy)	4.92	$P = 40$ MW
	eoe	58	0.05	–	10 (energy)	4.93	
DKDP	eoe	59.5	8	20	17 (energy)	4.7	$I_O = 0.25$ GW cm^{-2}
RDA	ooe	66.2	8	14.8	12 (power)	4.12	$\Delta\theta L = 1.0$ mrad cm
RDP	ooe	61.2	–	15.3	44 (power)	4.13	$I_O = 0.2$ GW cm^{-2}
	ooe	61.2	8	15.3	21 (power)	4.94	
LiIO$_3$	ooe	47	0.8	8	0.7 (power)	4.95	$P_{av} = 4.5$ mW
	ooe	47.5	–	4	4 (power)	4.96	
	eoe	61.7	–	4.65	10^{-3} (power)	4.96	
BBO	eoe	64	8	5.5	23 (energy)	4.30	$I_O = 0.25$ GW cm^{-2}
	ooe	31.3	8	7.5	20 (energy)	4.7	$I_O = 0.19$ GW cm^{-2}
	ooe	31.3	9	6	35 (quantum)	4.97	Intracavity THG, $P = 0.2$ W
LBO	Type I	38.1 (φ_{pm})	8	12.2	22 (energy)	4.7	$I_O = 0.19$ GW cm^{-2}
	Type II	41	8	12.6	60 (energy)	4.98	
LFM	eoe	8.2 (φ_{pm})	12	–	15	4.99	$P = 0.6$ MW, XY plane
NaCOOH	oee	2.2 (φ_{pm})	8	15	46 (power)	4.100	XY plane, $P_{av} = 1.9$ W, $P = 23$ MW, $\Delta\varphi = 8.7$ mrad

[a] Neodymium silicate glass laser.

0.2% at $I_0 = 50$ MWcm^{-2}. Direct THG of Nd:phosphate glass picosecond laser radiation ($\lambda = 1.054\,\mu$m, $\tau_p = 5$ ps) was realized in a $\beta - BaB_2O_4$ crystal ($\theta_{\text{ooee}} = 47.4°$, $\varphi = 90°$, $L = 0.72$cm) [4.102]. The conversion efficiency $\eta = 0.8\%$ at $I_0 = 50$ GW cm^{-2}; $\chi^3_{\text{eff}} = (6.4 \pm 2.8) \times 10^{-23}$ m^2/V^2.

4.1.5 Fourth-Harmonic Generation

Fourth-harmonic generation (FOHG) of Nd:YAG laser radiation at ($\lambda = 0.266\,\mu$m was obtained in KDP, DKDP, ADP, KB5, LFM, $\beta - BaB_2O_4$, and BeSO$_4 \cdot$ 4H$_2$O crystals (Table 4.7). DKDP and ADP crystals operating at $90°$ phase matching and KDP and BBO crystals are most suitable for this purpose. They have sufficiently large nonlinear coefficients, small coefficients of linear and two-photon absorption at the fourth-harmonic frequency, and high optical breakdown thresholds. An 85% conversion efficiency to radiation at $\lambda = 266$ nm was attained in an ADP crystal 4 mm in length [4.104].

4.1.6 Fifth-Harmonic Generation

Fifth-harmonic generation (FIHG) of Nd:YAG and neodymium silicate glass laser radiation was realized in KDP and ADP crystals upon cooling, and in $\beta - BaB_2O_4$, KB5, urea, and CaCO$_3$ crystals at room temperature (Table 4.8). Average powers of nanosecond radiation of 2–3 mW were attained at a high repetition frequency (120 kHz) [4.112, 113] and 5–7 mW at 10 Hz [4.114, 116]. In a KB5 crystal the peak power for radiation at $\lambda = 212.8$ nm was 11 MW for $\tau_p = 30$ ps [4.108]. First-and fourth-harmonic radiation waves propagated in the XY plane and were polarized in the same plane; a fifth-harmonic radiation wave was polarized along the Z axis (eeo interaction).

Unlike the foregoing cases, where the fundamental radiation was mixed with the fourth-harmonic, in the CaCO$_3$ crystal, for realization of FIHG, use made of four-photon parametric interaction of the oooe type: $\omega + \omega + 3\omega = 5\omega$ was used, i.e., two 1.06 μm photons were mixed with one 0.353 μm photon [4.119]. Here FIHG is realized by means of cubic nonlinear susceptibility $\chi^{(3)}$, whose tensor components are $\chi^{(3)}_{1111} = 2.5 \times 10^{-22}$ m^2/V^2 ($\lambda = 0.53\,\mu$m), $\chi^{(3)}_{3333} = 1.4 \times 10^{-22}$ m^2/V^2 ($\lambda = 0.53\,\mu$m), and $\chi^{(3)}_{3222} = 0.06 \times 10^{-22}$ m^2/V^2 ($\lambda = 0.69\,\mu$m). A train consisting of 20 pulses of neodymium laser radiation was used as a pump source ($\tau = 3$ ps). The CaCO$_3$ crystal was 0.5 cm long and linear absorption at $\lambda = 212$ nm amounted to $\alpha = 3.4$ cm^{-1}.

FIHG of Nd:YAG laser radiation [4.30, 109, 120, 121] was realized in a $\beta - BaB_2O_4$ crystal by mixing both the first-and fourth-harmonic radiations ($\theta_{\text{ooe}} = 51$–$55°$, $\theta_{\text{ooe}} = 57.2°$) and the third-and second-harmonic radiations ($\theta_{\text{ooe}} = 69.3°$). Overall conversion efficiencies of 15% and 4% to the fifth harmonic were achieved in a 6.4 mm long BBO crystal for mode-locked and Q-switched Nd:YAG lasers, respectively [4.30].

Table 4.7. Fourth–harmonic generation of Nd:YAG laser radiation ($1.064 \rightarrow 0.266$ µm)

Crystal	Type of interaction	θ_{pm} [deg]	I_O [W cm^{-2}]	τ_p [ns]	L [mm]	Conversion efficiency (from 532 nm) [%]	Refs.	Notes
KDP	ooe	78	–	7	–	30–35	4.103	
DKDP	ooe	90	2×10^7	–	40	40	4.5	$T = 40.6\,°C$
	ooe	90	8×10^9	0.03	4	75	4.104	
	ooe	90	5×10^7	25	20	40	4.105	$T = 60\,°C$ $P=2.5$ MW
	ooe	90	–	600	50	3.4	4.106	$T = 49.8\,°C$ $P_{av}=0.5$ W
ADP	ooe	90	8×10^9	0.03	4	85	4.104	
	ooe	90	–	8	30	15[a]	4.107	$T = 51.2\,°C$ $P_{av}=5$ W
KB5	eeo	47.2 (φ_{pm})	–	0.03	10	6[a]	4.108	$P=60$ MW
LFM	ooe	–	10^6	–	15	6–7	4.17	
BBO	ooe	48	–	5	–	16	4.109	$E=80$ mJ
	ooe	48	–	8	5	18	4.30	
	ooe	48	1.6×10^8	1	5	52	4.30	
	ooe	57.8	–	8×10^4	6.6	0.17	4.4	Nd:YAG laser, $T =$ 253 K, $\lambda=946$ nm
BeSO$_4 \cdot$ 4H$_2$O	ooe	81.6	2.2×10^8	10	10	30	4.110	$P=1.7$ W

[a]Efficiency of conversion from 1.064 µm.

Table 4.8. Fifth-harmonic generation of Nd:YAG and neodymium silicate glass laser radiation

Crystal	θ_{pm} [deg]	Type of interaction	Crystal temperature [°C]	Output parameters	τ_p [ns]	Refs.
KDP	90	ooe[a]	−70	10^{-4}J	–	4.111
	90	ooe	−35	$P_{av}=2.6$ mW, $f=120$ kHz	30	4.112
	90	ooe	−40	$P_{av}=2$ mW, $f=6$ kHz	30	4.113
ADP	90	ooe	−40	$P_{av}=5\text{–}7$ mW, $f=10$ Hz	10	4.114
	90	ooe	−55	$E=0.1$ mJ $f=10$ Hz	10	4.115
KB5	$53\pm1(\varphi)$	eeo	20	$E=0.7$ mJ	6	4.116
	$53\pm1(\varphi)$	eeo	20	$E=0.1$ mJ	0.02	4.117
	$52.1(\varphi)$	eeo	20	$E=0.3$ mJ	0.03	4.108
Urea	72	eeo	20	$E=30$ mJ	10	4.118
CaCO$_3$	51.5	oooe[a]	20	$E=0.6$ μJ	0.003	4.119
β-BaB$_2$O$_4$	55 ± 1	ooe	20	–	–	4.120
	51.1	ooe	20	–	10	4.121
	69.3	ooe[b]	20	–	10	4.121
	55	ooe	20	$E=20$ mJ	5	4.30,109
	55	ooe	20	$E=5$ mJ	1	4.30

[a] Neodimium silicate glass laser.
[b] $2\omega+3\omega=5\omega$.

Urea crystals are also suitable for FIHG of Q-switched Nd:YAG laser radiation due to a high nonlinear coefficient [d_{36} (urea) $= 3d_{36}$(KDP)], high transparency in the UV region (up to 200 nm), and high breakdown threshold (5 GW cm^{-2} at $\lambda = 1.06\,\mu$m). For instance, with a 5 mm long urea crystal, a radiation energy of up to 30 mJ was attained at $\lambda = 212.8$ nm [4.118].

Three different methods for fifth-harmonic generation, with $\lambda = 216$ nm, of YAlO$_3$:Nd^{3+} laser radiation ($\lambda = 1.08\,\mu$m, $\tau_p = 15$ps) have been reported [4.122]. In the first, FIHG in KB5, $2\omega + 3\omega = 5\omega$, interaction is of the eeo type, $\theta = 90°$, and $\varphi = 80°$. In the second, FIHG in KB5, $\omega + 4\omega = 5\omega$, interaction is of the eeo type, $\theta = 90°$, and $\varphi = 50.4°$. In the third, FIHG in KDP, $\omega + 4\omega = 5\omega$, interaction is of the ooe type, and $\theta = 84°$. Fifth harmonic energies amounted to 40, 50, and 450 µJ, respectively. Fulfillment of phase-matching conditions in KDP at room temperature [4.123], as well as higher effective nonlinearity and larger angular bandwidth, make it possible to attain great output energies in the third case.

4.1.7 Harmonic Generation of 1.318 µm Radiation

In some papers harmonic generation of Q-switched Nd:YAG laser radiation was studied with the use of fundamental radiation at $\lambda = 1.318$ µm. Generation of higher (up to fifth) harmonics with a peak power of 0.2–85 kW was attained with LiNbO$_3$ and KDP (Table 4.9) [4.124]. In all cases the ooe interaction was used. Researchers obtained the sixth harmonic at $\lambda = 219.3$ nm in a potassium pentaborate crystal by doubling the third harmonic [4.101]. The fundamental radiation propagated in the XY (ab) and eeo interaction was used. Spectral and angular bandwidths amounted to 1.2 nm and 1.5 mrad, respectively. LiNbO$_3$ is the most suitable material for doubling 1.318 µm radiation [4.127]. For instance, at a pump intensity $I_0 = 100$ MW cm^{-2}, 48% energy conversion efficiency to the second-harmonic was achieved for 90° phase matching. Without focusing of the fundamental radiation, the efficiency drops to 21% for a crystal 20 mm long [4.127]. The regime of 100% conversion for ICSHG of 1.318 µm fundamental radiation was realized with a LiIO$_3$ crystal, the output power being 1 W [4.126]. *Lin* et al. [4.128] studied the LBO crystal for doubling radiations of Nd:YAG (1.32 µm), Nd:YLF (1.31 µm), and Nd:YAP (1.34 µm) lasers. Due to a high effective nonlinear coefficient and large angular bandwidth, LBO was considered as very suitable for SHG of radiations near 1.3 µm.

Table 4.9. Generation of harmonics of Nd:YAG laser radiation with $\lambda = 1.318\,\mu m$

Number of harmonic	λ [nm]	Crystal	θ_{ooe}[deg]	L [mm]	τ_p [ns]	Output parameters	Energy conversion efficiency [%]	Refs.
2	659.4	$LiNbO_3$	44.67	16	40	85 kW	10	4.124
3	439.4	KDP	42.05	30	40	3.4 kW	0.4	4.124
4	329.7	KDP	53.47	30	40	6 kW	0.6	4.124
5[a]	263.8	KDP	55.33	30	30	0.2 kW	0.02	4.124
6[b]	219.3	KB5	78 (eeo)	15	45	3 kW	0.5	4.101
2	659.4	DCDA	70.38	13.5	25	1.4 MW	40	4.125
2[c]	659.4	$LiIO_3$	22	10	30	$P_{av} = 1$ W	100	4.126
2	659.4	$LiNbO_3$	90 ($T = 300\,°C$)	19	50	60 mJ	48	4.127
2	659.4	$LiNbO_3$	90	20	50	10 mJ	21	4.127
3	439.6	KDP	42.05 ($T = 300\,°C$)	40	50	1.4 mJ	3	4.127
3	439.6	$LiIO_3$	–	8	50	1.4 mJ	1.2	4.127

[a] $\omega + 4\omega = 5\omega$.
[b] $3\omega + 3\omega = 6\omega$.
[c] Intracavity SHG.

4.2 Harmonic Generation of High-Power Large-Aperture Neodymium Glass Laser Radiation

Recently, considerable advances have been achieved in harmonic generation of high-power neodymium glass laser radiation. In particular, such lasers are used in experiments on laser thermonuclear fusion. Frequency converters of this type use KDP crystals exclusively, which have certain advantages: high breakdown threshold for subnanosecond and picosecond radiation (> 10 GW cm^{-2}); low linear and two-photon absorption at wavelengths 0.27, 0.35, 0.53, and 1.06 µm; high threshold of picosecond continuum generation and SRS (> 100 GW cm^{-2}); the absence of optical damage; and the possibility of growing large single crystals (several tens of centimeters) possessing high optical quality. Also, small dispersive birefringence derivative with respect to temperature allows the use of KDP crystals without thermal stabilization. Maximum conversion efficiencies attained up to now are as follows: 90%, to the second harmonic at $\lambda = 0.53$ µm [4.129]; 80–81%, to the third harmonic at $\lambda = 0.36$ µm [4.129, 130]; 51% to the fourth harmonic at $\lambda = 0.27$ µm (with allowance for crystal antireflection coating $\eta = 70\%$ [4.131]; 92% (from the second harmonic), to the fourth harmonic at $\lambda = 0.264$ µm [4.132], and 19%, to the fifth harmonic at $\lambda = 0.211$ µm [4.132] (Table 4.10).

Tripling of neodymium glass laser radiation frequency has been thoroughly studied [4.136,137] and realized experimentally on a large-aperture laser setup at the University of Rochester in the USA [4.130]. Three schemes of tripling were compared, as described in the following subsections.

4.2.1 "Angle-Detuning" Scheme

Type II SHG is used with the angle $\alpha = 45°$ between the fundamental wave polarization vector and o-ray polarization vector. For THG, type II interaction is also used, with the parameters of both KDP crystals being $\theta = 59°$ and $\varphi = 0°$. The method is disadvantageous in that a special crystalline quartz plate must be placed between the KDP crystals to attain an angle of $90°$ between the polarization vectors of waves at fundamental and second-harmonic frequencies. If type I interaction is used for SHG, there is no need for a special polarization rotation plate. In this case, however, the scheme is 2.7 times more sensitive to the angle α.

4.2.2 "Polarization-Mismatch" Scheme

Similar to the first scheme, type II interaction ($\theta = 59°$, $\varphi = 0°$) is used for SHG and THG, but the rotation angle of the wave ω polarization vector with respect to the o-wave polarization vector is $\alpha = \arctan 1/\sqrt{2} = 35.3°$. It is

Table 4.10. Generation of harmonics of high–power Nd:glass laser radiation in KDP

Fundamental radiation			Second harmonic					Third and fourth harmonics					Refs.	Notes
λ[μm]	I_O [10^9 Wcm^{-2}]	τ_p [ns]	λ [μm]	Type of inter-action	η [%]	Crystal length [mm]	E [J]	λ [μm]	Type of inter-action	η [%]	Crystal length [mm]	E [J]		
1.054	2.5	0.14	0.53	eoe	67	12	9	0.35	eoe	80	12	11	4.130	The angle α between polarization vectors of 1.054 μm radiation and o–ray is 35°
1.054	3.5	0.7	0.53	eoe	67	12	25	0.35	eoe	80	12	30	4.130	
1.064	2.5	0.1	0.532	eoe	67	8	17	0.266	ooe	30	7	4	4.133	$\tau_{4\omega} = 50$ ps
1.064	9.5	0.7	0.532	ooe	83	10	346						4.131	
1.064	2.0	0.7	0.532	eoe	67	12	–	0.355	eoe	55	10	41	4.131	
1.064	1.2	0.7	0.532	ooe	–	10	–	0.266	ooe	51	10	50	4.131	
1.06	0.2	25	0.53	ooe	80	40	60						4.134	
1.06	2.7	0.5	0.53	ooe	90	30	10–20						4.129	
1.06	2.7	0.5	0.53	eoe	67	18	–	0.35	eoe	81	18	10–20	4.129	The angle between polarization vectors of 1.06 μm radiation and o–ray is 35.2°
1.053	1.5	0.6	0.53	eoe	70	16	70–80	0.26	ooe	46	7	53	4.135	
1.054	5	0.5	0.53	eoe	87	17.5	–	0.264	ooe	92[a]	10	–	4.132	

[a] Conversion efficiency from 0.527 μm.

known that, to attain maximum conversion to the third harmonic in the plane-wave approximation, the number of photons at the frequencies ω and 2ω must be equal, which results in the energy ratio 1:2. Hence, to attain maximum conversion to the third harmonic, the SHG efficiency must be 67%. In this scheme, when $\alpha = 35.3°$, one of two o-photons at the frequency ω is mixed with the e-photon at the frequency ω with the formation of an e-photon of 2ω, whereas the second o-photon at the frequency ω remains unconverted. Here 67% conversion to the second harmonic is attained and the unconverted radiation at frequency ω has a suitable polarization for the following cascade of tripling. This scheme has low sensitivity to angle detuning, which may reach 300 μrad; there is no need for an additional crystalline quartz plate for rotating the fundamental wave polarization vector. Besides, no birefringence problem arises, which strongly affects the parameters of the angle-detuning scheme with $\alpha = 45°$. This scheme is ideal for high-power Nd:glass laser systems.

4.2.3 "Polarization-Bypass" Scheme

This scheme is similar to the polarization-mismatch scheme. The only difference is the use of the type I interaction in both cascades. In the first cascade (SHG) maximum possible conversion to the second harmonic is attained ($\alpha = 35.3°, \theta = 41°$) by a proper choice of a nonlinear crystal of the required length. This scheme is especially suitable for SHG and THG cascades of CDA-type crystals, allowing 90° phase matching with temperature tuning, since they are less sensitive to angle detuning of wave polarization vectors than KDP.

4.2.4 Comparison of Schemes

All three schemes discussed ensure 80% conversion to the third harmonic (absorption at $\lambda = 1.06\,\mu\text{m}$, $\alpha = 0.04\,\text{cm}^{-1}$ is taken into account). However, the second scheme is best since the conversion efficiency (80%) is retained within a wide range of the fundamental radiation intensities: 2–5 GWcm^{-2}.

4.2.5 Experimental Results

For THG the scheme of "polarization mismatch" with $\eta = 80\%$ has been realized experimentally [4.130]. The laser setup parameters were as follows: wavelength 1.06 μm, beam diameter 60 mm, pulse length 140 and 700 ps, and pulse energy 25 and 40J, respectively. KDP crystals with $\theta_{\text{eoe}} = 59°$ and $\varphi = 0°$ were used. Maximum conversion efficiency was attained at $I_0 = 2$–3 GW cm^{-2}. The generation of second (532 nm), third (355 nm), and fourth (266 nm) harmonics was studied on a large-aperture noedymium laser Argus setup at the Lawrence Livermore National Laboratory [4.131]. The following mechanisms of nonlinear losses in frequency conversion were analyzed in detail: SRS in antireflection coatings of KDP crystals at intensities $>$ GW cm^{-2}; nonlinear losses (two-photon absorption) in KDP at $\lambda = 266$ nm at $I_0 > 1.5$ GWcm^{-2} (two-photon absorption coefficient $\beta = 2.7 \cdot 10^{-10}$ cmW^{-1}); volume breakdown of optical elements from fused silica due to self-focusing when $I > 1.5$ GWcm^{-2} at wavelengths of the third and fourth harmonics (355–266 nm); and the damage of the antireflection coatings at energy densities exceeding 1 Jcm^{-2}. When a KDP crystal (type I) 1.0 cm long was used (aperture 10×10 cm) at $I_0 = 9.5$ GWcm^{-2}, conversion efficiency to the second harmonic was 83% (outside the crystal). With antireflection coatings, η increased to 89%. For a KDP crystal (type I) 2.29 cm in length at $I_0 = 2.7$ GWcm^{-2} in length at $I_0 = 2.7$ GW cm^{-2}, η was 71%, the output energy being 65 J. At $I_0 = 5.7$ GW cm^{-2} in a type II KDP crystal 1.19 cm long, $\eta = 74\%$ and the output energy $E = 121$ J at $\lambda = 532$ nm were attained. Three doubling and tripling schemes were experimentally studied for THG: type I/type II, crystal lengths 2.3/1.2 cm; type I/type II, crystal lengths 1.3/1.2 cm; and type I/type II, crystal lengths 1.2/1.0 cm. The conversion efficiencies were 50%, 53%, and 55%, respectively. As was already mentioned, in the FOHG process nonlinear losses play an important role at $I_0 > 1.5$ GW cm^{-2}. Therefore, to obtain the effective FOHG, the pump intensity was 1–1.2 GW cm^{-2}. With the type I interaction (ooe) maximum conversion efficiencies to 266 nm radiation were 64% ($E = 50$ J) and 55% ($E = 44$ J) for crystals 1 cm and 1.5 cm in length, respectively. For antireflection coated crystals, η rises to 70% and 60%, respectively. These results [4.131] were theoretically interpreted by *Craxton* [4.138].

Ibragimov et al. [4.134] have theoretically analyzed the frequency-doubling process under the conditions applicable to large noedymium glass laser systems to evaluate the limiting conversion efficiency. Experimental investigation of SHG was performed on a multicascade neodymium glass laser ($\lambda = 1.06\,\mu$m) with the 45 mm aperture of the end cascade. Maximum second-harmonic radiation energy attained 90 J at 25 ns pulse duration; the fundamental laser beam intensity distribution corresponded to a hypergaussian function with $N = 5$. For doubling, KDP crystals were used with an aperture of 20 and 50 mm, the interaction type being ooe. Maximum energy-conversion efficiency to the second harmonic was obtained for KDP crystals with dimensions $50 \times 50 \times 40$ mm : $\eta = 80\%$ at an incident radiation energy of 70 J and divergence 6×10^{-5} rad.

Gulamov et al. [4.129] obtained maximum conversion efficiencies to second (90%) and third (81%) harmonics of high-power neodymium phosphate glass laser radiation. For doubling, KDP crystals 18, 30, and 40 mm in length were used with an aperture 50×50 mm. The conversion efficiency to 527 nm radiation amounted to 75%, 90%, and 80%, respectively. The beam diameter was 32 mm, the divergence 5.5×10^{-5} rad, and the depolarized fraction of the radiation did not exceed 3%. The polarization mismatch scheme with the 35% rotation of the fundamental wave polarization vector with respect to the o-wave polarization vector was used for THG. In both cascades KDP crystals

17.5 mm long and with an aperture of 50×50 mm were used; the interaction type was eoe.

The fourth-harmonic generation of radiation of a large-aperture neodymium laser consisting of a $LiYF_4:Nd^{3+}$ oscillator and neodymium-phosphate glass amplifiers ($\lambda = 1.053$ μm, $E = 115$ J) has been studied [4.135]. The conversion efficiency to $\lambda = 0.26$ μm was 46.5% at $I_0 = 1.5$ GW cm^{-2}. High efficiency FOHG with $\lambda = 0.264$ μm was realized by *Begishev* et al. [4.132] with conversion efficiency from 0.527 μm, $\eta = 92\%$. Further mixing of obtained radiation (0.264 μm) with the fundamental radiation (1.054 μm) in an ADP crystal ($\theta_{ooe} = 90°$, $T = -67.5\,°C$, $L = 10$ mm) allows us to generate the fifth harmonic (211 nm) with overall efficiency of 19% [4.132].

4.2.6 "Quadrature" Scheme

An interesting scheme for frequency conversion, introduced by *Eimerl* [4.139], called the "quadrature" scheme, uses two (instead of one) crystals at each conversion step. We shall illustrate the application of this scheme to SHG (Fig. 4.1). The quadrature scheme of SHG comprises two crystals oriented for type

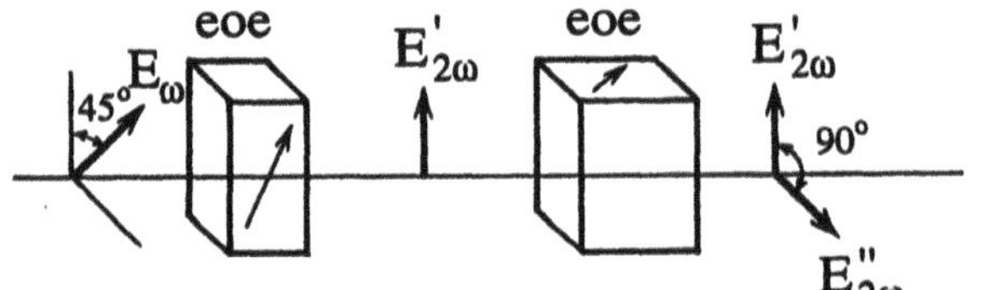

Fig. 4.1. Quadrature scheme of SHG

II interaction (eoe) and positioned so that the principal planes of these crystals (where the optic axes and beam propagation direction are arranged) are orthogonal. The scheme has two specific features. First, the fundamental radiation after the first conversion cascade has polarization suitable for the second cascade, so that both crystals participate effectively in the conversion. Second, the second harmonic generated in the first crystal has polarization unsuitable for the second conversion step, and therefore, is not converted in the second crystal. The total conversion efficiency for two crystals is

$$\eta = \eta_1 + (1 - \eta_2)\eta_2, \tag{4.2}$$

where η_1 and η_2 are the conversion efficiencies in the first and second crystals, respectively.

The quadrature scheme of conversion has an undoubted advantage over a scheme where only one crystal is used: the dynamic range for the pump intensity variation in the quadrature scheme (100–1000) greatly exceeds that in a one-crystal scheme (~ 10). The advantage of this scheme was illustrated experimentally for SHG of radiation of a Nd:YLF laser with three neodymium phosphate glass amplifiers ($\lambda = 1.053\,\mu\text{m}, \tau = 1.2$ ns). Two DKDP crystals 12 and 44 mm in length were used, the absorption at the fundamental frequency was $\alpha = 0.006\,\text{cm}^{-1}$, and Fresnel reflection amounted to 15.5–18.5%. With a change in the pump intensity from 0.2 to 9.6 GW cm^{-2} (the dynamic range 45) the conversion efficiency remained unchanged at 80%. When the crystals were antireflection coated, η increased to 95%. At low intensities the conversion takes place basically in the second long crystal, and at high intensities, in the first crystal. The dependences of η on the pump intensity for SHG were calculated also for KTP, $\beta - \text{BaB}_2\text{O}_4$ ($\lambda = 1.06\,\mu\text{m}$), $\text{CdGeAs}_2(10.6\,\mu\text{m})$. Quadrature schemes were also described for THG, FOHG, and SFG [4.139].

4.3 Harmonic Generation for Other Laser Sources

4.3.1 Ruby Laser

For SHG of ruby laser radiation ($\lambda = 694.3$ nm), KDP, DKDP, ADP, RDA, RDP, LiIO$_3$, and KB5 crystals have been used (Table 4.11). Maximum conversion efficiencies were attained in RDA, RDP, and LiIO$_3$ crystals. In a 1.45 cm long RDA crystal, power-conversion efficiency was 58% both at room temperature ($T = 20\,°\text{C}$, $\theta_{\text{ooe}} = 80\,°$) and at 90° phase matching ($T = 90\,°\text{C}$). The output power was 62 MW. The RDA crystal is suitable for this purpose because of a large angular bandwidth (Table 4.11). The third harmonic of ruby laser radiation ($\lambda_{3\omega} = 231.4$ nm) was obtained in a KB5 crystal by mixing its first (694.3 nm) and second (347.1 nm) harmonics [4.146]. Interacting waves propagated in the XY plane at an angle $\varphi = 57\,° \pm 1\,°$ to the X axis. The eeo interaction was used. The conversion efficiency calculated relative to the fundamental radiation was 0.2%; the output power was 40 kW at $\tau_\text{p} = 6$ ns.

Table 4.11. Second-harmonic generation of ruby laser radiation ($\lambda = 694.3$ nm)

Crystal	Type of interaction	θ_{pm}[deg]	I_0[Wcm^{-2}]	L [mm]	Power conversion efficiency [%]	Refs.	Notes
KDP	ooe	50.5	–	–	–	4.140	$L\Delta\theta = 1.75$ mrad cm
DKDP	ooe	52	–	–	–	4.141	$L\Delta\theta = 1.46$ mrad cm
ADP	ooe	52	–	–	–	4.140	$L\Delta\theta = 1.63$ mrad cm
RDA	ooe	80.3 (90)	1.5×10^8	1.45	58	4.142	$T = 20\,°\text{C}$ (90 °C), $L\Delta\theta = 4.37$ mrad cm
RDP	ooe	67	1.8×10^8	1.0	37	4.143	$T = 20\,°\text{C}$, $L\Delta\theta = 2.4$ mrad cm
LiIO$_3$	ooe	52	1.3×10^8	1.1	40	4.144	$L\Delta\theta = 0.2$ mrad cm
KB5	eeo	26.5 (φ_{pm})	–	1.0	10^{-3}	4.145	XY plane

4.3.2 Ti:sapphire Laser

Second harmonic of Ti:sapphire (Ti : Al_2O_3) laser radiation with $\lambda = 700$–900 nm has been realized in $LiIO_3$, BBO, LBO, and $KNbO_3$ crystals (Table 4.12); two organic crystals, 3-methoxy-4-hydroxy-benzaldehyde (MHBA) and 8-(4'-acetylphenyl)-1,4-dioxa-8-azaspiro [4.5] decane (APDA) were also used for this purpose. for continuous wave and cw pumping regimes of operation of Ti:-sapphire laser most suitable are the schemes with frequency doubling inside the laser cavity (ICSHG) or in an external ring resonator (ERR). Note that $KNbO_3$ can be used at noncritical phase-matching conditions (propagation direction along the a axis); by changing the temperature of the crystal between 20–180 °C the wavelength range of 860–940 nm can be frequency-doubled. By use of a 55 μm thickness BBO crystal ICSHG of Ti:sapphire laser radiation was realized with pulse-width as short as 54 fs [4.150]. Maximum second harmonic powers were achieved in continuous wave and mode-locked regimes with high repetition rate ($\tau = 1.5$ ps) : $P_{2\omega} = 0.7$ W [4.147, 156].

Third harmonic (272 nm) of mode-locked Ti: sapphire laser radiation was generated in a BBO crystal of 6.5–12 mm in length ($\theta = 50°$) with output power $P_{3\omega} = 150$ mW and $\tau = 1$ ps used [4.147, 158]. Conversion efficiency was 30%. For fourth-harmonic generation (210 nm) a BBO crystal ($\theta_{ooe} = 75°$, $L = 7$–8 mm was also used, the scheme of mixing of the fundamental radiation with the third harmonic $\omega + 3\omega = 4\omega$ was employed [4.147, 158]. Maximum average output power was about 10 mW ($\tau = 1$ ps).

4.3.3 Semiconductor Lasers

A $KNbO_3$ crystal is most convenient for SHG of semiconductor laser radiation (Table 4.13). Along with a very high nonlinear coefficient [$d_{32} = 50d_{36}$ (KDP) $= 2.1 \times 10^{-11}$ m/V], this crystal has 90° phase matching at room temperature at the wavelength of a diode laser ($\lambda = 860$ nm). The spectral bandwidth for a crystal length 9 mm is $\Delta\lambda = 0.056$ nm, which makes it possible to double the GaAlAs laser radiation with $\Delta\lambda = 0.02$ nm. The angular bandwidth at 90° phase matching is 51 mrad, which exceeds the divergence of the fundamental radiation beam under focusing into the crystal (12 mrad) [4.159].

Second-harmonic generation of pulsed $Ga_{1-x}Al_xAs$ laser radiation (860 nm) was realized in a 6 mm $KNbO_3$ crystal when the fundamental radiation propagated along the a axis ($T = 31°C$) [4.160]. The fundamental radiation was polarized along the b axis and the second harmonic along the c axis. At a pump intensity of 6 kW cm^{-2} the conversion efficiency attained 1.8×10^{-3}. The output power was 0.35 mW. Efficient frequency doubling of a 856 nm diode laser was realized by use of a monolithic ring resonator of $KNbO_3$; optical conversion efficiency was 39% and conversion from electrical power was $\approx$ 10% [4.165]. Continuous wave radiation at 429 nm with $P = 62$ mW was generated by frequency doubling in the $KNbO_3$ crystal of the emission of a

Table 4.12. Second-harmonic generation of Ti:sapphire (Ti:Al$_2$O$_3$) laser radiation

Crystal	λ_ω [nm]	τ_ω	θ_{pm} [deg]	L [mm]	Output power $P_{2\omega}$ [mW]	η [%]	Refs.	Notes
LiIO$_3$	720–850	1.5 ps	43	10	700	50	4.147	$f = 82$ MHz
LiIO$_3$	720–800	–	–	7	23	0.38[a]	4.148	External ring resonator (ERR)
BBO	720–850	1.5 ps	30	8	450	27	4.147	$f = 82$ MHz
BBO	760–865	134 fs	ooe	1	–	2.1	4.149	Dispersive frequency doubler
BBO	860	54 fs	27.5	55 μm	230	75 (5.2[a])	4.150	ICSHG, $f = 72$ MHz
BBO	766–814	–	ooe	5	170	7.4[a]	4.151	ICSHG
LBO	700–900	12–25 ns	90 (θ), 22–40 (φ)	5	25 mJ	30	4.152	$I_0 = 0.9$ GWcm^{-2}
LBO	720–850	1.5 ps	90 (θ), 32 (φ)	8	350	20	4.147	$f = 82$ MHz
LBO	720–800	–	–	10	10–60	1.0[a]	4.148	ERR
LBO	820	cw	90 (θ), 31.8 (φ)	10.7	410	21.6	4.153	ERR
LBO	740–900	–	90 (θ), 37–23 (φ)	6	–	–	4.154	$\Delta TL = 7.8$–15.3 °C cm, $\Delta\lambda L = 0.6$–1.25 nm cm
KNbO$_3$	860–940	35 ns	along a axis	7.9	7.8 kW (peak)	45 (2[a])	4.155	ICSHG, $T = 2$–180 °C
KNbO$_3$	860	cw	–	6	650	48	4.156	ERR
MHBA	800–900	10 ns	–	5	0.03 mJ	6	4.64	
APDA	760–900	cw	Type I	3	0.8 μW	0.0003	4.157	

[a]Total conversion efficiency from the pump source.

Table 4.13. Second-harmonic generation of semiconductor laser radiation

Crystal	λ_ω [nm]	Phase-matching conditions	L [mm]	$P_{2\omega}$ [mW]	η [%]	Refs.	Notes
KNbO$_3$	860	along a-axis	8.97	0.00028	0.005	4.159	
	860	along a-axis $T = 31\,°C$	5.74	0.35	0.04	4.160	$\tau = 10$ ns
	842	$T = -23\,°C$	5	0.72	0.27	4.161	
	842	$T = -23\,°C$	5	24	14	4.162	Crystal in an external resonator
	865	along a-axis	5	0.215	1.7	4.163	External ring resonator (ERR)
	842	along a-axis	5	6.7	0.57	4.164	ERR, cw regime
	856	along a-axis $T = 15\,°C$	7	41	39	4.165	External resonator
	972	along b-axis	5	1.2	4.8	4.166	Distributed Bragg reflection semiconductor laser
	862	$T = 34\,°C$	14	400 (peak)	6.3	4.167	GaAIAs amplifier injected by 5 µs-Ti:sapphire laser
	858	–	12.4	62	1.1	4.168	
KTP	1500	$\theta = 54\,°$, $\varphi = 0\,°$, type II	10	0.001	–	4.169	ERR
LiIO$_3$	740	$\theta_{ooe} = 45\,°$	6	0.018	0.18	4.170	ERR
K$_3$Li$_{1.97}$Nb$_{5.03}$O$_{15.06}$	820	90 °	2.4	0.36	3.1	4.171	External resonator

GaAlAs amplifier seeded by a laser diode [4.168]. Sum-frequency generation in a KTP crystal by mixing outputs of two diode lasers operating at wavelengths of 1.5 and $0.78 - 0.82\,\mu m$, allows us to generate radiation at $0.52 - 0.54\,\mu m$ with $P = 0.2 - 0.3\,\mu W$ [4.169, 172].

4.3.4 Dye Lasers

Table 4.14 shows some characteristics of nonlinear crystals used for doubling dye laser radiation: nonlinear coefficient d_{eff} for minimum wavelength attained by SHG at room temperature, the d_{eff}^2/n^3 ratio proportional to the conversion efficiency, the minimum wavelength attained by SHG, and the "walk-off" angle ρ at different wavelengths. For all crystals under consideration (except LFM) this wavelength corresponds to $90°$ phase matching when radiation propagates in the direction orthogonal to the optic axis ($\theta = 90°$) for uniaxial crystals, and along the Y axis ($\theta = 90°$, $\varphi = 90°$) for biaxial crystals. For lithium formate (LFM) the limiting wavelength 230 nm corresponds to the boundary of the absorption band, whereas the phase-matching conditions allow shorter wavelengths to be attained. Upon cooling the crystals, smaller wavelengths can be achieved with the aid of SHG; for instance, in ADP $\lambda_{2\omega\,\mathrm{min}} = 250\,\mathrm{nm}$ at $T = 200\,\mathrm{K}$ [4.173]. Since $90°$ phase matching has some advantages, nonlinear crystals which possess $90°$ phase matching at a given pump wavelength are generally used for SHG. For example, for SHG of 860

Table 4.14. Parameters of crystals doubling dye laser radiation frequency

Crystal	$d_{\mathrm{eff}}{}^a$	$d_{\mathrm{eff}}^2/n^{3\ a}$	$\lambda_{2\omega}$ min [nm]	"Walk-off" angle ρ[deg] at different λ_ω				
				500 nm	600 nm	700 nm	800 nm	900 nm
BBO	0.3	0.06	204.8	4.96	4.71	4.28	3.89	3.57
DKB5	0.1	0.01	216.2	–	–	–	–	–
KB5	0.1	0.01	217.1	1.99	1.96	1.56	1.05	0.11
LFM	1.4	2.1	230	–	7.22	6.76	6.43	6.19
KDP	1	1	258.5	–	1.51	1.69	1.69	1.65
ADP	1.2	1.5	262	–	1.57	1.81	1.82	1.79
DKDP	0.9	0.9	265.5	–	1.41	1.59	1.57	1.51
LiIO$_3$	12.7	107	293.2	–	3.34	4.98	5.00	4.74
ADA	–	–	294	–	0.80	1.88	1.42	2.03
DADA	–	–	296	–	–	–	–	–
DKDA	–	–	310	–	–	–	–	–
RDP	0.9	0.9	313.5	–	–	0.87	1.06	1.10
RDA	0.9	0.8	342	–	–	0.65	1.22	1.35
KNbO$_3$	30.3	390	430	–	–	–	–	0.94
DCDA	0.9	0.8	517	–	–	–	–	–
CDA	0.9	0.8	525	–	–	–	–	–

^{a}Values of d_{eff} and d_{eff}^2/n^3 are calculated relative to d_{eff} and d_{eff}^2/n^3 for KDP.

nm radiation, $KNbO_3$ is most suitable, and for 592 nm radiation, a DADA crystal is used.

Minimum wavelengths by SHG process were obtained in crystals of $\beta - BaB_2O_4$ (205 nm), potassium pentaborate (KB5), and its deuterated analog (DKB5) (217 nm). A KB5 crystal has been used for SHG of dye laser radiation at 434–630 nm [4.174–176] (Table 4.15). The dye laser radiation propagated in the XY (ab) plane and was polarized in the same plane. The second harmonic was polarized along the Z axis (the eeo interaction). The above spectral range was covered by varying the phase-matching angle φ_{ooe} from 90° to 30°. If interaction takes place in the YZ plane ($\varphi = 90°$), a much smaller spectral range (217.1–240 nm) is covered as the phase-matching angle θ_{ooe} varies from 90° to 0° [4.175]. In the YZ plane the effective nonlinearity is much less than in

Table 4.15. Second-harmonic generation of dye-laser radiation

Crystal	$\lambda_{2\omega}$ [nm]	Parameters of output radiation (energy, power, pulse duration); conversion efficiency	Refs.	Notes
KDP	267.5–310	0.1 kW, $\eta = 1\%$	4.177	
KDP	280–385	–	4.178,179	$\theta_{ooe} = 66$–$45°$
KDP	280–310	50 mJ	4.180	
ADP	280–315	–	4.181–183	$\theta_{ooe} = 70$–$58°$, $T = 20\,°C$
ADP	280–310	50 mJ, $\eta = 8.4\%$	4.180	
ADP[a]	290–315	up to 1 mW, $\eta = 3 \times 10^{-4}$	4.184	
ADP[a]	250–260	120 µW	4.173	$\theta_{ooe} = 90°$, $T = 200$–$280\,K$
ADP[a]	293	0.13 mW, $\eta = 0.08\%$, $\tau = 3$ ps	4.185	$L = 3$ mm
ADP[a]	295	$\eta = 10^{-4}$, $\tau = 3 - 4$ ps	4.186	$L = 1 - 3$ mm
RDP	313.8–318.5	3.6 MW, $\eta = 52\%$ in power, $\tau = 8$ ns	4.187	$\theta = 90°$, $T = 20° - 98\,°C$, $I_0 = 36\ MWcm^{-2}$ $L = 25$ mm
RDP	310–335	3.2 MW, $\eta = 36\%$, $\tau = 10$ ns $f = 10$ Hz	4.188	$\theta = 90°$
ADA	292–302	30 mW	4.189	$\theta = 90°$
ADA[a]	285–315	400 mW (single-mode regime), 50 mW (multimode regime)	4.190	$\theta = 90°$, temperature tuning, $L = 30$ mm
DKDA	310–355	0.8–3.2 MW, $\eta = 9$–36%, $\tau = 10$ ns, $f = 10$ Hz	4.188	$\theta = 90°$, $L = 15$ mm
LiIO$_3$[a]	295	$\eta = 10^{-4}$, $\tau = 2.1$ ps	4.186	$L = 0.3$ mm
LiIO$_3$[a]	293–312	0.37 mW, cw regime	4.191	$L = 10$ mm
LiIO$_3$	293–330	15 mW, cw regime	4.192	$L = 1$ mm
LiIO$_3$	293	3 kW, $\eta = 30\%$	4.177	$L = 6$ mm
LiIO$_3$	293–310	4 mW, $\eta = 0.4\%$, cw regime	4.193	$L = 6$ mm, $\Delta\lambda = 0.03$ nm
LiIO$_3$	293–310	21 mW, $\eta = 2\%$, cw regime	4.193	$L = 6$ mm, $\Delta\nu = 30$ MHz
BBO	204.8–215	100 kW, 4–17%, 8 ns	4.121	$\theta = 70°$–$90°$
BBO	205–310	50 kW, 1–36%, 9 - 22 ns	4.194	$L = 6$ and 8 mm

Table 4.15 (*Contd.*)

Crystal	$\lambda_{2\omega}$ [nm]	Parameters of output radiation (energy, power, pulse duration); conversion efficiency	Refs.	Notes
BBO[a]	315	20 mW (average), 43 fs	4.195	$\theta = 38°$, $\varphi = 90°$, $L = 55$ μm
BBO	230–303	0.02–0.18 mJ, 17 ns	4.196	$\theta_{ooe} = 40° - 60°$, $L = 7$ mm
BBO[a]	243	30 mW, cw regime	4.197	$\theta_{ooe} = 55°$, $L = 8$ mm, $\Delta v = 200$ Hz
KB5	217.3–234.5	0.3 kW, 1%, 7 ns	4.174	XY plane, eeo
KB5	217.1–240	$5 - 6$ μJ, 10%, $3 - 4$ ns	4.175	YZ plane, $\theta_{ooe} = 90°–0°$
KB5	217.1–315.0	$5 - 6$ μJ, 10%, 5 ns	4.175	XY plane, $\varphi_{eeo} = 90°–31°$, $L = 10$ mm
KB5	217.0–250	0.1–5 μJ, 0.2%–5%	4.176	XY plane, $\varphi_{eeo} = 90°–65°$
DKB5	216.15	2 μJ, 5%, 3 ns	4.198	$\theta = 90°$, $\varphi = 90°$
LFM	230–300	2%	4.199	XZ plane, $\theta_{ooe} = 35°–45°$, $L = 10$ mm
LFM[a]	290–315	$\eta = 10^{-4}$	4.184	XZ plane, $\theta_{ooe} = 45°$ (590 nm)
LFM[a]	238–249	70 μW (244 nm), cw regime	4.200	XZ plane, $\theta_{ooe} = 39°$ (486 nm)
LFM	237.5–260	20 W, nanosecond regime, $\eta = 0.7\%$	4.177	
LFM[a]	243	1.4 mW, cw regime	4.201	$\theta_{ooe} = 36.8°$, $L = 15$ mm
LFM	285–310	4 μW, cw regime	4.193	
KNbO₃	425–468	400 kW, 43%	4.202	Angular tuning in planes XY and YZ, temperature tuning $(20 - 220\,°C)$ along the a axis
KNbO₃	419–475	12 μW, $\eta = 6.5 \times 10^{-4}$, cw regime	4.159	Along the a axis, T from −36 to $+180\,°C$, $L = 9$ mm
KNbO₃[a]	425–435	21 mW, $\eta = 1.1\%$, cw regime	4.203	Along the a axis, $T = 0–50°C$, $L = 9$ mm
urea	238–300	–	4.204	$\theta_{eeo} = 90 - 45°$, $L = 2$ mm
urea	298–370	–	4.204	$\theta_{eoo} = 90 - 50°$, $L = 2$ mm

[a] Intracavity SHG.

the XY plane, since for KB5 $d_{31} \sim 10d_{32}$; therefore, in KB5 crystals the interactions in the XY plane are mainly used.

Kato [4.121] used a $\beta - BaB_2O_4$ crystal for SHG of dye laser radiation. The following parameters were obtained: $P = 1\,MW$, $\tau = 8$ ns, λ up to 204.8 nm (90° phase matching of the ooe type). The fundamental radiation was focused on the crystal by a lens with $F = 50$ cm; the conversion efficiency to $\lambda = 204.8$ nm was 4% and to $\lambda = 205.8$ nm, 17%. *Miyazaki* et al. [4.194] attained $\eta = 36\%$ in a BBO crystal for SHG of dye laser radiation at $I_0 = 423\,MW\,cm^{-2}$. The fundamental radiation was focused by a lens with $F = 50$ cm. The conversion efficiency obtained in BBO was 4–6 times that in ADP. Due to ICSHG of femtosecond dye laser radiation, UV radiation at $\lambda = 315$ nm with $\tau = 43$ fs was obtained in a BBO crystal 55 μm in length [4.195].

ADA crystals have been used for SHG of rhodamine 6G laser radiation with $\eta = 5 \times 10^{-3}$ [4.190].

Generation of cw UV radiation in the 299–330 nm range with $P = 2$–15 mW was achieved in LiIO$_3$ because of ICSHG of dye laser radiation [4.192]. Argon laser radiation at $\lambda = 514.5$ nm and $P = 2.5$ W was used as a pump for a rhodamine 6G laser. The UV radiation bandwidth was 180–500 kHz. By SHG under 90° phase matching in LiIO$_3$, *Buesener* et al. [4.191] obtained the wavelength $\lambda_{2\omega} = 293.15$ nm.

With the help of ICSHG of coumarin 102 laser radiation in a lithium formate crystal (LFM), UV radition at $\lambda = 243$ nm was attained [4.201]. The fundamental radiation in the crystal propagated in the XZ plane at $\theta = 36.8°$. The ooe type interaction was used, and the length of the crystal was 15 mm. The cross-sectional diameter of the focused fundamental radiation beam in the crystal was 20 μm. The conversion efficiency was 1.5×10^{-4}. Radiation with $\lambda = 243$ nm was also generated in a ADP crystal [4.205, 206] due to ICSFG of argon and dye laser radiations. Although the nonlinear coefficient of LFM exceeds that of ADP, ICSFG in ADP is more effective than ICSHG in LFM, since ADP crystals can be used at 90° phase matching by proper choice of the interacting wavelengths.

Third-harmonic generation has been obtained in potassium pentaborate (KB5) crystals [4.188]. Tunable UV radiation in the 207.3–217.4 nm region was attained at a peak power of 25 kW and an average power of 15 mW. Interactions of the eeo type (in the XY plane) and of the ooe type (in the YZ plane) were used. Third-harmonic generation of dye laser radiation in urea has been obtained [4.118]: $\lambda_{3\omega} = 231$ nm, $\theta_{\mathrm{eeo}} = 77°$.

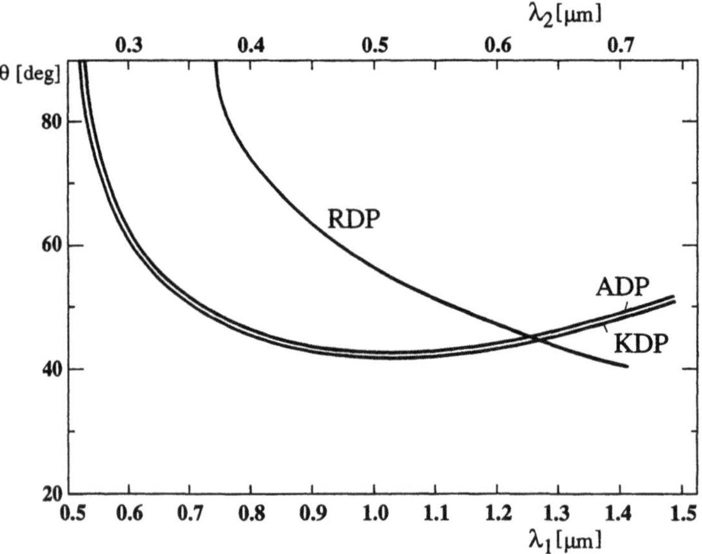

Fig. 4.2. Tuning curves for SHG in KDP, ADP, and RDP crystals (ooe interaction)

Figures 4.2–7 illustrate the tuning of the phase-matching angle versus the fundamental wavelength for SHG in crystals of KDP, ADP, RDP, $LiIO_3$, $LiNbO_3$, $\beta - BaB_2O_4$, KB5 (planes XY and YZ), LFM (XZ), KTP (XY), $KNbO_3$ (XY, YZ, XZ), and urea.

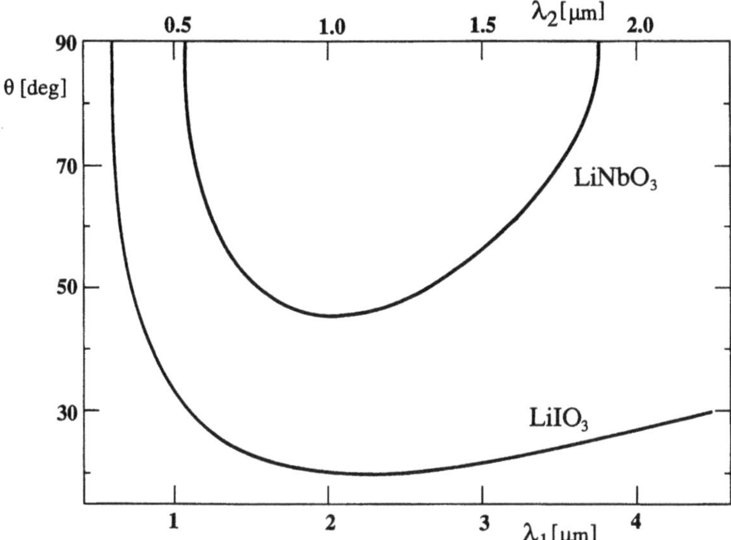

Fig. 4.3. Tuning curves for SHG in $LiIO_3$ and $LiNbO_3$ crystals (ooe interaction)

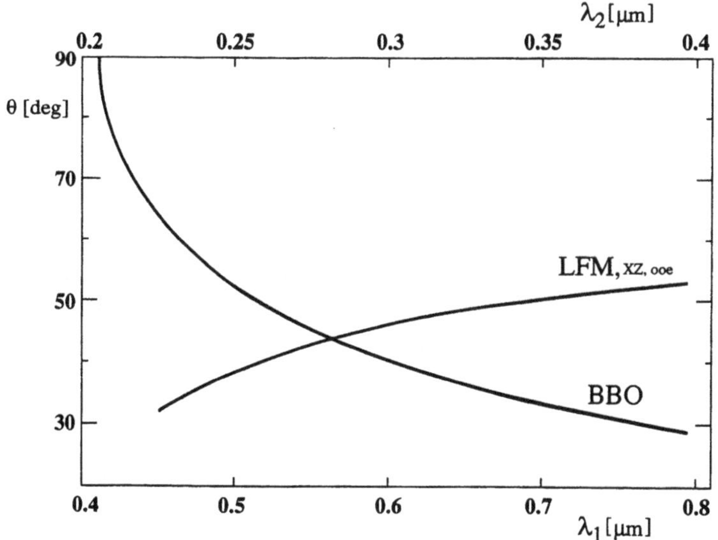

Fig. 4.4. Tuning curves for SHG in LFM (XZ plane, ooe interaction) and BBO (ooe interaction)

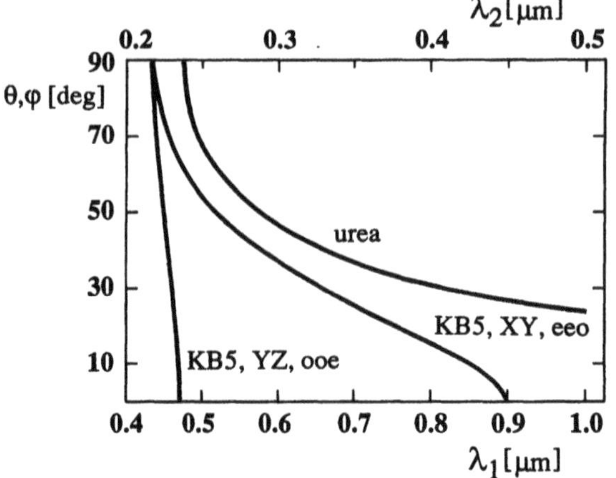

Fig. 4.5. Tuning curves for SHG in crystals of KB5 (XY, eeo; YZ, ooe) and urea (eeo)

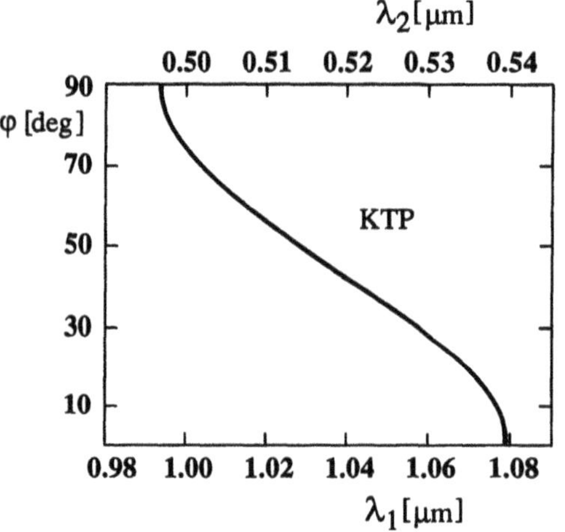

Fig. 4.6. Tuning curve for SHG in KTP (XY, eoe)

4.3.5 Gas Lasers

Second-harmonic generation of argon laser radiation (458–515 nm) has been realized in crystals of KDP, ADP, BBO, and KB5; SHG of He-Ne laser radiation ($1.15 - 3.39\,\mu m$) has been obtained in crystals of $LiIO_3$, $LiNbO_3$, and $AgGaS_2$; and that of NH_3 laser radiation ($\lambda = 12.8\,\mu m$) in Te and $CdGeAs_2$ crystals. Table 4.16 lists the crystals used, the corresponding phase-matching angles, and phase-matching temperatures. For SHG of argon laser radiation ($\lambda = 514.5\,nm$) in an ADP crystal placed in an external cavity the output power was 80 mW. For KDP and ADP crystals placed in the argon laser cavity (514.5 nm), $\eta = 50\%$ was achieved [4.220]. The second-harmonic generation power was 0.415 W. Continuous wave 257 nm radiation with a power of 1.2 W was generated by ICSHG of argon laser in a 6.5 long BBO crystal [4.212]. In

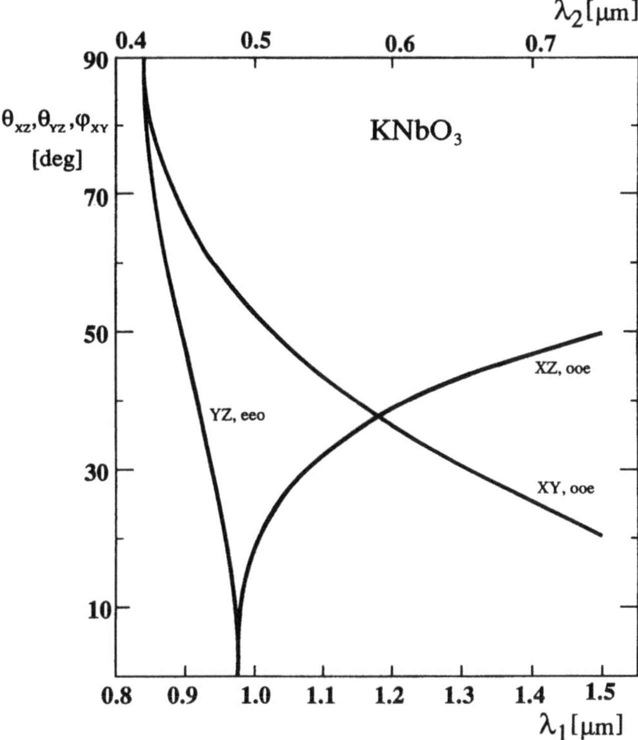

Fig. 4.7. Tuning curves for SHG in KNbO$_3$ (XY, ooe; YZ, eeo; XZ, ooe). The calculations were performed under assignment X, Y, $Z \Rightarrow c$, a, b ($n_X < n_Y < n_Z$)

the case of KB5 crystals, the conversion efficiency for SHG of argon laser radiation amounted to 10^{-5}. KB5 is suitable since it does not require thermal stabilization, but low nonlinearity [$d_{31} = 0.1 d_{36}$(KDP)] does not allow high conversion efficiencies. High-efficiency SHG of NH$_3$ laser radiation ($\tau_p = 1$ns) with $\eta = 60\%$ was realized in Te [4.217]. The power at $\lambda_{2\omega} = 6.4\,\mu$m amounted to 300 kW. Second-harmonic generation of CO laser radiation was obtained in a 7 mm long ZnGeP$_2$ crystal [4.219]. The average power at $\lambda = 2.6$–3.2 μm was 2.45–4.0 mW (peak power 350–500 W) at a pulse repetition rate $f = 100$ Hz. The conversion efficiency is $\eta = 3\%$.

4.3.6 Iodine Laser

In experiments on laser thermonuclear fusion high-power iodine lasers ($\lambda = 1.315\,\mu$m) are used with radiation-frequency conversion up to the sixth harmonic ($\lambda = 219.2$ nm) [4.221–224] (Table 4.17). Second-harmonic generation of iodine laser radiation was studied in KDP, DKDP, and LiIO$_3$ crystals. For SHG in KDP, η amounted to 16% and 12% for the eoe and ooe interactions, respectively. DKDP proved to be the best material for SHG of iodine

Table 4.16. Second-harmonic generation of gas laser radiation

Type of laser	Crystal	λ [μm]	θ_{pm} [deg]	T [°C]	Refs.
Argon laser	KDP[a]	0.5145	90	−13.7	4.207
	ADP	0.4965	90	−93.2	4.208
	ADP	0.5017	90	−68.4	4.208
	ADP	0.5145	90	−10.2	4.208
	ADP[a]	0.5145	90	−10	4.209
	KB5	0.4579	67.2 (φ_{pm})	20	4.145
	KB5	0.4765	60.2 (φ_{pm})	20	4.145
	KB5	0.4880	56.6 (φ_{pm})	20	4.145
	KB5	0.5145	50.2 (φ_{pm})	20	4.145
	BBO	0.5145	49.5	–	4.210
	BBO	0.4965	52.5	–	4.210
	BBO	0.4880	54.5	–	4.210
	BBO	0.4765	57.0	–	4.210
	BBO[a]	0.488	55	–	4.211
	BBO[a]	0.5145	–	–	4.212
He–Ne laser	LiIO$_3$[a]	1.152–1.198	25	20	4.213
	LiNbO$_3$	1.152	90	169	4.214
	LiNbO$_3$	1.152	90	281	4.215
	AgGaS$_2$	3.39	33	20	4.216
NH$_3$ laser	Te	12.8	–	–	4.217
	CdGeAs$_2$	11.7	35.7	–	4.218
CO laser	ZnGeP$_2$	5.2–6.3	47.5	–	4.219

[a]Intracavity SHG.

laser radiation. Whereas the values of damage thresholds are identical for KDP and DKDP, the latter has an absorption coefficient at $\lambda = 1.315\,\mu$m, an order of magnitude ($\alpha = 0.025\,\mathrm{cm}^{-1}$) smaller than KDP ($\alpha = 0.3\,\mathrm{cm}^{-1}$). In the single-pulse regime [4.223], 87–90% deuterated DKDP can be used ($\alpha = 0.06\,\mathrm{cm}^{-1}$). In the multipulse regime the maximum possible deuteration degree ($\sim 99\%$) is required.

The results of experiments on generation of iodine laser radiation harmonics are given in Table 4.17. Beams with diameters from 5 to 8 mm and pump intensity (1–3) GW cm^{-2} were used. In the beam 4 mm in diameter the efficiency of conversion to the second harmonic was 70%. For THG the "polarization mismatch" scheme [4.130, 136] was used. The conversion efficiency achieved for the third harmonic of iodine laser radiation amounted to 50%. For FOHG and FIHG, KDP crystals with the ooe interaction were used. In KB5 crystals SIHG of iodine laser radiation was attained upon doubling the third harmonic of the fundamental radiation frequency. The eeo interaction was used and third harmonic radiation propagated in the XY plane at an angle $\varphi = 80.5°$ to the X axis. Because of a small nonlinear coefficient of KB5 [$0.1d_{36}$ (KDP)], the efficiency of conversion to $\lambda = 219.2\,$nm was only 3%. Note that all the values of η are given with respect to the fundamental frequency.

Intracavity SHG of a chemical oxygen iodine laser ($\lambda = 1.315\,\mu$m) with conversion efficiency 1% [4.225] and 8% [4.226] was realized in a 10 mm long

Table 4.17. Generation of harmonics of iodine laser radiation: $\lambda = 1.315\,\mu m$, $E = 600$ mJ, $\tau_p = 1$ ns

	SHG $\omega + \omega = 2\omega$			THG $\omega + 2\omega = 3\omega$		FOHG $2\omega + 2\omega = 4\omega$	FIHG $2\omega + 3\omega = 5\omega$	SIHG $3\omega + 3\omega = 6\omega$
Wavelength [nm]	657.6			438.4		328.8	263.0	219.2
Crystal	DKDP	KDP	KDP	DKDP	KDP	KDP	KDP	KB5
Crystal length [mm]	19	20	10	20	10	40	–	10
Type of interaction	eoe	eoe	ooe	eoe	ooe	ooe	ooe	eeo
θ_{pm} [deg]	51.3	61.4	44.3	48	42.2	53.6	74	80.5 (φ_{pm})
Conversion efficiency [%]								
$I_O = (1\text{–}1.5) \times 10^9$ Wcm^{-2}	30	16	12	30	6	15	–	–
$I_O = 3 \times 10^9$ Wcm^{-2}	70	–	–	50	–	30	9	3

$LiIO_3$ crystal ($\theta_{ooe} = 23.5°$); achieved output powers in the latter case were 60 mW. Extracavity SHG of the same laser in 22 mm long $LiIO_3$ allows us to achieve cw powers up to 700 W with $\eta = 8\%$ [4.227].

4.3.7 CO_2 Laser

Second harmonic generation of CO_2 laser radiation has been realized in both pulsed and cw regimes with the use of crystals of proustite, pyrargyrite, silver thiogallate, $AgGaSe_2$, $ZnGeP_2$, $CdGeAs_2$, Tl_3AsSe_3 (TAS), GaSe, HgS, Se, and Te. The largest efficiencies for SHG with $\lambda = 5.3\,\mu m$ in a pulsed regime were attained with crystals of $AgGaSe_2$, $ZnGeP_2$, $CdGeAs_2$, and Tl_3AsSe_3 (Table 4.18). Second harmonic generation of pulsed CO_2 laser radiation with $\lambda = 10.25\,\mu m$ in a $AgGaSe_2$ crystal with $\eta = 35\%$ at $\theta_{ooe} = 52.7°$ has been reported [4.232]. Second-harmonic generation of CO_2 laser radiation was obtained in $CdGeAs_2$ crystals at 77 K with an average conversion efficiency of 21% in a pulsed regime ($f = 1.5$ kHz) and 0.44% in a cw regime [4.241]. The average power at $\lambda = 5.3\,\mu m$ amounted to 0.79 W for $f = 47$ kHz (pulsed regime) and 73 mW (cw regime).

In $ZnGeP_2$ high-efficiency SHG of CO_2 laser radiation was attained: $\eta = 49\%$ outside the crystal and $\eta = 80\%$ inside the crystal [4.236]. The energy of radiation with $\lambda_{2\omega} = 5.3\,\mu m$ and $\tau_p = 2$ ns was 5–10 mJ. Then the radiation was doubled again in another crystal of $ZnGeP_2$ ($\theta = 47°40'$, $\varphi = 0°$, $L = 10$ mm) with $\eta = 14\%$ (internal efficiency was 22%).

High-efficiency THG of pulsed CO_2 laser radiation ($\tau = 90\,ns$) in two $CdGeAs_2$ crystals has been realized [4.253] with conversion efficiency of 1.5% (inside the crystal $\eta = 3\%$). The phase-matching angle θ was 45° for the oeo interaction. The crystal lengths were 12 mm and 4.5 mm, respectively.

Fourth-harmonic generation was realized in $ZnGeP_2$ crystal ($\theta_{eeo} = 47.5°$, $L = 5\,mm$) by frequency doubling of the second harmonic of CO_2 laser radiation ($\tau = 170\,ns$); conversion efficiency was 2%, E = 0.2 mJ [4.254]. Second harmonic ($\lambda = 4.8\,\mu m$), third ($3.2\,\mu m$), fourth ($2.4\,\mu m$), and fifth harmonics ($1.9\,\mu m$) of CO_2 laser were obtained with the aid of Tl_3AsSe_3 crystals 5–6 mm in length [4.243]. In all cases ooe interaction was used with phase-matching angles 19°, 21°, 27°, and 28°, respectively. Conversion efficiency to the fourth harmonic was 27% (from 2ω), and to the fifth harmonic $\eta = 45\%$ (from 4ω).

4.3.8 Other Lasers

Isaev et al. [4.255] obtained SHG of copper-vapor laser radiation ($\lambda = 510.6\,nm$) in ADP crystals at $T = -30°$ and $\theta = 90°$. Ultraviolet radiation was obtained at $\lambda = 271$ and 289 nm and powers of 600 and 120 mW, respectively, in the case of SFG and SHG of copper-vapor laser radiation at

Table 4.18. Second-harmonic generation of CO_2 laser radiation

Crystal	$\lambda[\mu m]$	Type of interaction	θ_{pm} [deg]	$I_O[\mathrm{Wcm}^{-2}]$	L [mm]	η (power) [%]	Refs.
Ag_3AsS_3	9.2	ooe	19.9	–	10	4×10^{-4a}	4.228
Ag_3AsS_3	10.6	ooe	22.5	1.1×10^7	4.4	2.2	4.229
Ag_3SbS_3	10.6	ooe	24.2–30	–	–	–	4.230
$AgGaS_2$	10.6	ooe	67.5	–	–	–	4.216
$AgGaSe_2$	10.6	ooe	57.5	1.7×10^6	15.3	2.7	4.231
$AgGaSe_2$	10.25	ooe	52.7	$<10^7$	21	35	4.232
$AgGa_{0.65}In_{0.35}Se_2$	9.66	ooe	90	–	11	–	4.233
$ZnGeP_2$	9.19–9.7	eeo	76	–	–	5	4.234
	10.15–10.8					0.6^a	
$ZnGeP_2$	8.6	eeo	55.8	–	–	10.1	4.235
$ZnGeP_2$	10.6	eeo	76	10^9	3	49	4.236
$ZnGeP_2$	9.18–9.6	eeo	63–67	–	3.8	$-^a$	4.237
$ZnGeP_2$	9.3–9.6	eeo	–	2×10^7	4	2	4.238
$ZnGeP_2$	10.26–10.61	eeo	–	4.4×10^7	7.2	11.3	4.239
$CdGeAs_2$	10.6	oeo	48.4	1.4×10^7	9	15	4.240
$CdGeAs_2$	10.6	eeo	32.5	–	13	21	4.241
$CdGeAs_2$	10.6	eeo	32.5	–	13	0.44^a	4.241
Tl_3AsSe_3	9.6	–	–	–	3.7	10.9	4.242
Tl_3AsSe_3	9.6	ooe	19	10^7	5–6	28	4.243
Tl_3AsSe_3	10.6	ooe	–	6.3×10^8	4.57	57	4.244
GaSe	9.3–10.6	ooe	12.8–14.4	2×10^7	6.5	9	4.245
HgS	10.6	eeo	20.8	–	8.5	$-^a$	4.246
Se	10.6	eeo	5.5	–	8	10^{-5a}	4.247
Te	10.6	eeo	14.2	–	9	0.01	4.248
Te	10.6	eeo	15	10^6	4	0.25	4.249
Te	10.6	eeo	14.5	7×10^5	2.1	0.5	4.250
Te	10.6	oeo	20.4	–	–	–	4.251
Te	9–11	eeo	2-18	–	7	10^{-4a}	4.252

^{a}Continuous-wave regime.

510.6 and 578.2 nm [4.256]. Here a KDP crystal 4 cm in length was used, and $\eta = 3\%$ was obtained. For SHG and SFG of copper-vapor laser radiation BBO crystals are very opportune [4.257–259]. Thus, with the aid of 7–8 mm long BBO crystals ($\theta_{\mathrm{ooe}} = 51°$) output powers in UV (255, 271, 289 nm) up to 1.3 W were achieved with conversion efficiencies of 20–30% [4.259].

Second-harmonic generation of SOAP:Ho laser radiation (2.06 μm) [4.260] and of HF laser radiation (2.7–2.9 μm) [4.261] has been realized in proustite. Crystal of $ZnGeP_2$ ($\theta = 70°$) 13.6 mm in length, was used for SHG of DF laser radiation with $\eta = 6.2\%$ [4.262]. Intracavity SHG of a Q-switched cw pumped Cr^{3+}:$LiSrAlF_6$ laser radiation in $LiIO_3$ allows generation of UV radiation in the range of 395–435 nm with $P = 7$ mW [4.263]. *Chinn* [4.264] realized ICSHG of cw Nd^{3+}:NdP_5O_{14} laser radiation ($\lambda = 1.05$ μm) with $\eta = 1\%$ using a 1 mm long "banana" crystal. The fundamental radiation propagated along the crystallographic a axis and was polarized along the b axis; $T = 72.3\,°C$.

To study the possibility of SHG of Er^{3+}:YLF laser radiation in $KNbO_3$, SHG of a dye laser radiation at $\lambda = 850.2$ nm was realized with $\eta = 43\%$ at a fundamental power of 0.9 MW [4.202]. The $KNbO_3$ crystal, 5 mm in length, was kept at 18.0 °C and fundamental radiation propagated along the a axis. The third harmonic of alexandrite laser radiation ($\lambda_\omega = 730$–780 nm) has been obtained in a BBO crystal ($\theta_{\mathrm{ooe}} = 48.6°, L = 7.5$ mm) with $\eta = 24\%$ and energy $E = 7.5$ mJ [4.265].

4.3.9 Frequency Conversion of Femtosecond Pulses

Frequency conversion, in particular second-harmonic generation, of ultrashort laser pulses ($\tau < 1$ps) deliver certain additional limitations on the properties of applied nonlinear crystals. Radiation with ultrashort pulse-width has a broad frequency bandwidth. Thus the spectral bandwidth ($\Delta\nu$ or $\Delta\lambda$) and, as a result, also the angular bandwidth ($\Delta\theta$) of the nonlinear process should have a strong affect on the efficiency of conversion and on the pulse duration of converted radiation. This phenomena can also be characterized by the group-velocity mismatch between the fundamental and second-harmonic pulses determined from

$$\Delta v_{\mathrm{g}}^{-1} = \frac{\mathrm{d}k_{2\omega}}{\mathrm{d}(2\omega)} - \frac{\mathrm{d}k_\omega}{\mathrm{d}\omega} \tag{4.3}$$

where k_ω and $k_{2\omega}$ are absolute values of corresponding wave vectors at fundamental and second harmonic waves, respectively. An other factor which limits efficiency of SHG is the walk-off angle ρ between ordinary and extraordinary beams. Nonlinearity and the minimum wavelength attained by SHG should also be considered. *Pelouch* et al. [4.266] present the values of mentioned parameters for the case of SHG radiation with $\lambda = 630$ nm in KDP, ADP, BBO, and LBO crystals (Table 4.19).

Compared with those commonly applied for these purposes, KDP and ADP crystals, the BBO crystal has higher nonlinearity, however, as a draw-

Table 4.19. Optical properties of KDP, ADP, BBO, and LBO crystals for SHG ($\lambda = 630$ nm) and the minimum λ attained by SHG

Parameter	KDP	ADP	BBO	LBO
$d_{\text{eff}}/d_{\text{eff}}$ (KDP)	1.0	1.2	3.9	1.5
Δv_{g}^{-1} [fs/mm]	185	205	360	235
$L\Delta\lambda$ [mm nm]	3.2	2.9	1.6	2.5
$L\Delta\theta$ [mm mrad]	6.6	6.1	2.1	9.4
ρ [deg]	1.6	1.7	4.5	1.0
$\lambda_{2\omega}$ min [nm]	258.5	262	204.8	277.4

back, a very large walk-off angle should be mentioned. Theoretical considerations show that BBO crystal is most advantageous for SHG of pulses in the range of 0.1–1.0 ps [4.267]. Due to its small group-velocity mismatch and ρ-value and large angular bandwidth the LBO crystal is advantageous under focusing conditions and thicker crystals can be used compared with KDP, ADP, and BBO.

In a series of works [4.195, 266–271] SHG of femtosecond radiation in LBO, BBO, and KDP crystals was studied experimentally. By use of the autocorrelation method with noncollinear SHG in LBO 4.165 mm in length it is possible to measure pulsewidths as short as 40 fs [4.266]. *Ishida* and *Yajima* [4.268] performed measurements of femtosecond pulse-widths by use of 0.1 mm and 0.45 mm long KDP, and of 0.9 mm and 1.9 mm long BBO crystals. It was demonstrated that the BBO crystal is very useful for both pulse-width measurements and efficient conversion of femtosecond pulses into the UV. The possibility of pulse-width measurements up to 50 fs with a BBO crystal 2 mm in length was shown by *Cheng* et al. [4.269]. By use of a KDP crystal of 1 mm in length, SHG of mode-locked colliding-pulse dye laser radiation was realized with conversion efficiency of 15% and 21% for pulses of 100 fs and 180 fs duration, respectively [4.270]. Efficient SHG of a dye laser radiation with $\lambda = 496$ nm and $\tau_{\text{p}} = 300$ fs was realized in 7 mm BBO crystal: $\eta = 15\%$ [4.271]. Intracavity SHG of fs pulses from a ring dye laser is reported by *Edelstein* et al. [4.195]. An applied 55 µm thick BBO crystal allows to generate UV pulses (315 nm) as short as 43 fs with average power of 20 mW.

4.4 Sum-Frequency Generation

Sum-frequency generation greatly increases the spectral range emitted by tunable lasers. At present, with the aid of SFG in crystals, radiation with λ up to 185 nm has been obtained in the VUV spectral region. Dye laser radiation is most often used as a source of continuously tunable radiation for mixing. Ti:sapphire lasers and tunable OPO radiation, as well as ND:YAG laser radiation harmonics, Stokes components of SRS generated in various organic liquids and gases are also used for this purpose.

Sum-frequency generation is also used for effective conversion of IR radiation to the visible range (up-conversion), which greatly facilitates the detection of IR radiation. For instance, the IR radiation of CO_2 laser ($\lambda = 10.6\,\mu m$) is up-converted into the visible region with a conversion efficiency of 30–40% upon mixing with dye or Nd:YAG laser radiation (a pump source).

4.4.1 Up-Conversion to the UV Region

KDP, ADP, BBO, LBO, and KB5 crystals are usually used for up-conversion into the UV and VUV regions. Due to a high transparency within a wide wavelength range (up to 177 nm) and fairly high optical-damage threshold, KDP crystals have been successfully used for generation of picosecond and nanosecond UV pulses. Because of up-conversion in KDP, UV radiation was obtained within a range from 190 to 432 nm [4.114, 117, 272–281]. Table 4.20 lists the characteristics of the lasers used in SFG experiments with KDP, up-

Table 4.20. Sum-frequency generation of UV radiation in KDP

λ_{SF} [nm]	Sources of interacting radiation	τ_p [ns]	Conversion efficiency, power, energy	Refs.
190–212	SRS of 1.064 μm + sum frequency radiation (220–250 nm) [4.273]	0.02	20–40 μJ	4.272
215–223	2ω of dye laser + Nd:YAG laser	10	10 kW	4.114
215–245	SRS of 266 nm (4ω of Nd:YAG laser) + OPO (0.9–1.4 μm)	0.02	100 μJ	4.273
217–275	2ω of dye laser + Nd:YAG laser (1.064 μm)	25–30	50–55%, 10 mW (average)	4.274
217–226	OPO (1.1–1.5 μm) + 4ω of Nd:YAG laser (266 nm)	0.02	100 kW	4.117
218–244	(269–315) nm [4.275] + Nd:YAG laser	0.03	0.1 mJ	4.275
239	Nd:YAG laser (1.064 μm) + XeCl laser (308 nm)	0.7	50%	4.276
240–242	2ω of ruby laser (347 nm) + dye laser	30	1 MW	4.277
257–320	Dye laser + argon laser	cw regime	0.2 mW	4.278
269–315	SRS of 532 nm (2ω of Nd:YAG laser) + 532 nm	0.03	1–3 mJ	4.275
269–287	OPO (1.29–3.6 μm) + 3ω of Nd:YAG laser (355 nm)	0.02	100 kW	4.117
271	Two copper vapor lasers (511 and 578 nm	35	1.5%, 100 mW (average)	4.279
288–393[a]	OPO (0.63–1.5 μm) + 2ω of Nd:YAG laser (0.532 nm)	0.02	100 kW	4.117
360–415	Dye laser + Nd:YAG laser	25–30	60–70%	4.280
362–432	Dye laser + Nd:YAG laser	0.03	20%	4.281

[a]DKDP crystal was used.

conversion efficiencies, and output power or energy. In nearly all experiments angular tuning of phase matching was used, but *Stickel* and *Dunning* [4.277] attained the 240–242 nm region by mixing the second harmonic of ruby laser radiation and IR dye laser radiation in a temperature-tuned KDP crystal ($\theta_{ooe} = 90°, T = -20° \ldots + 80\,°C$). Mixing of Nd:YAG ($\tau_p = 0.7\,ns$) and XeCl ($\tau_p = 12\,ns$) laser radiation by SFG resulted in a conversion efficiency equal to 50%, measured with respect to the energy of the XeCl laser radiation ($\lambda_{SF} = 239$ nm; $\theta_{ooe} = 68.3°$) [4.276]. *Dudina* et al. [4.280] converted the dye laser radiation (545–680 nm) to the UV region ($\lambda_{SF} = 360–415$ nm and $\tau_p = 25–30\,ns$) with $\eta = 60–70\%$, using Nd:YAG laser radiation as a pump source (1.064 µm). The eoe interaction was used and the phase-matching angle changed from 56° to 58°. Continuous-wave radiation in the 257–320 nm region with a power of up to 0.2 mW was generated upon mixing argon ion laser radiation at wavelengths of 458, 488 and 515 nm with dye laser radiation [4.278]. The shortest wavelength in KDP (190 nm) was obtained [4.272] by mixing radiation at 1.415 µm (generated by SRS of 1.064 µm radiation in nitrogen) with radiation at $\lambda = 220–250$ nm (obtained in its turn by mixing the OPO radiation with SRS Stokes components induced by the fourth harmonic of Nd:YAG laser radiation in nitrogen and hydrogen, $\lambda = 283$ and 299 nm, respectively). Generated picosecond pulse energies amounted to 20–40 µJ within the 212–190 nm region ($\theta_{ooe} = 70 - 90°$).

Some papers report SFG realized in the UV region using ADP crystals (Table 4.21). Continuous wave radiation up to 243 nm with a power of 4 mW

Table 4.21. Sum-frequency generation of UV radiation in ADP

λ_{SF} [nm]	Sources of interacting radiation, phase-matching angle, crystal temperature	τ_p [ns]	Conversion efficiency, power, energy	Refs.
208–214	2ω dye laser + Nd:YAG laser, $\theta = 90°, T = -120° \ldots 0\,°C$	10	1.7 µJ	4.114
222–235	2ω dye laser + Nd:YAG laser	10	10%	4.114
240–248	Dye laser + 2ω of ruby laser, $\theta = 90°$, $T = -20° \ldots + 80\,°C$	30	1 MW, 4%	4.277
243–247[a]	Dye laser + argon laser (363.8 nm)	cw regime	4 mW,	4.197, 205
243[a]	Dye laser + argon laser (351 nm), $\theta = 90°, T = 8\,°C$	cw regime	0.3 mW,	4.282
247.5	Dye laser + krypton laser (431.1 nm), $\theta = 90°, T = -103\,°C$	cw regime	–	4.283
246–259	Dye laser + 2ω of Nd:YAG laser, $\theta = 90°, T = -120° \ldots 0\,°C$	10	1%, 3 µJ	4.114
252–268[a]	Dye laser + argon laser (477, 488, 497 nm), $\theta_{ooe} = 90°$	cw regime	8 mW	4.284
270–307	Dye laser + 2ω of Nd:YAG laser, $\theta_{ooe} = 81°$	ps regime	–	4.281

[a]ADP crystal was placed in an external resonator.

has been obtained by mixing dye and argon laser radiation [4.197, 205]. For a Q-switched ruby laser tuning was obtained in the 240–248 nm region with a pulse peak power of 1 MW [4.277]. The length of ADP crystal was 25 mm and $\theta = 90°$. The shortest wavelength (208 nm) for ADP was attained when Nd:YAG laser radiation was mixed with the second harmonic of dye laser radiation (246–302 nm) [4.114]. The ADP crystal was cooled to –120 °C at $\theta = 90°$. The peak power at $\lambda = 208$ nm was 500 W.

The use of BBO and LBO crystals made it possible to generate UV radiation down to 188.9 and 187.7 nm, respectively (Tables 4.22, 23). Due to their wide transparency range in UV (up to 160 nm for LBO), fulfillment of phase-matching conditions, and high nonlinearity [4.44a, 44b, 302] these crystals are very useful for nonlinear conversion in the UV. Ultraviolet radiation up to 197.4 nm ($\tau = 17$ ns) has been obtained in BBO crystal by THG of dye laser radiation [4.291]. The crystal was cut at an angle $\theta_C = 90°$ and had $L = 8.2$ mm. At $I_o = 19 - 29 \, \mathrm{MW \, cm^{-2}}$, η reached almost 5%. Upon cooling the crystal to 95 K, the minimum wavelength of the converted radiation was 195.3 nm. The experimental dependence of $\lambda_{3\omega}$ on the crystal temperature is given as $\lambda_{3\omega} = 194.34 + 10.3 \times 10^{-3} \, T(\mathrm{K})$. High-efficiency THG of dye laser radiation has been reported in an 8.2 mm BBO crystal ($\theta = 80°$); within most of the

Table 4.22. Sum-frequency generation of UV radiation in BBO

λ_{SF} [nm]	Sources of interacting radiation	τ_p [ns]	Conversion efficiency, power energy	Refs.
188.9–197	Dye laser (780–950 nm) + 2ω of another dye laser (248.5 nm)	10	up to 0.1 mJ	4.285
190.8–196.1	Ti:sapphire laser (738–825 nm) + 2ω of Ar laser (257 nm)	–	tens of nW	4.286
193	Dye laser + KrF laser (248.5 nm)	9	0.2%, 2 µJ	4.285
193	Dye laser (707 nm) + 4ω of Nd:YAG laser	90–250 fs	10 µJ (250 fs)	4.287
193.3	Dye laser (724 nm, 5 ps) + 4ω of Nd:YLF laser (263 nm, 25 ps)	0.01	1.7%, 4 µJ (2.5 mJ)[a]	4.288
193.4	FOHG of dye laser radiation (774 nm, 300 fs), $\omega + 3\omega = 4\omega$	800 fs	0.5 µJ (1.5 mJ)[a]	4.289
194	Ti:sapphire laser + 2ω of Ar laser (257 nm), three crystal configuration with external cavity	–	0.016 mJ	4.290
195.3	THG of dye laser ($T = 95$ K)	17	5%, 8 µJ	4.291
196–205	Dye laser + 2ω of another dye laser	5	0.1 mJ	4.292
197.7–202	THG of dye laser	0.008	1%, 1–4 mW	4.293
198–204	THG of dye laser	5	20%, 1.7 mJ	4.294
271	Two copper vapor lasers (511 and 578 nm)	35	0.9%, 64 mW	4.279
362.6–436.4	Dye laser + Nd:YAG laser, noncollinear SFG (NCSFG), $\alpha = 4.8$–$21.3°$	–	1%, 0.065 mJ	4.295
369	Diode laser (1310 nm) + Ar laser (515 nm)	–	1.3 µW	4.296
370.6	Dye laser (568.6 nm) + Nd:YAG laser, NCSFG, $\alpha = 6.3°$	–	8–18%	4.297

[a] After amplification in an ArF excimer gain module.

Table 4.23. Sum-frequency generation of UV radiation in LBO

λ_{SF} [nm]	Sources of interacting radiation	Conversion efficiency, power energy	Refs.
188–195	OPO (1.6–2.3 µm) + 5ω of Nd:YAG laser (212.8 nm) $\theta = 90°, \varphi = 90–52°$, ooe	0.2–2%, 2–40 µJ	4.298
187.7–195.2	OPO (1.591–2.394 µm) + 5ω of Nd:YAG laser $\theta = 90°, \varphi = 88–50°$, ooe	3 kW (peak)	4.299
191.4	SRS in H_2 (1.908 µm) + 5ω of Nd:YAG laser $\theta = 90°, \varphi = 88–50°$, ooe	10%, 67 kW (peak) 2 mW (average)	4.299
218–242	OPO (1.2–2.6 µm) + 4ω of Nd:YAG laser (266 nm) $\theta = 90°, \varphi = 90–33°$, ooe	0.2–2%, 20–400 µJ	4.298
232.5–238	Nd:YAG laser + 2ω of dye laser	–	4.300
240–255	Nd:YAG laser + 2ω of dye laser, NCSFG	8%, 0.12 mJ	4.301

tuning region (199–203.3 nm), $\eta > 20\%$ [4.294]. Tuning in the region 188.9–197 nm was attained by mixing in the SFG process dye laser radiation (780–950 nm) and the second harmonic of another dye laser (497 nm) in a BBO crystal [4.285]. The conversion efficiency was 0.2% and 7% in the nanosecond (9 ns) and picosecond (30 ps) regimes, respectively,

UV radiation with $\lambda = 193$ nm of femtosecond duration is of particular interest, since it can further be amplified in ArF excimer gain modules to mJ energies, corresponding to the GW cm^{-2} intensities [4.288, 289]. The minimum 193 nm pulse duration of 90 fs was achieved by mixing picosecond pulses of the fourth harmonic of an Nd:YAG laser with femtosecond dye laser output [4.287]. A dispersively compensated scheme used for SFG in a 1 mm thick BBO crystal results in output bandwidth of 0.6 nm [4.287]. Note that in all cases of SFG with a BBO crystal (Table 4.22), ooe interaction was used.

The UV range of 188–225 nm has been achieved by SFG in LBO crystal (Table 4.23). Generally type I (ooe) interaction in the XY plane was employed in these cases. Minimum wavelengths are observed when radiations propagate along the Y axis. *Liu* and *Kato* [4.303] discussed a proposed setup for generation of UV as short as 160 nm by SFG in an LBO crystal by mixing 190 nm radiation with IR of the range 1–2.5 µm.

To cover the 185–269 nm spectral region, KB5 crystals have been widely used for SFG (Table 4.24). Dye laser second-harmonic radiation has been mixed with radiations of ruby (694.3 nm) and Nd:YAG (1.064 µm) lasers, respectively [4.188, 306]. Nanosecond radiation was generated in the region 202–212 nm with an energy of 2–10 µJ [4.306] and in the region 207.3–217.4 nm with an energy of up to 2.5 mJ [4.188]. Temperature tuning was realized at 90° phase matching [4.311]: $\lambda_{SF} = 201.2–201.8$ nm at temperature varying from $-20°C$ to $40°C$. The mixing of the fourth harmonic of Nd:YAG laser radiation (266 nm) with dye laser radiation generated radiation at $\lambda_{SF} = 196.6$ nm with $\tau_p = 8$ ns and $P = 40$ kW [4.117]. Approximately 10 µJ radiation with λ up to 185 nm ($\tau_p = 30$ ns) was obtained by SFG process when mixing SH of dye laser radiation with radiation of another dye laser generating in the IR

Table 4.24. Generation of UV radiation in KB5

λ_{UV} [nm]	τ_p [ns]	Energy, power	Conversion efficiency	Process	Refs.
217.3–234.5	7	0.5 μJ	2×10^{-3}	SHG	4.174
217.1–315.0	2–5	5 μJ	0.1	SHG	4.175
217.0–250.0	5	Up to 5 μJ	$(2–5) \times 10^{-2}$	SHG	4.176
231.4	6	0.2 mJ	2×10^{-3}	THG	4.146
229–347	18	Up to 20 μJ	10^{-5}	SHG	4.145
212.8	6	110 kW	2×10^{-3}	FIHG	4.116
208–217	10	1 W	2.5×10^{-4}	SFG	4.304
196.6	8	0.5 mJ	10^{-3}	SFG	4.305
207.3–217.4	3	0.8 mJ	3×10^{-3}	SFG	4.188
201–212	20	2–10 μJ	0.1	SFG	4.306
185–200	30	Up to 10 μJ	0.1	SFG	4.307
211–216	cw regime	50–100 nW	10^{-6}	SFG	4.308
266	0.045	60 MW	0.05	FOHG	4.117
212.8	0.045	11 MW	0.01	FIHG	4.108
196.7–226	0.02	20 kW	–	SFG	4.117
212.8	0.02	5 MW	–	FIHG	4.117
219	0.75	100 MW	0.03	SIHG	4.222
194.1–194.3	cw regime	2 μW	–	SFG	4.309
200–222	0.045	1 μJ	2×10^{-5}	SFG	4.310
220–266	0.045	1 μJ	–	SHG	4.311

region 740–910 nm) [4.307]. The interacting radiations propogated along the Y axis and were polarized along the X axis. The up-conversion efficiency was 8–12% for IR radiation of intensity 500 MW cm^{-2}. Up to now 185 nm was the minimum wavelength achieved by frequency conversion in nonlinear crystals. The SFG process – by mixing dye laser radiation with that of an argon laser at wavelengths 351.1 and 334.5 nm, respectively [4.308, 312], or with SH of Ar laser radiation (257 nm) [4.309] – produces continuous-wave radiation at $\lambda = 211–216$ nm and 194 nm, respectively. Radiation power at $\lambda = 194$ nm was 2 μW. A temperature change from 20° to 45°C at 90° phase matching gave rise to tuning in the 194.1–194.3 nm region [4.309]. Generation of picosecond UV radiation in the 196–269 nm region has been achieved [4.117, 310] by mixing the third (355 nm) and fourth (266 nm) harmonics of mode-locked Nd:YAG laser radiation with OPO radiation. Powers up to 20 kW and 100 kW have been attained at $\lambda_{SF} = 197$ nm and 225–269 nm, respectively, for $\tau_p = 20$ ps and crystal length 7 mm [4.117]. The interaction was of eeo type in the XY plane: the crystal cut angle φ_C was 65°.

Note that minimum wavelengths are observed at 90° phase matching, i.e., when radiations propagate in the direction normal to the optic axis for the uniaxial crystals KDP and ADP or along the $Y(b)$ axis for a biaxial LBO and KB5 crystals. Figures 4.8–12 show for KDP, ADP, BBO, LBO, and KB5 crystals the SFG phase-matching curves for type I interaction; from these curves we can obtain λ_1 and λ_2 of the radiations being mixed to generate the

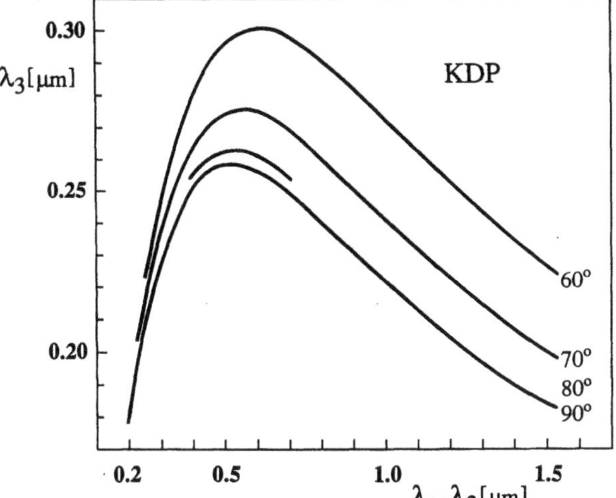

Fig. 4.8. Tuning curves for SFG in KDP at $\theta = 60°$, $70°$, $80°$, and $90°$ (ooe interaction)

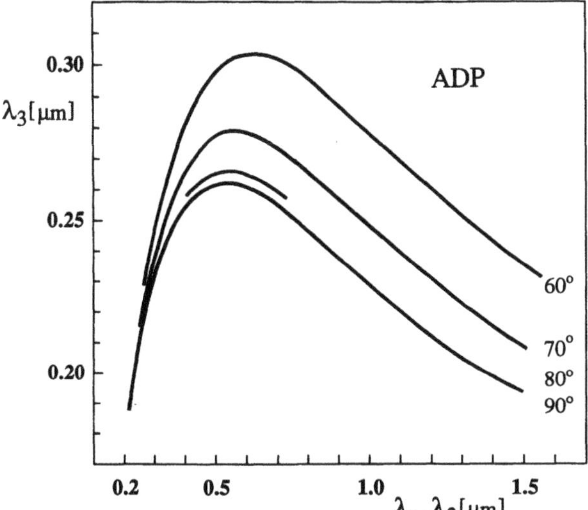

Fig. 4.9. Tuning curves for SFG in ADP at $\theta = 60°, 70°, 80°$, and $90°$ (ooe interaction)

radiation with λ_3. The tops of the curves correspond to the case of SHG and indicate the minimum wavelength attained by SHG.

4.4.2 Infrared Up-Conversion

$LiIO_3$ has been used for up-conversion of IR radiation with $\lambda = 1$–$5\,\mu m$ to visible light [4.313.–319] (Table 4.25). *Gursky* [4.313] obtained 100% conversion efficiency for radiation at $3.39\,\mu m$ and pump intensity $6.3 \times 10^4\,W\,cm^{-2}$ in

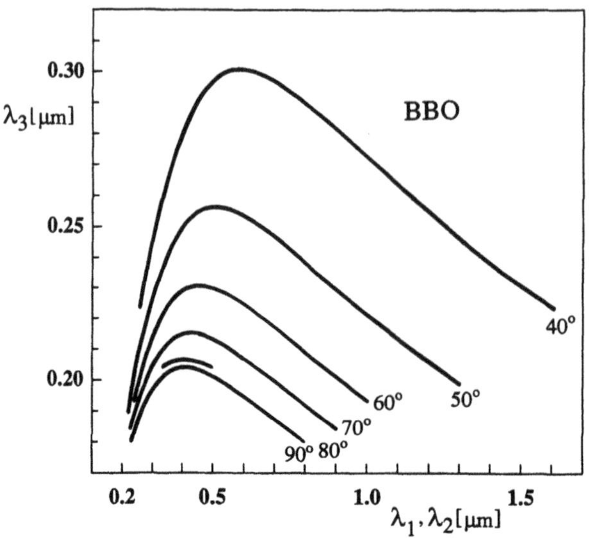

Fig. 4.10. Tuning curves for SFG in BBO at $\theta = 40°$, $50°$, $60°$, $70°$, $80°$, $90°$ (ooe interaction)

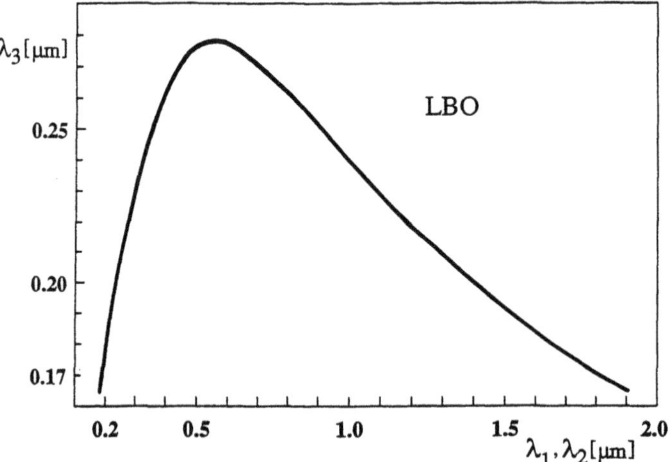

Fig. 4.11. Tuning curve for SFG in LBO at $\theta = 90°$, $\varphi = 90°$ (ooe interaction)

a crystal 5 cm long. LiNbO$_3$ crystals are also widely used for covering 1.5–4.5 μm radiation to the visible (Table 4.26). Argon-laser radiation has served as a pump and 90° phase matching of the ooe type has been used [4.214, 326, 327]. By varying the crystal temperature from 180° to 400 °C, the IR spectral regions from 2.7 to 4.5 μm were converted to the visible region [4.327, 328]. Ruby laser radiation has also served as a pump [4.320–322, 329]; 100% quantum efficiency was obtained in the 2 cm long crystal at a pump intensity of $5 \times 10^6 \, \mathrm{W\,cm^{-2}}$ [4.321]. Also, He-Ne and Nd:YAG laser radiations are frequently used as pump sources. Blackbody radiation at $\lambda = 6.5 - 12.5$ μm has been converted to

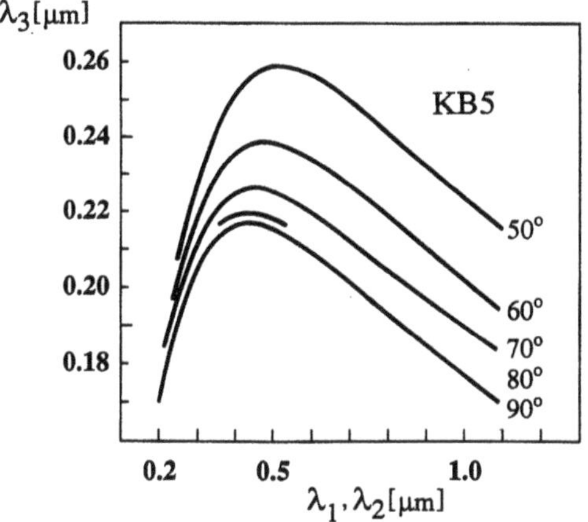

Fig. 4.12. Tuning curves for SFG in KB5 at $\theta = 90°, \varphi = 50°, 60°, 70°, 80°, 90°$ (eeo interaction)

Table 4.25. Up-conversion in LiIO$_3$

λ_{IR} [μm]	Pump source	η [%]	Refs.
3.39	0.694 μm, mode-locked ruby laser	100	4.313
3.2-5	1.064 μm, Nd:YAG laser	0.001	4.315
2.38	0.488 μm, argon laser	4×10^{-8}	4.316
1.98, 2.22, 2.67	0.694 μm, mode-locked ruby laser	0.14–0.28	4.317
3.39	0.5145 μm, argon laser	2.4×10^{-2}	4.318
1--2	0.694 μm, ruby laser	18	4.319

Table 4.26. Up-conversion in LiNbO$_3$

λ_{IR} [μm]	Pump source	η [%]	Refs.
1.69–1.71	0.694 μm, Q-switched ruby laser	1	4.320
1.6–3.0	0.694 μm, Q-switched ruby laser	100	4.321
1.6	0.694 μm, ruby laser	10^{-5}	4.322
3.3913	0.633 μm, cw He-Ne laser	10^{-5}	4.323
3.3922	0.633 μm, cw He-Ne laser	5×10^{-5}	4.324
2–4	1.064 μm, Q-switched Nd:YAG laser	–	4.325
3.39	0.515 μm, argon laser	–	4.326

the near IR region in a proustite crystal 6 mm long using Nd:YAG laser radiation as a pump source ($P = 1$ kW, $f = 2$ kHz) [4.330].

Up-conversion was used for converting IR radiation of astronomical objects to the visible spectrum [4.315]. Infrared radiation from the Moon, Venus, Mars, and some bright stars was converted to visible radiation at $\lambda = 0.76 - 0.38$ μm with η up to 0.01%. For a precise measurement of the wavelength, the radiation of a methane-stabilized He-Ne laser was converted to $\lambda = 0.533$ μm using I$_2$-stabilized He-Ne laser radiation as a pump and a LiNbO$_3$ crystal [4.324]. The IR wavelength measured in this way was 3.39223140 μm. Up-conversion also greatly widens the OPO tuning range. For

Table 4.27. Up-conversion in KTP

λ_{IR} [µm]	Pump source	η [%]	Refs.
1.064	0.809 µm, diode laser	68	4.331
1.54	0.78 µm, diode laser	7×10^{-4}	4.172
1.064	0.824 µm, dye laser(intracavity SFG)	0.26	4.332
1.064	0.809 µm, diode laser	55	4.333
1.064	0.805 µm, diode laser	24	4.334
1.319; 1.338	0.532 µm, 2ω of Q-switched Nd:YAG laser	10	4.335

instance, OPO radiation at $\lambda = 1$–2 µm has been converted to 0.42–0.51 µm radiation in a LiIO$_3$ crystal upon mixing with ruby laser radiation ($\eta = 18\%$) [4.319]. The interaction was of the ooe type and the phase-matching angle changed from 27° to 37°. The output power amounted to 10 kW.

Crystals of KTP are mainly used for up-conversion of Nd:YAG laser radiation (Table 4.27). Thus, by use of diode lasers as pump sources up to 50–70% conversion efficiencies were achieved [4.331, 333]. With aid of a BBO crystal, IR radiation with $\lambda = 2.45$ and 2.69 µm was converted into the visible with $\eta = 2.3\%$; dye laser with $\lambda = 571$ nm served as a pump source [4.336].

The signal radiation of a Nd:YAG laser ($\lambda = 1.064$ µm, P = 2.1 W) has been mixed with the pump radiation of a Kr laser ($\lambda = 676.4$ nm, P = 26.2 mW) in KNbO$_3$ [4.337]. The radiation propagated along the a-axis, the crystal temperature T was –4 °C, and the crystal length was 8.97 mm. The radiation power at a sum frequency with $\lambda = 413.6$ nm, was 0.1 mW.

4.4.3 Up-conversion of CO$_2$ Laser Radiation to the Near IR and Visible Regions

Nonlinear conversion of CO$_2$ laser radiation (10.6 µm) to the near IR and visible regions is performed in crystals of proustite (Ag$_3$AsS$_3$), Ag$_3$SbS$_3$, silver thiogallate (AgGaS$_2$), mercury thiogallate (HgGa$_2$S$_4$), ZnGeP$_2$, GaSe, CdSe, and HgS. Table 4.28 lists the data on the pump sources (pump wavelengths, operation regimes), phase-matching angles, type of interaction, pump intensities, crystal lengths, and quantum conversion efficiencies. In some studies up-conversion has been realized by means of difference frequency generation (DFG) [4.246, 339, 343], and by means of both SFG and DFG with interaction types ooe, eoe and oee [4.360]. However, in all other cases listed in Table 4.28 the SFG process has been used.

Proustite is widely used for up-conversion. In some research ruby laser radiation was used [4.338, 341, 345, 364, 365]. Maximum quantum efficiency for conversion to the visible spectrum was 10.7%, which corresponded to a power-conversion efficiency of 175% [4.357]. In other works Nd:YAG [4.340, 342, 344, 346, 347, 366–368], Kr [4.343, 369], and He–Ne (0.6328 µm) [4.348, 350, 353, 370] lasers were also used as sources of pump radiation. In one work [4.348] the proustite crystal was cooled to 77 K, which decreased the absorption at 633 and 593 nm to 0.8–0.9 cm^{-1}.

Table 4.28. Up-conversion of CO_2 laser radiation

Crystal	Pump source	λ_{pump} [μm]	Type of inter-action	θ_{pm} [deg]	I_0 [W cm^{-2}]	L [mm]	η [%]	Refs.
Ag_3AsS_3	Ruby laser, 300 μs, 20 mJ	0.694	eoe	27.7	–	4.5	1.4×10^{-4}	4.338
	cw He–Ne laser, 0.25 mW	0.633	ooe	25.4–27.8	–	10	2×10^{-5a}	4.339
	ns Nd:YAG laser, 740 W	1.064	eoe	20	–	6	0.84	4.340
	Ruby laser, 1 ms	0.694	–	–	10^4	10	0.14	4.341
	ns Nd:YAG laser	1.064	eoe	20	400	6	0.5	4.342
	cw Kr laser, 60 mW	0.647	eoe	–	–	9	2×10^{-4a}	4.343
	Nd:YAG laser	1.064	eoe	20	–	14	1.5	4.344
	Ruby laser, 25 ps	0.694	ooe	25.2	10^8	5	10.7	4.345
	ns Nd:YAG laser	1.064	eoe	20	–	–	30^b	4.346
	ns Nd:YAG laser	1.064	eoe	20	$(0.5–1.2) \times 10^6$	–	8^c	4.347
	cw He–Ne laser	0.633	eoe	29	–	–	4×10^{-4}	4.348
Ag_3SbS_3	cw Nd:YAG laser	1.064	ooe	27.3	86	3.5	4×10^{-6}	4.349
$AgGaS_2$	Nd:YAG laser	1.064	oee	40	6×10^5	3	40^b	4.350
	Dye laser, 3 ns	0.598	ooe	90	–	5	40	4.351
	Ruby laser, 30 ns	0.694	eoe	55	–	3.3	9	4.352
	Ruby laser, free-running generation	0.694	eoe	55	–	3.3	0.01	4.352
$AgGaS_2$	ns Nd:YAG laser	1.064	oee	40	–	–	30	4.353
	ns Nd:YAG laser	1.064	oee	40	$(0.5–1.2) \times 10^6$	–	14^c	4.347
	Dye laser	0.598	ooe	90	9×10^4	–	0.1	4.354
	Dye laser	0.597	ooe	90	–	8	–	4.355
$HgGa_2S_4$	ns Nd:YAG laser	1.064	ooe	41.6	$(0.5–1.2) \times 10^6$	3.6	$60(20)^c$	4.347,356
$ZnGeP_2$	Nd:YAG laser	1.064	oeo	82–89	–	10	1.4	4.357
	ns Nd:YAG laser	1.064	oeo	82.9	$(0.5–1.2) \times 10^6$	–	6^c	4.347
	CO laser, 4.7 W	5	eeo	56	–	–	0.01	4.358
	Nd:YAG laser, 30 ns	1.064	oeo	82.5	3×10^6	3	5	4.359

Table 4.28 (*Contd.*)

Crystal	Pump source	λ_{pump} [μm]	Type of inter-action	θ_{pm}[deg]	I_0[W cm^{-2}]	L [mm]	η[%]	Refs.
GaSe	CO laser	5–6	ooe, eoe, oee	10–18	–	1.5	2.5×10^{-2}	4.360
	Nd:YAG laser	1.064	ooe, eoe, oee	13–15	–	2.5	1.6×10^{-3}	4.360
	Nd:YAG laser	1.064	ooe	13.6	10^4	3.5	2×10^{-2}	4.361
CdSe	Nd:YAG laser	1.833	oeo	77	2.4×10^7	10	35[b]	4.362
	HF laser, 250 ns	2.72	oeo	70.5	6×10^6	30	40	4.363
HgS	cw He-Ne laser, 1 mW	0.633	eeo	25.3	–	8	4×10^{-8a}	4.246

[a]Difference frequency generation.
[b]Power-conversion efficiency.
[c]Power-conversion efficiency for two cascades:

$10.6 + 1.064 \rightarrow 0.967$ μm
$0.967 + 1.064 \rightarrow 0.507$ μm

The highest quantum efficiency of converting CO_2 laser radiation to the visible region (40%) was attained by *Jantz* and *Koidl* [4.351], who used a nanosecond dye laser as a pump source ($\lambda = 598$ nm, $\theta_{ooe} = 90°$, $AgGaS_2$ crystal length 5 mm). *Voronin* et al. [4.353] realized the scheme of two-cascade IR -to-visible up-conversion by pumping the radiation of an Nd:YAG laser operating in the Q-switched regime: $10.6 + 1.064 \Rightarrow 0.976$ µm, $0.967 + 1.064 \Rightarrow 0.507$ µm. At the first conversion step silver thiogallate was used. Quantum conversion efficiency was 30%. A further conversion of radiation with $\lambda = 0.967$ µm was performed with the same pump in a $LiIO_3$ crystal with $\theta_{ooe} = 21.6°$. To obtain the required polarization a crystalline quartz plate was placed between two crystals, which rotated the pump wave polarization,vector by $90°$ and the 0.967 µm wave polarization vector by $180°$. In the second cascade the quantum conversion efficiency η was 80%. A complete quantum efficiency of 10.6 µm $\Rightarrow$ 0.507 µm conversion was 24%. Efficient up-conversion of 10.6 µm radiation into the near IR in $HgGa_2S_4$ crystal with a Q-switched Nd:YAG laser a as pump source was realized with power conversion efficiency up to 60% [4.357].

Table 4.28 shows that proustite and silver thiogallate are the most promising nonlinear materials for up-conversion of CO_2 laser radiation to the visible range: quantum conversion efficiencies were 10–40% in the pulse regime and 4×10^{-6} in the cw regime.

Up-conversion of CO laser radiation to the visible region has also been reported. Researchers obtained a quantum conversion efficiency of 2.8×10^{-8} in a proustite crystal 0.8 cm long using He-Ne laser radiation (0.633 µm) as a pump source, θ_{ooe} being $33 - 36°$ [4.370]. In other work [4.361], a GaSe crystal was used for this purpose with Nd:YAG laser radiation (1.064 µm) as a pump source, $\theta_{ooe} = 18.8°$.

4.5 Difference-Frequency Generation

Difference-frequency generation or "down-conversion" is generally used for obtaining radiation in the middle and far IR regions as well as in the millimeter range of wavelengths. In some cases DFG is used for tuning high-power laser radiation in the visible region.

4.5.1 DFG in the Visible Region

Lyutskanov et al. [4.276] have reported the effective conversion of high-power XeCl laser radiation ($\lambda = 308.0, 308.2, 308.5$ nm; $\tau_p = 12$ ns) to the region with $\lambda = 434$ nm using Nd:YAG laser radiation ($\lambda = 1.064$ µm, $\tau_p = 0.7$ ns) as a pump. A KDP crystal 43 mm in length was used; the phase-matching angle was $\theta = 53°$ and interaction was of the ooe type. High-efficiency conversion of rhodamine 6G laser radiation at $\lambda = 555\text{--}580$ nm to $\lambda = 490\text{--}510$ nm was obtained with a DKDP crystal 5 cm long [4.371] (Table 4.29).

4.5.2 DFG in the Mid IR Region

To obtain IR radiation in the 1–6 µm region, $LiIO_3$ [4.281, 373–385] and $LiNbO_3$ [4.328, 386–391] crystals are mainly used. In this spectral region the mentioned crystals have high transparency, relatively high optical breakdown thresholds, and high nonlinear coefficients. Recently for a 1–3 µm region BBO crystals have also found application [4.392–395]. Crystals of proustite [4.396, 397], silver thiogallate [4.398–412], and GaSe [4.413, 410, 414–416] as well as $AgGaSe_2$ [4.231, 413, 417, 418], $CdGeAs_2$ [4.240, 419], CdSe [4.401, 420], and Te [4.421] are most often used in the wavelength range from 4 to 23 µm.

Table 4.30 lists the data on the crystals used for DFG in the mid IR range, corresponding sources of interacting radiation, and some output parameters of the generated IR radiation. Generation of cw IR radiation in the 2.3–4.6 µm region has been demonstrated [4.378]. With the aid of noncollinear DFG in $LiIO_3$ crystal the radiation from a rhodamine 6G laser ($\lambda = 570\text{--}630$ nm) and from an argon ion laser ($\lambda = 514$ or 488 nm) was mixed. A $LiIO_3$ crystal 10 mm in length was placed inside the dye laser cavity; the phase-matching angle was varied from 40° to 50°, and the angle α between the interacting beams amounted to $4 - 5°$. The conversion efficiency was $\sim 10^{-6}$; IR radiation power reached 0.5 µW in a single-mode regime and 4 µW in a multimode, the bandwidth being 5 cm^{-1}.

Generation of cw radiation in the 11.4–16.8 µm range with ~ 4µW power has been attained in a $CdGeAs_2$ crystal with the use of DFG between CO_2 and CO laser radiations [4.240]. For the type II interaction the phase-matching angle was 46–47°. An effective generation of nanosecond IR radiation in the regions 2–4 µm, 4.4–5.7 µm, and 5–11 µm has been reported [4.384, 389, 399] in $LiIO_3$, $LiNbO_3$, and $AgGaS_2$ crystals, respectively. In all three cases radiation from a dye laser and a Nd:YAG laser radiation was mixed. In a $LiNbO_3$ crystal

Table 4.29. Difference frequency generation in the visible region

Crystal	λ_{DF}[nm]	Sources of interacting radiations	Conversion efficiency [%]	Refs.
KDP	434	Nd:YAG laser ($\tau = 0.7$ ns) + XeCl laser (308 nm, 12 ns)	25	4.276
DKDP	490–510	Dye laser + 4ω of Nd:YAG laser (266 nm)	87	4.371
ADP	680–1100	Nitrogen laser (337 nm) + dye laser $\theta_{ooe} = 49\text{--}53°$	–	4.372
ADP	490–510	Dye laser + 4ω of ND:YAG laser (266 nm)	80	4.371

Table 4.30. Generation of IR radiation by DFG

Crystal	$\lambda[\mu m]$	Sources of interacting radiations, crystal parameters	Conversion efficiency, energy, power, τ_p	Refs.
$LiIO_3$	4.1–5.2	Dye laser + ruby laser, ICDFG, $L = 12$ mm	100 W (peak)	4.373
	1.25–1.60;	Dye laser + Q-switched Nd:YAG laser	0.5–70 W (peak), $\Delta\nu = 0.1$ cm^{-1}, 60 ns	4.374
	3.40–5.65	(1.064 and 0.532 μm, ICDFG, $\theta_{ooe} = 21$–28.5°)		
	2.8–3.4	Dye laser + Q-switched ND:YAG laser, $L = 10$ mm	80 mW (peak)	4.375
	1.1–5.6	Dye laser + Nd:YAG laser (1.064 and 0.532 μm), $\theta_c = 23°$	–	4.280
	2.3–4.6	Dye laser + argon laser (514 and 488 nm)	0.5–4 μW, cw	4.378
	4.3–5.3	Dye laser + 2ω of Nd:YAG laser, $\theta_{ooe} = 24.3°$	–	4.379
	0.7–2.2	Dye laser + nitrogen laser, $\theta_{ooe} = 51$–31°	3 ns	4.380
	3.8–6.0	Dye laser + copper vapor laser (511nm), $\theta_c = 21$–24°	10–100 μW, 20 ns	4.381
	3.5–5.4	Dye laser + 2ω of Nd:YAG laser, $\theta_{ooe} = 20°$	0.8 mJ, 10 ns	4.382
	1.2–1.6	Two dye lasers, $\theta_{ooe} = 29°$	1.5–5 ps	4.383
	4.4–5.7	Dye laser + Nd:YAG laser, $\theta_{ooe} = 20$–22°	550 kW, 8 ns	4.384
	~5	Two dye lasers, $\theta_{ooe} = 20°$, $L = 3$ mm	10%, 10 nJ, 400 fs	4.385
$LiNbO_3$	3–4	Dye laser + ruby laser	1%, 6 kW	4.386
	2.2–4.2	Dye laser + argon laser	1 μW, cw	4.328
	2–4.5	Dye laser(1.2 ps) + argon laser (100 ps), $\theta = 90°$, $T = 200$–400°C	25 μW(average), 1.2 ps, $f = 138$ MHz	4.387, 388
	2–4	Dye laser + Nd:YAG laser, $\theta_{ooe} = 46$–57°	60%, 1.6 MW	4.389
	2.04	Two dye lasers, $\theta_{ooe} = 90°$	50%, $\Delta\lambda = 0.03$ nm	4.390
$LiNbO_3$	1.7–4.0	CPM dye laser + subpicosecond continuum, $\theta_c = 55°$, $L = 1$mm	10 kW (peak), 0.2 ps, $\Delta\nu = 100$ cm^{-1}	4.391
BBO	2.5	Dye laser (620 nm) + picosecond continuum (825 nm), $\theta_{ooe} = 20.3°$, $L = 5$ mm	5%, 4 μJ, 0.5 ps	4.392
	0.9–1.5	Dye laser + Nd:YAG laser, $\theta_{ooe} = 20.5$–24.5°, $L = 10$ mm	23%, 4.5 mJ, 8 ns	4.393
	2.04–3.42	Two dye lasers, NCDFG, $\theta_{ooe} = 12$–17°, $L = 6$ mm	300–400 W (peak)	4.394
	1.23–1.76	Dye laser + Ti:sapphire laser	10 μW(average), 150 fs, $f = 80$ MHz	4.395
KTP	1.4–1.6	Dye laser + Nd:YAG laser, $\theta_{eoe} = 76$–78°, $\psi = 0°$	8.4 kW, $f = 76$ MHz, 94 fs	4.421
Ag_3AsS_3	11–23	Two dye lasers	3 W (peak), 30 ns	4.396
	3.7–10.2	OPO (1.60–1.67 μm) + 2ω of phosphate glass laser (527 nm)	25–50 μJ, 10 ps	4.397
$AgGaS_2$	5.5–18.3	Two dye lasers, $\theta = 90°$	4 W, 4 ns	4.398
	5–11	Dye laser + Nd:YAG laser, $\theta_{eoe} = 38$–52°	180 kW, 12 ns	4.399
	3.9–9.4	Dye laser + Nd:YAG laser	1%, 8 ps	4.400

Table 4.30 (*Contd.*)

Crystal	λ[μm]	Sources of interacting radiations, crystal parameters	Conversion efficiency, energy, power, τ_p	Refs.
AgGaS$_2$	4–11	OPO(2–4 μm) + radiation at λ =1.4–2.13 μm	1 kW, 8 ns	4.401
	8.7–11.6	Two dye lasers, $\theta_{ooe} = 65$–$85°$	0.1 mW, 500 ns	4.402
	4.6–12	Two dye lasers, $\theta_{ooe} = 45$–$83°$	300 mW, 10 ns	4.403
	7–9	Dye laser + Ti:sapphire laser, $\theta_{ooe} = 90°$	1μW, cw, $\Delta\nu = 0.5$ MHz	4.404
	4.76–6.45	Dye laser + Ti:sapphire laser, $\theta_{ooe} = 90°$, $L = 45$ mm	20 μW, cw, $\Delta\nu = 1$ MHz	4.405
	~ 4.26	GaAlAs laser (858 nm) + Ti: sapphire laser (715 nm, $\theta_{ooe} = 90°$)	47 μW(cw), 89 μW (50 μs)	4.406
	4.73; 5.12	Diode laser + Ti:sapphire laser, $\theta_{ooe} = 90°$	1μW, cw	4.407
	5.2–6.4	Nd:YAG laser + near IR(DFG in LiIO$_3$)	35%, 23 ps	4.408
	3.4–7.0	Dye laser + Nd:YAG laser, $\theta_{c} = 53.2°$	17 μW(average), 2.16 ps, $f = 76$ MHz	4.409
	4–10	Dye laser (1.1–1.4 μm) + Nd:glass laser (1.053 μm)	2%,10 nJ – 1 μJ, 1 ps	4.410
	4.5–11.5	Dye laser (870–1000 nm) + Ti:sapphire laser (815 nm), $\theta_{c} = 45°$, $L = 1$ mm	10 nJ, 400 fs, $f = 1$ kHz	4.411
	9	Ti:sapphire laser with dual wavelength output (50–70 fs), $\theta_{c} = 44°, L = 1$ mm	0.03 pJ, f= 85 MHz	4.412
AgGaSe$_2$	7–15	OPO (1.5–1.7 μm) + Nd:YAG laser (1.32 μm), $\theta_{ooe} = 90$–$57°$	1.2%	4.231
	12.2–13	CO laser(5.67–5.85 μm) + CO$_2$ laser, $\theta = 61°$	0.2 μW, cw	4.417
	8–18	Idler and signal waves of OPO	0.1 mJ, 3–6 ns	4.418
CdGeAs$_2$	11.4–16.8	CO laser + CO$_2$ laser	4 μW, cw	4.240
GaSe	9.5–18	Dye laser + ruby laser	300 W, 20 ns	4.414
	4–12	Idler and signal waves of OPO	60 W	4.415
	7–16	Nd:YAG laser + laser on F$_2^-$ colour centers, $\theta_{ooe} =13$–$15°$, $\theta_{eoe} =12$–$16°$	0.1–1 kW, 10 ns	4.416
CdSe	6–18	Dye laser(1.1–1.4 μm) + Nd:glass laser (1.053 μm)	10 nJ–1 μJ, 1 ps	4.410
	16	OPO singal wave(1.995 μm) + OPO idler wave(2.28 μm), $\theta = 62.22°$	0.5 kW, 20 Hz, 10 ns	4.420
	9–22	OPO (2–4 μm) + radiation at λ = 1.4–2.13 μm	10–100 W, 8 ns	4.401
Te	10.9–11.1	CO$_2$ laser(10.2 μm) + cw spin-flip laser(5.3 μm), $\theta_{eeo} = 14°$	10 μW	4.421

25 mm long, maximum IR power at $\lambda = 2 - 4$ μm amounted to 1.6 MW and the average power to 130 mW. The pulse duration at a difference frequency was 8 ns; the phase-matching angle varied from 46° to 57°. For the 4.4–5.7 μm region covered by DFG in the 19 mm long $LiIO_3$ crystal, the peak IR power was 550 kW at $\tau = 8$ ns; average power amounted to 45 mW at $\lambda = 4.9$ μm. The phase-matching angle θ_{ooe} was 20–22°. The IR radiation bandwidth at $\lambda = 4.9$ μm was evaluated as 0.1 cm^{-1}. For the region 5–11 μm covered by DFG in a 10 mm $AgGaS_2$ crystal, the peak power was 180 kW at $\tau = 8$ ns with an average power of ~ 14 mW. The phase-matching angle θ_{ooe} varied from 38° to 52° with a simultaneous tuning of the dye laser radiation wavelength from 1.35 to 1.17 μm. The IR radiation bandwidth was evaluated as 0.1–0.2 cm^{-1}.

In a proustite crystal DFG between radiation from two dye lasers covered the wavelength range from 11 to 23 μm [4.396]. The peak IR power at $\lambda = 16$–20 μm amounted to 1–3 W (pulse duration: 3 ns). An Ag_3AsS_3 crystal 4 mm long cut at the angle $\theta_c = 20°$ to the optic axis was used. The above mentioned spectral range was covered when the crystal was rotated by 7°. The transmittance of proustite at λ up to 24 μm has also been measured [4.396]: $\alpha \sim 70$ cm^{-1} at $\lambda = 24$ μm.

The down-conversion process in $LiNbO_3$, $AgGaS_2$, and CdSe crystals has been used to cover the 1.4–22 μm spectral range [4.401]. In the $LiNbO_3$ crystal the radiation from a Nd:YAG laser and a dye laser (610–710 nm) was mixed; as a result of DFG, tuning in the 1.4–2.13 μm region (IR 1) was possible. In another $LiNbO_3$ crystal, OPO in the 2.13–4 μm spectral region (IR 2) was realized. In $AgGaS_2$ and CdSe crystals DFG was realized upon mixing IR 1 and IR 2, which makes it possible to cover the 4–11 μm and 9–22 μm spectral regions, respectively. In the region 1.4–4 μm the output power was several MW, in the region 4–10 μm several kW, and in the region 9–22 μm, 10–100 W for $\tau = 8$ ns and $f = 10$Hz.

Some papers have demonstrated generation of picosecond IR radiation at $\lambda = 1$–9.4 μm by means of DFG. The mixing of radiation from two mode-locked dye lasers in a $LiIO_3$ crystal 1 mm long produced radiation at $\lambda = 1.2$–1.6 μm, with a peak power up to 10 W and pulse duration 1.5–5 ps [4.383]. The pulse length was measured by the correlation method using SFG between IR and dye laser radiation in a $LiIO_3$ crystal 1 mm long ($\theta_{ooe} = 37°$). Difference frequency generation in a 5 mm long $LiNbO_3$ crystal covered the IR spectral range 2–4.5 μm [4.388]. The radiation of an acousto-optically mode-locked argon laser ($\lambda = 514.4$ nm, $\tau = 10$ ps) was mixed with rhodamine 6G laser radiation ($\tau = 1.2$ ps) pumped by the same argon laser. The spectral bandwidth of IR pulses at $\lambda = 2.43$ μm was 2.6 nm; other characteristics are given in Table 4.30.

Elsaesser et al. [4.400] realized DFG of IR pulses in the 3.9–9.4 μm region with $\tau_p = 8$ ps when Nd:YAG laser radiation ($\lambda = 1.064$ μm, $\tau = 21$ ps) was mixed with dye laser radiation ($\lambda = 1.2$–1.46 μm) in a 15 mm $AgGaS_2$ crystal. The phase-matching angle varied from 36° to 48°. The IR radiation bandwidth was 6.5 cm^{-1} over the whole tuning range. The quantum efficiency of down-conversion to IR radiation was several percent with respect to Nd:YAG laser pulse energy. The generation of IR radiation at $\lambda = 1.4$–$1.6\,\mu$m, $f = 3.8$ MHz, and $\tau_p = 94$ fs has been reported [4.422]. Radiation of an acousto-optically mode-locked cw Nd:YAG laser ($\tau = 100$ ps) was mixed in a KTP crystal ($\theta_c = 76°$, XZ plane, $L = 3.4$ mm) with radiation of a cavity-dumped dye laser synchronously pumped by the second harmonic of Nd:YAG laser radiation. An average power of IR radiation was 3 mW. If a BBO crystal was used instead of KTP, an average power at $\lambda = 1.42\,\mu$m was 50 μW.

Difference-frequency generation in AgGaS$_2$ and GaSe 1 cm in length by mixing the output of a mode-locked Nd:glass laser (1.053 μm, 2 ps) with the travelling wave dye laser radiation (1.1–1.4 μm) allows to obtain ultrashort (1 ps) IR pulses in the range of 4–18μm [4.410]. The limiting wavelength corresponds to the absorption edge of the respective crystal: 10 μm AgGaS$_2$ and 18 μm for GaSe. IR pulses as short as 400 fs in the range of 4.5–11.5 μm were generated by mixing the Ti:sapphire laser and travelling wave dye laser outputs in AgGaS$_2$ crystal [4.411]. The duration of the IR pulses was measured by means of the pump-probe technique in silicon plate. The radiation-induced generation of hot carriers in Si by $\lambda_{ex} = 815$ nm results in the increase of IR absorption, which was monitored at $\lambda_{probe} = 8.0\,\mu$m.

The two-cascade method of shortening the CO$_2$ laser pulse duration has been proposed and realized [4.423]. The CO$_2$ laser radiation ($\lambda = 10.6\,\mu$m, $\tau = 150$ ns was mixed with Nd:YAG laser radiation ($\lambda = 1.064\,\mu$m, $\tau = 20$ ns), in the first proustite crystal. The difference-frequency radiation at $\lambda = 1.2\,\mu$m, was mixed once more with 1.064 μm, radiation in the second proustite crystal and became down-converted to 10.6 μm radiation ($\tau = 20$ ns). The power-conversion efficiency from $\lambda = 1.064\,\mu$m radiation was 0.05%, which made it possible to obtain IR radiation intensitites of about 10 kW cm^{-2}. In both cascades of nonlinear conversion 1 cm proustite crystals were used with $\theta_{eeo} = 20°$.

4.5.3 DFG in the Far IR Region

Difference-frequency generation between the radiations of two lasers generating at close frequencies is one of the methods of producing far IR radiation ($\lambda = 50\,\mu$m–20 mm). For instance, the mixing of frequencies of two temperature-tunable ruby lasers in LiNbO$_3$ and quartz gave rise to far IR radiation with the frequency 1.2–8.1 cm^{-1} [4.424]. One laser with a wide spectrum of radiation can also be used as a pump source. Then frequency components inside the generation spectrum interact and, as a result, the bandwidth-determined difference frequency is generated. This method was used for generating IR radiation at a fixed frequency of 100 cm^{-1} in LiNbO$_3$ pumped by neodymium silicate glass laser radiation [4.425].

Table 4.31. Difference frequency generation in the far IR region

Pump sources	Crystal	$v[\text{cm}^{-1}]$	$\lambda[\mu\text{m}]$	Power, energy	Refs.
Nd:glass laser (1.06 µm)	LiNbO$_3$	100	100	–	4.425
Ruby laser(0.694 µm)	LiNbO$_3$	29	330	–	4.426
Two ruby lasers(0.694 µm), 1 MW, 30 ns	LiNbO$_3$ quartz	1.2–8.0	1250–8330	20 mW	4.424
Nd:glass laser (1.06 µm), 50 mJ, 10 ps	ZnTe, LiNbO$_3$	8-30	330–1250	20 mW/cm^{-1}	4.427
Nd:glass laser (1.06 µm)10 ps	LiIO$_3$	–	–	–	4.428
Dye laser (0.73–0.93 µm), 11–15 ns, 4–13 MW	ZnTe, ZnSe, LiNbO$_3$	5–30	330–2000	1W (ZnTe)	4.429
Nd:glass laser (1.064 µm), 10 ps	LiNbO$_3$	0.4–2.5	4000–25000	60 W	4.430
Two ruby lasers(0.694 µm) 20 ns	LiNbO$_3$	1–3.3	3000–10000	0.5 W	4.431
Ruby laser(0.694 µm)	LiNbO$_3$	1.67–3.3	3000–6000	–	4.432
Two dye lasers: $\tau_1 = 1$–2 ps, $\lambda_1 = 589$ nm, $E_1 = 0.2$ mJ; $\tau_2 = 20$ ns, $\lambda_2 = 590$–596 nm, $E_2 = 20$ mJ	LiNbO$_3$	20–200	50–500	3 nJ	4.433
CO$_2$ laser at two frequencies	GaAs	2–100	100–5000	–	4.434
Two CO$_2$ lasers	ZnGeP$_2$	70–110	90–140	1.7 µW	4.435

LiNbO$_3$ is mainly used as the nonlinear material for the IR region, since it is fairly transparent in this region. Some isotropic crystals (GaAs, ZnTe, and ZnSe) possessing high nonlinearities are also used (Table 4.31). Down-conversion to $v = 20$–200 cm^{-1} with quantum efficiency 0.1–0.3% was attained [4.433]. Two dye lasers were used with nanosecond and picosecond pulse durations. The amplifiers of two lasers were pumped with the second harmonic of Nd:YAG laser radiation, which ensured synchronization between the interacting pulses. The two interacting beams were focused into a 4 mm LiNbO$_3$ crystal at a small angle α. Tuning of the far IR radiation frequency was attained by simultaneously varying the angle α from 5 to 50 mrad and changing the frequency of the nanosecond dye laser. The generated energy was 3 nJ at a pulse duration of 10 ps.

4.6 Optical Parametric Oscillation

4.6.1 OPO in the UV, Visible, and Near IR Spectral Regions

Optical parametric oscillation (OPO) in nonlinear crystals makes it possible to obtain radiation with a tunable frequency. The methods of angular and temperature phase-matching tuning are used for a smooth change of the wavelength in parametric light oscillators. Along with the advantages, both methods have certain drawbacks. Angular tuning is rather simple and more rapid than temperature tuning. Temperature tuning is generally used in the case of 90° phase matching, i.e., when the birefringence angle is zero. This method is mainly used in crystals with a strong temperature dependence of phase matching: ADP ($\lambda_{pump} = 266$ nm), LiNbO$_3$ ($\lambda_{pump} = 530$ nm), LBO ($\lambda_{pump} = 266$, 355 and 530 nm), Ba$_2$NaNb$_5$O$_{15}$ ($\lambda_{pump} = 530$ nm), KNbO$_3$ ($\lambda_{pump} = 532$ nm), and DKDP ($\lambda_{pump} = 266$ nm). At present, optical parametric oscillation makes it possible to obtain continuously tunable radiation from the UV (300 nm) to middle IR range (18 μm). Minimum pulse durations in the near IR region are as short as 57–65 fs (in visible, less than 100 fs), and the OPO radiation bandwidths are down to 0.02 cm^{-1}. Maximum efficiencies of OPO operation up to 50%, corresponding to 70–80% *pump depletions* (see below), were observed in femtosecond, picosecond, nanosecond, and continuous wave regimes by use of KTP, LBO, BBO, and LiNbO$_3$:MgO crystals, respectively.

Since the excitation of parametric oscillation requires high intensities of radiation (10^7–10^{10} W cm^{-2}), nanosecond and picosecond pump sources are usually used for OPO. All OPO schemes can be reduced to two schemes: the travelling-wave OPO (without a cavity) and the resonant OPO.

The travelling-wave OPO scheme (TWOPO) usually consists of one or two nonlinear crystals. TWOPO is simple and can be realized within the whole transparency range of the crystal; however, it has certain disadvantages. For instance, to attain high conversion efficiencies, high pump intensities are required (up to 30 GW cm^{-2}) close to the damage threshold of the crystal. Maximum conversion efficiencies in TWOPO schemes, were attained with crystals of KDP (67–74%) and ADP (60%) at total OPO pulse energies up to 2.3 J.

Singly-resonant OPO, or SROPO, uses resonant feedback at only the signal or idler frequency. Doubly-resonant OPO, or DROPO, uses resonant feedback of both signal and idler frequencies. Exotic triply-resonant OPO, with resonant feedback also at pump frequency, and intracavity OPO, with the crystal placed inside the laser cavity, e.g., CPM dye laser, are used very seldomly. Quadruply-resonant OPO, with SHG inside the OPO cavity and resonant feedback also at the second harmonic, can be mentioned as well.

Picosecond and femtosecond OPO with synchronous pumping is the most promising type of resonant OPO. A nonlinear crystal is placed in the cavity (or in two cavities), which ensures a positive feedback at one or two frequencies, and is pumped by a train of ultrashort pulses. The time period between pulses is equal to the double passing time of the cavity (axial period). The cavity generally consists of two broadband mirrors with reflection $R_1 = 99\%$ and $R_2 = 4$–80% at the OPO wavelengths. Synchronously pumped OPO is advantageous in that the generation threshold here is low ($I < 100$ MW cm^{-2}) and space and time pulse coherences are close to limiting. That is, in the synchronously pumped OPO scheme the shortest femtosecond pulses (60 fs) are attained. The drawback of this scheme is the necessity for special dielectric mirrors and its complexity as compared with the traveling-wave OPO scheme.

Injection seeding from an external source of radiation, mainly from other OPO, or from of the narrow-bandwidth laser radiation source, e.g., a single-frequency dye laser, significantly enhances the reproducibility and efficiency of parametric generators. Operating in this way, optical parametric amplifiers (OPA) ensure narrow-band output without using wavelength-selective elements. In the case of the seed at a fixed frequency, the tunability of the OPO-OPA system is achieved by changing the pump wavelength (dye-laser or Ti:-sapphire laser radiation).

Mode-locked or Q-switched Nd:YAG $(\lambda = 1.064\,\mu\text{m})$, Nd^{3+} phosphate glass $(\lambda = 1.054\,\mu\text{m})$, and Nd:YLF $(\lambda = 1.047\,\mu\text{m})$ lasers, as well as their second, third, and fourth harmonics, are generally used as an OPO pump source. A Nd:YAG laser operates with high reliability in the mode-locked regime at a high repetition rate. Pulse durations of passively mode-locked Nd:YAG lasers are about 25–45 ps. Currently, Nd:YAG laser systems can deliver 1 GW powers in a single picosecond pulse at a pulse repetition rate of more than 10 Hz. Nd^{3+}: phosphate glass lasers can deliver shorter pulses (1–2 ps); however, their operation is much less stable, and pulse repetition is low because of the low heat conductivity of the active elements.

As a pump source for OPO, the XeCl lasers $(\lambda = 308\,\text{nm})$ are also often used. Recently, very promising Ti:sapphire lasers $(\lambda = 700 - 900\,\text{nm})$ have found wide application in OPO devices. Compact schemes of OPO are realized with the aid of diode-laser-pumped Nd:YAG lasers as pump sources. Crystals with high nonlinearity, i.e., LiNbO_3 and KTP, are used in these devices.

Different OPO schemes and their energetic, temporal, spectral, and spatial characteristics are considered in detail in [4.38, 436–438]. A large variety of useful information on the OPO and their applications can be found in two special issues of the Journal of the Optical Society of America, B (vol. 10, No 9 and 11, 1994) devoted to optical parametric oscillators. In this handbook we list only the main output OPO parameters realized in practice.

The inorganic crystals KDP, DKDP, ADP, CDA, LiIO_3, LiNbO_3, BBO, LBO, KTP, KTA, "banana", $\alpha - \text{HIO}_3$, and KNbO_3 and the organic crystals of urea, NPP, and DLAP have been used as nonlinear materials for OPO in the 0.3–5 μm spectral range. Table 4.32 lists pump wavelengths, phase-matching angles, pump thresholds (peak intenstity and/or average power), tuning ranges, OPO pulse durations, and conversion efficiencies for OPO experiments in the UV, visible, and near IR spectral ranges. The column headed "notes" gives data on the OPO type, pump intensities, crystal lengths, phase-matching temperatures, and output characteristics of OPO radiation (energy, power, bandwidth). Note, that for the KTP crystal in the XY plane $(\theta = 90°)$ eoe interaction occurs, and in the XZ plane $(\varphi = 0°)$, it is oeo interaction. For the LBO crystal in the XY plane $(\theta = 90°)$, XZ plane $(\varphi = 0°)$, and YZ plane $(\varphi = 90°)$, respectively, ooe, oee, and eoo interactions take place.

Picosecond optial parametric oscillators are most thoroughly described in [4.439, 444, 483], Travelling-wave OPO in KDP, LiIO_3, LiNbO_3, and $\alpha - \text{HIO}_3$, crystals has been realized [4.439]. High-efficiency (10–12%) con-

Table 4.32. OPO in the UV, visible, and IR regions

Crystal	Phase-matching angle, type of interaction	λ_{pump} [µm]	Pump threshold, I_{thr} [MW cm^{-2}]	λ_{OPO} [µm]	τ_p	Conversion efficiency [%]	Refs.	Notes
KDP	eoe	0.532	–	0.8–1.7	35 ps	6–8	4.439	TWOPO, $I_0 = 15\,\mathrm{GW\ cm^{-2}}$, $L_1 = 2.5$ cm, $L_2 = 4\,\mathrm{cm}$
	eoe	0.532	–	0.8–1.67	40 ps	25	4.440	TWOPO, $E = 1\,\mathrm{mJ}$, $L_1 = L_2 = 4cm$
	eoe	0.532	–	0.9–1.3	30 ps	51	4.441, 442	TWOPO, $\Delta\nu\Delta\tau = 0.7$, $L_1 = 4\,\mathrm{cm}$, $L_2 = 6\,\mathrm{cm}$, $I_0 = 15$–$20\ \mathrm{GW\ cm^{-2}}$
	eoe	0.527	–	0.82–1.3	0.3–0.5 ps	2	4.443, 444	Synchronously pumped OPO, $E = 20\,\mathrm{µJ}$
	eoe	0.532	1000–2000	–	–	67–74[a]	4.445	TWOPO, $L_1 = 4\,\mathrm{cm}$, $L_2 = 6\,\mathrm{cm}$, $E = 2\,\mathrm{J}$
	eoe	0.355	–	0.45–0.64 0.79–1.69	45 ps	15	4.446	TWOPO, $L_1 = L_2 = 4\,\mathrm{cm}$
	eoe	0.35	1000	0.45–0.6	0.5 ps	70[a]	4.447	TWOPO, $L_1 = 2\,\mathrm{cm}$, $L_2 = 6\,\mathrm{cm}$, $E = 0.35\,\mathrm{J}$, $I_0 = 6 - 8\,\mathrm{GW\ cm^{-2}}$
	eoe	0.35	–	–	0.5 ns	67[a]	4.448	TWOPO, $L = 5\,\mathrm{cm}$, injection-seeding, $I_0 = 0.3\,\mathrm{GW\ cm^{-2}}$
KDP + BBO	eoe (KDP) ooe (BBO)	0.5275	–	0.75–1.77	0.6 ns	13	4.449, 450	TWOPO, $L(\mathrm{KDP}) = 4\,\mathrm{cm}$, $L(\mathrm{BBO}) = 1\,\mathrm{cm}$, $I_0 = 60\,\mathrm{GW\ cm^{-2}}$
DKDP	$\theta_{ooe} = 90°$	0.266	–	0.47–0.61	–	–	4.451	TWOPO, $T = 40$–$100\,°\mathrm{C}$
	$\theta_{ooe} = 90°$	0.266	–	0.37–0.6	–	–	4.452	
ADP	–	0.527	1500	0.93–1.21	–	60[a]	4.453	TWOPO, $E = 2.3\,\mathrm{J}$, $I_0 = 10\,\mathrm{GW\ cm^{-2}}$
	$\theta_{ooe} = 51\text{–}45°$	0.352	–	0.44–1.75	5 ps	0.1–1.0	4.454	TWOPO, $L_1 = 2.5\,\mathrm{cm}$, $L_2 = 3\,\mathrm{cm}$
	ooe	0.266	–	0.42–0.73	2 ns	25	4.455	TWOPO, $T = 50$–$105\,°\mathrm{C}$
	ooe	0.266	–	–	–	30	4.456	$L = 6\,\mathrm{cm}$, $I_0 = 1\,\mathrm{GW\ cm^{-2}}$
	$\theta_{ooe} = 90°$	0.266	–	0.44–0.68	–	10	4.457	TWOPO, $T = 50$–$110\,°\mathrm{C}$, $L = 5\,\mathrm{cm}$
CDA	$\theta_{ooe} = 90°$	0.532	–	0.854–1.41	10 ps	30–60	4.458	$L = 3\,\mathrm{cm}$, $T = 50$–$70\,°\mathrm{C}$, $I_0 = 0.3\,\mathrm{GW\ cm^{-2}}$
	$\theta_{ooe} = 90°$	0.53	1000	0.8–1.3	10 ps	12.5	4.459	synchronously pumped OPO, $L = 4\,\mathrm{cm}$, $I_0 = 3\,\mathrm{GW\ cm^{-2}}$

LiIO$_3$	$\theta_{ooe} = 21°$	1.06	80	–	10 ns	2	4.460	SROPO, $I_0 = 250\,\text{MW cm}^{-2}$, vector phase matching
	–	1.06	50	1.4–2.7	0.01–1 ns	–	4.461	SROPO, $P = 30$–50 MW
	$\theta_{ooe} = 24°$	1.06	50	2.5–3.2	–	15	4.462	SROPO, $L = 6\,\text{cm}$, $E = 0.1\,\text{J}$
	$\theta_{ooe} = 23.1$–22.4°	0.694	5	1.15–1.9	20 ns	50[b]	4.319, 463	DROPO, $L = 0.85\,\text{cm}$, $P = 10\,\text{kW}$
	$\theta_{ooe} = 21.8$–19.3°	0.694	–	0.95–0.84, 2.5–4.0	–	–	4.464	
	$\theta_{ooe} = 25$–30°	0.53	10	0.68–2.4	15 ns	8	4.465	SROPO, $L = 1.6\,\text{cm}$
	$\theta_{ooe} = 29.5°$	0.53	10	0.61–2.7	0.01–1 ns	–	4.461	SROPO, $P = 12\,\text{MW}$
	$\theta_{ooe} = 22$–34°	0.53	–	1.4–3.8	6 ps	–	4.466	Synchronously pumped OPO
	$\theta_{ooe} = 26°$	0.532	–	0.63–3.4	–	–	4.467	SROPO, $P = 100\,\text{kW}$, $\Delta v = 0.1\,\text{cm}^{-1}$
	$\theta_{ooe} = 23$–30°	0.532	10	0.63–3.35	30 ns	20	4.468	SROPO
	ooe	0.532	–	0.61–4.25	6 ps	4	4.439, 469	TWOPO, $L_1 = 1\,\text{cm}$, $L_2 = 2.5\,\text{cm}$, $I_0 = 2\,\text{GW cm}^{-2}$
	$\theta_{ooe} = 25$–30°	0.53	3000	0.68–2.4	–	5	4.470	TWOPO, $L_1 = L_2 = 4\,\text{cm}$, $I_0 = 6$ GW cm^{-2}
	–	0.53	10	0.74–1.85	10 ns	–	4.471	$E = 0.5\,\text{J}$
	$\theta_{ooe} = 22.5°$	0.532	–	4.1	50 ps	0.4	4.472	Injection seeding, $L = 3\,\text{cm}$, $E = 3\,\mu\text{J}$
	$\theta_{ooe} = 53$–37°	0.347	–	0.41–2.1	–	–	4.473	
LiNbO$_3$	$\theta_{ooe} = 90°$	1.06	–	2.13	100 ns	8	4.474	DROPO, $L = 3\,\text{mm}$
	$\theta_{ooe} = 90°$	1.06	–	1.43–4.0	6 ps	3	4.475	TWOPO, $L = 2\,\text{cm}$, $I_0 = 8\,\text{GW cm}^{-2}$
	$\theta_{ooe} = 90°$	1.06	–	1.1–4.45	20 ns	15	4.476	SROPO, $I_0 = 10\,\text{MW cm}^{-2}$
	ooe	1.06	–	1.4–4.0	3.5 ns	10	4.477	TWOPO, $\Delta v = 6.5\,\text{cm}^{-1}$, $I_0 = 1\,\text{GW cm}^{-2}$
	ooe	1.06	–	–	0.5 ns	5–20	4.478	TWOPO
	–	1.064	–	1.55–3.5	20 ns	–	4.420	SROPO, $L = 5\,\text{cm}$
	45–51°	1.064	–	1.37–4.83	40 ps	17	4.479	TWOPO
	47°	1.054	100	1.35–2.11	0.5 ps	15	4.480	Synch. pumped OPO, $L = 18\,\text{mm}$, $I_0 = 0.14\,\text{GW cm}^{-2}$
	47°	1.064	–	2.5–4.0	10 ns	–	4.481	Injection seeding, $L = 5\,\text{cm}$, $E = 4\,\text{mJ}$, $\Delta v = 0.2\,\text{cm}^{-1}$

Table 4.32 (*Contd.*)

Crystal	Phase-matching angle, type of interaction	λ_{pump} [μm]	Pump threshold, I_{thr} [MW cm^{-2}]	λ_{OPO} [μm]	τ_p	Conversion efficiency [%]	Refs.	Notes
LiNbO$_3$	47°	1.064	–	1.50–1.58, 3.27–3.65	–	–	4.482	Injection seeding, $L = 5$ cm, $E = 20$ mJ,
	90°	0.53	–	0.75–0.64 1.8–3.1	–	–	4.215	$T = 180$–400 °C
	50–90°	0.53	–	0.59–3.7	5 ps	–	4.483	TWOPO, $L_1 = L_2 = 15$ mm
	84°	0.53	–	0.66–2.7	40 ps	17	4.479	TWOPO, $T = 46$–360 °C
	90°	0.532	–	0.68–0.76	20 ps	9	4.484	Synch. pumped OPO
	90°	0.532	–	0.93–1.3	20 ps	2–3	4.444	Synch. pumped OPO
	90°	0.532	–	0.63–3.6	30 ps	–	4.485	$L = 5$ cm, $T = 50$–450 °C
	90°	0.532	8	0.85–1.4	15 ps	17.5	4.486	SROPO, $P = 30$ kW, $f = 10$ kHz
	90°	0.532	< 30	0.65–3.0	10 ps	7.2	4.459	Synch. pumped OPO, $L = 25$ mm
	90°	0.473–0.659	–	0.55–3.65	130–700 ns	46(67^b)	4.487	SROPO, $T = 110 - 430$ °C, $P_{av} = 105$ mW
MgO : LiNbO$_3$	–	1.06	0.4 mW	1–1.14	cw	–	4.488	Quadruply resonant OPO
	$\theta_{ooe} = 90°$	0.532	35 mW	1.01–1.13	cw	40(60^b)	4.489	DROPO, $T = 107$–110 °C
	90°	0.532	12 mW	1.007–1.129	cw	34(78^b)	4.490	DROPO, $T = 107 - 111$ °C, $P = 8.15$ mW
	90°	0.532	13 mW	0.966–1.185	cw	38(73^b)	4.491	DROPO, $L = 15$ mm, $T = 113$–126 °C, $P = 100$ mW
	$\theta_{ooe} = 60$–84°	0.532	–	0.7–2.2	30 ps	5.4	4.492	TWOPO, $\Delta\lambda = 0.3$ nm(0.7 μm) and 1.4 nm(2μm)
	90°	0.532	20 mW	–	cw	–	4.493	DROPO, $L = 12.5$ mm, $T = 107$ °C,
BBO	ooe	0.62	–	0.75–2.8	200 fs	15	4.494	TWOPO, $L_1 = 5$ mm, $L_2 = 7$ mm, $E = 20$ μJ
	ooe, ooe	0.6	–	0.75–3.1	–	20–25	4.450	TWOPO-OPA, $L_1 = L_2 = 8$ mm, $I_0 = 70$ GW cm^{-2}
	ooe, eoe	0.6	–	0.75–3.1	180–250 fs	23	4.450	
	eoe, ooe	0.6	–	0.75–3.1	200–250 fs	20–25	4.450	

BBO								
$\theta_{\text{ooe}} = 21.7-21.9°$	0.532	278	0.94–1.22	12 ns	10	4.496		SROPO, $L = 9\,\text{mm}$, $E = 1\,\text{mJ}$
$\theta_{\text{ooe}} = 20.7-22.8°$	0.532	–	0.67–2.58	18 ps	13	4.497		TWOPO, $L_1 = L_2 = 9\,\text{mm}$, $I_0 = 2.5-3.8\,\text{GW cm}^{-2}$, $E = 0.1-0.5\,\text{mJ}$
ooe	0.527	–	0.7–1.8	65–260 fs	3	4.498–500		Synch. pumped SROPO, $L = 5.8\,\text{mm}$
ooe	0.527	–	1.04–1.07	70 fs	–	4.501		OPA with gain ratio 2×10^4
ooe	0.53	–	0.63–3.2	1.3 ps	25	4.450		TWOPO-OPA, $L_1 = L_2 = 8\,\text{mm}$
ooe	0.36	500	0.406–3.17	20 ps	30	4.502		Synch. pumped OPO, $L = 12\,\text{mm}$, $I_0 = 2\,\text{GWcm}^{-2}$, $E = 3\,\text{mJ}$, $\Delta\lambda = 0.24\,\text{nm}$
ooe	0.355	130	0.45–1.68	8 ns	9.4	4.503		SROPO, $L = 11.5\,\text{mm}$, $E = 15\,\text{mJ}$
$\theta_{\text{ooe}} = 26-33°$	0.355	–	0.43–2.0	–	–	4.504		$L = 7.6\,\text{mm}$
$\theta_{\text{ooe}} = 25-55°$	0.355	–	0.41–2.6	–	–	4.504		$L = 6.5\,\text{mm}$, SHG of OPO radiation to 205 nm in BBO
$\theta_{\text{ooe}} = 24-33°$	0.355	20	0.412–2.55	2.5 ns	24	4.505		SROPO, $L = 12\,\text{mm}$, $P_{\text{av}} = 140\,\text{mW}$
ooe	0.355	27	0.42–2.3	8 ns	32	4.506		SROPO, $L_1 = 11.5\,\text{mm}$, $L_2 = 9.5\,\text{mm}$ $\Delta\lambda = 0.03\,\text{mm}$
$\theta_{\text{ooe}} = 33.7-44.4°$	0.355	38	0.48–0.63; 0.81–1.36	8 ns	12	4.507		SROPO, $L_1 = 17\,\text{mm}$, $L_2 = 10\,\text{mm}$ $\Delta\lambda = 0.05-0.3$ nm
ooe	0.355	39	0.59–0.89	20–30 ps	2	4.508		Synch. pumped OPO, $L = 11.8\,\text{mm}$, $P = 15\,\text{kW}$
$\theta_{\text{ooe}} = 26-33°$	0.355	–	0.4–2.0	15 ps	30	4.509		OPO-OPA, $L_1 = 12\,\text{mm}$, $L_2 = 6\,\text{mm}$, $L_3 = 15\,\text{mm}$, $I_0 = 3\,\text{GW cm}^{-2}$, $\Delta\lambda = 0.3\,\text{nm}$
ooe	0.355	–	0.4–2.86	24 ps	6.5	4.510		TWOPO, $L_1 = L_2 = L_3 = 8\,\text{mm}$, $I_0 = 5\,\text{GW cm}^{-2}$, $\Delta\nu = 10\,\text{cm}^{-1}$
$\theta_{\text{ooe}} = 27-33°$	0.355	–	0.45–1.768	–	2	4.511		SROPO, $L = 10\,\text{mm}$, $E = 0.2\,\text{mJ}$
$\theta_{\text{ooe}} = 23-33°$	0.355	20–40	0.402–3.036	7 ns	40–61	4.512, 513		SROPO, $L = 15\,\text{mm}$, $E = 0.1-0.2\,\text{J}$
ooe	0.355	300	0.407–2.78	9 ps	–	4.514		DROPO, $L = 7\,\text{mm}$
ooe	0.355	–	0.43–2.1	15 ps	30	4.515		Injection seeding, $L = 15\,\text{mm}$
$\theta_{\text{ooe}} = 35.5-37°$	0.308	150	0.422–0.477	8 ns	10	4.516		SROPO, $L = 7\,\text{mm}$, $E = 0.26\,\text{mJ}$
ooe	0.308	18	0.354–2.37	–	64[b]	4.517		SROPO, $L = 20\,\text{mm}$, $E = 20\,\text{mJ}$
ooe	0.308	–	0.4–0.56	–	15	4.518		SROPO, $L = 20\,\text{mm}$, $\Delta\nu = 0.07\,\text{cm}^{-1}$ (with intracavity etalon)

Table 4.32 (*Contd.*)

Crystal	Phase-matching angle, type of interaction	λ_{pump} [μm]	Pump threshold, I_{thr} [MW cm^{-2}]	λ_{OPO} [μm]	τ_p	Conversion efficiency [%]	Refs.	Notes
BBO	$\theta_{ooe} = 36.5–47.5°$	0.266	23	0.33–1.37	9 ns	–	4.519	SROPO, $L = 20.5$ mm, $I_0 = 23$ MW cm^{-2}
	$\theta_{ooe} = 30–48°$	0.266	–	0.302–2.248	7 ns	6.3	4.512	SROPO
LBO	$\theta = 90°, \varphi = 0°$	0.78–0.81	(360 mW)	1.49–1.70	cw	40^b	4.520	DROPO, $L = 2$ cm, $T = 130–185$ °C, $P = 30$ mW
	$\theta = 86°, \varphi = 0°$	0.6515	–	1.2–1.4	20 ps	0.8	4.521	TWOPO, $L_1 = 8$ mm, $L_2 = 17$ mm, $I_0 = 0.8$ GW cm^{-2}, $E = 10$ μJ
	$\theta = 90°, \varphi = 0°$	0.605	–	0.85–0.97; 1.6–2.1	200 fs	10–15	4.450, 522	TWOPO, $L_1 = L_2 = L_3 = 15$ mm, $T = 30–85$ °C, $I_0 = 25$ GW cm^{-2}
	$\theta = 81°, \varphi = 5°$	0.57–0.63	–	1.2–1.5	580 fs	10	4.523	Injection seeding by 1.08 μm
	$\theta = 85°, \varphi = 9°$	0.57–0.63	–	1.2–1.5	400 fs	25	4.524	Injection seeding by 1.08 μm (40 ps, $L = 9$ mm, $I_0 = 1$ TW cm^{-2})
	$\theta = 0°, \varphi = 0°$	0.532	220	0.95–1.006; 1.13–1.21	10 ns	0.5	4.525	SROPO, $T = 20–120$ °C, $I_0 = 250$ MW cm^{-2} $\Delta\lambda = 0.4$ nm
	$\theta = 90°, \varphi = 0°$	0.532	–	0.75–1.8	–	20	4.33	Injection seeding from OPO (0.72–2 μm), $T = 106.5–148.5$ °C, $I_0 = 3.1$ GW cm^{-2}
	$\theta = 90°$, ooe	0.53	–	0.65–2.5	–	24	4.526	OPA, angle ($\varphi = 8.7–15.9°$) and temperature ($T = 103–210$ °C) tuning, $E = 0.45$ mJ
	$\theta = 90°, \varphi = 0°$	0.532	1500	0.77–1.7	100 ps	30	4.527	Synch. pumped SROPO, $L = 15$ mm $T = 105–137$ °C, $\Delta\lambda = 0.14$ nm
	$\theta = 90°, \varphi = 0°$	0.5235	2500 (10 mW)	0.652–2.65	12 ps	13	4.528–530	SROPO, $L = 12$ mm, $T = 125–190$ °C
	$\theta = 90°, \varphi = 0°$	0.5235	1100 (4.5 mW)	0.909–1.235	33 ps	50	4.530, 531	DROPO, $T = 167–180$ °C
	$\theta = 90°, \varphi = 0°$	0.5235	15 (30 mw)	0.65–2.65	1.7 ps	50	4.530	DROPO, $L = 12$ mm, $P = 0.21$ W
	$\theta = 90°, \varphi = 0°$	0.5235	700	0.924–1.208	12 ns	45	4.532	DROPO, $L = 12$ mm, $T = 156–166$ °C

LBO	$\theta = 90°, \varphi = 0°$	0.523	100	0.72–1.91	1 ps	34	4.533	Synch. pumped SROPO, $L = 13$ mm $T = 125$–175 °C, $P_{av} = 89$ mW
	$\theta = 90°, \varphi = 0°$	0.523	80 (70 mW)	0.8–1.5	1.2–1.5 ps	27(75[b])	4.534	SROPO, $L = 12$ mm, $P_{av} = 78$ mW
	$\theta = 90°, \varphi = 0°$	0.5145	(50 mW)	0.966–1.105	cw	10	4.535	TROPO, $L = 20$ mm, $T = 183° \pm 3$ °C, $P = 90$ mW
	$\theta = 0°, \varphi = 90°$	0.364	(115 mW)	0.494–0.502; 1.32–1.38	cw	9.4	4.536, 537	SROPO and DROPO, $L = 20$ mm, $T = 18$–86 °C, $P = 103$ mW
	$\theta = 0°, \varphi = 0°$	0.355	–	0.47–0.487	10 ns	9	4.538	SROPO, $L = 12$ mm, $T = -35° + 100$ °C, $E = 4.5$ mJ
	$\theta = 90°, \varphi = 24$–42°	0.355	14	0.435–1.922	10 ns	22	4.539	DROPO, $I_0 = 40$ MW cm^{-2}, $E = 2.7$ mJ
	$\theta = 90°, \varphi = 27$–43°	0.355	–	0.46–1.6	15 ps	30	4.540	Injection seeding from OPO, $L = 16$ mm, $I_0 = 2.8$ GWcm^{-2}, $E = 0.3$ mJ
	$\theta = 0°, \varphi = 0°$	0.355	15	0.48–0.457; 1.355–1.59	12 ns	27	4.541	SROPO, $T = 20$–200 °C
	$\theta = 90°, \varphi = 27$–42°	0.355	60	0.455–0.655; 0.76–1.62	10 ns	35[b]	4.542	SROPO, $L = 16$ mm
	$\theta = 90°, \varphi = 18$–42°	0.355	–	0.403–2.58	12 ps	28	4.543	TWOPO, $L_1 = L_2 = 15$ mm, $I_0 = 5$ GW cm^{-2}, E=0.1–1 mJ
	$\theta = 0°, \varphi = 0°$	0.355	2300	0.4159–0.4826	30 ps	37.6	4.544, 545	TWOPO, $L = 10$ mm, $T = 21°$–450 °C, $I_0 = 18$ GW cm^{-2}, $\Delta\lambda = 0.15$ nm
	$\theta = 90°, \varphi = 30$–42°	0.355	1000	0.452–1.65	9 ps	26	4.514	DROPO, $L = 10.5$ mm, $E = 0.15$ mJ
	$\theta = 90°, \varphi = 26$–52°	0.308	26	0.355–0.497; 0.809–2.34	–	28–40[b]	4.517, 546	SROPO, $L = 15$ mm
	type II in XZ and YZ planes, $\theta = 0$–9°	0.3078	30	0.381–0.387; 1.5–1.6	5 ns	35	4.547	$L = 16$ mm, $I_0 = 0.1$ GW cm^{-2}
	$\theta = 0°, \varphi = 0°$	0.266	10	0.314–1.74,	10 ns	10	4.548	SROPO, $L = 16$ mm, $T = 20$ °C

Table 4.32 (*Contd.*)

Crystal	Phase-matching angle, type of interaction	λ_{pump} [μm]	Pump threshold, I_{thr} [MW cm^{-2}]	λ_{OPO} [μm]	τ_p	Conversion efficiency [%]	Refs.	Notes
KTP	$\theta = 50–58°$, $\varphi = 0°$	1.064	–	1.8–2.4	10 ns	10	4.549	DROPO, E = 0.1–0.5 mJ
	$\theta = 90°$, $\varphi = 53°$	1.064	80	3.2	10 ns	5	4.550	SROPO, $L = 15$ mm, $P = 0.2$ W
	$\theta = 82–90°$, $\varphi = 0°$	1.064	(0.8 W)	1.57–1.59; 3.21–3.30	2–3 ps	15	4.551	SROPO, $L = 10$ mm, $f = 75$ MHz, $\Delta\lambda = 1.5$ nm
	$\theta = 90°$, $\varphi = 0°$	1.06	–	1.61	15 ns	47(66^b)	4.552	Diode-pumped Nd:YAG laser
	–	1.064	–	2.128	–	25	4.553	Synch. pumped OPO with 6 KTP (total length 58 mm), $P = 14$ W
	$\theta = 81–90°$, $\varphi = 0°$	1.053	(5.8 W)	1.55–1.56; 3.22–3.28	12 ps	21	4.554	Synch. pumped OPO, $L = 6$ mm, $P = 2$ W
	$\theta = 90°$, $\varphi = 0°$	0.7–0.95	70	1.04–1.38; 2.15–3.09	10 ns	20	4.555	SROPO, $L = 15$ mm
	$\varphi = 0°$	0.765–0.815	–	1.22–1.37; 1.82–2.15	57–135 fs	55^b	4.556	$L = 1.15$ mm, $f = 90$ MHz, $P = 340$ mW (135 fs) and 115 mW (57 fs)
	$\theta = 90°$, $\varphi = 0°$	0.72–0.853	150	1.052–1.214; 2.286–2.871	1.2 ps	42	4.557	Synch. pumped OPO, $L = 6$ mm, $P = 0.7$ W
	$\theta = 54°$, $\varphi = 0°$		–	1.38–1.67	cw	0.001	4.558	$L = 10$ mm, $P = 2$ μW
	$\theta = 67°$, $\varphi = 0°$	0.73–0.80 0.765	40000; (180 mW)	1.2–1.34; 1.78–2.1	62 fs	–	4.559	Synch. pumped OPO, $L = 1.5$ mm, $f = 76$ MHz, $P = 175$ mW
	$\theta = 45°$, $\varphi = 0°$	0.68	–	1.16–2.2; 0.58–0.657	57 fs	60^b	4.560	$L = 1.5$ mm, $P = 0.68$ W, ICSHG in BBO ($L = 47$ μm)
	$\varphi = 0°$	0.645	(110 mW)	1.2–1.34	220 fs	13	4.561	Synch. pumped OPO, $P = 30$ mW
	$\theta = 53°$, $\varphi = 0°$	0.61	–	0.755–1.04; 1.5–3.2	105–120 fs	–	4.562, 563	Synch. pumped OPO in CPM dye laser cavity, $L = 1.4$ mm
	$\theta = 40–70°$, $\varphi = 90°$	0.526	–	0.6–2.0	30 ps	10	4.564	$L = 20$ mm
	$\theta = 40–80°$, $\varphi = 0°$	0.526	–	0.6–4.3	30 ps	10	4.564	$L = 20$ mm

KTP	$\theta = 90°,\ \varphi = 0°$	0.532	1.4 W (SROPO) 30 mW (DROPO)	1.039; 1.09	cw	35	4.493, 565	SROPO and DROPO, $L = 10\,\text{mm}$, $P = 1.07\,\text{W}$
	$\theta = 90°,$ $\varphi = 10\text{--}35°$	0.523	57(61 mW)	1.002–1.096	2.2 ps	16(79[b])	4.566	Synch. pumped OPO, $L = 5\,\text{mm},\ P = 42\,\text{mW}$
	$\varphi = 0°$	0.532	80	0.7–0.9; 1.3–2.2	3.5 ns	12	4.418, 567	$L = 15\,\text{mm},\ E = 3\,\text{mJ},\ \Delta v = 0.02\,\text{cm}^{-1}$
	$\theta = 90°,$ $\varphi = 25.3°$	0.531	(40 mW)	1.0617	cw	30	4.568	DROPO, $L = 8\,\text{mm}$
	–	0.5235	1000(2 mW)	0.946–1.02; 1.075–1.172	8 ps	10(56[b])	4.530, 569	Synch. pumped SROPO, $L = 5\,\text{mm},\ P = 2\,\text{mW}$
	$\theta = 90°$	0.523	60(61 mW)	0.938–1.184	1–2 ps	16	4.570	SROPO, $L = 5\,\text{mm},\ f = 125\,\text{MHz}$, $P = 40\,\text{mW}$
	$\theta = 90°,$ $\varphi = 0 - 33°$	0.526	(0.5 W)	1.01–1.1	14 ps	44	4.554	Synch. pumped OPO, $L = 6\,\text{mm}$, $P = 0.58\,\text{W}$
	$\theta = 53 - 72°$	0.5235	–	1.2–1.9	1.5 ps	–	4.500	Synch. pumped OPO, $L = 9\,\text{mm}$, $\Delta v = 10\,\text{cm}^{-1}$
		0.5235	4 (150 mW)	1.02; 1.075	–	7	4.530	cw mode-locked DROPO
	$\theta = 69°,\ \varphi = 0°$	0.532	7	0.76–1.04	6 ns	30	4.571	$L = 15\,\text{mm}$, ICSHG in BBO with $\eta = 40\%$ (380–520 nm)
KTA	$\varphi = 0°,\ \text{oeo}$	0.78	–	1.29–1.44; 1.83–1.91	85–150 fs	10–15	4.572	$L = 1.47\,\text{mm},\ \text{P} = 75\,\text{mW}$
	$\theta = 53°,\ \varphi = 0°$	0.773-0.792	–	1.45; 1.7	300 ns	0.3	4.573	DROPO, $L = 7\,\text{mm}$
In:KTA	Type II	0.77	–	1.435;1.662	–	–	4.560	$L = 1.5\,\text{mm}$
"Banana"	$\theta_{\text{ooe}} = 90°$	0.532	–	0.75–1.82	10 ns	5	4.574	SROPO, $\text{T} = 80 - 220\,°\text{C}$
	$\theta_{\text{ooe}} = 90°$	0.532	5	0.8–1.6	10 ps	25	4.575	Synch. pumped OPO
	$\theta_{\text{ooe}} = 90°$	0.53	50	0.65–3	10 ps	5.3	4.459	Synch. pumped OPO, $I_0 = 250\ \text{MWcm}^{-2}$

Table 4.32 (*Contd.*)

Crystal	Phase-matching angle, type of interaction	λ_{pump} [μm]	Pump threshold, I_{thr} [MW cm^{-2}]	λ_{OPO} [μm]	τ_p	Conversion efficiency [%]	Refs.	Notes
"Banana"	$\theta_{ooe} = 90°$	0.532	7–9	0.672–2.56	15–45ps	8.1	4.576, 577	Synch. pumped SROPO, $L = 10$ mm, $f = 139$ MHz, $T = 75$–350 °C
$\alpha - HIO_3$	eoe	0.532	–	1–1.1	–	57	4.578	$L = 23$ mm, $I_0 = 20$ MWcm^{-2}
	eoe	0.532	–	0.7–2.2	30–45ps	10–12	4.439	TWOPO, $L_1 = L_2 = 2$ cm, $I_0 = 4$–5 GWcm^{-2}
	eoe	0.527	60	–	5–6ps	10	4.579	Synch. pumped OPO, $\Delta\nu\Delta\tau = 0.7$
$KNbO_3$	type II	1.064	240	1.87;2.47	–	–	4.580	$L = 9$ mm
		0.532	2.2	0.86–1.4	–	5	4.581	DROPO, $L = 19$ mm, $T = 184$–220 °C
	along the b axis	0.532	3.5	0.88–1.35	10 ns	32	4.582	DROPO, $T = 180$–200 °C, $P = 12$ MW
Urea	$\theta_{oeo} = 81$–90°	0.355	55 (45 mW)	0.5–0.51; 1.17–1.22	7 ns	20	4.583	SROPO, $L = 12.7$ mm, $I_0 = 90$ MWcm^{-2}
	$\theta_{oeo} = 50$–90°	0.355	–	0.5–1.23	7ns	23	4.584, 585	SROPO, $L = 23$ mm
	$\theta_{oeo} = 64$–90°	0.308	16–20	0.537–0.72	4–6 ns	37	4.586	$L = 15$ mm
	eeo	0.266	–	0.33–0.42	7ns	–	4.585	
NPP	–	0.62	–	0.8–1.6	150–290 fs	–	4.587, 588	$L = 1.5$ mm
	$\theta = 9.5$–13°, $\varphi = 0°$	0.5927	30	0.9–1.7	1 ns	5	4.589, 590	$L = 1.9$ mm
DLAP	type I	0.308	18	0.415–0.526; 0.743–1.194	–	5[b]	4.591	SROPO, $L = 25$ mm

[a]Conversion efficiency was determined from Eq. 4.4.
[b]Pump depletion.

version to parametric radiation was attained in an $\alpha - HIO_3$ crystal at pump intensity 3–4 GW cm^{-2} without focussing. For KDP and LiNbO$_3$ crystals, cylindrical telescoping was used with optimum conditions found experimentally. For LiNbO$_3$ a one-crystal scheme and 2:1 spherical telescoping were used. In α-HIO$_3$ an effective SRS was observed, which competed with OPO and consumed up to 30% of the pump energy. The SRS threshold was very low and amounted to 0.3 GW cm^{-2}. In LiIO$_3$ SRS was less effective: up to 5% of the pump energy was consumed for stimulated scattering. Study of the parametric pulse shape has shown that in KDP the parametric pulse duration decreases to 17 ps, and in LiIO$_3$ to 6 ps in comparison with pump pulse duration 45 ps.

Danelyus et al. [4.444] realized OPO with synchronous pumping by a train of picosecond pulses of the second harmonic of Nd:phosphate glass laser radiation ($\lambda = 527$ nm). A KDP crystal (L = 4 cm, eoe interaction) was placed in a resonator with an optical length of 130 cm, equal to the axial period of the pumping laser. The shortest OPO pulses were 0.3–0.5 ps at an energy of 20 µJ (the tuning range was from 0.8 to 1.5 µm). Then, the OPO pulses were amplified to 1 mJ in F_2^+:LiF crystals (L = 2 cm) pumped with the second harmonic of electro-optically mode-locked Nd:YAG laser radiation. The pulses of parametric radiation can be considerably shortened in a two-cascade TWOPO by introducing the corresponding time delay between the pump and signal (or idler) waves. For this purpose, for instance, a CaCO$_3$ crystal several millimeters in length can be placed between the TWOPO crystals, which ensures the temporal delay between the signal and pump waves with different polarizations and, hence, different refractive indices in a CaCO$_3$ crystal [4.478, 592]. This method shortened the OPO pulses to 4 ps when the pump pulse duration (τ_{pump}) was 21 ps [4.592], and to 0.5 ps when $\tau_{pump} = 8$ ps [4.478]. In the latter case the temporal delay amounted to 8.5 ps.

Maximum OPO efficiency in traveling-wave OPO schemes $\eta_{eff} = 60–70\%$ has been attained with two KDP or ADP crystals spaced at a great distance from each other (up to 1 m) [4.445, 447, 453]. The efficiency η_{eff} was calculated by the equation

$$\eta_{eff} = E_{OPO}/(E_{OPO} + E_{unc}) \tag{4.4}$$

where E_{OPO} is the total OPO radiation energy (signal + idler) and E_{unc} is the energy of unconverted pump radiation measured after second crystal. The value η_{eff} is greater than the ordinary η value calculated from the equation $\eta = E_{OPO}/E_{pump}$, since

$$(E_{OPO} + E_{unc})/E_{pump} = 50–80\%. \tag{4.5}$$

This is because the pump and OPO radiations are always partially lost due to scattering and absorption in the crystals [4.453]. Conversion efficiency can also be determined in terms of pump depletion:

$$\eta = 1 - E_{unc}/E_{pump}. \tag{4.6}$$

Pump depletions are usually much greater than the ordinary η values.

Generation of ultrashort OPO pulses ($\tau < 100\mathrm{fs}$) was reported in a number of articles [4.498–501, 556, 559, 560, 572]. Synchronously pumped OPO schemes are mainly used in these devices. *Laenen* et al. [4.499] pumped BBO (L = 5.8 mm, $\theta_c = 23°$) based SROPO by the train of 300 pulses with 0.8 ps duration from a frequency-doubled Nd:glass laser. Near the degeneracy point ($\lambda = 1.0796$ mm) OPO pulse durations were 65 fs (FWHM). With a KTP crystal (L = 1.5 mm, $\theta_c = 67°$, $\varphi = 0°$) and additional external pulse compression, 175 mW IR radiation near 1.3 μm was generated with $\tau = 62$ fs and $f = 76$ MHz [4.559]. As a pump source a Ti:sapphire laser (765 nm, 800 mW, 110 fs) was used. The measurements of pulse duration were carried out by the auto-correlation method with 1 mm thick $LiIO_3$ [4.499] or 0.3 mm thick KDP [4.559]. Minimum OPO pulse durations obtained up to now are as short as 57 fs [4.556, 560]. Here the Ti:sapphire laser (2.5 W, 125 fs) was also employed for synchronous pumping of KTP (1.15 mm) based OPO. The use of an intracavity dispersion compensation allows generation of 57 fs unchirped pulses with a high repetition rate (90 MHz): average power was 115 mW. Output OPO powers up to 1 W were attained at $\tau = 135$ fs. When a BBO crystal (L = 47 μm) was placed inside the ring OPO cavity the tuning range of OPO was shifted into the visible by ICSHG: $\lambda = 580–657$ nm, $\tau < 100$ fs, P = 240 mW [4.560].

Of special interest is the use of the first, second, or third harmonics of *diode-pumped* Nd:YAG or Nd:YLF laser radiations with relatively low average powers ($\sim 10–100$ mW) as OPO pump sources [4.489–491, 528, 531, 532, 541, 542, 552, 566, 569, 571]. Crystals of LBO, KTP, and $LiNbO_3$:MgO are used for this purpose. An IR parametric power of 100 mW was obtained in such a continuous-wave DROPO scheme using a $LiNbO_3$:MgO crystal [4.491]. *Marshall* and *Kaz* [4.552] achieved an absolute electrical (wall-plug) efficiency of 2.1% with pump-to-OPO power-conversion efficiency of 70%. ICSHG with a BBO crystal placed inside the cavity of KTP based OPO allows frequency tuning in the ranges 380–520 nm and 760–1040 nm by use of a frequency-doubled Nd:YAG laser as a pump source [4.571]. Wall-plug efficiency was 0.4% and 0.9%, respectively.

Choice of the OPO crystal is similar to that for other nonlinear frequency conversions described above. "Banana" and potassium niobate crystals have maximum nonlinearities and, hence, minimum OPO thresholds (for instance for "banana" $I_{\mathrm{thr}} = 5$ MW cm^{-2} [4.575], and for $KNbO_3$, $I_{\mathrm{thr}} = 2.2$ MW cm^{-2} [4.581]. These crystals are nonhygroscopic and ensure noncritical (90°) phase matching with temperature tuning. At the same time they are unstable under radiation and cannot be grown to large sizes. The minimum OPO threshold down to 0.4 mW was achieved in a cw regime in a quadruply resonant MgO:LiNbO$_3$ OPO scheme by use of a monolithic total-internal-reflection resonator, in which doubly-resonant second-harmonic generation and DRO-PO occur simultaneously [4.488]. KTP crystals are useful for near IR region (1–3 μm) in the resonant schemes of OPO. $LiIO_3$, $LiNbO_3$, and α-HIO_3 crystals are effective in the visible and IR regions (up to 5 μm); BBO and LBO crystals in the UV, visible, and near IR; and KDP, ADP, and urea crystals, in the UV and visible regions when they are pumped by the second, third, or fourth harmonics of Nd:YAG laser radiation. Because of their high optical quality, KDP, ADP, and $LiIO_3$ crystals can be used in TWOPO schemes, which require high pump intensities (up to 10 GW cm^{-2}) but at the same time are technologically simple. $LiNbO_3$ crystals are advantageous in cw diode-pumped Nd:YAG laser devices. Attractive KTP and LBO crystals are very suitable, in particular, for synchronously pumped picosecond and femtosecond OPO schemes. Among the advantageous parameters of these crystals, the small walk-off angle should be mentioned. For nanosecond OPO, BBO crystals represent an optimal choice due to their large nonlinearity, high optical damage threshold, and fulfillment of phase-matching conditions in a wide spectral range (0.23–3.2 μm). By changing the phase-matching angle of the BBO crystal by only 2°, the near IR region from 0.7 to 2.6 μm can be covered. However, it should be mentioned that the small angular bandwidth of BBO based OPO necessitates the use of single-mode diffraction limited laser radiation as a pump source. The detailed comparison of BBO and LBO crystals for OPO applications is presented in a series of articles [4.169, 450, 512, 514, 515].

4.6.2 OPO in the Mid IR Region

For OPO in the mid IR region (5–16 μm), Ag_3AsS_3, $AgGaS_2$, $AgGaSe_2$, $ZnGeP_2$, GaSe, and CdSe crystals are used. *Elsaesser* et al. [4.593] have reported OPO in the range of 1.2–8 μm with a proustite crystal and Nd:YAG laser radiation as a pump source ($\lambda = 1.064$ μm, $I_0 = 6$ GW cm^{-2}, $\tau_p = 21$ ps, $\Delta v = 2$ cm^{-1}. The energy-conversion efficiency amounted to 10^{-2}–10^{-4}. The OPO pulse spectral bandwidth was 10–40 cm^{-1} and OPO pulse duration ∼8 ps. When an Nd^{3+}:phosphate glass laser was used as a pump source, the conversion efficiency to OPO radiation with $\lambda = 1.4$–5 μm was 1% [4.594].

Traveling-wave OPO has been realized [4.595] with two $AgGaS_2$ crystals (L = 1.5 cm and 3 cm) with output radiation in the range of 1.2–10 μm. Here Nd:YAG laser radiation ($\lambda = 1.064$ μm, $\tau = 20$ ps, $I_0 = 3$ GW cm^{-2}) was a pump source. The phase-matching angle (θ_{ooe}) varied from 35° to 55°. The OPO pulse duration (8 ps) was determined by the cross-correlation method using noncollinear SFG. The OPO radiation (5 μm) and second harmonic of the Nd:YAG laser radiation (0.53 μm) were mixed in a $LiIO_3$ crystal 0.1 cm in length ($\theta = 25°$). The OPO pulse spectral bandwidth was 10 cm^{-1} (6–10 μm).

Optical parametric amplification up to 13 μm in $AgGaS_2$ crystals has been demonstrated by injection seeding from OPO based on $LiNbO_3$ ($\lambda = 1.4$–4.5 μm and its second harmonic) [4.596] or on LBO crystal (1.16–2.13 μm) [4.597]. The energies of single picosecond pulses (20 ps) reach 90 μJ [4.596], and spectral widths were $\Delta V = 3$–8 cm^{-1} [4.597].

Effective generation of parametric radiation in an $AgGaSe_2$ crystal has been reported for the wavelength ranges 1.6–1.7, 6.7–6.9 and 2.65–9.02 μm. [4.598]. Here Q-switched Nd:YAG ($\lambda = 1.34$ μm) and Ho:YLF ($\lambda = 2.05$ μm) laser radiations were used as pump sources. The crystal lengths were 18–21 mm; the conversion efficiency amounted to 18% for the output power P = 100 kW and τ = 30 ns. Picosecond OPO in $ZnGeP_2$ was realized with pumping by a train of pulses of erbium laser radiation at λ = 2.94 μm and τ = 80 ps [4.599, 600]. The efficiency of conversion to parametric radiation with λ = 5.51–5.38 μm and 6.29–6.46 μm was 5.3% for I_0 = 4 GW cm^{-2}, a 42 mm crystal, and θ_{oeo} = 84.5–79.3° [4.599]. On increasing I_0 to 16 GW cm^{-2} the quantum conversion efficiency was 17%; the OPO output power amounted to 1 MW. High-efficiency OPO was obtained with a CdSe crystal [4.601]; the efficiency of power conversion to parametric radiation with λ_{signal} = 2.26–2.23 μm and λ_{idler} = 9.8 − 10.4 μm reached 40%. As a pump source, Nd:YAG laser radiation (λ = 1.833 μm, τ = 30 ns, and $I_0 = 2 \times 10^7$ W cm^{-2} was used. The phase-matching angle θ_{oeo} changed from 90° to 78°. With pumping by CaF_2:Dy^{2+} laser radiation (λ = 2.36 μm; τ = 40 ns), OPO in a CdSe crystal was obtained in the region 7.9–13.7 μm at θ_{oeo} = 90–65° [4.602, 603]. The energy-conversion efficiency was 15% at $I_0 = 10^7$ W cm^{-2} and crystal length 30 mm. With the same crystal and HF laser radiation (λ = 2.87 μm) as a pump source, OPO was realized in the ranges 4.3–4.5 μm, 8.1–8.3 μm, and 14.1–16.4 μm with a peak power of up to 800 W [4.604, 605].

Table 4.33 contains the results of OPO experiments in the mid IR range and lists the crystals, pump wavelengths, OPO tuning ranges, pulse durations and conversion efficiencies. The table demonstrates that $AgGaSe_2$ and CdSe crystals are most promising in a nanosecond regime; and proustite, silver thiogallate and $ZnGeP_2$, in a picosecond regime.

4.6.3 Conversion of OPO Radiation to the UV Region

The range of wavelengths achieved by OPO can be considerably widened by the generation of harmonics (SHG, FOHG) and also by sum-frequency mixing (SFM) of the OPO radiation with the harmonics of pump laser radiation. For example, *Kryukov* et al. [4.440] used a 30 mm $LiIO_3$ crystal cut at the angle θ_{ooe} = 30° for SHG of OPO radiation. The energy of the second harmonic at λ = 420–700 nm was 0.1–0.4 mJ at η = 5%–47%. The second-harmonic radiation was, in its turn, doubled in a KDP or an ADP crystal 40 mm in length; the tuning range was 0.26–0.4 μm for an output pulse energy of several μJ and τ = 30 ps. Second-harmonic generation of OPO radiation up to 313 nm at τ = 5 ps and Δv = 12 cm^{-1} has been attained [4.483, 592] with $LiIO_3$ (L = 1.4 mm, θ = 52°) and KDP (L = 8 mm, θ = 41.5° or 52°) crystals. The SH radiation energy was 50 mJ.

Singly resonant OPO radiation obtained in urea (λ = 0.498–1.23 μm, τ = 7 ns) was doubled in another urea crystal (L = 6 mm, eeo interaction) with η = 8.3%, as result of which the spectral range from 249 to 320 nm was covered [4.584]. Also used for SHG of OPO radiation were crystals of BBO ($\lambda_{2\omega}$ = 205–305 nm, η = 3.5–8.1% [4.614]; 220 nm, η = 10% [4.515]; 205 nm [4.504],

Table 4.33. Optical parametric oscillation in the mid IR region

Crystal	λ_{pump} [μm]	λ_{OPO} [μm]	τ_p	Conversion efficiency [%]	Refs.
Ag_3AsS_3	1.06	2.1	100 ns	–	4.606
	1.065	1.82–2.56	26 ns	1	4.607
	1.065	1.22–8.5	25 ns	–	4.608
	1.064	1.2–8	8 ps	0.01–1	4.593
	1.055	1.4–5	10 ps	1	4.594
$AgGaS_2$	1.064	1.2–10	8 ps	0.1–10	4.595
	1.06	1.4–4.0	18 ns	16	4.609
	1.064	4.5–8.7	15–20 ps	5.4	4.596
	1.064	1.16–12.9	19 ps	25	4.597
$AgGaSe_2$	1.34	1.6–1.7, 6.7–6.9	30 ns	>18	4.598
	2.05	2.65–9.02	30 ns	>18	4.598
	2.06	~ 4.1	~ 30 ns	23	4.610
$ZnGeP_2$	2.94	5.51–5.38 6.29–6.46	80 ps	5.3	4.599
	2.94	5–5.3, 5.9–6.3	150 ps	17	4.600
	2.79	5.3; 5.9	~ 100 ps	10	4.611
	2.94	4–10	~ 100 ps	17.6	4.612, 613
GaSe	2.94	3.5–18	~ 100 ps	1	4.612, 613
CdSe	1.833	9.8–10.4, 2.26–2.23	300 ns	40	4.601
	2.36	7.9–13.7	40 ns	15	4.602, 603
	2.87	4.3–4.5, 8.1–8.3	140 ns	15	4.604
	2.87	14.1–16.4	–	–	4.605

$LiIO_3$ (560–915 nm, $\eta = 10\%$ [4.463]), and $KNbO_3$ (427–470 nm, $\eta = 40\%$ [4.615]). Use of a BBO crystal for ICSHG of KTP based OPO allows generation of nanosecond radiation in the range of 380–580 nm with $\eta = 40\%$ [4.571] and femtosecond radiation ($\tau = 57$ fs) in the 580–657 nm range with average power 680 mW [4.560]. In the latter case the thickness of applied BBO was 47 μm.

The generation of 197–393 nm picosecond radiation has been reported [4.117], obtained by mixing the second, third, and fourth harmonics of Nd: YAG laser radiation ($\lambda = 532$, 355, and 266 nm, respectively) with OPO radiation in $KB_5O_8 \cdot 4H_2O$, KDP, and DKDP crystals. The UV radiation power amounted to 20 kW at $\lambda = 197$ nm and 100 kW at $\lambda = 225$–393 nm at a pulse duration of 20 ps. *Petrosyan* et al. [4.310] covered the 200–222 nm spectral region, mixing OPO radiation with 266 nm radiation in a KB5 crystal (E = 1 μJ, $\tau = 45$ ps). In both papers [4.117, 310] the interaction in KB5 was of the eeo type in the *XY* plane. By SFM between OPO and 532 nm radiations in DKDP a UV range of 302–392 nm was covered with energies up to 3 mJ and $\eta = 28\%$ [4.614]. A wavelength as short as 215 nm with E = 30 mJ and $\eta = 20\%$ was achieved in ADP ($\theta = 90°$, L = 10 mm) by SFM of TWOPO output with the fourth harmonic of a neodimium laser [4.616]. The shortest wavelengths in this way were obtained by mixing OPO radiation with fourth (266 nm) and fifth

(212.8 nm) harmonics of Nd:YAG laser in an LBO crystal: 188–242 nm, η = 0.2–2%, E = 2–400 μJ [4.299], 188–195 nm, E = 9 μJ [4.300].

4.7 Stimulated Raman Scattering and Picosecond Continuum Generation in Crystals

Nonlinear optical losses of pump radiation energy are the factors limiting the energy-conversion efficiency for OPO in $LiIO_3$, $LiNbO_3$ $\alpha - HIO_3$, and Ag_3AsS_3 crystals [4.437]. In $LiIO_3$ and $\alpha - HIO_3$ crystals pumped at λ_{pump} = 530 nm, stimulated Raman scattering (SRS) is the main mechanism of nonlinear losses, and it decreases significantly the OPO conversion efficiency. For instance, in $LiIO_3$ at λ_{pump} = 527 nm and τ = 5 ps, a maximum value of η is 2% at I = 7 GW cm^{-2}. With a subsequent increase of the intensity, the value of η decreases. Here SRS consumes more than 70% of the pump energy [4.437].

Table 4.34 lists the data on the SRS thresholds in KDP, $LiIO_3$, $LiNbO_3$ and α-HIO_3 crystals. As is known, the SRS threshold corresponds to the pump power density at which a sharp jump of the Stokes wave intensity is observed. It is seen that the KDP crystal has the highest SRS threshold; therefore, in this crystal maximum conversion efficiencies to parametric radiation are obtained. *Dzhotyan* et al. [4.468] studied SRS in $LiIO_3$ which was pumped by the second harmonic of a single-frequency Q-switched neodymium glass laser radiation. A peak pulse pump power at λ = 530 nm was 15 MW at τ = 30 ns. A 3 cm $LiIO_3$ crystal cut at an angle θ_C = 26° to the optic axis was used for SRS. The measured Stokes shift was 822±2 cm^{-1}. Without SRS, the efficiency of conversion to parametric radiation was almost 20%, whereas in the presence of SRS the OPO radiation energy did not depend on the pump energy and maximum conversion efficiency was only 3%.

The process of SRS in α-HIO_3 and its competition with parametric amplification has been thoroughly studied [4.439]. The second harmonic of mode-

Table 4.34. Stimulated Raman scattering thresholds in crystals at $\lambda = 532$ nm

Crystal	$I_{thr}[10^9 \text{ W cm}^{-2}]$	$\tau_p[ps]$	Refs.
KDP	22	30	4.617
	44	4	4.617
$LiIO_3$	0.015	30 000	4.468
	0.7	30	4.439
$LiNbO_3$	1.2	30	4.437
	5	4	4.437
α–HIO_3	0.3	35	4.439
	1.4	30	4.437
	6	4	4.437

Table 4.35. Stimulated Raman scattering in $\alpha - HIO_3$

SRS components	$\Delta v\,[\mathrm{cm}^{-1}]$	$\lambda\,[\mathrm{nm}]$	η
1st Stokes	790	555.2	0.21
2nd Stokes	1580	580.7	0.07
3rd Stokes	2370	608.6	0.0095
4th Stokes	3160	640.2	0.001
1st anti-Stokes	790	510.5	0.01
2nd anti-Stokes	1580	490.7	0.001

locked Nd:YAG laser radiation ($\lambda = 532$ nm, $\tau = 35$ ps) was used as pump radiation. Maximum pulse-repetition rate was 25 Hz. The SRS threshold intensity in α-HIO$_3$ was rather low, 0.3 GW cm^{-2}. Table 4.35 gives the efficiencies of conversion to different Stokes and anti-Stokes SRS components in an α-HIO$_3$ crystal when pumped with $\lambda_{\mathrm{pump}} = 532$ nm at pump power density $I_0 = 1$ GW cm^{-2} and $\tau = 35$ ps [4.439]. The crystal was oriented for parametric interaction of the eoe type near the degeneracy point ($\lambda_{\mathrm{signal}} = \lambda_{\mathrm{idler}}$). Dependence of the SRS threshold on the pump radiation polarization was observed. The transition from extraordinary to ordinary pump polarization halves the SRS threshold. This is accompanied by the appearance of a fine structure of SRS components that gradually fill the spectrum between the main components with increasing pump power.

Thus, with increasing pump intensity the parametric signal is inhibited and the SRS Stokes components sharply rise. Note, however, that the process of SRS in the crystals can itself be successfully used for discrete frequency tuning.

Along with stimulated Raman scattering in crystals, the generation of wide-band radiation (picosecond continuum) is also observed. The picosecond continuum was first detected in media with cubic nonlinearity: D$_2$O, H$_2$O, NaCl, and others. It has been successfully used in kinetic spectroscopy as probe radiation. Among crystals with square nonlinearity, picosecond continuum was observed in KDP, LiIO$_3$, LiNbO$_3$, GaAs, AgBr, ZnSe, and CdS. Table 4.36 illustrates some main characteristics of the picosecond continuum and

Table 4.36. Picosecond continuum generation

Crystal	$\lambda_{\mathrm{pump}}\,[\mu m]$	$I_{\mathrm{pump}}\,[10^9\,\mathrm{W\,cm}^{-2}]$	$\lambda_{\mathrm{cont}}\,[\mu m]$	Conversion efficiency[%]	Cut angle θ [deg]	Refs.
KDP	1.054	50	0.3–1.1	10	49	4.618
	0.527	30–40	0.84–1.4	15	42	4.619
LiIO$_3$	0.355	–	0.46–1.55	–	90	4.620
	0.532	0.3	0.67–2.58	–	90	4.620
	1.064	–	1.72–3.0	–	90	4.620
LiNbO$_3$	1.064	–	1.92–2.38	3	44.7	4.621
GaAs	9.3	100	3–14	–	–	4.622

conditions for its generation. *Bareika* et al. [4.619] used the single pulse of the second harmonic of Nd^{3+} phosphate glass laser radiation ($\lambda = 0.527$ μm, $\tau = 3$–6 ps, E = 1 mJ) as a pump source for picosecond continuum generation. The pump radiation was focused to the 4 cm KDP crystal cut for the ooe interaction ($\theta = 42°$, $\varphi = 45°$). The efficiency of conversion to the picosecond continuum under saturation conditions was 15% and spectral conversion efficiency amounted to $10^{-4}(cm^{-1})^{-1}$. The divergence of the picosecond continuum beam was ~ 5 mrad and the spectral bandwidth due solely to collinear generation was 3000 cm^{-1}. When the crystal was rotated by 10′, two maxima in the continuum spectrum, corresponding to the signal and idler radiations, appeared. The time delay between the boundary continuum frequencies did not exceed 0.2 ps. Then the continuum radiation was doubled in another 3 cm KDP crystal (ooe interaction) with an efficiency of 10% [4.619].

Picosecond continuum generation was also realized in KDP pumped with radiation at $\lambda_{pump} = 1.054$ μm and I = 50 GW cm $^{-2}$ [4.618]. The continuum obtained extended from 0.3 μm to 1.1 μm. Time delay of blue components compared with red ones was observed. In $LiIO_3$, the picosecond continuum was realized by *Pokhsraryan* [4.620] upon pumping with the first, second, and third harmonics of picosecond Nd:YAG laser radiation. The pumping radiation propagated in a 3 cm crystal along the X axis. Noncollinear generation of the picosecond continuum was observed along the circular cone generatrix; the cone axis corresponded to the X axis, and the angle was found from the phase-matching condition $K_3 = K_1 + K_2$. At $\lambda_{pump} = 1.064$ μm, the generated picosecond continuum covered the 1.7–3.0 μm spectral range; at $\lambda_{pump} = 0.532$ μm, the range was 0.67–2.58 μm; and at $\lambda_{pump} = 0.355$ μm , the spectral range was 0.46–1.55 μm.

Generation of the picosecond continuum in a 4.5 cm $LiNbO_3$ crystal has been demonstrated [4.621]. Type I interaction (ooe) was used at $\theta = 44.7°$. The continuum energy amounted to 0.2 mJ. The generation of the picosecond continuum in the 3–14 μm region was reported [4.622] when CO_2 laser radiation ($\lambda_{pump} = 9.3$ μm, $\tau = 2.5$ and 8 ps) was focused in GaAs, AgBr, ZnSe, and CdS crystals. The contribution of different nonlinear processes – phase self-modulation, the four-photon parametric process, high-order harmonic generation, and SRS – to the formation of the picosecond continuum was studied. GaAs crystals have very high nonlinearity: $d_{14} = (134$–$189) \times 10^{-12}$ m/V. However, they cannot be used in nonlinear SHG, SFG, or OPO processes because of zero birefringence and hence the absence of phase-matching.

The generated ps continuum can be further up- or down-converted by mixing with the pump in SFG or DFG processes respectively. Thus, continuously tunable picosecond radiation in the range 350–680 nm [4.623] and 330–700 nm [4.624] with linewidth of 10–30 cm^{-1} was obtained by mixing the ps continuum generated, respectively, in the laser active element (Nd:glass) or in D_2O, with the fundamental frequency (1.06 μm). A crystal of $LiIO_3$ 3 cm in length was used for SFG; output powers attained 1–2 MW ($\tau = 10$ ps). By mixing dye laser output with the ps continuum in DFG process, IR radiation of subpicosecond duration (1.7 – 4.0 µm, 0.2 ps [4.392] and 2.5 µm, 0.5 ps [4.393]) was generated. Crystals of $LiNbO_3$ and BBO, respectively, were used for this purpose.

Second harmonic generation of a broadband continuum (0.6–1.4 µm) generated by 620 nm, 100 fs radiation from CPM dye laser in D_2O was realized in an ultrathin ($L = 1$ µm) organic crystal of MNA* in a non phase-matched configuration [4.625]. Conversion efficiency η was comparable to that of a 100 µm KDP crystal under phase-matched conditions.

References

Chapter 1

1.1   T.H. Maiman: Nature **187**, 493–494 (1960)
1.2   L.F. Johnson, K. Nassau: Proc. IRE **49**, 1704–1706 (1961)
1.3   E. Snitzer: Phys. Rev. Lett. **7**, 444–446 (1961)
1.4   J.E. Geusic, H.M. Marcos, L.G. Van Uitert: Appl. Phys. Lett. **4**, 182–184 (1964)
1.5   A. Javan, W.R. Bennett, D.R. Herriot: Phys. Rev. Lett. **6**, 106–110 (1961)
1.6   E. Gordon, E. Labuda, W. Bridges: Appl. Phys. Lett. **4**, 178–180 (1964)
1.7   C.K.N. Patel: Phys. Rev. Lett. **13**, 617–619 (1964)
1.8   P.P. Sorokin, J.R. Lankard: IBM J. Res. Dev. **10**, 162–163 (1966)
1.9   F.P. Schafer, W. Schmidt, J. Volse: Appl. Phys. Lett. **9**, 306–309 (1966)
1.10 R.N. Hall, G.E. Fenner, J.D. Kingsley, T.J. Soltys, R.O. Carlson: Phys. Rev. Lett. **9**, 366–368 (1962)
1.11 M.I. Nathan, W.P. Dumke, G. Burns, F.H. Dill, G. Lasher: Appl. Phys. Lett. **1**, 62–64 (1962)
1.12 T.M. Quist, R.H. Rediker, R.J. Keyes, W.E. Krag, B. Lax, A.L. McWhorter, H.E. Zeiger: Appl. Phys. Lett. **1**, 91–93 (1962)
1.13 P.A. Franken, A.E. Hill, C.W. Peters, G. Weinreich: Phys. Rev. Lett. **7**, 118–119 (1961)
1.14 J.A. Giordmaine: Phys. Rev. Lett. **8**, 19–20 (1962)
1.15 P.D. Maker, R.W. Terhune, M. Nicenoff, C.M. Savage: Phys. Rev. Lett. **8**, 21–22 (1962)
1.16 R.C. Miller, A. Savage: Phys. Rev. **128**, 2175–2179 (1962)
1.17 A.W. Smith, N. Braslau: IBM J. Res. Dev. **6**, 361–362 (1962)
1.18 F. Zernike, P.R. Berman: Phys. Rev. Lett. **15**, 999–1001 (1965)
1.19 G.D. Boyd, A. Ashkin: Phys. Rev. **146**, 187–198 (1966)
1.20 J.A. Giordmaine, R.C. Miller: Phys. Rev. Lett. **14**, 973–976 (1965)
1.21 F. Jona, G. Shirane: *Ferroelectric Crystals* (Pergamon, Oxford, 1962)
1.22 R.C. Miller: Appl. Phys. Lett. **5**, 17–19 (1964)
1.23 J.E. Geusic, H.J. Levinstein, J.J. Rubin, S. Singh, L.G. Van Uitert: Appl. Phys. Lett. **11**, 269–271 (1967)
1.24 K.F. Hulme, O. Jones, P.H. Davies, M.V. Hobden: Appl. Phys. Lett. **10**, 133–135 (1967)
1.25 G. Nath, S. Haussühl: Appl. Phys. Lett. **14**, 154–156 (1969)
1.26 F.C. Zumsteg, J.D. Bierlein, T.E. Gier: J. Appl. Phys. **47**, 4980–4985 (1976)
1.27 I.S. Rez: Usp. Fiz. Nauk **93**, 633–674 (1967) [English transl. : Sov. Phys. - Usp. **10**, 759–782 (1968)]
1.28 D.N. Nikogosyan: Kvantovaya Elektron. **4**, 5–26 (1977) [English transl. : Sov. J. Quantum Electron. **7**, 1–13 (1977)]
1.29 C. Chen, B. Wu, A. Jiang, G. You: Scientia Sinica, Ser. B **28**, 235–243 (1985)
1.30 C. Chen, Y. Wu, A. Jiang, B. Wu, G. You, R. Li, S. Lin: J. Opt. Soc. Am. B **6**, 616–621 (1989)

Chapter 2

2.1   S.A. Akhmanov, R.V. Khokhlov: *Problems of Nonlinear Optics* (VINITI, Moscow 1964)
       (In Russian)
2.2   N. Bloembergen: *Nonlinear Optics* (Benjamin, New York 1965)
2.3   F. Zernicke, J.E. Midwinter: *Applied Nonlinear Optics* (Wiley, New York 1973)
2.4   V.G. Dmitriev, L.V. Tarasov: *Optique Non Lineaire Appliquee* (MIR, Moscow 1987).
       [Translated from Russian: V.G. Dmitriev, L.V. Tarasov, *Prikladnaya Nelineinaya Optika*
       (Radio i Sviyaz, Moscow 1982)]
2.5   Yu.V. Voroshilov, D.N. Nikogosyan: Kvantovaya Elektron. **3**, 608–610 (1976) [English
       transl.: Sov. J. Quantum Electron. **6**, 326–327 (1976)]
2.6   A. Yariv, P. Yeh: *Optical Waves in Crystals* (Wiley, New York 1984)
2.7   D.N. Nikogosyan, G.G. Gurzadyan: Kvantovaya Elektron. **13**, 2519–2520 (1986) [English
       transl.: Sov. J. Quantum Electron. **16**, 1663–1664 (1986)]
2.8   F. Brehat, B. Wyncke: J. Phys. B **22**, 1891–1898 (1989)
2.9   J.P. Feve, B. Boulanger, G. Marnier: Opt. Commun. **99**, 284–302 (1993)
2.10   D.Yu. Stepanov, V.D. Shigorin, G.P. Shipulo: Kvantovaya Elektron. **11**, 1957–1964 (1984)
       [English transl.: Sov. J. Quantum Electron. **14**, 1315–1320 (1984)]
2.11   M.V. Hobden: J. Appl. Phys. **38**, 4365–4372 (1967)
2.12   D.N. Nikogosyan, G.G. Gurzadyan: Kvantovaya Elektron. **14**, 1529–1541 (1987) [English
       transl.: Sov. J. Quantum Electron. **17**, 970–977 (1987)]
2.13   M. Kashke, C. Koch: Appl. Phys. B **49**, 419–423 (1989)
2.14   N.K. Sidorov: Kvantovaya Elektron. **19**, 880–881 (1992) [English transl.: Sov. J. Quantum
       Electron. **22**, 818–819 (1992)]
2.15   D.A. Kleinman: Phys. Rev. **126**, 1977–1979 (1962)
2.16   V.G. Dmitriev, I.Ya. Itskhoki: "Optical Frequency Multipliers", in *Handbook of Lasers*,
       Vol. 2, ed. by A.M. Prokhorov (Sovetskoe Radio, Moscow 1978) pp. 292–319 (In Russian)
2.17   H. Ito, H. Inaba: IEEE J. QE-**8**, 612 (1972)
2.18   H. Ito, H. Naito, H. Inaba: IEEE J. QE-**10**, 247–252 (1974)
2.19   H. Ito, H. Naito, H. Inaba: J. Appl. Phys. **46**, 3992–3998 (1975)
2.20   K. Kato: Opt. Quant. Electron. **8**, 261–262 (1976)
2.21   O.I. Lavrovskaya, N.I. Pavlova, A.V. Tarasov: Kristallografiya **31**, 1145–1151 (1986)
       [English transl.: Sov. Phys. -Crystallogr. **31**, 678–682 (1978)]
2.22   V.G. Dmitriev, D.N. Nikogosyan: Opt. Commun. **95**, 173–182 (1992)
2.23   M.A. Dreger, J.H. Erkkila: Opt. Lett. **17**, 787–788 (1992)
2.24   J.F. Nye: *Physical Properties of Crystals* (Clarendon, Oxford 1957)
2.25   J.L. Oudar, R. Hierle: J. Appl. Phys. **48**, 2699–2704 (1977)
2.26   J.-M. Halbout, C.L. Tang: IEEE J. QE-**18**, 410–415 (1982)
2.27   D. Eimerl, S. Velsko, L. Davis, F. Wang, G. Loiacono, G. Kennedy: IEEE J. QE-**25**, 179–
       193 (1989)
2.28   I. Ledoux, C. Lepers, A. Perigaud, J. Badan, J. Zyss: Opt. Commun. **80**, 149–154 (1990)
2.29   D. Roberts: IEEE J. QE-**28**, 2057–2074 (1992)
2.30   A.P. Sukhorukov: *Nonlinear Wave Interactions in Optics and Radiophysics* (Nauka, Mos-
       cow 1988) (In Russian)
2.31   E. Yanke, F. Emde, F. Lösch: *Tafeln höherer Funktionen* (Teubner, Stuttgart 1960)
2.32   S.A. Akhmanov, V.A. Vysloukh, A.S. Chirkin: *Optics of Femtosecond Laser Pulses*
       (American Institute of Physics, New York 1992) [Translated from Russian: S.A. Akhma-
       nov, V.A. Vysloukh, A.S. Chirkin, *Optics of Femtosecond Laser Pulses* (Nauka, Moscow
       1988)]
2.33   R. Danelyus, A. Piskarskas, V. Sirutkaitis, A. Stabinis, Ya. Yasevichyute: *Optical Para-
       metric Oscillators and Picosecond Spectroscopy* (Mosklas, Vilnyus 1983) (In Russian)

2.34   S.L. Shapiro (ed.): *Ultrashort Light Pulses. Picosecond Technques and Applications*, Topics in Applied Physics, Vol. 18 (Springer-Verlag, Berlin, Heidelberg, 1977)

2.35   J. Herrmann, B. Wilhelmi: *Laser für ultrakurze Lichtimpulse. Grundlagen und Anwendungen* (Akademie, Berlin 1984)

2.36   J.D. Boyd, D.A. Kleinman: J. Appl. Phys. **39**, 3597–3639 (1968)

2.37   V.G. Dmitriev, L.A. Kulevskii: "Optical Parametric Oscillators", in *Handbook of Lasers*, Vol. 2, ed. by A.M. Prokhorov (Sovetskoe Radio, Moscow 1978), pp. 319–348 (In Russian)

2.38   D.S. Chemla, J. Zyss (eds.): *Nonlinear Optical Properties of Organic Molecules and Crystals*, Vols. 1, 2 (Academic, New York 1987)

2.39   I.S. Rez, Yu. M. Poplavko: *Dielectrics. Main Properties and Applications in Electronics* (Radio i Sviyaz, Moscow 1989) (In Russian)

Chapter 3

3.1   C. Chen, Y. Wu, A. Jiang, B. Wu, G. You, R. Li, S. Lin: J. Opt. Soc. Am. B **6**, 616–621 (1989)

3.2   K. Kato: IEEE J. **26**, 1173–1175 (1990)

3.3   S.P. Velsko, M. Webb, L. Davis, C. Huang: IEEE J. **27**, 2182–2192 (1991)

3.4   S. Lin, Z. Sun, B. Wu, C. Chen: J. Appl. Phys. **67**, 634–638 (1990)

3.5   S. Lin, B. Wu, F. Xie, C. Chen: Appl. Phys. Lett. **59**, 1541–1543 (1991)

3.6   B. Wu, F. Xie, C. Chen, D. Deng, Z. Xu: Opt. Commun. **88**, 451–454 (1992)

3.7   T. Ukachi, R.J. Lane, W.R. Bosenberg, C.L. Tang: Appl. Phys. Lett. **57**, 980–982 (1990)

3.8   T. Ukachi, R.J. Lane, W.R. Bosenberg, C.L. Tang: J. Opt. Soc. Am. B **9**, 1128–1133 (1992)

3.9   K. Kato: IEEE J. **26**, 2043–2045 (1990)

3.10   J.T. Lin, J.L. Montgomery, K. Kato: Opt. Commun. **80**, 159–165 (1990)

3.11   D.-W. Chen, J.T. Lin: IEEE J. **29**, 307–310 (1993)

3.12   Y. Wang, Z. Xu, D. Deng, W. Zheng, B. Wu, C. Chen: Appl. Phys. Lett. **59**, 531–533 (1991)

3.13   H.-J. Krause, W. Daum: Appl. Phys. Lett. **60**, 2180–2182 (1992)

3.14   J.Y. Zhang, J.Y. Huang, Y.R. Shen, C. Chen, B. Wu: Appl. Phys. Lett. **58**, 213–215 (1991)

3.15   W.S. Pelouch, T. Ukachi, E.S. Wachman, C.L. Tang: Appl. Phys. Lett. **57**, 111–113 (1990)

3.16   A. Borsutzky, R. Brunger, Ch. Huang, R. Wallenstein: Appl. Phys. B **52**, 55–62 (1991)

3.17   V.A. Dyakov, M.Kh. Dzhafarov, A.A. Lukashev, A.A. Podshivalov, V.I. Pryalkin: Kvantovaya Elektron. **18**, 339–341 (1991) [English transl.: Sov. J. Quantum Electron. **21**, 307–308 (1991)]

3.18   B. Wu, N. Chen, C. Chen, D. Deng, Z. Xu: Opt. Lett. **14**, 1080–1081 (1989)

3.19   F. Hanson, D. Dick: Opt. Lett. **16**, 205–207 (1991)

3.20   S. Lin, J.Y. Huang, J. Ling, C. Chen, Y.R. Shen: Appl. Phys. Lett. **59**, 2805–2807 (1991)

3.21   H. Mao, B. Wu, C. Chen, D. Zhang, P. Wang: Appl. Phys. Lett. **62**, 1866–1868 (1993)

3.22   G.C. Bhar, P.K. Datta, A.M. Rudra: Appl. Phys. B **57**, 431–434 (1993)

3.23   T. Schröder, K.-J. Boller, A. Fix, R. Wallenstein: Appl. Phys. B **58**, 425–438 (1994)

3.24   J.Y. Huang, Y.R. Shen, C. Chen, B. Wu: Appl. Phys. Lett. **58**, 1579–1581 (1991)

3.25   I. Gontijo: Opt. Commun. **108**, 324–328 (1994)

3.26   S.T. Yang, C.C. Pohalski, E.K. Gustafson, R.L. Byer, R.S. Feigelson, R.J. Raymakers, R.K. Route: Opt. Lett. **16**, 1493–1495 (1991)

3.27   G.J. Hall, A.I. Ferguson: Opt. Lett. **18**, 1511–1513 (1993)

3.28   G.P.A. Malcolm, M. Ebrahimzadeh, A.I. Ferguson: IEEE J. **28**, 1172–1178 (1992)

3.29   A. Robertson, A.I. Ferguson: Opt. Lett. **19**, 117–119 (1994)

3.30   S.D. Butterworth, M.J. McCarthy, D.C. Hanna: Opt. Lett. **18**, 1429–1431 (1993)

3.31   M.J. McCarthy, S.D. Butterworth, D.C. Hanna: Opt. Commun. **102**, 297–303 (1993)

3.32   M. Ebrahimzadeh, G.J. Hall, A.I. Ferguson: Appl. Phys. Lett. **60**, 1421–1423 (1992)

3.33   C. Chen: Laser Focus World **25**, No. **11**, 129–137 (1989)

3.34   G.A. Skripko, S.G. Bartoshevich, I.V. Mikhnyuk, I.G. Tarazevich: Opt. Lett. **16**, 1726–1728 (1991)

3.35   B.V. Bokut: Zh. Prikl. Spektrosk. **7**, 621–624 (1967) [English transl.: J. Appl. Spectrosc. **7**, 425–429 (1967)]

3.36   V.G. Dmitriev, D.N. Nikogosyan: Opt. Commun. **95**, 173–182 (1993)

3.37   D.A. Roberts: IEEE J. **28**, 2057–2074 (1992)

3.38   Y. Tang, Y. Cui, M.H. Dunn: Opt. Lett. **17**, 192–194 (1992)

3.39   G. Robertson, A. Henderson, M.H. Dunn: Appl. Phys. Lett. **60**, 271–273 (1992)

3.40   G. Robertson, A. Henderson, M. Dunn: Opt. Lett. **16**, 1584–1586 (1991)

3.41   M. Ebrahimzadeh, G. Robertson, M.H. Dunn: Opt. Lett. **16**, 767–769 (1991)

3.42   I.M. Bayanov, V.M. Gordienko, M.S. Djidjoev, V.A. Dyakov, S.A. Magnitskii, V.I. Pryalkin, A.P. Tarasevitch: Proc. SPIE **1800**, 2–17 (1991)

3.43   Y. Cui, M.H. Dunn, C.J. Norrie, W. Sibbett, B.D. Sinclair, Y. Tang, J.A.C. Terry: Opt. Lett. **17**, 646–648 (1992)

3.44   Y. Cui, D.E. Withers, C.F. Rae, C.J. Norrie, Y. Tang, B.D. Sinclair, W. Sibbett, M.H. Dunn: Opt. Lett. **18**, 122–124 (1993)

3.45   A. Fix, T. Schröder, R. Wallenstein: Laser und Optoelektronik **23**, No. **3**, 106–110 (1991)

3.46   F. Huang, L. Huang: Appl. Phys. Lett. **61**, 1769–1771 (1992)

3.47   F. Huang, L. Huang, B.-I. Yin, Y. Hua: Appl. Phys. Lett. **62**, 672–674 (1993)

3.48   H.-J. Krause, W. Daum: Appl. Phys. B **56**, 8–13 (1993)

3.49   F.G. Colville, A.J. Henderson, M.J. Padgett, J. Zhang, M.H. Dunn: Opt. Lett. **18**, 205–207 (1993)

3.50   M. Ebrahimzadeh, G.J. Hall, A.I. Ferguson: Opt. Lett. **17**, 652–654 (1992)

3.51   F. Hanson, P. Poirier: Opt. Lett. **19**, 1526–1528 (1994)

3.52   H. Zhou, J. Zhang, T. Chen, C. Chen, Y.R. Shen: Appl. Phys. Lett. **62**, 1457–1459 (1993)

3.53   G.P. Banfi, R. Danielius, A. Piskarskas, P.Di Trapani, P. Foggi, R. Righini: Opt. Lett. **18**, 1633–1635 (1993)

3.54   S.A. Akhmanov, I.M. Bayanov, V.M. Gordienko, V.A. Dyakov, S.A. Magnitskii, V.I. Pryalkin, A.P. Tarasevitch: "Parametric generation of femtosecond pulses by LBO crystal in the near IR", in *Ultrafast Processes in Spectroscopy 1991*, Inst. Phys. Conf. Ser. No. 126, ed. by A. Laubereau, A. Seilmeier (IOP, Bristol, 1992) pp. 67–70

3.55   V.M. Gordienko, S.A. Magnitskii, A.P. Tarasevitch: "Injection-locked femtosecond parametric oscillators on LBO crystal; towards 10^{17} W cm^{-2}", in *Frontiers in Nonlinear Optics. The Sergei Akhmanov Memorial Volume*, ed. by H. Walther, N. Koroteev, M.O. Scully (IOP, Bristol, 1993) pp. 286–292

3.56   A. Nebel, R. Beigang: Opt. Lett. **16**, 1729–1731 (1991)

3.57   F. Xie, B. Wu, G. You, C. Chen: Opt. Lett. **16**, 1237–1239 (1991)

3.58   J.D. Beasley: Appl. Opt. **33**, 1000–1003 (1994)

3.59   E.M. Voronkova, B.N. Grechushnikov, G.I. Distler, I.P. Petrov: *Optical Materials for Infrared Technique* (Nauka, Moscow 1965) (In Russian)

3.60   W.L. Smith: Appl. Opt. **16**, 798 (1977)

3.61   A.G. Akmanov, S.A. Akhmanov, B.V. Zhdanov, A.I. Kovrigin, N.K. Podsotskaya, R.V. Khokhlov: Pisma Zh. Eksp. Teor. Fiz. **10**, 244–249 (1969) [English transl].: JETP Lett. **10**, 154–156 (1969)]

3.62   M.W. Dowley, E.B. Hodges: IEEE J. QE-4, 552–558 (1968)

3.63   E.F. Labuda, A.M. Johnson: IEEE J. QE-3, 164–167 (1967)

3.64   A.A. Blistanov, V.S. Bondarenko, N.V. Perelomova, F.N. Strizhevskaya, V.V. Chkalova, M.P. Shaskolskaya: *Acoustic Crystals* (Nauka, Moscow 1982) (In Russian)

3.65   P.J. Wegner, M.A. Henesian, D.R. Speck, C. Bibeau, R.B. Ehrlich, C.W. Laumann, J.K. Lawson, T.L. Weiland: Appl. Opt. **31**, 6414–6426 (1992)

3.66   A. Yokotani, T. Sasaki, K. Yoshida, S. Nakai: Appl. Phys. Lett. **55**, 2692–2693 (1989)

3.67   E.N. Volkova, V.V. Fadeev: In *Nonlinear Optics*, ed. by R.V. Khokhlov (Nauka, Novosibirsk, 1968) pp. 185–187 (In Russian)

3.68   G. Dikchyus, E. Zhilinskas, A. Piskarskas, V. Sirutkaitis: Kvantovaya Elektron. **6**, 1610–1619 (1979) [English transl.: Sov. J. Quantum Electron. **9**, 950–955 (1979)]

3.69   E.E. Fill: Opt. Commun. **33**, 321–322 (1980)

3.70   G.G. Gurzadyan, R.K. Ispiryan: Appl. Phys. Lett. **59**, 630–631 (1991)

3.71   P. Liu, W.L. Smith, H. Lotem, J.H. Bechtel, N. Bloembergen, R.S. Adhav: Phys. Rev. B **17**, 4620–4632 (1978)

3.72   G.J. Linford, B.C. Johnson, J.S. Hildum, W.E. Martin, K. Snyder, R.D. Boyd, W.L. Smith, C.L. Vercimak, D. Eimerl, J.T. Hunt: Appl. Opt. **21**, 3633–3643 (1982)

3.73   F. Zernike, Jr.: J. Opt. Soc. Am. **54**, 1215–1220 (1964)

3.74   N.P. Barnes, D.J. Gettemy, R.S. Adhav: J. Opt. Soc. Am. **72**, 895–898 (1982)

3.75   R.A. Philips: J. Opt. Soc. Am. **56**, 629–632 (1966)

3.76   M. Yamazaki, T. Ogawa: J. Opt. Soc. Am. **56**, 1407–1408 (1966)

3.77   G.C. Ghosh, G.C. Bhar: IEEE J. **QE-18**, 143–145 (1982)

3.78   D. Eimerl: Ferroelectrics **72**, 95–139 (1987)

3.79   V.S. Suvorov, A.S. Sonin: Kristallografiya **11**, 832–848 (1966) [English transl.: Sov. Phys. - Crystallogr. **11**, 711–723 (1966)]

3.80   F.M. Johnson, J.A. Duardo: Laser Focus **3**, No. 6, 31–37 (1967)

3.81   W.F. Hagen, P.C. Magnante: J. Appl. Phys. **40**, 219–224 (1969)

3.82   A.P. Sukhorukov, I.V. Tomov: Opt. Spektrosk. **28**, 1211–1213 (1970) [English transl.: Opt. Spectrosc. USSR **28**, 651–653 (1970)]

3.83   Y. Takagi, M. Sumitani, N. Nakashima, K. Yoshihara: IEEE J. **QE-21**, 193–195 (1985)

3.84   G.A. Massey, J.C. Johnson: IEEE J. **QE-12**, 721–727 (1976)

3.85   M. Okada, S. Ieiri: Jpn. J. Appl. Phys. **10**, 808 (1971)

3.86   E. Fill, J. Wildenauer: Opt. Commun. **47**, 412–413 (1983)

3.87   V.I. Bredikhin, V.N. Genkin, S.P. Kuznetsov, M.A. Novikov: Pisma Zh. Tekh. Fiz. **3**, 407–409 (1977) [English transl.: Sov. Tech. Phys. Lett. **3**, 165–166 (1977)]

3.88   V.I. Bredikhin, G.L. Gaushkina, V.N. Genkin, S.P. Kuznetsov: Pisma Zh. Tekh. Fiz. **5**, 505–508 (1979) [English transl.: Sov. Tech. Phys. Lett. **5**, 207–208 (1977)]

3.89   M.D. Jones, G.A. Massey: IEEE J. **QE-15**, 204–206 (1979)

3.90   G.A. Massey, M.D. Jones, J.C. Johnson: IEEE J. **QE-14**, 527–532 (1978)

3.91   A. Ashkin, G.D. Boyd, J.M. Dziedzic: Phys. Rev. Lett. **11**, 14–17 (1963)

3.92   R.C. Eckardt, H. Masuda, Y.X. Fan, R.L. Byer: IEEE J. **26**, 922–933 (1990)

3.93   M. Webb: IEEE J. **30**, 1934–1942 (1994)

3.94   U. Deserno, S. Haussühl: IEEE J. **QE-9**, 598–601 (1973)

3.95   R.S. Craxton, S.D. Jacobs, J.E. Rizzo, R. Boni: IEEE J. **QE-17**, 1782–1786 (1981)

3.96   R.B. Andreev, V.D. Volosov, A.G. Kalintsev: Opt. Spektrosk. **37**, 294–299 (1974) [English transl.: Opt. Spectrosc. USSR **37**, 169–171 (1974)]

3.97   R.B. Andreev, V.D. Volosov, V.N. Krylov: Zh. Tekh. Fiz. **47**, 1977–1978 (1977) [English transl.: Sov. Phys. - Tech. Phys. **22**, 1146 (1977)]

3.98   A. Yokotani, T. Sasaki, T. Yamanaka, C. Yamanaka: Jpn. J. Appl. Phys. **25**, 161–162 (1986)

3.99   M.W. Dowley: Opto-electron. **1**, 179–181 (1969)

3.100   J.E. Midwinter, J. Warner: Brit. J. Appl. Phys. **16**, 1135–1142 (1965)

3.101   L. Armstrong, S.E. Neister, R. Adhav: Laser Focus **18**, No. 12, 49–53 (1982)

3.102   B.F. Bareika, I.A. Begishev, Sh. A. Burdulis, A.A. Gulamov, E.A. Erofeev, A.S. Piskarskas, V.A. Sirutkaitis, T. Usmanov: Pisma Zh. Tekh. Fiz. **12**, 186–189 (1988) [English transl.: Sov. Tech. Phys. Lett. **12**, 78–79 (1988)]

3.103   V.D. Volosov, V.N. Krylov, V.A. Serebryakov, D.V. Sokolov: Pisma Zh. Eksp. Teor. Fiz. **19**, 38–41 (1974) [English transl.: JETP Lett. **19**, 23–25 (1974)]

3.104 K.P. Burneika, M.V. Ignatavichyus, V.I. Kabelka, A.S. Piskarskas, A.Yu. Stabinis: Pisma Zh. Eksp. Teor. Fiz. **16**, 365–367 (1972) [English transl.: JETP Lett. **16**, 257–258 (1972)]

3.105 V. Kabelka, A. Kutka, A. Piskarskas, V. Smilgiavichyus, Ya. Yasevichyute: Kvantovaya Elektron. **6**, 1735–1739 (1979) [English transl.: Sov. J. Quantum Electron. **9**, 1022–1024 (1979)]

3.106 V.D. Volosov, Yu.E. Kamach, E.N. Kozlovsky, V.M. Ovchinnikov: Opt. Mekh. Promyshl. **36**, No. **10**, 3–4 (1969) [English transl.: Sov. J. Opt. Technol. **36**, 656–657 (1969)]

3.107 Y. Nishida, A. Yokotani, T. Sasaki, K. Yoshida, T. Yamanaka, C. Yamanaka: Appl. Phys. Lett. **52**, 420–421 (1988)

3.108 W. Seka, S.D. Jacobs, J.E. Rizzo, R. Boni, R.S. Craxton: Opt. Commun. **34**, 469–473 (1980)

3.109 V.D. Volosov, E.V. Nilov: Opt. Spektrosk. **21**, 715–719 (1966) [English transl.: Opt. Spectrosc. USSR **21**, 392–394 (1966)]

3.110 S.A. Akhmanov, I.A. Begishev, A.A. Gulamov, E.A. Erofeev, B.V. Zhdanov, V.I. Kuznetsov, L.N. Rashkovich, T.V. Usmanov: Kvantovaya Elektron.: **11**, 1701–1702 (1984) [English transl.: Sov. J. Quantum Electron. **14**, 1145–1146 (1984)]

3.111 J.E. Swain, S.E. Stokowski, D. Milam, G.C. Kennedy: Appl. Phys. Lett. **41**, 12–14 (1982)

3.112 D. Eimerl, S. Velsko, L. Davis, F. Wang, G. Loiacono, G. Kennedy: IEEE J. QE-**25**, 179–193 (1989)

3.113 A.S. Sonin, A.S. Vasilevskaya: *Electrooptic Crystals* (Atomizdat, Moscow 1971) (In Russian)

3.114 T.R. Sliker, S.R. Burlage: J. Appl. Phys. **34**, 1837–1840 (1963)

3.115 J. Reintjes, R.C. Eckardt: IEEE J. QE-**13**, 791–793 (1977)

3.116 J.P. Machewirth, R. Webb, D. Anafi: Laser Focus: **12**, No. **5**, 104–107 (1976)

3.117 G. Brederlow, E. Fill, K.J. Witte: *The High-Power Iodine Laser*, Springer Ser. Opt. Sci., Vol. 34 (Springer, Berlin, Heidelberg, 1983)

3.118 J. Reintjes, R.C. Eckardt: Appl. Phys. Lett. **30**, 91–93 (1977)

3.119 R.S. Adhav: Laser Focus **19**, No. **6**, 73–78 (1983)

3.120 T.A. Rabson, H.J. Ruiz, P.L. Shah, F.K. Tittel: Appl. Phys. Lett. **20**, 282–284 (1972)

3.121 P.E. Perkins, T.S. Fahlen: IEEE J. QE-**21**, 1636–1638 (1985)

3.122 Y.S. Liu, W.B. Jones, J.P. Chernoch: Appl. Phys. Lett. **29**, 32–34 (1976)

3.123 R.M. Kogan, T.G. Crow: Appl. Opt. **17**, 927–930 (1978)

3.124 R.S. Adhav, S.R. Adhav, J.M. Pelaprat: Laser Focus **23**, No. **9**, 88–100 (1987)

3.125 H. Nakatani, W.R. Bosenberg, L.K. Cheng, C.L. Tang: Appl. Phys. Lett. **53**, 2587–2589 (1988)

3.126 A.S. Vasilevskaya, E.N. Volkova, V.A. Koptsik, L.N. Rashkovich, T.A. Regulskaya, I.S. Rez, A.S. Sonin, V.S. Suvorov: Kristallografiya **12**, 518–519 (1967) [English transl.: Sov. Phys. - Crystallogr. **12**, 446 (1967)]

3.127 B.V. Zhdanov, V.V. Kalitin, A.I. Kovrigin, S.M. Pershin: Pisma Zh. Tekh. Fiz. **1**, 847–851 (1975) [English transl.: Sov. Tech. Phys. Lett. **1**, 368–369 (1975)]

3.128 Y.P. Kim, M.H.R. Hutchinson: Appl. Phys. B **49**, 469–478 (1989)

3.129 F. Zernike, Jr.: J. Opt. Soc. Am. **55**, 210–211 (1965)

3.130 F. Wondrazek, A. Seilmeier, W. Kaiser: Appl. Phys. B **32**, 39–42 (1983)

3.131 R.E. Stickel, Jr., F.B. Dunning: Appl. Opt. **17**, 1313–1314 (1978)

3.132 R.K. Jain, T.K. Gustafson: IEEE J. QE-**12**, 555–556 (1976)

3.133 R.K. Jain, T.K. Gustafson: IEEE J. QE-**9**, 859–861 (1973)

3.134 R.W. Wallace: Opt. Commun. **4**, 316–318 (1971)

3.135 G.V. Venkin, L.L. Kulyuk, D.I. Maleev: Kvantovaya Elektron.: **2**, 2475–2480 (1975) [English transl.: Sov. J. Quantum Electron. **5**, 1348–1351 (1975)]

3.136 B.G. Huth, Y.C. Kiang: J. Appl. Phys. **40**, 4976–4977 (1969)

3.137 D.P. Schinke: IEEE J. QE-**8**, 86–87 (1972)

3.138 A.H. Kung: Appl. Phys. Lett. **25**, 653–655 (1974)

3.139 K. Kato: Opt. Commun. **13**, 361–362 (1975)
3.140 J.M. Yarborough, G.A. Massey: Appl. Phys. Lett. **18**, 438–440 (1971)
3.141 G.A. Massey: Appl. Phys. Lett. **24**, 371–373 (1974)
3.142 T. Sato: J. Appl. Phys. **44**, 2257–2259 (1973)
3.143 C. Chen, B. Wu, A. Jiang, G. You: Scientia Sinica, Ser. B **28**, 235–243 (1985)
3.144 L.J. Bromley, A. Guy, D.C. Hanna: Opt. Commun. **67**, 316–320 (1988)
3.145 K. Kato: IEEE J. QE-**22**, 1013–1014 (1986)
3.146 Y.X. Fan, R.C. Eckardt, R.L. Byer, C. Chen, A.D. Jiang: IEEE J. **25**, 1196–1199 (1989)
3.147 Y.X. Fan, R.C. Eckardt, R.L. Byer, J. Nolting, R. Wallenstein: Appl. Phys. Lett. **53**, 2014–2016 (1988)
3.148 D. Eimerl, L. Davis, S. Velsko, E.K. Graham, A. Zalkin: J. Appl. Phys. **62**, 1968–1983 (1987)
3.149 G.C. Bhar, S. Das, U. Chatterjee: Appl. Opt. **28**, 202–204 (1989)
3.150 K. Miyazaki, H. Sakai, T. Sato: Opt. Lett. **11**, 797–799 (1986)
3.151 X. Xinan, Y. Shuzhong: Chinese Phys. – Lasers **13**, 892–894 (1986)
3.152 L.K. Cheng, W.R. Bosenberg, C.L. Tang: Appl. Phys. Lett. **53**, 175–177 (1988)
3.153 W.R. Bosenberg, L.K. Cheng, C.L. Tang: Appl. Phys. Lett. **54**, 13–15 (1989)
3.154 C. Chen, Y.X. Fan, R.C. Eckardt, R.L. Byer: Proc. SPIE **681**, 12–19 (1987)
3.155 H.J. Müschenborn, W. Theiss, W. Demtröder: Appl. Phys. B **50**, 365–369 (1990)
3.156 M. Ebrahimzadeh, A.J. Henderson, M.H. Dunn: IEEE J. **26**, 1241–1252 (1990)
3.157 J.Y. Huang, J.Y. Zhang, Y.R. Shen, C. Chen, B. Wu: Appl. Phys. Lett. **57**, 1961–1963 (1990)
3.158 J.Y. Zhang, J.Y. Huang, Y.R. Shen, C. Chen: J. Opt. Soc. Am. B **10**, 1758–1764 (1993)
3.159 H. Vanherzeele, C. Chen: Appl. Opt. **27**, 2634–2636 (1988)
3.160 H. Komine: J. Opt. Soc. Am. B **10**, 1751–1757 (1993)
3.161 G.C. Bhar, S. Das, U. Chatterjee: J. Appl. Phys. **66**, 5111–5113 (1989)
3.162 G.C. Bhar, S. Das, U. Chatterjee: Appl. Phys. Lett. **54**, 1383–1384 (1989)
3.163 X.D. Zhu, L. Deng: Appl. Phys. Lett. **61**, 1490–1492 (1992)
3.164 M. Watanabe, K. Hayasaka, H. Imajo, J. Umezu, S. Urabe: Appl. Phys. B **53**, 11–13 (1991)
3.165 I.V. Tomov, T. Anderson, P.M. Rentzepis: Appl. Phys. Lett. **61**, 1157–1159 (1992)
3.166 W.L. Glab, J.P. Hessler: Appl. Opt. **26**, 3181–3182 (1987)
3.167 U. Heitmann, M. Kötteritzsch, S. Heitz, A. Hese: Appl. Phys. B **55**, 419–423 (1992)
3.168 D.W. Coutts, M.D. Ainsworth, J.A. Piper: IEEE J. **25**, 1985–1987 (1989)
3.169 S. Lu, Y. Yuan, Y. Tang, W. Xu, C. Wu: "Mixing Frequency Generation of 271.0–291.5 nm in β-BaB$_2$O$_4$", in *Proceedings of the Topical Meeting on Laser Materials and Laser Spectroscopy*, ed. by Z. Wang, Z. Zhang (World Scientific, Singapore 1989) pp. 77–79
3.170 M. -H. Lu, Y. -M. Liu: Opt. Commun. **84**, 193–198 (1991)
3.171 K. Kurokawa, M. Nakazawa: Appl. Phys. Lett. **55**, 7–9 (1989)
3.172 G.C. Bhar, S. Das, U. Chatterjee: J. Phys. D **22**, 562–563 (1989)
3.173 H. Komine: Opt. Lett. **13**, 643–645 (1988)
3.174 K. Kuroda, T. Omatsu, T. Shimura, M. Chihara, I. Ogura: Opt. Commun. **75**, 42–46 (1990)
3.175 D.W. Coutts, J.A. Piper: IEEE J. **28**, 1761–1764 (1992)
3.176 D.W. Coutts, M.D. Ainsworth, J.A. Piper: IEEE J. **26**, 1555–1558 (1990)
3.177 Y. Taira: Jpn. J. Appl. Phys. **31**, L682–L684 (1992)
3.178 G.G. Gurzadyan, A.S. Oganesyan, A.V. Petrosyan, R.O. Sharkhatunyan: Zh. Tekh. Fiz. **61**, 152–154 (1991) [English transl.: Sov. Phys. -Tech. Phys. **36**, 341–342 (1991)]
3.179 W. Joosen, H.J. Bakker, L.D. Noordam, H.G. Muller, H.B. van Linden van den Heuvell: J. Opt. Soc. Am. B **8**, 2087–2093 (1991)
3.180 T.R. Zhang, H.R. Choo, M.C. Downer: Appl. Opt. **29**, 3927–3933 (1990)
3.181 P. Qiu, A. Penzkofer: Appl. Phys. B **45**, 225–236 (1988)
3.182 G.D. Hager, S.A. Hanes, M.A. Dreger: IEEE J. **28**, 2573–2576 (1992)
3.183 G. Nath, S. Haussühl: Appl. Phys. Lett. **14**, 154–156 (1969)

3.184   F.R. Nash, J.G. Bergman, G.D. Boyd, E.H. Turner: J. Appl. Phys. **40**, 5201–5206 (1969)

3.185   G. Nath, H. Mehmanesch, M. Gsänger: Appl. Phys. Lett. **17**, 286–288 (1970)

3.186   D.J. Gettemy, W.C. Harker, G. Lindholm, N.P. Barnes: IEEE J. **24**, 2231–2237 (1988)

3.187   J. Jerphagnon: Appl. Phys. Lett. **16**, 298–299 (1970)

3.188   N.M. Bityurin, V.I. Bredikhin, V.N. Genkin: Kvantovaya Elektron. **5**, 2453–2457 (1978) [English transl.: Sov. J. Quantum Electron. **8**, 1377–1379 (1978)]

3.189   K. Takizawa, M. Okada, S. Ieiri: Opt. Commun. **23**, 279–281 (1977)

3.190   S. Umegaki, S. I. Tanaka, T. Uchiyama, S. Yabumoto: Opt. Commun. **3**, 244–245 (1971)

3.191   J.M. Crettez, J. Comte, E. Coquet: Opt. Commun. **6**, 26–29 (1972)

3.192   M.M. Choy, R.L. Byer: Phys. Rev. B **14**, 1693–1706 (1976)

3.193   Z.B. Perekalina, G.F. Dobrzhansky, I.A. Spilko: Kristallografiya **15**, 1252–1253 (1970) [English transl.: Sov. Phys. – Crystallogr. **15**, 1095 (1970)]

3.194   V.A. Kizel, V.I. Burkov: *Gyrotropy of Crystals* (Nauka, Moscow 1980) (In Russian)

3.195   I.M. Beterov, V.I. Stroganov, V.I. Trunov, B.Ya. Yurshin: Kvantovaya Elektron. **2**, 2440–2443 (1975) [English transl.: Sov. J. Quantum Electron. **5**, 1329–1331 (1975)]

3.196   H. Buesener, A. Renn, M. Brieger, F. von Moers, A. Hese: Appl. Phys. B **39**, 77–81 (1986)

3.197   G. Nath, S. Haussühl: Phys. Lett. **29** A, 91–92 (1969)

3.198   J.E. Pearson, G.A. Evans, A. Yariv: Opt. Commun. **4**, 366–367 (1972)

3.199   A.I. Izrailenko, A.I. Kovrigin, P.V. Nikles: Pisma Zh. Eksp. Teor. Fiz. **12**, 475–478 (1970) [English transl.: JETP Lett. **12**, 331–333 (1970)]

3.200   M. Okada, S. Ieiri: Phys. Lett. **34** A, 63–64 (1971)

3.201   R.B. Chesler, M.A. Karr, J.E. Geusic: J. Appl. Phys. **41**, 4125–4127 (1970)

3.202   K. Kato: IEEE J. QE-**21**, 119–120 (1985)

3.203   A.J. Campillo, C.L. Tang: Appl. Phys. Lett. **19**, 36–38 (1971)

3.204   A.J. Campillo: IEEE J. QE-**8**, 809–811 (1972)

3.205   D.W. Meltzer, L.S. Goldberg: Opt. Commun. **5**, 209–211 (1972)

3.206   L.S. Goldberg: Appl. Phys. Lett. **17**, 489–491 (1970)

3.207   T.M. Jedju, L. Rothberg: Appl. Opt. **27**, 615–618 (1988)

3.208   F. Huisken, A. Kulcke, D. Voelkel, C. Laush, J.M. Lisy: Appl. Phys. Lett. **62**, 805–807 (1993)

3.209   G. Nath, G. Pauli: Appl. Phys. Lett. **22**, 75–76 (1973)

3.210   D. Malz, J. Bergmann, J. Heise: Exp. Techn. Phys. **23**, 379–388 (1975)

3.211   Y.C. See, J. Falk: Appl. Phys. Lett. **36**, 503–505 (1980)

3.212   D. Malz, J. Bergmann, J. Heise: Exp. Techn. Phys. **23**, 495–498 (1975)

3.213   S.G. Karpenko, N.E. Kornienko, V.L. Strizhevskii: Kvantovaya Elektron. **1**, 1768–1779 (1974) [English transl.: Sov. J. Quantum Electron. **4**, 979–985 (1974)]

3.214   V.I. Kabelka, V.G. Kolomiets, A.S. Piskarskas, A.Yu. Stabinis: Zh. Prikl. Spektrosk. **21**, 947–950 (1974) [English transl.: J. Appl. Spectrosc. **21**, 582–585 (1974)]

3.215   V.I. Kabelka, A.S. Piskarskas, A.Yu. Stabinis, R.L. Sher: Kvantovaya Elektron. **2**, 434–436 (1975) [English transl.: Sov. J Quantum Electron. **5**, 255–246 (1975)]

3.216   B.I. Kidyarov, I.V. Nikolaev, E.V. Pestryakov, V.M. Tarasov: Izv. Akad. Nauk, Ser. Fiz. **58**, 131–134 (1994) [English transl.: Bull. Acad. Sci., Phys. Ser. **58**, No. **2** (1994)]

3.217   M. Webb, S.P. Velsko: IEEE J. **26**, 1394–1398 (1990)

3.218   H. Gerlach: Opt. Commun. **12**, 405–408 (1974)

3.219   R.B. Andreev, V.D. Volosov, V.N. Krylov: Pisma Zh. Tekh. Fiz. **4**, 256–258 (1978) [English transl.: Sov. Tech. Phys. Lett. **4**, 105–106 (1978)]

3.220   E.W. Van Stryland, W.E. Williams, M.J. Soileau, A.L. Smirl: IEEE J. QE-**20**, 434–439 (1984)

3.221   A. Arutunyan, G. Arzumanyan, R. Danielius, V. Kabelka, R. Sharkhatunyan, Ya. Yasevichyute: Litovskii Fizicheskii Sbornik **18**, 255–263 (1978) (In Russian)

3.222   R. Danielius, G. Dikchyus, V. Kabelka, A. Piskarskas, A. Stabinis, Ya. Yasevichyute: Kvantovaya Elektron. **4**, 2379–2395 (1977) [English transl. : Sov. J. Quantum Electron. **7**, 1360–1368 (1977)]

3.223 V.G. Dmitriev, V.N. Krasnyanskaya, M.F. Koldobskaya, I.S. Rez, E.A. Shalaev, E.M. Shvom: Kvantovaya Elektron. No. 2 (14), 64–66 (1973) [English transl. : Sov. J. Quantum Electron. **3**, 126–127 (1973)]
3.224 A. Koeneke, A. Hirth: Opt. Commun. **34**, 245–248 (1980)
3.225 T.Y. Fan, C.E. Huang, B.Q. Hu, R.C. Eckardt, Y.X. Fan, R.L. Byer, R.S. Feigelson: Appl. Opt. **26**, 2390–2394 (1987)
3.226 Y.S. Liu, D. Dentz, R. Belt: Opt. Lett. **9**, 76–78 (1984)
3.227 D.N. Dovchenko, V.A. Dyakov, V.I. Pryalkin: Izv. Akad. Nauk SSSR, Ser. Fiz. **52**, 225–230 (1988) [English transl. : Bull. Acad. Sci. USSR, Phys. Ser. **52**, No. 2, 13–17 (1988)]
3.228 J.C. Jacco: Proc. SPIE **968**, 93–99 (1988)
3.229 R.F. Belt, G. Gashurov, Y.S. Liu: Laser Focus **21**, No. **10**, 110–124 (1985)
3.230 Y. Kitaoka, T. Sasaki, S. Nakai, Y. Goto: Appl. Phys. Lett. **59**, 19–21 (1991)
3.231 A.L. Aleksandrovskii, S.A. Akhmanov, V.A. Dyakov, N.I. Zheludev, V.I. Pryalkin: Kvantovaya Elektron. **12**, 1333–1334 (1985) [English transl. : Sov. J. Quantum Electron. **15**, 885–886 (1985)]
3.232 K. Kato: IEEE J. **27**, 1137–1140 (1991)
3.233 P.F. Bordui, R. Blachman, R.G. Norwood: Appl. Phys. Lett. **61**, 1369–1371 (1992)
3.234 P.E. Perkins, T.S. Fahlen: J. Opt. Soc. Am. B **4**, 1066–1071 (1987)
3.235 J.D. Bierlein, H. Vanherzeele: J. Opt. Soc. Am. B **6**, 622–633 (1989)
3.236 H.Y. Shen, Y.P. Zhou, W.X. Lin, Z.D. Zeng, R.R. Zeng, G.F. Yu, C.H. Huang, A.D. Jiang, S.Q. Jia, D.Z. Shen: IEEE J. **28**, 48–51 (1992)
3.237 K. Kato: IEEE J. **28**, 1974–1976 (1992)
3.238 D.W. Anthon, C.D. Crowder: Appl. Opt. **27**, 2650–2652 (1988)
3.239 H. Vanherzeele, J.D. Bierlein, F.C. Zumsteg: Appl. Opt. **27**, 3314–3316 (1988)
3.240 H. Vanherzeele: Appl. Opt. **27**, 3608–3615 (1988)
3.241 R.A. Stolzenberger, C.C. Hsu, N. Peyghambarian, J.J.E. Reid, R.A. Morgan: IEEE Photon. Technol. Lett. **1**, 446–448 (1989)
3.242 V.A. Dyakov, V.V. Krasnikov, V.I. Pryalkin, M.S. Pshenichnikov, T.B. Razumikhina, V.S. Solomatin, A.I. Kholodnykh: Kvantovaya Elektron. **15**, 1703–1704 (1988) [English transl. : Sov. J. Quantum Electron. **18**, 1059–1060 (1988)]
3.243 O.I. Lavrovskaya, N.I. Pavlova, A.V. Tarasov: Kristallografiya **31**, 1145–1151 (1986) [English transl. : Sov. Phys. – Crystallogr. **31**, 678–682 (1986)]
3.244 R.A. Stolzenberger: Appl. Opt. **27**, 3883–3886 (1988)
3.245 L.J. Bromley, A. Guy, D.C. Hanna: Opt. Commun. **70**, 350–354 (1989)
3.246 Yu.A. Galaichuk, V.A. Dyakov, N.I. Likholit, V.S. Ovechko, R.A. Petrenko, T.V. Rozhdestvenskaya, V.L. Strizhevskii, A.I. Khilchevskii, Yu.N. Yashkir: Izv. Akad. Nauk SSSR, Ser. Fiz. **52**, 560–563 (1988) [English transl. : Bull. Akad. Sci. USSR, Phys. Ser. **52**, No. 3, 131–133 (1988)]
3.247 T. Nishikawa, N. Uesugi, H. Ito: Appl. Phys. Lett. **55**, 1943–1945 (1989)
3.248 V.M. Garmash, G.A. Ermakov, N.I. Pavlova, A.V. Tarasov: Pisma Zh. Tekh. Fiz. **12**, 1222–1225 (1986) [English transl. : Sov. Tech. Phys. Lett. **12**, 505–506 (1986)]
3.249 R. Burnham, R.A. Stolzenberger, A. Pinto: IEEE. Photon. Technol. Lett. **1**, 27–28 (1989)
3.250 W. Wang, K. Nakagawa, Y. Toda, M. Ohtsu: Appl. Phys. Lett. **61**, 1886–1888 (1992)
3.251 J.T. Lin, J.L. Montgomery: Opt. Commun. **75**, 315–320 (1990)
3.252 W.X. Lin, H.Y. Shen, Y.P. Zhou, R.R. Zeng, G.F. Yu, C.H. Huang, Z.D. Zeng, W.J. Zhang: Opt. Commun. **82**, 333–336 (1991)
3.253 W. Wang, M. Ohtsu: Opt. Commun. **102**, 304–308 (1993)
3.254 K. Kato: IEEE J. QE-**24**, 3–4 (1988)
3.255 H. Liao, H. Shen, T. Lian, Y. Zhou, C. Huang, R. Zeng, G. Yu: Optics and Laser Technology **20**, 103–104 (1988)
3.256 W.P. Risk, R.N. Payne, W. Lenth, C. Harder, H. Meier: Appl. Phys. Lett. **55**, 1179–1181 (1989)

3.257   J.-C. Baumert, F.M. Schellenberg, W. Lenth, W.P. Risk, G.C. Bjorklund: Appl. Phys. Lett. **51**, 2192–2194 (1987)

3.258   Z.Y. Ou, S.F. Pereira, E.S. Polzik, H.J. Kimble: Opt. Lett. **17**, 640–642 (1992)

3.259   S.T. Yang, R.C. Eckardt, R.L. Byer: J. Opt. Soc. Am. B **10**, 1684–1695 (1993)

3.260   S.T. Yang, R.C. Eckardt, R.L. Byer: Opt. Lett. **18**, 971–973 (1993)

3.261   K. Kato, M. Masutani: Opt. Lett. **17**, 178–179 (1992)

3.262   J.A.C. Terry, Y. Cui, Y. Yang, W. Sibbett, M.H. Dunn: J. Opt. Soc. Am. B **11**, 758–769 (1994)

3.263   G.I. Dyakonov, V.A. Maslov, V.A. Mikhailov, S.K. Pak, V.N. Semenenko, I.A. Shcherbakov: Kvantovaya Elektron. **16**, 1601–1603 (1989) [English transl. : Sov. J. Quantum Electron. **19**, 1031–1032 (1989)]

3.264   J.-J. Zondy, M. Abed, A. Clairon: J. Opt. Soc. Am. B **11**, 2004–2015 (1994)

3.265   B. Boulanger, J.P. Feve, G. Marnier, B. Menaert, X. Cabirol, P. Villeval, C. Bonnin: J. Opt. Soc. Am. B **11**, 750–757 (1994)

3.266   S.E. Moody, J.M. Eggleston, J.F. Seamans: IEEE J. QE-**23**, 335–340 (1987)

3.267   T.A. Driscoll, H.J. Hoffman, R.E. Stone, P.E. Perkins: J. Opt. Soc. Am. B **3**, 683–686 (1986)

3.268   F.C. Zumsteg, J.D. Bierlein, T.E. Gier: J. Appl. Phys. **47**, 4980–4985 (1976)

3.269   F. Ahmed: Appl. Opt. **28**, 119–122 (1989)

3.270   P. Yankov, D. Schumov, A. Nenov, A. Monev : Opt. Lett. **18**, 1771–1773 (1993)

3.271   J.C. Jacco, D.R. Rockafellow, E.A. Teppo: Opt. Lett. **16**, 1307–1309 (1991)

3.272   R.J. Bolt, M. van der Mooren: Opt. Commun. **100**, 399–410 (1993)

3.273   L.G. Van Uitert, J.J. Rubin, W.A. Bonner: IEEE J. QE-**4**, 622–627 (1968)

3.274   G.D. Boyd, R.C. Miller, K. Nassau, W.L. Bond, A. Savage: Appl. Phys. Lett. **5**, 234–236 (1964)

3.275   G.V. Ageev, R.P. Bashuk, A.S. Bebchuk, N.S. Voidetskaya, D.A. Gromov, Yu.N. Solovieva, A.V. Chesnokov: In *Nonlinear Optics*, ed. by R.V. Khokhlov (Nauka, Novosibirsk 1968) pp. 211–217 (In Russian)

3.276   Y.C. See, S. Guha, J. Falk: Appl. Opt. **19**, 1415–1418 (1980)

3.277   D.H. Jundt, M.M. Fejer, R.L. Byer, R.G. Norwood, P.F. Bordui: Opt. Lett. **16**, 1856–1858 (1991)

3.278   A. Seilmeier, W. Kaiser: Appl. Phys. **23**, 113–119 (1980)

3.279   D. von der Linde, A.M. Glass, K.F. Rodgers: Appl. Phys. Lett. **25**, 155–157 (1974)

3.280   D.H. Jundt, M.M. Fejer, R.L. Byer: IEEE J. **26**, 135–138 (1990)

3.281   D.S. Smith, H.D. Riccius, R.P. Edwin: Opt. Commun. **17**, 332–335 (1976)

3.282   D.F. Nelson, R.M. Mikulyak: J. Appl. Phys. **45**, 3688–3689 (1974)

3.283   J.E. Midwinter: J. Appl. Phys. **39**, 3033–3038 (1968)

3.284   M.V. Hobden, J. Warner: Phys. Lett. **22**, 243–244 (1966)

3.285   G.J. Edwards, M. Lawrence: Opt. Quant. Electron. **16**, 373–375 (1984)

3.286   A.M. Prokhorov, Yu.S. Kuzminov: *Physics and Chemistry of Crystalline Lithium Niobate* (Hilger, Bristol, 1990)

3.287   D.S. Moore, S.C. Schmidt: Opt. Lett. **12**, 480–482 (1987)

3.288   A. Laubereau, L. Greiter, W. Kaiser: Appl. Phys. Lett. **25**, 87–89 (1974)

3.289   R.L. Herbst, R.N. Fleming, R.L. Byer: Appl. Phys. Lett. **25**, 520–522 (1974)

3.290   Z.I. Ivanova, V. Kabelka, S.A. Magnitskii, A. Piskarskas, V. Smilgiavichyus, N.M. Rubinina, V.G. Tunkin: Kvantovaya Elektron. **4**, 2469–2472 (1977) [English transl. : Sov. J. Quantum Electron. **7**, 1414–1416 (1977)]

3.291   K. Kato: IEEE J. QE-**16**, 1017–1018 (1980)

3.292   P.M. Bridenbaugh, J.R. Carruthers, J.M. Dziedzic, F.R. Nash: Appl. Phys. Lett. **17**, 104–106 (1970)

3.293   R.C. Miller, G.D. Boyd, A. Savage: Appl. Phys. Lett. **6**, 77-79 (1965)

3.294   H. Fay, W.J. Alfred, H.M. Dess: Appl. Phys. Lett. **12**, 89–92 (1968)

3.295   N.B. Angert, O.F. Butyagin, V.P. Zorenko, A.P. Kudryavtseva, V.R. Kushnir, S.R. Rustamov: Kvantovaya Elektron. No. **5**, 128–129 (1971) [English transl. : Sov. J. Quantum Electron. **1**, 542–543 (1971)]

3.296   J.C. Bergman, A. Ashkin, A.A. Ballman, J.M. Dziedzic, H.J. Levinstein, R.G. Smith: Appl. Phys. Lett. **12**, 92–94 (1968)

3.297   R.L. Byer, J.F. Young, R.S. Feigelson: J. Appl. Phys. **41**, 2320–2325 (1970)

3.298   T.R. Volk, N.M. Rubinina, A.I. Kholodnykh: Kvantovaya Elektron. **15**, 1705–1706 (1988) [English transl. : Sov. J. Quantum Electron. **18**, 1061–1062 (1988)]

3.299   F.R. Nash, G.D. Boyd, M. Sargent III, P.M. Bridenbaugh: J. Appl. Phys. **41**, 2564–2576 (1970)

3.300   V.A. Dyakov, V.I. Pryalkin, A.I. Kholodnykh: Kvantovaya Elektron. **8**, 715–721 (1981) [English transl. : Sov. J. Quantum Electron. **11**, 433–436 (1981)]

3.301   J.E. Midwinter, J. Warner: J. Appl. Phys. **38**, 519–523 (1967)

3.302   E.N. Antonov, V.G. Koloshnikov, D.N. Nikogosyan: Opt. Spektrosk. **36**, 768–722 (1974) [English transl. : Opt. Spectrosc. USSR **36**, 446–448 (1974)]

3.303   R.C. Miller, W.A. Nordland, P.M. Bridenbaugh: J. Appl. Phys. **42**, 4145–4147 (1971)

3.304   T. Kushida, Y. Tanaka, M. Ojima, Y. Nakazaki: Jpn. J. Appl. Phys. **14**, 1097–1098 (1975)

3.305   M. Berg, C.B. Harris, T.W. Kenny, P.L. Richards: Appl. Phys. Lett. **47**, 206–208 (1985)

3.306   J. Falk, J.E. Murray: Appl. Phys. Lett. **14**, 245–247 (1969)

3.307   G.M. Zverev, E.A. Levchuk, V.A. Pashkov, Yu.D. Poryadin: Kvantovaya Elektron. No. 2(8), 94–96 (1972) [English transl. : Sov. J. Quantum Electron. **2**, 167–169 (1972)]

3.308   G.M. Zverev, S.A. Kolyadin, E.A. Levchuk, L.A. Skvortsov: Kvantovaya Elektron. **4**, 1882–1889 (1977) [English transl. : Sov. J. Quantum Electron. **7**, 1071–1075 (1977)]

3.309   S.J. Brosnan, R.L. Byer: IEEE J. QE-15, 415–431 (1979)

3.310   M.J. Soileau: Appl. Opt. **20**, 1030–1033 (1981)

3.311   Y. Uematsu: Jpn. J. Appl. Phys. **13**, 1362–1368 (1974)

3.312   B. Zysset, I. Biaggio, P. Günter: J. Opt. Soc. Am. B **9**, 380–386 (1992)

3.313   W.R. Bosenberg, R.H. Jarman: Appl. Phys. Lett. **18**, 1323–1325 (1993)

3.314   K. Kato: IEEE J. QE-15, 410–411 (1979)

3.315   J.J.E. Reid: Appl. Phys. Lett. **62**, 19–21 (1993)

3.316   Y. Uematsu, T. Fukuda: Jpn. J. Appl. Phys. **12**, 841–844 (1973)

3.317   J.-C. Baumert, J. Hoffnagle, P. Günter: Proc. SPIE **492**, 374–385 (1984)

3.318   W.P. Risk, R.Pon, W. Lenth: Appl. Phys. Lett. **54**, 1625–1627 (1989)

3.319   Y. Lu, Q. Zhao, Y. Li, H. He, Q. Zou, Z. Lu, Z. Geng: Optical Engineering **32**, 713–716 (1993)

3.320   Y. Uematsu, T. Fukuda: Jpn. J. Appl. Phys. **4**, 507 (1971)

3.321   E. Wiesendanger: Ferroelectrics **1**, 141–148 (1970)

3.322   C. Zimmermann, T.W. Hänsch, R. Byer, S. O'Brien, D. Welch: Appl. Phys. Lett. **61**, 2741–2743 (1992)

3.323   I. Biaggio, P. Kerkoc, L.-S. Wu, P. Günter, B. Zysset: J. Opt. Soc. Am. **9**, 507–517 (1992)

3.324   P. Günter: Appl. Phys. Lett. **34**, 650–652 (1979)

3.325   W. Seelert, P. Kortz, D. Rytz, B. Zysset, D. Ellgehausen, G. Mizell: Opt. Lett. **17**, 1432–1434 (1992)

3.326   K. Kato: IEEE J. QE-18, 451–452 (1982)

3.327   I. Biaggio, H. Looser, P. Günter: Ferroelectrics **94**, 157–161 (1989)

3.328   J.-C. Baumert, P. Günter, H. Melchior: Opt. Commun. **48**, 215–220 (1983)

3.329   A. Hemmerich, D.H. McIntyre, C. Zimmermann, T.W. Hänsch: Opt. Lett. **15**, 372–374 (1990)

3.330   M.K. Chun, L. Goldberg, J.F. Weller: Appl. Phys. Lett. **53**, 1170–1171 (1988)

3.331   W.J. Kozlovsky, W. Lenth, E.E. Latta, A. Moser, G.L. Bona: Appl. Phys. Lett. **56**, 2291–2292 (1990)

3.332   J.-C. Baumert, J. Hoffnagle, P. Günter: Appl. Opt. **24**, 1299–1301 (1985)

3.333   P. Günter, P.M. Asbeck, S.K. Kurtz: Appl. Phys. Lett. **35**, 461–463 (1979)

3.334   L. Goldberg, L. Busse, D. Mehuys: Appl. Phys. Lett. **60**, 1037–1039 (1992)

3.335   J.-C. Baumert, P. Günter: Appl. Phys. Lett. **50**, 554–556 (1987)

3.336 D.H. Jundt, P. Günter, B. Zysset: Nonlinear Opt. **4**, 341–345 (1993)

3.337 U. Ellenberger, R. Weber, J.E. Balmer, B. Zysset, D. Ellgehausen, G.D. Mizell: Appl. Opt. **31**, 7563–7569 (1992)

3.338 *Physical-Chemical Properties of Semiconductors: Handbook.* (Nauka, Moscow 1979) (In Russian)

3.339 V.V. Badikov, O.N. Pivovarov, Yu.V. Skokov, O.V. Skrebneva, N.K. Trotsenko: Kvantovaya Elektron. **2**, 618–621 (1975) [English transl. : Sov. J. Quantum Electron. **5**, 350–351 (1975)]

3.340 E.S. Voronin, V.S. Solomatin, N.I. Cherepov, V.V. Shuvalov, V.V. Badikov, O.N. Pivovarov: Kvantovaya Elektron. **2**, 1090–1092 (1975) [English transl. : Sov. J. Quantum Electron. **5**, 597–598 (1975)]

3.341 P. Canarelli, Z. Benko, R. Curl, F.K. Tittel: J. Opt. Soc. Am. B **9**, 197–202 (1992)

3.342 A.H. Hielscher, C.E. Miller, D.C. Bayard, U. Simon, K.P. Smolka, R.F. Curl, F.K. Tittel: J. Opt. Soc. Am. B **9**, 1962–1967 (1992)

3.343 H. Matthes, R. Viehmann, N. Marschall: Appl. Phys. Lett. **26**, 237–239 (1975)

3.344 G.D. Boyd, H. Kasper, J.H. McFee: IEEE J. QE-7, 563–573 (1971)

3.345 V.V. Badikov, I.N. Matveev, S.M. Pshenichnikov, O.V. Skrebneva, N.K. Trotsenko, N.D. Ustinov: Kristallografiya **26**, 537–539 (1981) [English transl. : Sov. Phys. – Crystallogr. **26**, 304–305 (1981)]

3.346 G.C. Bhar, D.K. Ghosh, P.S. Ghosh, D. Schmitt: Appl. Opt. **22**, 2492–2494 (1983)

3.347 P. Canarelli, Z. Benko, A.H. Hielscher, R.F. Curl, F.K. Tittel: IEEE J. **28**, 52–55 (1992)

3.348 G.C. Bhar, R.C. Smith: IEEE J. QE-10, 546–550 (1974)

3.349 T. Itabe, J.L. Bufton: Appl. Opt. **23**, 3044–3047 (1984)

3.350 Y.X. Fan, R.C. Eckardt, R.L. Byer, R.K. Route, R.S. Feigelson: Appl. Phys. Lett. **45**, 313–315 (1984)

3.351 P.J. Kupecek, C.A. Schwartz, D.S. Chemla: IEEE J. QE-10, 540–545 (1974)

3.352 D.S. Chemla, P.J. Kupecek, D.S. Robertson, R.C. Smith: Opt. Commun. **3**, 29–31 (1971)

3.353 T. Elsaesser, A. Seilmeier, W. Kaiser, P. Koidl, G. Brandt: Appl. Phys. Lett. **44**, 383–385 (1984)

3.354 H.J. Bakker, J.T.M. Kennis, H.J. Kop, A. Lagendijk: Opt. Commun. **86**, 58–64 (1991)

3.355 T. Elsaesser, H. Lobentanzer, A. Seilmeier: Opt. Commun. **52**, 355–359 (1985)

3.356 A.G. Yodh, H.W.K. Tom, G.D. Aumiller, R.S. Miranda: J. Opt. Soc. Am. B **8**, 1663–1667 (1991)

3.357 D.C. Hanna, V.V. Rampal, R.C. Smith: Opt. Commun. **8**, 151–153 (1973)

3.358 D.C. Hanna, V.V. Rampal, R.C. Smith: IEEE J. QE-10, 461–462 (1974)

3.359 K. Kato: IEEE J. QE-20, 698–699 (1984)

3.360 G.C. Bhar, S. Das, U. Chatterjee, R.S. Feigelson, R.K. Route: Appl. Phys. Lett. **54**, 1489–1491 (1989)

3.361 S.A. Andreev, I.N. Matveev, I.P. Nekrasov, S.M. Pshenichnikov, N.P. Sopina: Kvantovaya Elektron. **4**, 657–659 (1977) [English transl. : Sov. J. Quantum Electron. **7**, 366–368 (1977)]

3.362 G.C. Bhar: Appl. Opt. **15**, 305–307 (1976)

3.363 W. Jantz, P. Koidl: Appl. Phys. Lett. **31**, 99–101 (1977)

3.364 A.P. Gorchakov, A.A. Popesku, V.S. Solomatin: Kvantovaya Elektron. **5**, 413–415 (1978) [English transl. : Sov. J. Quantum Electron. **8**, 236–237 (1978)]

3.365 D.C. Hanna, B. Luther-Davies, H.N. Rutt, R.C. Smith, C.R. Stanley: IEEE J. QE-8, 317–324 (1972)

3.366 K.G. Spears, X. Zhu, X. Yang, L. Wang: Opt. Commun. **66**, 167–171 (1988)

3.367 T. Dahinten, U. Plödereder, A. Seilmeier, K.L. Vodopyanov, K.R. Allakhverdiev, Z.A. Ibragimov: IEEE J. **29**, 2245–2250 (1993)

3.368 H. Kildal, G.W. Iseler: Appl. Opt. **15**, 3062–3065 (1976)

3.369 G.D. Boyd, E. Buehler, F.G. Storz: Appl. Phys. Lett. **18**, 301–304 (1971)

3.370   K.L. Vodopyanov: J. Opt. Soc. Am. B **10**, 1723–1729 (1993)
3.371   Yu.M. Andreev, S.D. Velikanov, A.S. Elutin, A.F. Zapolskii, D.V. Konkin, S.N. Mikshin, S.V. Smirnov, Yu.N. Frolov, V.V. Shchurov: Kvantovaya Elektron. **19**, 1110 (1992) [English transl. : Sov. J. Quantum Electron. **22**, 1035 (1987)]
3.372   Yu.M. Andreev, V.G. Voevodin, P.P. Geiko, A.I. Gribenyukov, V.V. Zuev, A.S. Solodukhin, S.A. Trushin, V.V. Churakov, S.F. Shubin: Kvantovaya Elektron. **14**, 2137–2138 (1987) [English transl. : Sov. J. Quantum Electron. **17**, 1362–1363 (1987)]
3.373   Yu.M. Andreev, V.Yu. Baranov, V.G. Voevodin, P.P. Geiko, A.I. Gribenyukov, S.V. Izyumov, S.M. Kozochkin, V.D. Pismenny, Yu.A. Satov, A.P. Streltsov: Kvantovaya Elektron. **14**, 2252–2254 (1987) [English transl. : Sov. J. Quantum Electron. **17**, 1435–1436 (1987)]
3.374   K.L. Vodopyanov, V.G. Voevodin, A.I. Gribenyukov, L.A. Kulevskii: Izv. Akad. Nauk SSSR, Ser. Fiz. **49**, 569–572 (1985) [English transl. : Bull. Acad. Sci. USSR, Phys. Ser. **49**, No. **3**, 146–149 (1985)]
3.375   Yu.M. Andreev, V.G. Voevodin, A.I. Gribenyukov, O.Ya. Zyryanov, I.I. Ippolitov, A.N. Morozov, A.V. Sosnin, G.S. Khmelnitskii: Kvantovaya Elektron. **11**, 1511–1512 (1984) [English transl. : Sov. J. Quantum Electron. **14**, 1021–1022 (1984)]
3.376   K.L. Vodopyanov, V.G. Voevodin, A.I. Gribenyukov, L.A. Kulevskii: Kvantovaya Elektron. **14**, 1815–1819 (1987) [English transl. : Sov. J. Quantum Electron. **17**, 1159–1161 (1987)]
3.377   Yu.M. Andreev, A.D. Belykh, V.G. Voevodin, P.P. Geiko, A.I. Gribenyukov, V.A. Gurashvili, S.V. Izyumov: Kvantovaya Elektron. **14**, 782–783 (1987) [English transl. : Sov. J. Quantum Electron. **17**, 490–491 (1987)]
3.378   Yu.M. Andreev, V.G. Voevodin, A.I. Gribenyukov, V.P. Novikov: Kvantovaya Elektron. **14**, 1177–1179 (1987) [English transl. : Sov. J. Quantum Electron. **17**, 748–749 (1987)]
3.379   G.D. Boyd, W.B. Gandrud, E. Buehler: Appl. Phys. Lett. **18**, 446–448 (1971)
3.380   V.E. Zuev, M.V. Kabanov, Yu.M. Andreev, V.G. Voevodin, P.P. Geiko, A.I. Gribenyukov, V.V. Zuev: Izv. Akad. Nauk SSSR, Ser. Fiz. **52**, 1142–1148 (1988) [English transl. : Bull. Acad. Sci. USSR, Phys. Ser. **52**, No. **6**, 87–92 (1988)]
3.381   A.A. Barykin, S.V. Davydov, V.P. Dorokhov, V.P. Zakharov, V.V. Butuzov: Kvantovaya Elektron. **20**, 794–800 (1993) [English transl. : Quantum Electron. **23**, 688–693 (1993)]
3.382   Yu.M. Andreev, A.N. Bykanov, A.I. Gribenyukov, V.V. Zuev, V.D. Karyshev, A.V. Kisletsov, I.O. Kovalev, V.I. Konov, G.P. Kuzmin, A.A. Nesterenko, A.E. Osorgin, Yu.M. Starodumov, N.I. Chapliev: Kvantovaya Elektron. **17**, 476–480 (1990) [English transl.: Sov. J. Quantum Electron. **20**, 410–414 (1990)]
3.383   P.D. Mason, D.J. Jackson, E.K. Gorton: Opt. Commun. **110**, 163–166 (1994)
3.384   G.B. Abdullaev, K.R. Allakhverdiev, M.E. Karasev, V.I. Konov, L.A. Kulevskii, N.B. Mustafaev, P.P. Pashinin, A.M. Prokhorov, Yu.M. Starodumov, N.I. Chapliev: Kvantovaya Elektron. **16**, 757–763 (1989) [English transl. : Sov. J. Quantum Electron. **19**, 494–498 (1989)]
3.385   G.C. Bhar, L.K. Samanta, D.K. Ghosh, S. Das: Kvantovaya Elektron. No. **14**, 1361–1363 (1987) [English transl. : Sov. J. Quantum Electron. **17**, 860–861 (1987)]
3.386   Yu.M. Andreev, T.V. Vedernikova, A.A. Betin, V.G. Voevodin, A.I. Gribenyukov, O.Ya. Zyryanov, I.I. Ippolitov, V.I. Masychev, O.V. Mitropolskii, V.P. Novikov, M.A. Novikov, A.V. Sosnin: Kvantovaya Elektron. **12**, 1535–1537 (1985) [English transl. : Sov. J. Quantum Electron. **15**, 1014–1015 (1985)]
3.387   G.C. Bhar, S. Das, U. Chatterjee, K.L. Vodopyanov: Appl. Phys. Lett. **54**, 313–314 (1989)
3.388   K.L. Vodopyanov, L.A. Kulevskii, V.G. Voevodin, A.I. Gribenyukov, K.R. Allakhverdiev, T.A. Kerimov: Opt. Commun. **83**, 322–326 (1991)
3.389   K.L. Vodopyanov, Yu.A. Andreev, G.C. Bhar: Kvantovaya Elektron. **20**, 879–881 (1993) [English transl. : Quantum Electron. **23**, 763–765 (1993)]
3.390   G.C. Bhar, G.C. Ghosh: IEEE J. QE-16, 838–843 (1980)
3.391   G.C. Bhar, G. Ghosh: J. Opt. Soc. Am. **69**, 730–733 (1979)

3.392 N.P. Andreeva, S.A. Andreev, I.N. Matveev, S.M. Pshenichnikov, N.D. Ustinov: Kvantovaya Elektron. **6**, 357–359 (1979) [English transl. : Sov. J. Quantum Electron. **9**, 208–210 (1979)]
3.393 K. Kato: Appl. Phys. Lett. **29**, 562–563 (1976)
3.394 W.R. Cook, H. Jaffe: Acta Crystallogr. **10**, 705–707 (1957)
3.395 J.A. Paisner, M.L. Spaeth, D.C. Gerstenberger, I.W. Ruderman: Appl. Phys. Lett. **32**, 476–478 (1978)
3.396 K.B. Petrosyan, A.L. Pogosyan, K.M. Pokhsraryan: Izv. Akad. Nauk SSSR, Ser. Fiz. **47**, 1619–1621 (1983) [English transl. : Bull. Acad. Sci. USSR, Phys. Ser. **47**, No. **8**, 155–157 (1983)]
3.397 K. Kato: Opt. Commun. **19**, 332–333 (1976)
3.398 K. Kato: IEEE J. QE-**13**, 544–546 (1977)
3.399 G.G. Gurzadyan, R.K. Ispiryan: Int. J. Nonl. Opt. Phys. **1**, 533–540 (1992)
3.400 H. Zacharias, A. Anders, J.B. Halpern, K.H. Welge: Opt. Commun. **19**, 116–119 (1976)
3.401 W.R. Cook, L.M. Hubby: J. Opt. Soc. Am. **66**, 72-73 (1976)
3.402 F.B. Dunning, R.E. Stickel, Jr. : Appl. Opt. **15**, 3131–3134 (1976)
3.403 H.J. Dewey: IEEE J. QE-**12**, 303–306 (1976)
3.404 R.E. Stickel, Jr., S. Blit, G.F. Hildenbrandt, E.D. Dahl, F.B. Dunning, F.K. Tittel: Appl. Opt. **17**, 2270 (1978)
3.405 C.F. Dewey, Jr., W.R. Cook, Jr., R.T. Hodgson, J.J. Wynne: Appl. Phys. Lett. **26**, 714–716 (1975)
3.406 R.E. Stickel, Jr., F.B. Dunning: Appl. Opt. **16**, 2356–2358 (1977)
3.407 K. Kato: Appl. Phys. Lett. **30**, 583–584 (1977)
3.408 H. Hemmati, J.C. Bergquist, W.M. Itano: Appl. Opt. **8**, 73–75 (1983)
3.409 R.E. Stickel, Jr., F.B. Dunning: Appl. Opt. **17**, 981–982 (1978)
3.410 A.G. Arutyunyan, V.G. Atanesyan, K.B. Petrosyan, K.M. Pokhsraryan: Pisma Zh. Tekh. Fiz. **6**, 277–280 (1980) [English transl. : Sov. Tech. Phys. Lett. **6**, 120–121 (1980)]
3.411 K. Kato: IEEE J. QE-**16**, 810–811 (1980)
3.412 D. Bauerle, K. Betzler, H. Hesse, S. Kapphan, P. Loose: Phys. Status Solidi A **42**, K119–K121 (1977)
3.413 K. Betzler, H. Hesse, P. Loose: J. Mol. Struct. **47**, 393–396 (1978)
3.414 J.-M. Halbout, S. Blit, W. Donaldson, C.L. Tang: IEEE J. QE-**15**, 1176–1180 (1979)
3.415 W.R. Donaldson, C.L. Tang: Appl. Phys. Lett. **44**, 25–27 (1984)
3.416 M.J. Rosker, K. Cheng, C.L. Tang: IEEE J. QE-**21**, 1600-1606 (1985)
3.417 M. Ebrahimzadeh, M.H. Dunn, F. Akerboom: Opt. Lett. **14**, 560–562 (1989)
3.418 M. Ebrahimzadeh, M.H. Dunn: Opt. Commun. **69**, 161–165 (1988)
3.419 M.J. Rosker, C.L. Tang: J. Opt. Soc. Am. B **2**, 691–696 (1985)
3.420 C. Cassidy, J.M. Halbout, W. Donaldson, C.L. Tang: Opt. Commun. **29**, 243–246 (1979)
3.421 M.J. Rosker: Proc. SPIE **681**, 10–11 (1986)
3.422 K. Kato: IEEE J. QE-**10**, 616–618 (1974)
3.423 V.S. Suvorov, I.S. Rez: Opt. Spektrosk. **27**, 181–183 (1969) [English transl. : Opt. Spectrosc. USSR **27**, 94–95 (1969)]
3.424 Yu.D. Golyaev, V.G. Dmitriev, I.Ya. Itskhoki, V.N. Krasnyanskaya, I.S. Rez, E.A. Shalaev: Kvantovaya Elektron. No. 1 (13), 122–123 (1973) [English transl. : Sov. J. Quantum Electron. **3**, 72–73 (1973)]
3.425 R.S. Adhav, A.D. Vlassopoulos: Laser Focus **10**, No. **5**, 47–48 (1974)
3.426 K.V. Vetrov, V.D. Volosov, A.G. Kalintsev: Izv. Akad. Nauk SSSR, Ser. Fiz. **52**, 301–303 (1988) [English transl. : Bull. Acad. Sci. USSR, Ser. Phys. **52**, No. **2**, 78–79 (1988)]
3.427 K. Kato: Opt. Commun. **9**, 249–251 (1973)
3.428 R.S. Adhav, R.W. Wallace: IEEE J. QE-**9**, 855–856 (1973)
3.429 G.A. Massey, R.A. Elliot: IEEE J. QE-**10**, 899–900 (1974)
3.430 J.D. Bierlein, H. Vanherzeele, A.A. Ballman: Appl. Phys. Lett. **54**, 783–785 (1989)
3.431 A.H. Kung: Appl. Phys. Lett. **65**, 1082–1084 (1994)
3.432 W.R. Bosenberg, L.K. Cheng, J.D. Bierlein: Appl. Phys. Lett. **65**, 2765–2767 (1994)

3.433   L.K. Cheng, L.T. Cheng, J.D. Bierlein, F.C. Zumsteg, A.A. Ballman: Appl. Phys. Lett. **62**, 346–348 (1993)
3.434   K. Kato: IEEE J. **30**, 881–883 (1994)
3.435   A.L. Aleksandrovskii, G.I. Ershova, G.Kh. Kitaeva, S.P. Kulik, I.I. Naumova, V.V. Tarasenko: Kvantovaya Elektron. **18**, 254–256 (1991) [English transl. : Sov. J. Quantum Electron. **21**, 225–227 (1991)]
3.436   Y. Chang, J. Wen, H. Wang, B. Li: Chinese Phys. Lett. **9**, 427–430 (1992)
3.437   J.L. Nightingale, W.J. Silva, G.E. Reade, A. Rybicki, W.J. Kozlovsky, R.L. Byer: Proc. SPIE **681**, 20–24 (1986)
3.438   W.J. Kozlovsky, C.D. Nabors, R.L. Byer: IEEE J. **24**, 913–919 (1988)
3.439   G.T. Maker, A.I. Ferguson: Opt. Lett. **15**, 375–377 (1990)
3.440   W.J. Kozlovsky, C.D. Nabors, R.C. Eckardt, R.L. Byer: Opt. Lett. **14**, 66–68 (1989)
3.441   C.D. Nabors, R.C. Eckardt, W.J. Kozlovsky, R.L. Byer: Opt. Lett. **14**, 1134–1136 (1989)
3.442   D.C. Gerstenberger, G.E. Tye, R.W. Wallace: Opt. Lett. **16**, 992–994 (1991)
3.443   D.C. Gerstenberger, R.W. Wallace: J. Opt. Soc. Am. B **10**, 1681–1683 (1993)
3.444   G.T. Maker, A.I. Ferguson: Opt. Commun. **76**, 369–375 (1990)
3.445   K.F. Hulme, O. Jones, P.H. Davies, M.V. Hobden: Appl. Phys. Lett. **10**, 133–135 (1967)
3.446   N. Ito: Opt. Lett. **7**, 63–65 (1982)
3.447   E.N. Antonov, V.R. Mironenko, D.N. Nikogosyan, M.I. Golovey: Kvantovaya Elektron. **1**, 1742–1746 (1974) [English transl. : Sov. J. Quantum Electron. **4**, 963–965 (1974)]
3.448   D.S. Hanna, A.J. Turner: Opt. Quant. Electron. **8**, 213–217 (1976)
3.449   V.V. Berezovskii, Yu.A. Bykovskii, S.N. Potanin, I.S. Rez: Kvantovaya Elektron. No. **2** (**14**), 74–75 (1973) [English transl. : Sov. J. Quantum Electron. **3**, 134–135 (1973)]
3.450   D.N. Nikogosyan, A.P. Sukhorukov, M.I. Golovey: Kvantovaya Elektron. **2**, 609–612 (1975) [English transl. : Sov. J. Quantum Electron. **5**, 344–346 (1975)]
3.451   V.V. Berezovskii, Yu.A. Bykovskii, M.I. Goncharov, I.S. Rez: Kvantovaya Elektron. No. **2** (**8**), 105–107 (1972) [English transl. : Sov. J. Quantum Electron. **2**, 180–182 (1972)]
3.452   G.J. Ernst, W.J. Witteman: IEEE J. QE-**8**, 382–383 (1972)
3.453   L.O. Hocker, C.F. Dewey: Appl. Phys. **11**, 137–140 (1976)
3.454   N.P. Barnes, R.C. Eckardt, D.J. Gettemy, L.B. Edgett: IEEE J. QE-**15**, 1074–1076 (1979)
3.455   D.S. Chemla, Ph.J. Kupecek, C.A. Schwartz: Opt. Commun. **7**, 225–228 (1973)
3.456   D. Cotter, D.C. Hanna, B. Luther-Davies, R.C. Smith: Opt. Commun. **11**, 54–56 (1974)
3.457   M.V. Hobden: Opto-electron. **1**, 159 (1969)
3.458   R.A. Andrews: IEEE J. QE-**6**, 68–80 (1970)
3.459   Yu.A. Gorokhov, D.P. Krindach, D.N. Nikogosyan, A.P. Sukhorukov: Kvantovaya Elektron. **1**, 679–683 (1974) [English transl. : Sov. J. Quantum Electron. **4**, 382–384 (1974)]
3.460   D.C. Hanna, B. Luther-Davies, H.N. Rutt, R.C. Smith: Appl. Phys. Lett. **20**, 34–36 (1972)
3.461   T. Elsaesser, A. Seilmeier, W. Kaiser: Opt. Commun. **44**, 293–296 (1983)
3.462   J. Falk, J.M. Yarborough: Appl. Phys. Lett. **19**, 68–70 (1971)
3.463   D.N. Nikogosyan: Kvantovaya Elektron. **2**, 2524–2525 (1975) [English transl. : Sov. J. Quantum Electron. **5**, 1378–1379 (1975)]
3.464   E.K. Pfitzer, H.D. Riccius, K.J. Siemsen: Opt. Commun. **3**, 277–278 (1971)
3.465   H.D. Riccius, K.J. Siemsen: Phys. Lett. **45 A**, 377–378 (1973)
3.466   A.J. Alcock, A.C. Walker: Appl. Phys. Lett. **23**, 467–468 (1973)
3.467   J. Warner: Appl. Phys. Lett. **12**, 222–224 (1968)
3.468   G.C. Bhar, D.C. Hanna, B. Luther-Davies, R.C. Smith: Opt. Commun. **6**, 323–326 (1972)
3.469   A.F. Milton: Appl. Opt. **11**, 2311–2330 (1972)
3.470   G.B. Abdullaev, L.A. Kulevskii, A.M. Prokhorov, A.D. Saveliev, E.Yu. Salaev, V.V. Smirnov: Pisma Zh. Eksp. Teor. Fiz. **16**, 130–133 (1972) [English transl. : JETP Lett. **16**, 90–92 (1972)]
3.471   G.B. Abdullaev, L.A. Kulevskii, P.V. Nikles, A.M. Prokhorov, A.D. Saveliev, E.Yu. Salaev, V.V. Smirnov: Kvantovaya Elektron. **3**, 163–167 (1976) [English transl. : Sov. J. Quantum Electron. **6**, 88–90 (1976)]
3.472   A. Bianchi, A. Ferrario, M. Musci: Opt. Commun. **25**, 256–258 (1978)

3.473 G.B. Abdullaev, K.R. Allakhverdiev, L.A. Kulevskii, A.M. Prokhorov, E.Yu. Salaev, A.D. Saveliev, V.V. Smirnov: Kvantovaya Elektron. **2**, 1228–1233 (1975) [English transl. : Sov. J. Quantum Electron. **5**, 665–668 (1975)]

3.474 A. Bianchi, M. Garbi: Opt. Commun. **30**, 122–124 (1979)

3.475 K.L. Vodopyanov, L.A. Kulevskii: Opt. Commun. **118**, 375–378 (1995)

3.476 Yu.A. Gusev, A.V. Kirpichnikov, S.N. Konoplin, S.I. Marennikov, P.V. Nikles, Yu.N. Polivanov, A.M. Prokhorov, A.D. Saveliev, R.Sh. Sayakhov, V.V. Smirnov, V.P. Chebotaev: Pisma Zh. Tekh. Fiz. **6**, 1262–1265 (1980) [English transl. : Sov. Tech. Phys. Lett. **6**, 541–542 (1980)]

3.477 R.L. Byer, M.M. Choy, R.L. Herbst, D.S. Chemla, R.S. Feigelson: Appl. Phys. Lett. **24**, 65–68 (1974)

3.478 N.P. Barnes, D.J. Gettemy, J.R. Hietanen, R.A. Iannini: Appl. Opt. **28**, 5162–5168 (1989)

3.479 V.V. Badikov, V.B. Laptev, V.L. Panyutin, E.A. Ryabov, G.S. Shevyrdyaeva, O.B. Scherbina: Kvantovaya Elektron. **19**, 782–784 (1992) [English transl. : Sov. J. Quantum Electron. **22**, 722–724 (1992)]

3.480 G.C. Catella, L.R. Shiozawa, J.R. Hietanen, R.C. Eckardt, R.K. Route, R.S. Feigelson, D.G. Cooper, C.L. Marquardt: Appl. Opt. **32**, 3948–3951 (1993)

3.481 P.A. Budni, M.G. Knights, E.P. Chicklis, K.L. Schepler: Opt. Lett. **18**, 1068–1070 (1993)

3.482 U. Simon, Z. Benko, M.W. Sigrist, R.F. Curl, F.K. Tittel: Appl. Opt. **32**, 6650–6655 (1993)

3.483 R.C. Eckardt, Y.X. Fan, R.L. Byer, C.L. Marquardt, M.E. Storm, L. Esterowitz: Appl. Phys. Lett. **49**, 608–610 (1986)

3.484 C.L. Marquardt, D.G. Cooper, P.A. Budni, M.G. Knights, K.L. Schepler, R. DeDomenico, G.C. Catella: Appl. Opt. **33**, 3192–3197 (1994)

3.485 B.C. Ziegler, K.L. Schepler: Appl. Opt. **30**, 5077–5080 (1991)

3.486 R.C. Eckardt, Y.X. Fan, R.L. Byer, R.K. Route, R.S. Feigelson, J. van der Laan: Appl. Phys. Lett. **47**, 786–788 (1985)

3.487 G.D. Boyd, H.M. Kasper, J.H. McFee, F.G. Storz: IEEE J. QE-8, 900–908 (1972)

3.488 H. Kildal, J.C. Mikkelsen: Opt. Commun. **9**, 315–318 (1973)

3.489 D.A. Russell, R. Ebert: Appl. Opt. **32**, 6638–6644 (1993)

3.490 A.A. Davydov, L.A. Kulevskii, A.M. Prokhorov, A.D. Saveliev, V.V. Smirnov: Pisma Zh. Eksp. Teor. Fiz. **15**, 725–727 (1972) [English transl. : JETP Lett. **15**, 513–514 (1972)]

3.491 A. Ferrario, M. Garbi: Opt. Commun. **17**, 158–159 (1976)

3.492 J.A. Weiss, L.S. Goldberg: Appl. Phys. Lett. **24**, 389–391 (1974)

3.493 R.G. Wenzel, G.P. Arnold: Appl. Opt. **15**, 1322–1326 (1976)

3.494 R.L. Herbst, R.L. Byer: Appl. Phys. Lett. **19**, 527–530 (1971)

3.495 F. Bryukner, V.S. Dneprovskii, V.U. Khattatov: Kvantovaya Elektron. **1**, 1360–1364 (1974) [English transl. : Sov. J. Quantum Electron. **4**, 749–751 (1974)]

3.496 J.M. Ralston, R.K. Chang: Opto-electron. **1**, 182–188 (1969)

3.497 W.L. Bond: J. Appl. Phys. **36**, 1674–1677 (1965)

3.498 D. Andreou: Opt. Commun. **23**, 37–43 (1977)

3.499 A.A. Davydov, L.A. Kulevskii, A.M. Prokhorov, A.D. Saveliev, V.V. Smirnov, A.V. Shirkov: Opt. Commun. **9**, 234–236 (1973)

3.500 R.L. Herbst, R.L. Byer: Appl. Phys. Lett. **21**, 189–191 (1972)

3.501 R.L. Byer, H. Kildal, R.S. Feigelson: Appl. Phys. Lett. **19**, 237–240 (1971)

3.502 D.S. Chemla, R.F. Begley, R.L. Byer: IEEE J.QE-10, 71–81 (1974)

3.503 H. Kildal, J.C. Mikkelsen: Opt. Commun. **10**, 306–309 (1974)

3.504 N. Menyuk, G.W. Iseler, A. Mooradian: Appl. Phys. Lett. **29**, 422–424 (1976)

3.505 Yu.M. Andreev, V.G. Voevodin, P.P. Geyko, A.I. Gribenyukov, A.P. Dyadkin, S.V. Pigulsky, A.I. Starodubtsev: Kvantovaya Elektron. **14**, 784–786 (1987) [English transl. : Sov. J. Quantum Electron. **17**, 491–493 (1987)]

3.506 G.D. Boyd, E. Buehler, F.G. Storz, J.H. Wernick: IEEE J. QE-8, 419–426 (1972)

3.507 Y. Wu, T. Sasaki, S. Nakai, A. Yokotani, H. Tang, C. Chen: Appl. Phys. Lett. **62**, 2614–2615 (1993)

3.508 M.P. Golovey, G.F. Dobrzhansky, G.I. Kosourov, I.N. Kalinkina, E.I. Kortukova, Yu.S. Likhacheva, V.V. Ogadzhanova: Kristallografiya **15**, 757–761 (1970) [English transl. : Sov. Phys. -Crystallogr. **15**, 651–654 (1970)]

3.509 G.F. Dobrzhansky, M.P. Golovey, G.I. Kosourov: Pisma Zh. Eksp. Teor. Fiz. **16**, 263–265 (1969) [English transl. JETP Lett. **10**, 167–168 (1969)]

3.510 K. Kato: Appl. Phys. Lett. **33**, 413–414 (1978)

3.511 K. Kato: IEEE J. **26**, 1455–1456 (1990)

3.512 P.S. Bechtold, S. Haussühl: Appl. Phys. **14**, 403–410 (1977)

3.513 J.C. Bergman, G.R. Crane, H. Guggenheim: J. Appl. Phys. **46**, 4645–4646 (1975)

3.514 V.S. Suvorov, A.S. Sonin, I.S. Rez: Zh. Exp. Teor. Fiz. **53**, 49–55 (1967) [English transl. : Sov. Phys. – JETP **26**, 33–37 (1968)]

3.515 K. Kato: Appl. Phys. Lett. **25**, 342–343 (1974)

3.516 A.S. Vasilevskaya, M.F. Koldobskaya, L.G. Lomova, V.P. Popova, T.A. Regulskaya, I.S. Rez, Yu.P. Sobesskii, A.S. Sonin, V.S. Suvorov: Kristallografiya **12**, 447–450 (1967) [English transl. : Sov. Phys. -Crystallogr. **12**, 383–385 (1967)]

3.517 S. Singh: "Nonlinear Optical Materials" in *Handbook of Lasers*, ed. by R.G. Pressley (Chemical Rubber Co., Cleveland 1971) pp. 489–525

3.518 E.N. Volkova, Sh.L. Faerman: Kvantovaya Elektron. **3**, 2508–2511 (1976) [English transl. : Sov. J. Quantum Electron. **6**, 1380–1382 (1976)]

3.519 K. Kato: J. Appl. Phys. **46**, 2721–2722 (1975)

3.520 M.P. Golovey, I.N. Kalinkina, G.I. Kosourov: Opt. Spektrosk. **28**, 991–992 (1970) [English transl. : Opt. Spectrosc. USSR **28**, 535–536 (1970)]

3.521 K. Kato, S. Nakao: Jpn. J. Appl. Phys. **13**, 1681–1682 (1974)

3.522 K. Kato, A.J. Alcock, M.C. Richardson: Opt. Commun. **11**, 5–7 (1974)

3.523 E.V. Nilov, I.L. Yachnev: Zh. Prikl. Spektrosk. **7**, 943–945 (1967) [English transl. : J. Appl. Spectrosc. **7**, 628–630 (1967)]

3.524 R.S. Adhav: J. Appl. Phys. **39**, 4095–4098 (1968)

3.525 R.C. Miller: Appl. Phys. Lett. **5**, 17–19 (1964)

3.526 W.J. Deshotels: J. Opt. Soc. Am. **50**, 865 (1960)

3.527 S. Blit, E.G. Weaver, T.A. Rabson, F.K. Tittel: Appl. Opt. **17**, 721–723 (1978)

3.528 R.S. Adhav: J. Phys. D **2**, 177 –182 (1969)

3.529 K. Kato: Opt. Commun. **13**, 93–95 (1975)

3.530 K. Kato: IEEE J. QE-**10**, 622–624 (1974)

3.531 V.S. Suvorov, A.A. Filimonov: Fiz. Tverd. Tela **9**, 2131–2132 (1967) [English transl. : Sov. Phys. – Solid State **9**, 1674–1675 (1968)]

3.532 S. Singh, W.A. Bonner, J.R. Potopowicz, L.G. Van Uitert: Appl. Phys. Lett. **17**, 292–294 (1970)

3.533 H. Ito, H. Naito, H. Inaba: IEEE J. QE-**10**, 247–252 (1974)

3.534 K. Kato: Opt. Quant. Electron. **8**, 261–262 (1976)

3.535 H. Naito, H. Inaba: Opto-electron. **5**, 256–259 (1973)

3.536 S.J. Bastow, M.H. Dunn: Opt. Commun. **35**, 259–263 (1980)

3.537 K. Kato: IEEE J. QE-**19**, 893–894 (1983)

3.538 P.V. Lenzo, E.G. Spencer, J.P. Remeika: Appl. Opt. **4**, 1036–1037 (1965)

3.539 R.C. Miller, W.A. Nordland, E.D. Kolb, W.L. Bond: J. Appl. Phys. **41**, 3008–3011 (1970)

3.540 S.K. Kurtz, T.T. Perry, J.G. Bergman, Jr. : Appl. Phys. Lett. **12**, 186–188 (1967)

3.541 V.I. Bespalov, I.A. Batyreva, L.A. Dmitrenko, V.V. Korolikhin, S.P. Kuznetsov, M.A. Novikov: Kvantovaya Elektron. **4**, 1563–1566 (1977) [English transl. : Sov. J. Quantum Electron. **7**, 885–887 (1977)]

3.542 H. Naito, H. Inaba: Opto-electron. **4**, 335–337 (1972)

3.543 S.K. Kurtz: "Nonlinear Optical Materials" in *Laser Handbook*, Vol. 1 ed. by F.T. Arecchi, E.O. Schulz-Dubois (North-Holland, Amsterdam 1972) pp. 923–974

3.544 V.A. Kiselev, V.F. Kitaeva, L.A. Kulevskii, Yu.N. Polivanov, S.N. Poluektov: Zh. Eksp. Teor. Fiz. **62**, 1291–1301 (1972) [English transl. : Sov. Phys. -JETP **35**, 687–691 (1972)]

384   References

3.545 H. Ito, H. Naito, H. Inaba: J. Appl. Phys. **46**, 3992–3998 (1975)
3.546 J.E. Bjorkholm: IEEE J. QE-**4**, 970–972 (1968)
3.547 G.A. Dikchyus, V.I. Kabelka, A.S. Piskarskas, A.Yu. Stabinis: Kvantovaya Elektron. **1**, 2513–2515 (1974) [English transl. : Sov. J. Quantum Electron. **4**, 1402–1403 (1974)]
3.548 G. Dikchyus, R. Danielius, V. Kabelka, A. Piskarskas, T. Tomkiavichyus, A. Stabinis: Kvantovaya Elektron. **3**, 779–784 (1976) [English transl. : Sov. J. Quantum Electron. **6**, 425–428 (1976)]
3.549 R. Danielius, G. Dikchyus, V. Kabelka, A. Piskarskas: Zh. Tekh. Fiz. **47**, 1075–1077 (1977) [English transl. : Sov. Phys. - Tech. Phys. **22**, 642–643 (1977)]
3.550 C.A. Ebbers, L.D. DeLoach, M. Webb, D. Eimerl, S.P. Velsko, D.A. Keszler: IEEE J. **29**, 497–507 (1993)
3.551 L.T. Cheng, L.K. Cheng, J.D. Bierlein, F.C. Zumsteg: Appl. Phys. Lett. **63**, 2618–2620 (1993)
3.552 B. Hofmann, H. Vogt: J. Phys. C **6**, 543–550 (1973)
3.553 K. Iio: J. Phys. Soc. Jpn. **34**, 138–147 (1973)
3.554 D.S. Chemla, E. Batifol, R.L. Byer, R.L. Herbst: Opt. Commun. **11**, 57–61 (1974)
3.555 S. Singh, D.A. Draegert, J.E. Geusic: Phys. Rev. B **2**, 2709–2724 (1970)
3.556 J.D. Barry, C.J. Kennedy: IEEE J. QE-**11**, 575–579 (1975)
3.557 J.E. Geusic, H.J. Levinstein, J.J. Rubin, S. Singh, L.G. Van Uitert: Appl. Phys. Lett. **11**, 269–271 (1967)
3.558 J.E. Murray, R.J. Pressley, J.H. Boyden, R.B. Webb: IEEE J. QE-**10**, 263–267 (1974)
3.559 J.E. Geusic, H.J. Levinstein, S. Singh, R.G. Smith, L.G. Van Uitert: Appl. Phys. Lett. **12**, 306–308 (1968)
3.560 F.R. Nash, E.H. Turner, P.M. Bridenbaugh, J.M. Dziedzic: J. Appl. Phys. **43**, 1–9 (1972)
3.561 R.G. Smith, J.E. Geusic, H.J. Levinstein, J.J. Rubin, S. Singh, L.G. Van Uitert: Appl. Phys. Lett. **12**, 308–310 (1968)
3.562 R.B. Chesler, M.A. Karr, J.E. Geusic: Proc. IEEE **58**, 1899–1914 (1970)
3.563 A. Piskarskas, V. Smilgevichius, A. Umbrasas: Opt. Commun. **73**, 322–324 (1989)
3.564 L.G. Van Uitert, S. Singh, H.J. Levinstein, J.E. Geusic, W.A. Bonner: Appl. Phys. Lett. **11**, 161–163 (1967)
3.565 L.G. Van Uitert, S. Singh, H.J. Levinstein, J.E. Geusic, W.A. Bonner: Appl. Phys. Lett. **12**, 224 (1968)
3.566 B.F. Levine, C.G. Bethea, H.M. Kasper, F.A. Thiel: IEEE J. QE-**12**, 367–368 (1976)
3.567 V.V. Badikov, I.N. Matveev, S.M. Pshenichnikov, O.V. Rychik, N.K. Trotsenko, N.D. Ustinov, S.I. Shcherbakov: Kvantovaya Elektron. **7**, 2235–2237 (1980) [English transl. : Sov. J. Quantum Electron. **10**, 1300–1301 (1980)]
3.568 S.A. Andreev, N.P. Andreeva, V.V. Badikov, I.N. Matveev, S.M. Pschenichnikov: Kvantovaya Elektron. **7**, 2003–2006 (1980) [English transl. : Sov. J. Quantum Electron.**10**, 1157–1158 (1980)]
3.569 V.V. Badikov, I.N. Matveev, V.L. Panyutin, S.M. Pshenichnikov, T.M. Repyakhova, O.V. Rychik, A.E. Rozenson, N.K. Trotsenko, N.D. Ustinov: Kvantovaya Elektron. **6**, 1807–1810 (1979) [English transl. : Sov. J. Quantum Electron. **9**, 1068–1069 (1979)]
3.570 W.L. Bond, G.D. Boyd, H.L. Carter: J. Appl. Phys. **38**, 4090–4091 (1967)
3.571 G.D. Boyd, T.J. Bridges, E.G. Burkhardt: IEEE J. QE-**4**, 515–519 (1968)
3.572 J. Jerphagnon, E. Batifol, G. Tsoucaris, M. Sourbe: C.R. Acad. Sci. **265 B**, 495–497 (1967)
3.573 J.D. Feichtner, R. Johannes, G.W. Roland: Appl. Opt. **9**, 1716–1717 (1970)
3.574 W.B. Gandrud, G.D. Boyd: Opt. Commun. **1**, 187–190 (1969)
3.575 W.B. Gandrud, G.D. Boyd, J.H. McFee, F.H. Wehmeier: Appl. Phys. Lett. **16**, 59–61 (1970)
3.576 J.H. McFee, G.D. Boyd, P.H. Schmidt: Appl. Phys. Lett. **17**, 57–59 (1970)
3.577 J. Jerphagnon, E. Batifol, M. Sourbe: C.R. Acad. Sci. **265 B**, 400–402 (1967)
3.578 S. Caldwell, H.Y. Fan: Phys. Rev. **114**, 664–675 (1959)
3.579 G.W. Day: Appl. Phys. Lett. **18**, 347–349 (1971)

3.580   G.H. Sherman, P.D. Coleman: J. Appl. Phys. **44**, 238–241 (1973)

3.581   L. Gampel, F.M. Johnson: J. Opt. Soc. Am. **59**, 72–73 (1969)

3.582   W. Henrion, F. Eckart: Z. Naturforsch. **19a**, 1024–1025 (1964)

3.583   C.K.N. Patel: Phys. Rev. Lett. **16**, 613–616 (1966)

3.584   R.C. Weast, M.J. Astle (eds.): *CRC Handbook of Chemistry and Physics* (CRC Press, Boca Raton, FL 1980)

3.585   M. Gottlieb, T.J. Isaacs, J.D. Feichtner, G.W. Roland: J. Appl. Phys. **45**, 5145–5151 (1974)

3.586   J.D. Feichtner, G.W. Roland: Appl. Opt. **11**, 993–998 (1972)

3.587   M.D. Ewbank, P.R. Newman, N.L. Mota, S.M. Lee, W.L. Wolfe, A.G. DeBell, W.A. Harrison: J. Appl. Phys. **51**, 3848–3852 (1980)

3.588   R.C.Y. Auyeung, D.M. Zielke, B.J. Feldman: Appl. Phys. B **48**, 293–297 (1989)

3.589   D.R. Suhre: Appl. Phys. B **52**, 367–370 (1991)

3.590   G.H. Sherman, P.D. Coleman: IEEE J. QE-**9**, 403–409 (1973)

3.591   D.E. McCarthy: Appl. Opt. **7**, 1997–2000 (1968)

3.592   W.B. Gandrud, R.L. Abrams: Appl. Phys. Lett. **17**, 302–305 (1970)

3.593   J.-L. Oudar, C.A. Schwartz, E.M. Batifol: IEEE J. QE-**11**, 623–629 (1975)

3.594   K.C. Nomura: Phys. Rev. Lett. **5**, 500-501 (1960)

3.595   C.K.N. Patel: Phys. Rev. Lett. **15**, 1027–1030 (1965)

3.596   J. Jerphagnon, M. Sourbe, E. Batifol: C.R. Acad. Sci. **263 B**, 1067–1070 (1966)

3.597   A. Delahaigue, C. Thiebeaux, P. Jouve: Appl. Phys. **24**, 21–22 (1981)

3.598   J. Jerphagnon, M. Bernard: IEEE J. QE-**4**, 395–396 (1968)

3.599   J.D. Taynai, R. Targ, W.B. Tiffany: IEEE J. QE-**7**, 412–416 (1971)

3.600   J.-M. Halbout, C.L. Tang: IEEE J. QE-**18**, 410–415 (1982)

3.601   I.M. Silvestrova, G.N. Nabakhtiani, V.B. Kozin, V.A. Kuznetsov, Yu.V. Pisarevskii: Kristallografiya **37**, 1535-1541 (1992) [English transl. : Sov. Phys. -Crystallogr. **37**, 831–834 (1992)]

3.602   H. Minemoto, Y. Ozaki, N. Sonoda, T. Sasaki: Appl. Phys. Lett. **63**, 3565–3567 (1993)

3.603   R.B. Andreev, K.V. Vetrov, V.N. Voitsechovskii, V.D. Volosov, I.V. Nikiforuk, B.P. Nikolaeva, V.E. Yakobson: Izv. Akad. Nauk. SSSR, Ser. Fiz. **54**, 2491–2493 (1990) [English transl. : Bull. Acad. Sci. USSR, Phys. Ser. **54**, No. **12**, 187–189 (1990)]

3.604   C.E. Barker, D. Eimerl, S.P. Velsko: J. Opt. Soc. Am. B **8**, 2481–2492 (1991)

3.605   G. Robertson, M.H. Dunn: Appl. Phys. Lett. **62**, 3405–3407 (1993)

3.606   M. Kitazawa, R. Higuchi, M. Takahashi, T. Wada, H. Sasabe: Appl. Phys. Lett. **64**, 2477–2479 (1994)

3.607   C. Medrano P., P. Günter, H. Abend: Phys. Status Solidi B **143**, 749–754 (1987)

3.608   M.V. Hobden: J. Appl. Phys. **38**, 4365–4372 (1967)

3.609   A.N. Izrailenko, R.Yu. Orlov, V.A. Koptsik: Kristallografiya **13**, 171 (1968) [English transl. : Sov. Phys. -Crystallogr. **13**, 136 (1968)]

3.610   R.Yu. Orlov: Izv. Vyssh. Ucheb. Zaved., Ser. Radiofiz. **12**, 1351–1353 (1969) [English transl. : Radiophysics. Quantum Electron. **12**, 1056–1058 (1969)]

3.611   V.D. Shigorin: "Second Harmonic Generation in Molecular Crystals", in Proceedings of P.N. Lebedev Physical Institute, USSR Academy of Sciences, Vol. 98 (Nauka, Moscow 1977) pp. 78–140 [English transl. : Proc. Lebedev Phys. Inst. Acad. Sci. USSR **98**, (1977)]

3.612   N. Zhang, D.R. Yuan, X.T. Tao, D. Xu, Z.S. Shao, M.H. Jiang, M.G. Liu: Opt. Commun. **99**, 247–251 (1993)

3.613   X.T. Tao, D.R. Yuan, N. Zhang, M.H. Jiang, Z.S. Shao: Appl. Phys. Lett. **60**, 1415–1417 (1992)

3.614   T. Konoshita, S. Horinouchi, K. Sasaki, H. Okamoto, N. Tanaka, T. Fukaya, M. Goto: J. Opt. Soc. Am. B **11**, 986–994 (1994)

3.615   J. Zyss, D.S. Chemla, J.F. Nicoud: J. Chem. Phys. **74**, 4800–4811 (1981)

3.616   J. Zyss, I. Ledoux, R.B. Hierle, R.K. Raj, J.-L. Oudar: IEEE J. QE-**21**, 1286–1295 (1985)

3.617   D. Josse, R. Hierle, I. Ledoux, J. Zyss: Appl. Phys. Lett. **53**, 2251–2253 (1988)

3.618   S.X. Dou, D. Josse, J. Zyss: J. Opt. Soc. Am. B **10**, 1708–1715 (1993)

3.619   I. Ledoux, J. Badan, J. Zyss, A. Migus, D. Hulin, J. Etchepare, G. Grillon, A. Antonetti: J. Opt. Soc. Am. B **4**, 987–997 (1986)

3.620   G. Puccetti, A. Perigaud, J. Badan, I. Ledoux, J. Zyss: J. Opt. Soc. Am. B **10**, 733-744 (1993)

3.621   K. Sutter, Ch. Bosshard, W.S. Wang, G. Surmely, P. Günter: Appl. Phys. Lett. **53**, 1779–1781 (1988)

3.622   K. Sutter, Ch. Bosshard, P. Günter: Ferroelectrics **92**, 395–401 (1989)

3.623   C. Bosshard, K. Sutter, P. Günter: Ferroelectrics **92**, 387–393 (1989)

3.624   P. Günter, Ch. Bosshard, K. Sutter, H. Arend, G. Chapuis, R.J. Twieg, D. Dobrowolski: Appl. Phys. Lett. **50**, 486–488 (1987)

3.625   T. Ukachi, T. Shigemoto, H. Komatsu, T. Sugiyama: J. Opt. Soc. Am. B **10**, 1372–1378 (1993)

3.626   K. Iio, Y. Kusuhara, K. Hamano, S. Sawada: Jpn. J. Appl. Phys. **13**, 1299–1300 (1974)

3.627   P. Kerkoc, M. Zgonik, K. Sutter, Ch. Bosshard, P. Günter: Appl. Phys. Lett. **54**, 2062–2064 (1989)

3.628   J.-C. Baumert, R.J. Twieg, G.D. Bjorklund, J.A. Logan, C.W. Dirk: Appl. Phys. Lett. **51**, 1484–1486 (1987)

3.629   P. Kerkoc, M. Zgonik, K. Sutter, Ch. Bosshard, P. Günter: J. Opt. Soc. Am. B **7**, 313–319 (1990)

3.630   J.L. Oudar, R. Hierle: J. Appl. Phys. **48**, 2699–2704 (1977)

3.631   K. Kato: IEEE J. QE-**16**, 1288–1290 (1980)

3.632   B.L. Davydov, L.G. Koreneva, E.A. Lavrovskii: Radiotech. Elektron. **19**, 1313–1315 (1974) [English transl. : Radio Eng. Electron. Phys. **19**, No. **6**, 130–131 (1974)

3.633   P.D. Southgate, D.S. Hall: Appl. Phys. Lett. **18**, 456–458 (1971)

3.634   P.V. Vidakovic, M. Coquillay, F. Salin: J. Opt. Soc. Am. B **4**, 998–1012 (1987)

3.635   R. Morita, P.V. Vidakovic: Appl. Phys. Lett. **61**, 2854–2856 (1992)

3.636   I. Ledoux, C. Lepers, A. Perigaud, J. Badan, J. Zyss: Opt. Commun. **80**, 149–154 (1990)

3.637   D. Josse, S.X. Dou, J. Zyss, P. Andreazza, A. Perigaud: Appl. Phys. Lett. **61**, 121–123 (1992)

3.638   I. Ledoux, J. Zyss, A. Migus, J. Etchepare, G. Grillon, A. Antonetti: Appl. Phys. Lett. **48**, 1564–1566 (1986)

3.639   J.D. Bierlein, L.K. Cheng, Y. Wang, W. Tam: Appl. Phys. Lett. **56**, 423–425 (1990)

3.640   J.G. Calvert, J.N. Pitts: *Photochemistry* (Wiley, New York 1966)

3.641   R.A. Smith, F.E. Jones, R.P. Chasmar: *The Detection and Measurement of Infrared Radiation* (Clarendon, Oxford 1957)

3.642   R.B. Sosman: *The Properties of Silica* (Chemical Catalog Co., New York 1927)

3.643   A. Smakula: *Einkristalle* (Springer, Berlin 1962)

3.644   M.J. Soileau, M. Bass: IEEE J. QE-**16**, 814 (1980)

3.645   R.S. Krishnan (ed.): *Progress in Crystal Physics*, Vol. 1 (Viswanathan, Madras 1958)

3.646   Y. Mori, I. Kuroda, S. Nakajima, T. Sasaki, S. Nakai: Appl. Phys. Lett. **67**, 1818–1820 (1995)

3.647   Y. Mori, S. Nakajima, A. Miyamoto, M. Inagaki, T. Sasaki, H. Yoshida, S. Nakai: Proc. SPIE, **2633**, 299–307 (1995)

3.648   K. Kato: IEEE J. **30**, 2950–2952 (1994)

3.649   Y. Tang, Y. Cui, M.H. Dunn: J. Opt. Soc. Am. B **12**, 638–643 (1995)

3.650   M. Oka, L.Y. Liu, W. Wiechmann, N. Eguchi, S. Kubota: IEEE J. Selected Topics Quant. Electron. **1**, 859–866 (1995)

3.651   K. Kato: IEEE J. **31**, 169–171 (1995)

3.652   W. Wiechmann, S. Kubota, T. Fukui, H. Masuda: Opt. Lett. **18**, 1208–1210 (1993)

3.653   D.L. Fenimore, K.L. Schepler, U.B. Ramabadran, S.R. McPherson: J. Opt. Soc. Am. B **12**, 794–796 (1995)

3.654   J. Han, Y. Liu, M. Wang, D. Nie: J. Crystal Growth **128**, 864–866 (1993)

3.655   D.T. Reid, M. Ebrahimzade, W. Sibbett: J. Opt. Soc. Am. B **12**, 2168–2179 (1995)

3.656   L.K. Cheng, L.T. Cheng, J. Galperin, P.A.M. Hotsenpiller, J.D. Bierlein: J. Crystal Growth **137**, 107–115 (1994)
3.657   H. Komine, J.M. Fukumoto, W.H. Long, Jr., E.A. Stappaerts: IEEE J. Selected Topics Quant. Electron. **1**, 44–49 (1995)

Chapter 4

4.1   D.T. Attwood, E.L. Pierce, L.W. Coleman: Opt. Commun. **15**, 10–12 (1975)
4.2   D.T. Attwood, E.S. Bliss, E.L. Pierce, L.W. Coleman: IEEE J. QE-**12**, 203–204 (1976)
4.3   Yu.A. Matveets, D.N. Nikogosyan, V. Kabelka, A. Piskarskas: Kvantovaya Elektron. **5**, 664–666 (1978) [English transl.: Sov. J. Quantum Electron. **8**, 386–388 (1978)]
4.4   S.S. Dimov: Opt. Quant. Electr. **25**, 545–550 (1993)
4.5   V.P. Machewirth, R. Webb, D. Anafi: Laser Focus **12**, No. **5**, 104–107 (1976)
4.6   R.M. Kogan, R.M. Pixton, T.G. Crow: Opt. Eng. **17**, 120–124 (1978)
4.7   A. Borsutzky, R. Brünger, Ch. Huang, R. Wallenstein: Appl. Phys. **B 52**, 55–62 (1991)
4.8   K. Kato: IEEE J. QE-**10**, 616–618 (1974)
4.9   T.A. Rabson, H.J. Ruiz, P.L. Shah, F.K. Tittel: Appl. Phys. Lett. **20**, 282–284 (1972)
4.10   E.O. Ammann, C.D. Decker, J. Falk: IEEE J. QE-**10**, 463–465 (1974)
4.11   D.T. Hon: IEEE J. QE-**12**, 148–151 (1976)
4.12   K. Kato: Opt. Commun. **13**, 93–95 (1975)
4.13   K. Kato, S. Nakao: Jpn. J. Appl. Phys. **13**, 1681–1682 (1974)
4.14   V.G. Dmitriev, V.N. Krasnyanskaya, M.F. Koldobskaya, I.S. Rez, E.A. Shalaev, E.M. Shvom: Kvantovaya Elektron. No. **2** (**14**), 64–66 (1973) [English transl.: Sov. J. Quantum Electron. **3**, 126–127 (1973)]
4.15   E.W. Van Stryland, W.E. Williams, M.J. Soileau, A.L. Smirl: IEEE J. QE-**20**, 434–439 (1984)
4.16   R.L. Byer, Y.K. Park, R.S. Feigelson, W.L. Kway: Appl. Phys. Lett. **39**, 17–19 (1981)
4.17   R.B. Andreev, V.D. Volosov, L.I. Kuznetsova: Kvantovaya Elektron. **2**, 420–421 (1975) [English transl.: Sov. J. Quantum Electron. **5**, 242–243 (1975)]
4.18   P. Hargis: Laser Focus **14**, No. **7**, 18–20 (1978)
4.19   A.M. Johnson, W.M. Simpson: Opt. Lett. **8**, 554–556 (1983)
4.20   A.L. Aleksandrovsky, S.A. Akhmanov, V.A. Dyakov, N.I. Zheludev, V.I. Pryalkin: Kvantovaya Elektron. **12**, 1333–1334 (1985) [English transl.: Sov. J. Quantum Electron. **15**, 885–886 (1985)]
4.21   R.F. Belt, G. Gashurov, Y.S. Liu: Laser Focus **21**, No. **10**, 110–124 (1985)
4.22   T.A. Driscoll, H.J. Hoffman, R.E. Stone, P.E. Perkins: J. Opt. Soc. Am. **B 3**, 683–686 (1986)
4.23   S.E. Moody, J.M. Eggleston, J.F. Seamans: IEEE J. QE-**23**, 335–340 (1987)
4.24   O.I. Lavrovskaya, N.I. Pavlova, A.V. Tarasov: Kristallografiya **31**, 1145–1151 (1986) [Engl. transl.: Sov. Phys–Crystallogr. **31**, 678–682 (1986)]
4.25   R.J. Bolt, M. van der Mooren: Opt. Commun. **100**, 399–410 (1993)
4.26   A.J.W. Brown, M.S. Bowers, K.W. Kangas, C.H. Fisher: Opt. Lett. **17**, 109–111 (1992)
4.27   J.E. Murray, R.J. Pressley, J.H. Boyden, R.B. Webb: IEEE J. QE-**10**, 263–267 (1974)
4.28   W. Seelert, P. Kortz, D. Rytz, B. Zysset, D. Ellgehausen, G. Mizell: Opt. Lett. **17**, 1432–1434 (1992)
4.29   R.S. Adhav, S.R. Adhav, J.M. Pelaprat: Laser Focus **23**, No. **9**, 88–100 (1987)
4.30   C. Chen, Y.X. Fan, R.C. Eckardt, R.L. Byer: Proc. SPIE, **681**, 12–19 (1986)
4.31   G.C. Bhar, U. Chatterjee, P. Datta: Appl. Phys. **B 51**, 317–319 (1990)
4.32   G.C. Bhar, S. Das, P.K. Datta: Phys. Stat. Sol. a, **119**, K173–K176 (1990)
4.33   J.Y. Huang, Y.R. Shen, C. Chen, B. Wu: Appl. Phys. Lett. **58**, 1579–1581 (1991)

4.34   J.T. Lin, J.L. Montgomery, K. Kato: Opt. Commun. **80**, 159–165 (1990)
4.35   V.A. Dyakov, M.Kh. Dzhafarov, A.A. Lukashev, A.A. Podshivalov, V.I. Pryalkin: Kvant. Elektron. **18**, 339–341 (1991) [English transl.: Sov. J. Quantum Electron. **21**, 307–308 (1991)]
4.36   F. Xie, B. Wu, G. You, C. Chen: Opt. Lett. **16**, 1237–1239 (1991)
4.37   H. Mao, F. Fu, B. Wu, C. Chen: Appl. Phys. Lett. **61**, 1148–1150 (1992)
4.38   V. Dmitriev, L. Tarasov: *Optique Non Lineaire Appliquee* (MIR, Moscow 1987)
4.39   D.A. Bryan, R. Gerson, H.E. Tomaschke: Appl. Phys. Lett. **44**, 847–849 (1984)
4.40   K.L. Sweeney, L.E. Halliburton, D.A. Bryan, R.R. Rice, R. Gerson, H.E. Tomaschke: Appl. Phys. Lett. **45**, 805–807 (1984)
4.41   T.R. Volk, N.M. Rubinina, A.I. Kholodnykh: Kvantovaya Elektron. **15**, 1705–1706 (1988) [English transl.: Sov. J. Quantum Electron. **18**, 1061–1062 (1988)]
4.42   E.O. Ammann, S. Gush, Jr.: Appl. Phys. Lett. **52**, 1374–1376 (1988)
4.43   S. Singh, D.A. Draegert, J.E. Geusic: Phys. Rev. **B 2**, 2709–2724 (1970)
4.44a   D.N. Nikogosyan: Appl. Phys. A **52**, 359–368 (1991)
4.44b   D.N. Nikogosyan: Appl. Phys. A **58**, 181–190 (1994)
4.45   G.C. Bhar, S. Das, U. Chatterjee: Appl. Phys. Lett. **54**, 1383–1384 (1989)
4.46   G.C. Bhar, U. Chatterjee, S. Das: J. Appl. Phys. **66**, 5111–5113 (1989)
4.47   V.D. Shigorin. "Second Harmonic Generation in Molecular Crystals", in Proc. of P.N. Lebedev Physical Institute, USSR Academy of Sciences, Vol.98 (Nauka, Moscow 1977) pp. 78–140 [English transl.: Proc. Lebedev Phys. Inst. Acad. Sci. USSR **98** (1977)]
4.48   M.J. Rosker, C.L. Tang: IEEE J. QE-**20**, 334–336 (1984)
4.49   B.L. Davydov, S.G. Kotovshchikov, V.A. Nefedov: Kvantovaya Elektron. **4**, 214–220 (1977) [English transl.: Sov. J. Quantum Electron. 7, 129–131 (1977)]
4.50   M. Halbout, C.L. Tang: IEEE J. QE-**18**, 410–415 (1982)
4.51   J. Zyss, D.S. Chemla, J.F. Nicoud: J. Chem. Phys. **74**, 4800–4811 (1981)
4.52   D. Josse, R. Hierle, I. Ledoux, J. Zyss: Appl. Phys. Lett. **53**, 2251–2253 (1988)
4.53   J.L. Oudar, R. Hierle: J. Appl. Phys. **48**, 2699–2704 (1977)
4.54   B.L. Davydov, L.G. Koreneva, E.A. Lavrovsky: Radiotekh. Elektron. **19**, 1313–1316 (1974) [English transl.: Radio Eng. Electron. Phys. **19** (6), 130–131 (1974)]
4.55   K. Kato: IEEE J. QE-**16**, 1288–1290 (1980)
4.56   B.F. Levine, C.G. Bethea, C.D. Thurmond, R.T. Lynch, J.L. Bernstein: J. Appl. Phys. **50**, 2523–2527 (1979)
4.57   G.S. Belikova, M.P. Golovey, V.D. Shigorin, G.P. Shipulo: Opt. Spektrosk. **38**, 779–783 (1975) [English transl.: Opt. Spectrosc. USSR **38**, 441–443 (1975)]
4.58   P. Günter, Ch. Bosshard, K. Sutter, H. Arend, G. Chapuis, R.J. Twieg, D. Dobrowolski: Appl. Phys. Lett. **50**, 486–488 (1987)
4.59   D. Eimerl, S. Velsko, L. Davis, F. Wang, G. Loiacono, G. Kennedy: IEEE J. **25**, 179–193 (1989)
4.60   P.A. Norman, D. Bloor, J.S. Obhi, S.A. Karaulov, M.B. Hursthouse, P.V. Kolinsky, R.J. Jones, S.R. Hall: J. Opt. Soc. Am. B **4**, 1013–1016 (1987)
4.61   R. Morita, P.V. Vidakovic: Appl. Phys. Lett. **61**, 2854–2856 (1992)
4.62   H. Minemoto, Y. Ozaki, N. Sonoda, T. Sasaki: Appl. Phys. Lett. **63**, 3565–3567 (1993)
4.63   T. Ukachi, T. Shigemoto, H. Komatsu, T. Sugiyama: J. Opt. Soc. Am. B **10**, 1372–1378 (1993)
4.64   N. Zhang, D.R. Yuan, X.T. Tao, D. Xu, Z.S. Shao, M.H. Jiang, M.G. Liu: Opt. Comm. **99**, 247–251 (1993)
4.65   M. Kitazawa, P. Higuchi, M. Takahashi, T. Wada, H. Sasabe: Appl. Phys. Lett. **64**, 2477–2479 (1994)
4.66   V. Deserno, G. Nath: Phys. Lett. **30A**, 483–484 (1969)
4.67   V.G. Dmitriev, P.G. Konvisar, I.B. Lyushnya, V.Yu. Mikhailov, S.R. Rustamov, M.F. Stelmakh: Kvantovaya Elektron. **8**, 906–907 (1981) [English transl.: Sov. J. Quantum Electron. **11**, 545–546 (1981)]

4.68   Yu.D. Golyaev, S.A. Grodsky, V.G. Dmitriev, P.G. Konvisar, S.V. Lantratov, V.Yu. Mikhailov, S.R. Rustamov: Kvantovaya Elektron. **9**, 2093–2095 (1982) [English transl.: Sov. J. Quantum Electron. **12**, 1360–1362 (1982)]
4.69   A. Koeneke, A. Hirth: Opt. Commun. **34**, 245–248 (1980)
4.70   V.G. Dmitriev, V.R. Kushnir, S.R. Rustamov, A.A. Fomichev: Kvantovaya Elektron. No. 2(**8**), 111–112 (1972) [English transl.: Sov. J. Quantum Electron. **2**, 188–189 (1972)]
4.71   R.G. Smith, J.E. Geusic, H.J. Levinstein, S. Singh, L.G. Van Uitert: J. Appl. Phys. **39**, 4030–4032 (1968)
4.72   J.E. Geusic, H.J. Levinstein, S. Singh, R.G. Smith, L.G. Van Uitert: Appl. Phys. Lett. **12**, 306–308 (1968)
4.73   R.G. Smith, J.E. Geusic, H.J. Levinstein, J.J. Rubin, S. Singh, L.G. Van Uitert: Appl. Phys. Lett. **12**, 308–310 (1968)
4.74   R.B. Chesler, M.A. Karr, J.E. Geusic: Proc. IEEE **58**, 1899–1914 (1970)
4.75   W. Gulshaw, J. Kannelaud, J.E. Peterson: IEEE J. **QE-10**, 253–263 (1974)
4.76   Y.S. Liu, D. Dentz, R. Belt: Opt. Lett. **9**, 76–78 (1984)
4.77   P.E. Perkins, T.S. Fahlen: J. Opt. Soc. Am. B **4**, 1066–1071 (1987)
4.78   D.W. Anthon, D.L. Sipes, T.J. Pier, M.R. Ressl: IEEE J. **28**, 1148–1157 (1992)
4.79   L.R. Marshall, A.D. Hays, A. Kaz, R.L. Burnham: IEEE J. **28**, 1158–1163 (1992)
4.80   T. Fukuda, Y. Uematsu: Izv. Akad. Nauk SSSR, Ser. Fiz. **41**, 548–554 (1977) [English transl.: Bull. Acad. Sci. USSR, Phys. Ser. **41**, No. 3, 73–77 (1977)]
4.81   W.P. Risk, R. Pon, W. Lenth: Appl. Phys. Lett. **54**, 1625–1627 (1989)
4.82   I. Biaggio, H. Looser, P. Günter: Ferroelectr. **94**, 157–161 (1989)
4.83   V.M. Garmash, G.A. Ermakov, N.I. Pavlova, A.V. Tarasov: Pisma Zh. Tekh. Fiz. **12**, 1222–1225 (1986) [English transl.: Sov. Tech. Phys. Lett. **12**, 505–506 (1986)]
4.84   A. Ashkin, G.D. Boyd, J.M. Dziedzic: IEEE J. **QE-2**, 109–124 (1966)
4.85   W.J. Kozlovsky, C.D. Nabors, R.L. Byer: IEEE J. **24**, 913–919 (1988)
4.86   D.C. Gerstenberger, G.E. Tye, R.W. Wallace: Opt. Lett. **16**, 992–994 (1991)
4.87   K. Fiedler, S. Schiller, R. Paschotta, P. Kürz, J. Mlynek: Opt. Lett. **18**, 1786–1788 (1993)
4.88   D.H. Jundt, M.M. Fejer, R.L. Byer, R.G. Norwood, P.F. Bordui: Opt. Lett. **16**, 1856–1858 (1991)
4.89   Z.Y. Ou, S.F. Pereira, E.S. Polzik, H.J. Kimble: Opt. Lett. **17**, 640–642 (1992)
4.90   S.T. Yang, C.C. Pohalski, E.K. Gustafson, R.L. Byer, R.S. Feigelson, R.J. Raymakers, R.K. Route: Opt. Lett. **16**, 1493–1495 (1991)
4.91   G.P.A. Malcolm, M. Ebrahimzadeh, A.I. Ferguson: IEEE J. **QE-28**, 1172–1178 (1992)
4.92   R.B. Andreev, V.D. Volosov, V.S. Gorshkov: Opt. Spektrosk. **46**, 376–381 (1979) [English transl.: Opt. Spectrosc. USSR **46**, 207–210 (1979)]
4.93   A.H. Kung, J.F. Young, G.C. Bjorklund, S.E. Harris: Phys. Rev. Lett. **29**, 985–988 (1972)
4.94   K. Kato: Appl. Phys. Lett. **25**, 342–343 (1974)
4.95   S.F. Bogdanov, P.G. Konvisar, S.R. Rustamov: Kvantovaya Elektron. **12**, 2143–2144 (1985) [English transl.: Sov. J. Quantum. Electron. **15**, 1409 (1985)]
4.96   M. Okada, S. Ieiri: Jpn. J. Appl. Phys. **10**, 808 (1971)
4.97   R. Wu: Appl. Opt. **32**, 971–975 (1993)
4.98   B. Wu, N. Chen, C. Chen, D. Deng, Z. Xu: Opt. Lett. **14**, 1080–1081 (1989)
4.99   K. Kato: Opt. Quantum Electron. **8**, 261–262 (1976)
4.100   K. Kato: IEEE J. **QE-19**, 893–894 (1983)
4.101   A.G. Arutyunyan, G.R. Buniatyan, A.A. Melkonyan, A.V. Mesropyan, G.A. Paityan: In *Nonlinear Optical Interactions* (Yerevan University Press, Yerevan, 1987) pp. 135–144
4.102   P. Qiu, A. Penzkofer: Appl. Phys. B **45**, 225–236 (1988)
4.103   G.A. Abakumov, Kh.S. Bagdasarov, V.V. Vetrov, S.A. Vorobev, V.P. Zakharov, V.F. Pikelni, A.P. Smirnov, V.V. Fadeev, E.A. Fedorov: Kvantovaya Elektron. **4**, 1152–1153 (1977) [English transl.: Sov. J. Quantum Electron. **7**, 656–657 (1977)]
4.104   J.J. Reintjes, R.C. Eckardt: Appl. Phys. Lett. **30**, 91–93 (1977)
4.105   Y.S. Liu, W.B. Jones, J.P. Chernoch: Appl. Phys. Lett. **29**, 32–34 (1976)

390   References

4.106 P.E. Perkins, T.S. Fahlen: IEEE J. QE-21, 1636–1638 (1985)
4.107 K. Kato: Opt. Commun. 13, 361–362 (1975)
4.108 A.G. Arutyunyan, V.G. Atanesyan, K.B. Petrosyan, K.M. Pokhsraryan: Pisma Zh. Tekh. Fiz. 6, 277–280 (1980) [English transl.: Sov. Tech. Phys. Lett. 6, 120–121 (1980)]
4.109 A. Lago, R. Wallenstein, C. Chen, Y.X. Fan, R.L. Byer: Opt. Lett. 13, 221–223 (1988)
4.110 K. Kato: IEEE J. QE-26, 1455–1456 (1990)
4.111 A.G. Akmanov, S.A. Akhmanov, B.V. Zhdanov, A.I. Kovrigin, N.K. Podsotskaya, R.V. Khokhlov: Pisma Zh. Eksp. Teor. Fiz. 10, 244–249 (1969) [English transl.: JETP Lett. 10, 154–156 (1969)]
4.112 G.A. Massey, M.D. Jones, J.C. Johnson: IEEE J. QE-14, 527–532 (1978)
4.113 M.D. Jones, G.A. Massey: IEEE J. QE-15, 204–206 (1979)
4.114 G.A. Massey, J.C. Johnson: IEEE J. QE-12, 721–727 (1976)
4.115 G.A. Massey: Appl. Phys. Lett. 24, 371–373 (1974)
4.116 K. Kato: Opt. Commun. 19, 332–333 (1976)
4.117 Y. Tanaka, H. Kuroda, S. Shionoya: Opt. Commun. 41, 434–436 (1982)
4.118 K. Kato: IEEE J. QE-16, 810–811 (1980)
4.119 V.G. Tunkin, T. Usmanov, V.A. Shakirov: Kvantovaya Elektron. No. 5 (11), 117–118 (1972) [English transl.: Sov. J. Quantum. Electron. 2, 487–488 (1972)]
4.120 C. Chen, B. Wu, A. Jiang, G. You: Sci. Sin. B 28, 235–243 (1985)
4.121 K. Kato: IEEE J. QE-22, 1013–1014 (1986)
4.122 A.G. Arutyunyan, G.G. Gurzadyan, R.K. Ispiryan: Kvantovaya Elektron. 16, 2493–2495 (1989) [English transl.: Sov. J. Quantum Electron. 19, 1602–1603 (1989)]
4.123 N.P. Garayanz, K.B. Petrosyan, K.M. Pokhsraryan: Izv. Akad. Nauk Arm. SSR, Ser. Fiz. 23, 109–111 (1988) [English transl.: Sov. J. Contemp. Phys. Armen. Acad. Sci. 23, No 2 (1988)]
4.124 A.G. Akmanov, A.M. Valshin, A.G. Yamaletdinov: Kvantovaya Elektron. 8, 408–410 (1981) [English transl.: Sov. J. Quantum Electron. 11, 247–248 (1981)]
4.125 K. Kato, R.S. Adhav: IEEE J. QE-12, 443–444 (1976)
4.126 A.A. Kazakov, S.V. Shavkunov, E.A. Shalaev: Kvantovaya Elektron. 8, 2259–2261 (1981) [English transl.: Sov. J. Quantum. Electron. 11, 1381–1383 (1981)]
4.127 A.A. Kazakov, V.A. Konovalov, S.V. Shavkunov, E.A. Shalaev: Kvantovaya Elektron. 10, 1603–1610 (1983) [English transl.: Sov. J. Quantum. Electron. 13, 1054–1058 (1983)]
4.128 S. Lin, B. Wu, F. Xie, C. Chen: Appl. Phys. Lett. 59, 1541–1543 (1991)
4.129 A.A. Gulamov, E.A. Ibragimov, V.I. Redkorechev, T. Usmanov: Kvantovaya Elektron. 10, 1305–1306 (1983) [English transl.: Sov. J. Quantum Electron. 13, 844–845 (1983)]
4.130 W. Seka, S.D. Jakobs, J.E. Rizzo, R. Boni, R.S. Craxton: Opt. Commun. 34, 469–473 (1980)
4.131 G.J. Linford, B.C. Johnson, J.S. Hildum, W.E. Martin, K. Snyder, R.D. Boyd, W.L. Smith, C.L. Vercimak, D. Eimerl, J.T. Hunt: Appl. Opt. 21, 3633–3643 (1982)
4.132 I.A. Begishev, R.A. Ganeev, A.A. Gulamov, E.A. Erofeev, Sh.R. Kamalov, T. Usmanov, A.D. Khadzhaev: Kvantovaya Elektron. 15, 353–361 (1988) [English transl.: Sov. J. Quantum Electron. 18, 224–228 (1988)]
4.133 C. Loth, D. Bruneau, E. Fabre: Appl. Opt. 19, 1022–1023 (1980)
4.134 E.F. Ibragimov, V.I. Redkorechev, A.P. Sukhorukov, T. Usmanov: Kvantovaya Elektron. 9, 1131–1140 (1982) [English transl.: Sov. J. Quantum Electron. 12, 714–719 (1982)]
4.135 D. Bruneau, A.M. Tournade, E. Fabre: Appl. Opt. 24, 3740–3745 (1985)
4.136 R.S. Craxton: Opt. Commun. 34, 474–478 (1980)
4.137 R.S. Craxton: IEEE J. QE-17, 1771–1782 (1981)
4.138 R.S. Craxton: Appl. Opt. 22, 2739–2742 (1983)
4.139 D. Eimerl: IEEE J. QE-23, 1361–1371 (1987)
4.140 S. Sullivan, E.L. Thomas: Opt. Commun. 25, 125–128 (1978)
4.141 V.S. Suvorov, A.S. Sonin, I.S. Rez: Zh. Eksp. Teor. Fiz. 53, 49–55 (1967) [English transl.: Sov. Phys.-JETP 26, 33–37 (1968)]

4.142 K. Kato: IEEE J. QE-**10**, 622–624 (1974)
4.143 K. Kato, A.J. Alcock, M.C. Richardson: Opt. Commun. **11**, 5–7 (1974)
4.144 G. Nath, H. Mehmanesch, M. Gsänger: Appl. Phys. Lett. **17**, 286–288 (1970)
4.145 T.S. Chen, W.P. White: IEEE J. QE-**12**, 436–437 (1976)
4.146 K. Kato: Appl. Phys. Lett. **29**, 562–563 (1976)
4.147 A. Nebel, R. Beigang: Opt. Lett. **16**, 1729–1731 (1991)
4.148 C.S. Adams, A.I. Ferguson: Opt. Commun. **90**, 89–94 (1992)
4.149 R.A. Cheville, M.T. Reiten, N.J. Halas: Opt. Lett. **17**, 1343–1345 (1992)
4.150 R.J. Ellingson, C.L. Tang: Opt. Lett. **17**, 343–345 (1992)
4.151 P. Poirier, F. Hanson: Opt. Lett. **18**, 1925–1927 (1993)
4.152 G.A. Skripko, S.G. Bartoshevich, I.V. Mikhnyuk, I.G. Tarazevich: Opt. Lett. **16**, 1726–1728 (1991)
4.153 S. Bourzeix, M.D. Plimmer, F. Nez, L. Julien, F. Biraben: Opt. Commun. **99**, 89–94 (1993)
4.154 D-W. Chen, J.T. Lin: IEEE J. QE-**29**, 307–310 (1993)
4.155 L.S. Wu, H. Looser, P. Günter: Appl. Phys. Lett. **56**, 2163–2165 (1990)
4.156 E.S. Polzik, H.J. Kimble: Opt. Lett. **16**, 1400–1402 (1991)
4.157 M. Sagawa, H. Kagawa, A. Kakuta, M. Kaji: Appl. Phys. Lett, **63**, 1877–1879 (1993)
4.158 A. Nebel, R. Beigang: Opt. Commun. **94**, 369–372 (1992)
4.159 J.C. Baumert, P. Günter, H. Melchior: Opt. Commun. **48**, 215–220 (1983)
4.160 P. Günter, P.M. Asbeck, S.K. Kurtz: Appl. Phys. Lett. **35**, 461–463 (1979)
4.161 M.K. Chun, L. Goldberg, J.F. Weller: Appl. Phys. Lett. **53**, 1170–1171 (1988)
4.162 L. Goldberg, M.K. Chun: Appl. Phys. Lett. **55**, 218–220 (1989)
4.163 G.J. Dixon, C.E. Tanner, C.E. Wieman: Opt. Lett. **14**, 731–733 (1989)
4.164 A. Hemmerich, D.H. McIntyre, C. Zimmermann, T.W. Hänsch: Opt. Lett. **15**, 372–374 (1990)
4.165 W.J. Kozlovsky, W. Lenth, E.E. Latta, A. Moser, G.L. Bona: Appl. Phys. Lett. **56**, 2291–2292 (1990)
4.166 C. Zimmermann, T.W. Hänsch, R. Byer, S. O'Brien, D. Welch: Appl. Phys. Lett. **61**, 2741–2743 (1992)
4.167 L. Goldberg, L. Busse, D. Mehuys: Appl. Phys. Lett. **60**, 1037–1039 (1992)
4.168 L. Goldberg, L.E. Busse, D. Mehuys: Appl. Phys. Lett. **63**, 2327–2329 (1993)
4.169 W. Wang, K. Nakagawa, Y. Toda, M. Ohtsu: Appl. Phys. Lett. **61**, 1886–1888 (1992)
4.170 Chr. Tamm: Appl. Phys. **B 56**, 295–300 (1993)
4.171 J.J.E. Reid: Appl. Phys. Lett, **62**, 19–21 (1993)
4.172 W. Wang, M. Ohtsu: Opt. Commun. **102**, 304–308 (1993)
4.173 C.R. Webster, L. Wöste, R.N. Zare: Opt. Commun. **35**, 435–440 (1980)
4.174 C.F. Dewey, Jr., W.R. Cook, Jr., R.T. Hodgson, J.J. Wynne: Appl. Phys. Lett. **26**, 714–716 (1975)
4.175 H.J. Dewey: IEEE J. QE-**12**, 303–306 (1976)
4.176 H. Zacharias, A. Anders, J.B. Halpern, K.H. Welge: Opt. Commun. **19**, 116–119 (1976)
4.177 V.I. Stroganov, V.I. Trunov, A.A. Chernenko, A.N. Izrailenko: Kvantovaya Elektron. **3**, 1122–1124 (1976) [English transl.: Sov. J. Quantum. Electron. **6**, 601–602 (1976)]
4.178 B.V. Bokut, N.S. Kazak, A.G. Maschenko, V.A. Mostovnikov, A.N. Rubinov: Pisma Zh. Eksp. Teor. Fiz. **15**, 26–30 (1972) [English transl.: JETP Lett. **15**, 18–20 (1972)]
4.179 B.G. Huth, G.I. Farmer, L.M. Taylor, M.R. Kagan: Spectrosc. Lett. **1**, 425–432 (1968)
4.180 A. Hirth, K. Vollrath, J.V. Allain: Opt. Commun. **20**, 347–349 (1977)
4.181 D.J. Bradley, J.V. Nicholas, J.R.D. Shaw: Appl. Phys. Lett. **19**, 172–173 (1971)
4.182 D.A. Jennings, A.J. Varga: J. Appl. Phys. **42**, 5171–5172 (1971)
4.183 R.W. Wallace: Opt. Commun. **4**, 316–318 (1971)
4.184 C. Gabel, M. Hercher: IEEE J. QE-**8**, 850–851 (1972)
4.185 M. Yamashita, W. Sibbett, D. Welford, D.J. Bradley: J. Appl. Phys. **51**, 3559–3562 (1980)
4.186 D. Welford, W. Sibbett, J.R. Taylor: Opt. Commun. **35**, 283–286 (1980)
4.187 K. Kato: J. Appl. Phys. **46**, 2721–2722 (1975)

392   References

4.188   K. Kato: IEEE J. QE-13, 544–546 (1977)
4.189   A.I. Ferguson, M.H. Dunn: Opt. Commun. 23, 177–182 (1977)
4.190   D. Frölich, L. Stein, H.W. Schröder, H. Welling: Appl. Phys. 11, 97–101 (1976)
4.191   H. Buesener, A. Renn, M. Brieger, F. Von Moers, A. Hese: Appl. Phys. B 39, 77–81 (1986)
4.192   W.A. Majewsky: Opt. Commun. 45, 201–206 (1983)
4.193   I.M. Beterov, V.I. Stroganov, V.I. Trunov, B.Ya. Yurshin: Kvantovaya Elektron. 2, 2440–2443 (1975) [English transl.: Sov. J. Quantum Electron. 5, 1329–1331 (1975)]
4.194   K. Miyazaki, H. Sakai, T. Sato: Opt. Lett. 11, 797–799 (1986)
4.195   D.C. Edelstein, E.S. Wachman, L.K. Cheng, W.R. Bosenberg, C.L. Tang: Appl. Phys. Lett. 52, 2211–2213 (1988)
4.196   H.J. Müschenborn, W. Theiss, W. Demtröder: Appl. Phys. B 50, 365–369 (1990)
4.197   R. Kallenbach, F. Schmidt-Kaler, M. Weitz, C. Zimmermann, T.W. Hänsch: Opt. Commun. 81, 63–66 (1991)
4.198   J.A. Paisner, M.L. Spaeth, D.C. Gerstenberger, I.W. Ruderman: Appl. Phys. Lett. 32, 476–478 (1978)
4.199   F.B. Dunning, F.K. Tittel, R.F. Stebbings: Opt. Commun. 7, 181–183 (1973)
4.200   S.J. Bastow, M.H. Dunn: Opt. Commun. 35, 259–263 (1980)
4.201   C.J. Foot, P.E.G. Baird, M.G. Boshier, D.N. Stacey, G.K. Woodgate: Opt. Commun. 50, 199–204 (1984)
4.202   K. Kato: IEEE J. QE-15, 410–411 (1979)
4.203   J.C. Baumert, J. Hoffnagle, P. Günter: Appl. Opt. 24, 1299–1301 (1985)
4.204   J.M. Halbout, S. Blit, W. Donaldson, C.L. Tang: IEEE J. QE-15, 1176–1180 (1979)
4.205   B. Couillaud, L.A. Bloomfield, T.W. Hansh: Opt. Lett. 8, 259–261 (1983)
4.206   B. Couillaud, T.W. Hansh, S.G. Mac Lean: Opt. Commun. 50, 127–129 (1984)
4.207   E.F. Labuda, A.M. Johnson: IEEE J. QE-3, 164–167 (1967)
4.208   R.K. Jain, T.K. Gustafson: IEEE J. QE-9, 859–861 (1973)
4.209   J.C. Bergquist, H. Hemmati, W.M. Itano: Opt. Commun. 43, 437–442 (1982)
4.210   X. Xinan, Y. Shuzhong: Chin. Phys. - Lasers 13, 892–894 (1986)
4.211   C. Zimmermann, R. Kallenbach, T.W. Hänsch, J. Sandberg: Opt. Commun. 71, 229–234 (1989)
4.212   Y. Taira: Jpn. J. Appl. Phys. 31, L682–L684 (1992)
4.213   F. Kaczmarek, A. Jendrzejczak: Opt. Quantum Electron. 15, 187–191 (1983)
4.214   E.N. Antonov, V.G. Koloshnikov, D.N. Nikogosyan: Opt. Spektrosk. 36, 768–772 (1974) [English transl.: Opt. Spectrosc. USSR 36, 446–448 (1974)]
4.215   T. Kushida, Y. Tanaka, M. Ojima, Y. Nakazaki: Jpn. J. Appl. Phys. 14, 1097–1098 (1975)
4.216   V.V. Badikov, O.N. Pivovarov, Yu.V. Skokov, O.V. Skrebneva, N.K. Trotsenko: Kvantovaya Elektron. 2, 618–621 (1975) [English transl.: Sov. J. Quantum Electron. 5, 350–351 (1975)]
4.217   E.D. Shaw, C.K.N. Patel, R.J. Chichester: Opt. Commun. 33, 221–224 (1980)
4.218   Yu.M. Andreev, V.G. Voevodin, P.P. Geiko, A.I. Gribenyukov, A.P. Dyadkin, S.V. Pigulsky, A.I. Starodubtsev: Kvantovaya Elektron. 14, 784–786 (1987) [English transl.: Sov. J. Quantum. Electron. 17, 491–493 (1987)]
4.219   Yu.M. Andreev, A.D. Belykh, V.G. Voevodin, P.P. Geiko, A.I. Gribenyukov, V.A. Gurashvili, S.V. Izyumov: Kvantovaya Elektron. 14, 782–783 (1987) [English transl.: Sov. J. Quantum Electron. 17, 490–491 (1987)]
4.220   M.W. Dowley: Appl. Phys. Lett. 13, 395 (1968)
4.221   E.E. Fill: Opt. Commun. 33, 321–322 (1980)
4.222   E. Fill, J. Wildenauer: Opt. Commun. 47, 412–413 (1983)
4.223   G. Brederlow, E. Fill, K.J. Witte: *The High-Power Iodine Laser*, Springer Ser. Opt. Sci., Vol. 34 (Springer, Berlin, Heidelberg 1983)
4.224   K.J. Witte, E. Fill, G. Brederlow, H. Baumhacker, R. Volk: IEEE J. QE-17, 1809–1816 (1981)
4.225   D.E. Johnson, R.H. Humphreys, Jr., P. Keating, G.D. Hager: Appl. Phys. B 48, 339–342 (1989)

4.226 D.V. Ishkov, P.G. Kryukov, V.S. Pazyuk, M.P. Frolov, N.N. Yuryshev: Kvantovaya Elektron. **19**, 407–409 (1992) [English transl.: Sov. J. Quantum Electron. **22**, 370–372 (1992)]

4.227 G.D. Hager, S.A. Hanes, M.A. Dreger: IEEE J. **28**, 2573–2576 (1992)

4.228 C.J. Ernst, W.J. Witteman: IEEE J. QE-**8**, 382–383 (1972)

4.229 D.N. Nikogosyan, A.P. Sukhorukov, M.I. Golovey: Kvantovaya Elektron. **2**, 609–612 (1975) [English transl.: Sov. J. Quantum Electron. **5**, 344–346 (1975)]

4.230 D.S. Chemla, P.I. Kupecek, C.A. Schwartz: Opt. Commun. **7**, 225–228 (1973)

4.231 R.L. Byer, M.M. Choy, R.L. Herbst, D.S. Chemla, R.S. Feigelson: Appl. Phys. Lett. **24**, 65–68 (1974)

4.232 R.C. Eckardt, Y.X. Fan, R.L. Byer, R.K. Route, R.S. Feigelson, J. Van der Laan: Appl. Phys. Lett. **47**, 786–788 (1985)

4.233 G.C. Bhar, S. Das, U. Chatterjee, P.K. Datta, Yu.M. Andreev: Appl. Phys. Lett. **63**, 1316–1318 (1993)

4.234 Yu.M. Andreev, V.G. Voevodin, A.I. Gribenyukov, O.Ya. Zyryanov, I.I. Ippolitov, A.N. Morozov, A.V. Soskin, G.S. Khmelnitsky: Kvantovaya Elektron. **11**, 1511–1512 (1984) [English transl.: Sov. J. Quantum Electron. **14**, 1021–1022 (1984)]

4.235 Yu.M. Andreev, V.G. Voevodin, P.P. Geiko, A.I. Gribenyukov, V.V. Zuev, A.S. Solodukhin, S.A. Trushin, V.V. Churakov, S.F. Shubin: Kvantovaya Elektron.**14**, 2137–2138 (1987) [English transl.: Sov. J. Quantum Electron. **17**, 1362–1363 (1987)]

4.236 Yu.M. Andreev, V.Yu. Baranov, V.G. Voevodin, P.P. Geiko, A.I. Gribenyukov, S.V. Izyumov, S.M. Kozochkin, V.D. Pismenny, Yu.A. Satov, A.P. Streltsov: Kvantovaya Elektron. **14**, 2252–2254 (1987) [English transl.: Sov. J. Quantum Electron. **17**, 1435–1436 (1987)]

4.237 G.C. Bhar, S. Das, U. Chatterjee, K.L. Vodopyanov: Appl. Phys. Lett. **54**, 313–314 (1989)

4.238 Yu.M. Andreev, A.N. Bykanov, A.I. Gribenyukov, V.V. Zuev, V.D. Karyshev, A.V. Kisletsov, I.O. Kovalev, V.I. Konov, G.P. Kuzmin, A.A. Nesterenko, A.E. Osorgin, Yu.M. Starodumov, N.I. Chapliev: Kvantovaya Elektron. **17**, 476–480 (1990) [English transl.: Sov. J. Quantum Electron. **20**, 410–414 (1990)]

4.239 A.A. Barykin, S.V. Davydov, V.P. Dorokhov, V.P. Zakharov, V.V. Butuzov: Kvant. Elektron. **20**, 794–800 (1993) [English transl.: Quantum Electron. **23**, 688–693 (1993)]

4.240 H. Kildal, J.C. Mikkelsen: Opt. Commun. **10**, 306–309 (1974)

4.241 N. Menyuk, G.W. Iseler, A. Mooradian: Appl. Phys. Lett. **29**, 422–424 (1976)

4.242 R.L. Pastel: Appl. Opt. **26**, 1574–1576 (1987)

4.243 R.C.Y. Auyeung, D.M. Zielke, B.J. Feldman: Appl. Phys. B **48**, 293–297 (1989)

4.244 D.R. Suhre: Appl. Phys. B **52**, 367–370 (1991)

4.245 G.B. Abdullaev, K.R. Allakhverdiev, M.E. Karasev, V.I. Konov, L.A. Kulevskii, N.B. Mustafaev, P.P. Pashinin, A.M. Prokhorov, Yu.M. Starodumov, N.I. Chapliev: Kvantovaya Elektron. **16**, 757–763 (1989) [English transl.: Sov. J. Quantum Electron. **19**, 494–498 (1989)]

4.246 G.D. Boyd, T.J. Bridges, E.G. Burkhardt: IEEE J. QE-**4**, 515–519 (1968)

4.247 G.W. Day: Appl. Phys. Lett. **18**, 347–349 (1971)

4.248 C.K.N. Patel: Phys. Rev. Lett. **15**, 1027–1030 (1965)

4.249 J.D. Taynai, R. Targ, W.B. Tiffany: IEEE J. QE-**7**, 412–416 (1971)

4.250 V.V. Berezovskii, Yu.A. Bykovskii, M.I. Goncharov, I.S. Rez: Kvantovaya Elektron. No. **2 (8)**, 105–107 (1972) [English transl.: Sov. J. Quantum Electron. **2**, 180–182 (1972)]

4.251 J. Jerphagnon, M. Bernard: IEEE J.QE-**4**, 395–396 (1968)

4.252 A. Delahaigne, C. Thiebeaux, P. Louve: Appl. Phys. **24**, 21–22 (1981)

4.253 N. Menyuk, G.W. Iseler: Opt. Lett. **4**, 55–57 (1979)

4.254 Yu.M. Andreev, T.V. Vedernikova, A.A. Betin, V.G. Voevodin, A.I. Gribenyukov, O.Ya. Zyryanov, I.I. Ippolitov, V.I. Masychev, O.V. Mitropolskii, V.P. Novikov, M.A. Novikov, A.V. Sosnin: Kvantovaya Elektron. **12**, 1535–1537 (1985) [English transl.: Sov. J. Quantum Electron. **15**, 1014–1015 (1985)]

4.255   A.A. Isaev, G.Yu. Lemmerman, G.L. Malafeeva: Kvantovaya Elektron. **7**, 1700–1704 (1980) [English transl.: Sov. J. Quantum Electron. **10**, 983–985 (1980)]

4.256   Yu.P. Polunin, V.O. Troitsky: Kvantovaya Elektron. **14**, 2249–2251 (1987) [English transl.: Sov. J. Quantum Electron. **17**, 1433–1434 (1987)]

4.257   K. Kuroda, T. Omatsu, T. Shimura, M. Chihara, I. Ogura: Opt. Commun. **75**, 42–46 (1990)

4.258   D.W. Coutts, M.D. Ainsworth, J.A. Piper: IEEE J. **26**, 1555–1558 (1990)

4.259   D.W. Coutts. J.A. Piper: IEEE J. **28**, 1761–1764 (1992)

4.260   T. Henningsen, J.D. Feichtner, N.T. Melamed: IEEE J. QE-7, 248–250 (1971)

4.261   D.P. Juyal, G.C. Thomas: Opt. Commun. **15**, 26–28 (1975)

4.262   Yu.M. Andreev, S.D. Velikanov, A.S. Yelutin, A.F. Zapolskii, D.V. Konkin, S.N. Mishkin, S.V. Smirnov, Yu.N. Frolov, V.V. Shchurov: Kvantovaya Elektron. **19**, 1110 (1992) [English transl.: Sov. J. Quant. Electron. **22**, 1035 (1992)]

4.263   F. Balembois, P. Georges, F. Salin, G. Roger, A. Brun: Appl. Phys. Lett. **61**, 2381–2382 (1992)

4.264   S.R. Chinn: Appl. Phys. Lett. **29**, 176–179 (1976)

4.265   S. Imai, T. Yamada, Y. Fujimori, K. Ishikawa: Appl. Phys. Lett. **54**, 1206–1208 (1989)

4.266   W.S. Pelouch, T. Ukachi, E.S. Wachman, C.L. Tang: Appl. Phys. Lett, **57**, 111–113 (1990)

4.267   T.R. Zhang, H.R. Choo, M. Downer: Appl. Opt. **29**, 3927–3933 (1990)

4.268   Y. Ishida, T. Yajima: Opt. Commun. **62**, 197–200 (1987)

4.269   K.L. Cheng, W. Bosenberg, F.W. Wise, I.A. Walmsley, C.L. Tang: Appl. Phys. Lett. **52**, 519–521 (1988)

4.270   D. Kühlke, U. Herpers: Opt. Commun. **69**, 75–78 (1988)

4.271   G. Szabo, Z. Bor: Appl. Phys. B **50**, 51–54 (1990)

4.272   Y. Takagi, M. Sumitani, N. Nakashima, K. Yoshihara: IEEE J. QE-21, 193–195 (1985)

4.273   Y. Takagi, M. Sumitani, N. Nakashima, D.V. O'Connor, K. Yoshihara: Appl. Phys. Lett. **42**, 489–491 (1983)

4.274   S.M. Kopylov, L.K. Mikhailov, O.B. Cherednichenko: Kvantovaya Elektron. **10**, 625–627 (1983) [English transl.: Sov. J. Quantum Electron. **13**, 375–376 (1983)]

4.275   D.A. Angelov, G.G. Gurzadyan, D.N. Nikogosyan: Kvantovaya Elektron. **6**, 2267–2269 (1979) [English transl.: Sov. J. Quantum Electron. **9**, 1334–1335 (1979)]

4.276   V.L. Lyutskanov, S.D. Savov, S.M. Saltiel, K.V. Stamenov, I.V. Tomov: Opt. Commun. **37**, 149–152 (1981)

4.277   R.E. Stickel, Jr., F.B. Dunning: Appl. Opt. **17**, 1313–1314 (1978)

4.278   S. Blit, E.G. Weaver, F.B. Dunning, F.K. Tittel: Opt. Lett. **1**, 58–60 (1977)

4.279   D.W. Coutts, M.D. Ainsworth, J.A. Piper: IEEE J. **25**, 1985–1987 (1989)

4.280   N.S. Dudina, S.M. Kopylov, L.K. Mikhailov, O.B. Cherednichenko: Kvantovaya Elektron. **6**, 2478–2481 (1979) [English transl.: Sov. J. Quantum Electron. **9**, 1468–1469 (1979)]

4.281   C.A. Moore, L.S. Goldberg: Opt. Commun. **16**, 21–25 (1976)

4.282   H. Hemmati, J.C. Bergquist: Opt. Commun. **47**, 157–160 (1983)

4.283   R.P. Mariella, Jr.: Opt. Commun. **29**, 100–102 (1979)

4.284   E. Liu, F.B. Dunning, F.K. Tittel: Appl. Opt. **21**, 3415–3416 (1982)

4.285   W. Mückenheim, P. Lokai, B. Burghardt, D. Basting: Appl. Phys. B **45**, 259–261 (1988)

4.286   M. Watanabe, K. Hayasaka, H. Imajo, J. Umezu, S. Urabe: Appl. Phys. B **53**, 11–13 (1991)

4.287   Th. Hofmann, K. Mossavi, F.K. Tittel, G. Szabo: Opt. Lett. **17**, 1691–1693 (1992)

4.288   I.V. Tomov, T. Anderson, P.M. Rentzepis: Appl. Phys. Lett. **61**, 1157–1159 (1992)

4.289   J. Ringling, O. Kittelmann, F. Noack: Opt. Lett. **17**, 1794–1796 (1992)

4.290   M. Watanabe, K. Hayasaka, H. Imajo, S. Urabe: Opt. Lett. **17**, 46–48 (1992)

4.291   P. Lokai, B. Burghardt, W Mückenheim: Appl. Phys. B **45**, 245–247 (1988)

4.292   U. Heitmann, M. Kötteritzsch, S. Heitz, A. Hese: Appl. Phys. B **55**, 419–423 (1992)

4.293   T.L. Gustafson: Opt. Commun. **67**, 53–57 (1988)

4.294   W.L. Glab, J.P. Hessler: Appl. Opt. **26**, 3181–3182 (1987)

4.295   G.C. Bhar, U. Chatterjee, S. Das: Jpn. J. Appl. Phys. **29**, L1127–L1129 (1990)

4.296   K. Sugiyama, J. Yoda, T. Sakurai: Opt. Lett. **16**, 449–451 (1991)

4.297 G.C. Bhar, P.K. Datta, U. Chatterjee: J. Phys. D: Appl. Phys. **25**, 1042–1047 (1992)
4.298 A. Borsutzky, R. Brünger, R. Wallenstein: Appl. Phys. **B 52**, 380–384 (1991)
4.299 B. Wu, F. Xie, C. Chen, D. Deng, Z. Xu: Opt. Commun. **88**, 451–454 (1992)
4.300 K. Kato: IEEE J. **26**, 1173–1175 (1990)
4.301 G.C. Bhar, P.K. Datta, A.M. Rudra: Appl. Phys. **B 57**, 431–434 (1993)
4.302 R.C. Eckardt, H. Masuda, Y.X. Fan, R.L. Byer: IEEE J. **QE-26**, 922–933 (1990)
4.303 J.T. Lin, K. Kato: Proc. SPIE **1220**, 58–63 (1990)
4.304 F.B. Dunning, R.E. Stickel, Jr.: Appl. Opt. **15**, 3131–3134 (1976)
4.305 K. Kato: Appl. Phys. Lett. **30**, 583–584 (1977)
4.306 R.E. Stickel, F.B. Dunning: Appl. Opt. **16**, 2356–2358 (1977)
4.307 R.E. Stickel, Jr., F.B. Dunning: Appl. Opt. **17**, 981–982 (1978)
4.308 R.E. Stickel, Jr., S. Blit, G.F. Hidebrandt, E.D. Dahl, F.B. Dunning, F.K. Tittel: Appl.
       Opt. **17**, 2270 (1978)
4.309 H. Hemmati, J.C. Bergquist, W.M. Itano: Opt. Lett. **8**, 73–75 (1983)
4.310 K.B. Petrosyan, A.L. Pogosyan, K.M. Pokhsraryan: Izv. Akad. Nauk SSSR, Ser. Fiz. **47**,
       1619–1621 (1983) [English transl.: Bull. Acad. Sci. USSR, Phys. Ser. **47** No. **8**, 155–157
       (1983)]
4.311 K.B. Petrosyan, K.M. Pokhsraryan: Izv. Akad. Nauk Arm. SSR, Ser. Fiz. **20**, 39–42 (1985)
       [English transl.: Sov. J. Contemp. Phys. Armen. Acad. Sci. **20** No. **1**, 43 (1985)]
4.312 F.B. Dunning: Laser Focus **14** (5), 72–76 (1978)
4.313 T.R. Gurski: Appl. Phys. Lett. **23**, 273–275 (1973)
4.314 S.N. Kosolobov, V.V. Lebedev, S.I. Marennikov, Yu.N. Popov, G.V. Krivoshchekov, P.L.
       Mitnitsky, B.I. Kidiyarov: Pisma Zh. Eksp. Teor. Fiz. **16**, 475–479 (1972) [English transl.:
       JETP Lett. **16**, 338–340 (1972)]
4.315 T.R. Gurski, H.W. Epps, S.P. Maran: Nature **249**, 638–639 (1974)
4.316 D. Malz, J. Bergmann, J. Heise: Exp. Tech. Phys. **23**, 495–498 (1975)
4.317 D. Malz, J. Bergmann, J. Heise: Exp. Tech. Phys. **23**, 379–388 (1975)
4.318 Y.C. See, J. Falk: Appl. Phys. Lett. **36**, 503–505 (1980)
4.319 A.J. Campillo, C.L. Tang: Appl. Phys. Lett. **19**, 36–38 (1971)
4.320 J.E. Midwinter, J. Warner: J. Appl. Phys. **38**, 519–523 (1967)
4.321 E.A. Arutyunyan, V.S. Mkrtchyan: Kvantovaya Elektron. **2**, 812–814 (1975) [English
       transl.: Sov. J. Quantum Electron. **5**, 450–451 (1975)]
4.322 J.E. Midwinter: Appl. Phys. Lett. **12**, 68–70 (1968)
4.323 R.C. Miller, W.A. Nordland: IEEE J. **QE-3**, 642–643 (1967)
4.324 K.M. Baird, D.S. Smith, W.E. Berger: Opt. Commun. **7**, 107–109 (1973)
4.325 J.E. Midwinter: Appl. Phys. Lett. **14**, 29–32 (1969)
4.326 Y.C. See, S. Guha, J. Falk: Appl. Opt. **19**, 1415–1418 (1980)
4.327 M.M. Abbas, T. Kostiuk, K.W. Ogilvie: Appl. Opt. **15**, 961–970 (1976)
4.328 A.S. Pine: J. Opt. Soc. Am. **64**, 1683–1690 (1974)
4.329 E.A. Arutyunyan, R.B. Kostanyan, V.S. Mkrtchyan, M.A. Mkrtchyan: Kvantovaya
       Elektron. **2**, 1811–1813 (1975) [English transl.: Sov. J. Quantum Electron. **5**, 985–987 (1975)]
4.330 J. Falk, J.M. Yarborough: Appl. Phys. Lett. **19**, 68–70 (1971)
4.331 P.N. Kean, R.W. Standley, G.J. Dixon: Appl. Phys. Lett. **63**, 302–304 (1993)
4.332 P.K. Benicewicz, D. McGraw: Opt. Lett. **15**, 165–167 (1990)
4.333 W.P. Risk, W.J. Kozlovsky: Opt. Lett. **17**, 707–709 (1992)
4.334 P.N. Kean, G.J. Dixon: Opt. Lett. **17**, 127–129 (1992)
4.335 R.A. Stolzenberger, C.C. Hsu, N. Peyghambarian, J.J.E. Reid, R.A. Morgan: IEEE
       Photon. Technol. Lett. **1**, 446–448 (1989)
4.336 M-H. Lu, Y-M. Liu: Opt. Commun. **84**, 193–198 (1991)
4.337 J.C. Baumert, P. Günter: Appl. Phys. Lett. **50**, 554–556 (1987)
4.338 J. Warner: Appl. Phys. Lett. **12**, 222–224 (1968)
4.339 E.K. Pfitzer, H.D. Riccius, K.J. Siemsen: Opt. Commun. **3**, 277–278 (1971)
4.340 D.Y. Tseng: Appl. Phys. Lett. **21**, 382–384 (1972)

4.341   R.F. Lucy: Appl. Opt. **11**, 1329–1336 (1972)

4.342   A.S. Alcock, A.C. Walker: Appl. Phys. Lett. **23**, 467–468 (1973)

4.343   G.P. Arumov, E.S. Voronin, Yu.A. Ilinsky, V.S. Solomatin, V.V. Shuvalov: Kvantovaya Elektron. **5**(17), 95–99 (1973) [English transl.: Sov. J. Quantum Electron. **3**, 421–423 (1973)]

4.344   E.S. Voronin, V.S. Solomatin, V.V. Shuvalov: Opto-electron. **6**, 189–190 (1974)

4.345   D.N. Nikogosyan: Kvantovaya Elektron. **2**, 2524–2525 (1975) [English transl.: Sov. J. Quantum Electron. **5**, 1378–1379 (1975)]

4.346   P.A. Jaanimagi, M.C. Richardson, N.R. Isenor: Opt. Lett. **4**, 45–47 (1979)

4.347   S.A. Andreev, N.P. Andreeva, I.N. Matveev, S.A. Pshenichnikov: Kvantovaya Elektron. **8**, 1361–1363 (1981) [English transl.: Sov. J. Quantum Electron. **11**, 821–822 (1981)]

4.348   N. Ito: Opt. Lett. **7**, 63–65 (1983)

4.349   W.B. Gandrud, G.D. Boyd: Opt. Commun. **1**, 187–190 (1970)

4.350   E.S. Voronin, V.S. Solomatin, N.I. Cherepov, V.V. Shuvalov, V.V. Badikov, O.N. Pivovarov: Kvantovaya Elektron. **2**, 1090–1092 (1975) [English transl.: Sov. J. Quantum Electron. **5**, 597–598 (1975)]

4.351   W. Jantz, P. Koidl: Appl. Phys. Lett. **31**, 99–101 (1977)

4.352   S.A. Andreev, I.N. Matveev, I.P. Nekrasov, S.M. Pshenichnikov, N.P. Sopina: Kvantovaya Elektron. **4**, 657–659 (1977) [English transl.: Sov. J. Quantum. Electron. **7**, 366–368 (1977)]

4.353   E.S. Voronin, V.S. Solomatin, V.V. Shuvalov: Kvantovaya Elektron. **5**, 2031–2032 (1978) [English transl.: Sov. J. Quantum Electron. **8**, 1145–1146 (1978)]

4.354   T. Itabe, J.L. Bufton: Appl. Opt. **23**, 3044–3047 (1984)

4.355   G.S. Bhar, U. Chatterjee, P.K. Datta, S. Das, R.S. Feigelson, R.K. Route: Appl. Phys. **B 53**, 19–22 (1991)

4.356   S.A. Andreev, N.P. Andreeva, V.V. Badikov, I.N. Matveev, S.M. Pshenichnikov: Kvantovaya Elektron. **7**, 2003–2006 (1980) [English transl.: Sov. J. Quantum Electron. **10**, 1157–1158 (1980)]

4.357   G.D. Boyd, W.B. Gandrud, E. Buechler: Appl. Phys. Lett. **18**, 446–448 (1971)

4.358   Yu.M. Andreev, V.G. Voevodin, A.I. Gribenyukov, V.P. Novikov: Kvantovaya Elektron. **14**, 1177–1179 (1987) [English transl.: Sov. J. Quantum Electron. **17**, 748–749 (1987)]

4.359   N.P. Andreeva, S.A. Andreev, I.N. Matveev, S.M. Pshenichnikov, N.D. Ustinov: Kvantovaya Elektron. **6**, 357–359 (1979) [English transl.: Sov. J. Quantum Electron. **9**, 208–210 (1979)]

4.360   Ph. Kupecek, E. Batifol, A. Kuhn: Opt. Commun. **11**, 291–295 (1974)

4.361   G.B. Abdullaev, K.R. Allakhverdiev, L.A. Kulevsky, A.M. Prokhorov, E.Yu. Salaev, A.D. Savelev, V.V. Smirnov: Kvantovaya Elektron. **2**, 1228–1233 (1975) [English transl.: Sov. J. Quantum. Electron. **5**, 665–668 (1975)]

4.362   R.L. Herbst, R.L. Byer: Appl. Phys. Lett. **19**, 527–530 (1971)

4.363   A. Ferrario, M. Garbi: Opt. Commun. **17**, 158–159 (1976)

4.364   E.S. Voronin, M.I. Divlekeev, Yu.A. Ilinsky, V.S. Solomatin, V.V. Badikov, A.A. Godovikov: Kvantovaya Elektron. **1**, 151–153 (1971) [English transl.: Sov. J. Quantum Electron. **1**, 115–116 (1971)]

4.365   P.G. Kryukov, Yu.A. Matveets, D.N. Nikogosyan: Kvantovaya Elektron. **2**, 2269–2275 (1975) [English transl.: Sov. J. Quantum. Electron. **5**, 1236–1238 (1975)]

4.366   G.P. Arumov, E.S. Voronin, Yu.A. Ilinsky, V.E. Prokopenko, V.S. Solomatin: Kvantovaya Elektron. **2**, 272–276 (1975) [English transl.: Sov. J. Quantum Electron. **5**, 153–155 (1975)]

4.367   Y. Klinger, F. Arams: Proc. IEEE **57**, 1797–1798 (1969)

4.368   J. Falk, J.M. Yarborough: Appl. Phys. Lett. **19**, 68–70 (1971)

4.369   E.N. Antonov, V.R. Mironenko, D.N. Nikogosyan, M.I. Golovey: Kvantovaya Elektron. **1**, 1742–1746 (1974) [English transl.: Sov. J. Quantum Electron. **4**, 963–965 (1974)]

4.370   H.D. Riccius, K.J. Siemsen: Phys. Lett. **45A**, 377–378 (1973)

4.371   G.A. Massey, J.C. Johnson: Appl. Opt. **17**, 3702–3703 (1978)

4.372   C.L. Sam, M.M. Choy: Appl. Phys. Lett. **30**, 199–201 (1977)

4.373   D.W. Meltzer, L.S. Goldberg: Opt. Commun. **5**, 209–211 (1972)
4.374   L. Goldberg: Appl. Opt. **14**, 653–656 (1975)
4.375   H. Tashiro, T. Yajima: Opt. Commun. **12**, 129–131 (1974)
4.376   T.M. Jedju, L. Rothberg, A. Labrie: Opt. Lett. **13**, 961–963 (1988)
4.377   P. Mutin, J.P. Boquillon: Appl. Phys. **B 48**, 411–416 (1989)
4.378   B. Wellegehausen, D. Friede, H. Vogt, S. Shahdin: Appl. Phys. **11**, 363–370 (1976)
4.379   G.F. Dobrzhansky, L.A. Kulevsky, Yu.N. Polivanov, R.Sh. Sayakhov, A.T. Sukhodolsky: Kvantovaya Elektron. **4**, 1794–1796 (1977) [English transl.: Sov. J. Quantum Electron. **7**, 1019–1020 (1977)]
4.380   R. Koenig, A. Rosenfeld, N. Tam, S. Mory: Opt. Commun. **24**, 190–194 (1978)
4.381   Sh. Atabaev, Yu.N. Polivanov, S.N. Poluektov: Kvantovaya Elektron. **9**, 378–380 (1982) [English transl.: Sov. J. Quantum Electron. **12**, 212–213 (1982)]
4.382   L. Mannik, S.K. Brown: Opt. Commun. **47**, 62–64 (1983)
4.383   D. Cotter, K.I. White: Opt. Commun. **49**, 205–209 (1984)
4.384   K. Kato: IEEE J. QE-**21**, 119–120 (1985)
4.385   T. Elsaesser, M.C. Nuss: Opt. Lett. **16**, 411–413 (1991)
4.386   C.F. Dewey, L.O. Hocker: Appl. Phys. Lett. **18**, 58–60 (1971)
4.387   I.S. Ruddock, R. Illingworth, L. Reekie: Opt. Quantum Electron. **16**, 87–88 (1984)
4.388   L. Reekie, I.S. Ruddock, R. Illingworth: Opt. Quantum Electron. **17**, 169–173 (1985)
4.389   K. Kato: IEEE J. QE-**16**, 1017–1018 (1980)
4.390   R.L. Seymour, M.M. Choy: Opt. Commun. **20**, 101–103 (1977)
4.391   D.S. Moore, S.C. Schmidt: Opt. Lett. **12**, 480–482 (1987)
4.392   P.C.M. Planken, E. Snoeks, L.D. Noordam, H.G. Muller, H.B. van Linden van den Heuvell: Opt. Commun. **85**, 31–35 (1991)
4.393   S. Ashworth, C. Iaconis, O. Votava, E. Riedle: Opt. Commun. **97**, 109–114 (1993)
4.394   G.C. Bhar, U. Chatterjee, S. Das: Appl. Phys. Lett. **58**, 231–233 (1991)
4.395   F. Seifert, V. Petrov: Opt. Commun. **99**, 413–420 (1993)
4.396   L.O. Hocker, C.F. Dewey, Jr.: Appl. Phys. **11**, 137–140 (1976)
4.397   B. Bareika, G. Dikchyus, E.D. Isianova, A. Piskarskas, V. Sirutkaitis: Pisma Zh. Tech. Fiz. **6**, 694–697 (1980) [English transl.: Sov. Tech. Phys. Lett. **6**, 301–302 (1980)]
4.398   R.J. Seymour, F. Zernike: Appl. Phys. Lett. **29**, 705–707 (1976)
4.399   K. Kato: IEEE J. QE-**20**, 698–699 (1984)
4.400   T. Elsaesser, H. Lobentanzer, A. Seilmeier: Opt. Commun. **52**, 355–359 (1985)
4.401   D.S. Bethune, A.C. Luntz: Appl. Phys. **B 40**, 107–113 (1986)
4.402   D.C. Hanna, V.V. Rampal, R.C. Smith: IEEE J. QE-**10**, 461–462 (1974)
4.403   D.C. Hanna, V.V. Rampal, R.C. Smith: Opt. Commun. **8**, 151–153 (1973)
4.404   P. Canarelli, Z. Benko, R. Curl, F.K. Tittel: J. Opt. Soc. Am. **B 9**, 197–202 (1992)
4.405   A.H. Hielscher, C.E. Miller, D.C. Bayard, U. Simon, K.P. Smolka, R.F. Curl, F.K. Tittel: J. Opt. Soc. Am. **B 9**, 1962–1967 (1992)
4.406   U. Simon, F.K. Tittel, L. Goldberg: Opt. Lett. **18**, 1931–1933 (1993)
4.407   U. Simon, C.E. Miller, C.C. Bradley, R.G. Hulet, R.F. Curl, F.K. Tittel: Opt. Lett. **18**, 1062–1064 (1933)
4.408   K.G. Spears, X. Zhu, X. Yang, L. Wang: Opt. Commun. **66**, 167–171 (1988)
4.409   A.G. Yodh, H.W.K. Tom, G.D. Aumiller, R.S. Miranda: J. Opt. Soc. Am. **B 8**, 1663–1667 (1991)
4.410   T. Dahinten, U. Plödereder, A. Seilmeier, K.L. Vodopyanov, K.R. Allakhverdiev, Z.A. Ibragimov: IEEE J. **29**, 2245–2250 (1993)
4.411   P. Hamm, C. Lauterwasser, W. Zinth: Opt. Lett. **18**, 1943–1945 (1993)
4.412   M.R.X. de Barros, P.C. Becker: Opt. Lett. **18**, 631–633 (1993)
4.413   A. Bianchi, M. Garbi: Opt. Commun. **30**, 122–124 (1979)
4.414   G.B. Abdullaev, L.A. Kulevskii, P.V. Nikles, A.M. Prokhorov, A.D. Savelev, E.Yu. Salaev, V.V. Smirnov: Kvantovaya Elektron. **3**, 163–167 (1976) [English transl.: Sov. J. Quantum Electron. **6**, 88–90 (1976)]

4.415 A. Bianchi, A. Ferrario, M. Musci: Opt. Commun. **25**, 256–258 (1978)

4.416 Yu.A. Gusev, A.V. Kirpichnikov, S.N. Konoplin, S.I. Marennikov, P.V. Nikles, Yu.N. Polivanov, A.M. Prokhorov, A.D. Savelev, R.Sh. Sayakhov, V.V. Smirnov, V.P. Chebotaev: Pisma Zh. Tekh. Fiz. **6**, 1262–1265 (1980) [English transl.: Sov. Tech. Phys. Lett. **6**, 541–542 (1980)]

4.417 H. Kildal, J.C. Mikkelsen: Opt. Commun. **9**, 315–318 (1973)

4.418 W.R. Bosenberg, D.R. Guyer: J. Opt. Soc. Am. **B 10**, 1716–1722 (1993)

4.419 M.S. Piltch, J. Rink, C. Tallman: Opt. Commun. **15**, 112–114 (1975)

4.420 D. Andreou: Opt. Commun. **23**, 37–43 (1977)

4.421 T.J. Bridges, V.T. Nguyen, E.G. Burkhardt, C.K.N. Patel: Appl. Phys. Lett. **27**, 600–602 (1975)

4.422 K. Kurokawa, M. Nakazawa: Appl. Phys. Lett. **55**, 7–9 (1989)

4.423 A.M. Valshin, V.M. Gordienko, E.O. Danilov, A.I. Kovrigin: Kvantovaya Elektron. **12**, 437–439 (1985) [English transl.: Sov. J. Quantum Electron. **15**, 291–293 (1985)]

4.424 D.W. Faries, K.A. Gehring, P.L. Richards, Y.R. Shen: Phys. Rev. **180**, 363–365 (1969)

4.425 F. Zernike, Jr., P.R. Berman: Phys. Rev. Lett. **15**, 999–1001 (1965)

4.426 T. Yajima, K. Inoue: Phys. Lett. **26A**, 281–282 (1968); IEEE J. QE-5, 140–146 (1969)

4.427 T. Yajima, N. Takeuchi: Jpn. J. Appl. Phys. **10**, 907–915 (1971)

4.428 N. Takeuchi, N. Matsumoto, T. Yajima, S. Kishida: Jpn. J. Appl. Phys. **11**, 268–269 (1972)

4.429 N. Matsumoto, T. Yajima: Jpn. J. Appl. Phys. **12**, 90–97 (1973)

4.430 Yu.O. Avetisyan, P.S. Pogosyan: Pisma Zh. Tech. Fiz. **2**, 1144–1146 (1976) [English transl.: Sov. Tech. Phys. Lett. **2**, 450 (1976)]

4.431 Yu.O. Avetisyan, A.O. Makaryan, K.M. Movsesyan, P.S. Pogosyan: Pisma Zh. Tech. Fiz. **5**, 233–235 (1979) [English transl.: Sov. Tech. Phys. Lett. **5**, 93 (1979)]

4.432 A.H. Makarian, K.M. Movsessian, P.G. Pogossian: Opt. Commun. **35**, 147–148 (1980)

4.433 M. Berg, C.B. Harris, T.W. Kenny, P.L. Richards: Appl. Phys. Lett. **47**, 206–208 (1985)

4.434 S.G. Ryabov, G.N. Toropkin, I.F. Usoltsev: *Instruments of Quantum Electronics* (Radio i Svyaz, Moscow 1985) [In Russian]

4.435 G.D. Boyd, T.J. Bridges, C.K.N. Patel, E. Buehler: Appl. Phys. Lett. **21**, 553–555 (1972)

4.436 V.G. Dmitriev, L.A. Kulevskii: "Parametric Generators of Light" in *Handbook of Lasers*, Vol. 2, ed. by A.M. Prokhorov (Sovetskoye Radio, Moscow 1978) pp. 319–348 [In Russian]

4.437 R. Danelyus, A. Piskarskas, V. Sirutkaitis, A. Stabinis, Ya. Yasevichyute: *Parametric Generators of Light and Picosecond Spectroscopy* (Mokslas, Vilnus 1983) [In Russian]

4.438 Y.R. Shen: *The Principles of Nonlinear Optics* (Wiley, New York, 1984)

4.439 R. Danelyus, G. Dikchyus, V. Kabelka, A. Piskarskas, A. Stabinis, Ya. Yasevichyute: Kvantovaya Elektron. **4**, 2379–2395 (1977) [English transl.: Sov. J. Quantum Electron. **7**, 1360–1368 (1977)]

4.440 P.G. Kryukov, Yu.A. Matveets, D.N. Nikogosyan, A.V. Sharkov: Kvantovaya Elektron. **5**, 2348–2353 (1978) [English transl.: Sov. J. Quantum Electron. **8**, 1319–1322 (1978)]

4.441 R. Danelyus, V. Kabelka, A. Piskarskas, A. Smilgyavichyus: Pisma Zh. Tech. Fiz. **4**, 765–769 (1978) [English transl.: Sov. Tech. Phys. Lett. **4**, 308–309 (1978)]

4.442 V. Kabelka, A. Kutka, A. Piskarskas, V. Smilgyavichyus, Ya. Yasevichyute: Kvantovaya Elektron. **6**, 1735–1739 (1979) [English transl.: Sov. J. Quantum Electron. **9**, 1022–1024 (1979)]

4.443 B. Bareika, G. Dikchyus, A. Piskarskas, V. Sirutkaitis, Ya. Yasevichyute: Kvantovaya Elektron. **10**, 2318–2324 (1983) [English transl.: Sov. J. Quantum Electron. **13**, 1507–1510 (1983)]

4.444 R. Danelyus, A. Piskarskas, V. Sirutkaitis: Kvantovaya Elektron. **9**, 2491–2501 (1982) [English transl.: Sov. J. Quantum Electron. **12**, 1626–1632 (1982)]

4.445 B.F. Bareika, I.A. Begishev, Sh.A. Burdulis, A.A. Gulamov, E.A. Erofeev, A.S. Piskarskas, V.A. Sirutkaitis, T. Usmanov: Pisma Zh. Tech. Fiz. **12**, 186–189 (1986) [English transl.: Sov. Tech. Phys. Lett. **12**, 78–79 (1986)]

4.446 R. Danelyus, V. Kabelka, A. Piskarskas, V. Smilgyavichyus: Kvantovaya Elektron. **5**, 679–682 (1978) [English transl.: Sov. J. Quantum. Electron. **8**, 398–400 (1978)]

4.447 I.A. Begishev, A.A. Gulamov, E.A. Erofeev, T. Usmanov: Pisma Zh. Tech. Fiz. **13**, 305–309 (1987) [English transl.: Sov. Tech. Phys. Lett. **13**, 125–126 (1987)]

4.448 I.A. Begishev, A.A. Gulamov, E.A. Erofeev, E.A. Ibragimov, Sh.R. Kamalov, T. Usmanov, A.D. Khadzhaev: Kvantovaya Elektron. **17**, 1196–1199 (1990) [English transl.: Sov. J. Quantum Electron. **20**, 1104–1106 (1990)]

4.449 R. Danielius, A. Piskarskas, D. Podenas, P. Di Trapani, A. Varanavicius, G.P. Banfi: Opt. Commun. **87**, 23–27 (1992)

4.450 R. Danielius, A. Piskarskas, A. Stabinis, G.P. Banfi, P. Di Trapani, R. Righini: J. Opt. Soc. Am. **B10**, 2222–2232 (1993)

4.451 G.A. Massey, J.C. Johnson: IEEE J. QE-**15**, 201–203 (1979)

4.452 G.C. Ghosh, G.C. Bhar: IEEE J. QE-**18**, 143–145 (1982)

4.453 S.A. Akhmanov, I.E. Begishev, A.A. Gulamov, E.A. Erofeev, B.V. Zhdanov, V.I. Kuznetsov, L.N. Rashkovich, T.B. Usmanov: Kvantovaya Elektron. **11**, 1701–1702 (1984) [English transl.: Sov. J. Quantum Electron. **14**, 1145–1146 (1984)]

4.454 F. Wondrazek, A. Seilmeier, W. Kaiser: Appl. Phys. **B 32**, 39–42 (1983)

4.455 J.M. Yarborough, G.A. Massey: Appl. Phys. Lett. **18**, 438–440 (1971)

4.456 B.V. Zhdanov, V.V. Kalitin, A.I. Kovrigin, S.M. Pershin: Pisma Zh. Tech. Fiz. **1**, 847–851 (1975) [English transl.: Sov. Tech. Phys. Lett. **1**, 368–369 (1975)]

4.457 G.A. Massey, J.C. Johnson, R.A. Elliott: IEEE J. QE-**12**, 143–147 (1976)

4.458 G.A. Massey, R.A. Elliott: IEEE J. QE-**10**, 899–900 (1974)

4.459 G. Ionushauskas, A. Piskarskas, V. Sirutkaitis, A. Yuozapavichyus: Kvantovaya Elektron. **14**, 2044–2045 (1987) [English transl.: Sov. J. Quantum Electron. **17**, 1303–1304 (1987)]

4.460 A.A. Babin, Yu.N. Belyaev, V.N. Petryakov, M.M. Sushchik, G.I. Freidman: Kvantovaya Elektron. **3**, 1138–1139 (1976) [English transl.: Sov. J. Quantum Electron. **6**, 613–614 (1976)]

4.461 A.A. Babin, Yu.N. Belyaev, Yu.K. Verevkin, G.I. Freidman: Kvantovaya Elektron. **6**, 1237–1246 (1979) [English transl.: Sov. J. Quantum. Electron. **9**, 728–733 (1979)]

4.462 I.I. Ashmarin, Yu.A. Bykovsky, V.A. Ukraintsev, A.A. Chistyakov, L.V. Shishonkov: Kvantovaya Elektron. **11**, 1847–1850 (1984) [English transl.: Sov. J. Quantum Electron. **14**, 1237–1239 (1984)]

4.463 A.J. Campillo: IEEE J. QE-**8**, 809–811 (1972)

4.464 L.S. Goldberg: Appl. Phys. Lett. **17**, 489–491 (1970)

4.465 A.I. Izrailenko, A.I. Kovrigin, P.V. Nikles: Pisma Zh. Exp. Teor. Fiz. **12**, 475–478 (1970) [English transl.: JETP Lett. **12**, 331–333 (1970)]

4.466 R.B. Weisman, S.A. Rice: Opt. Commun. **19**, 28–32 (1976)

4.467 S.A. Akhmanov, B.V. Zhdanov, A.I. Kovrigin, V.I. Kuznetsov, S.M. Pershin, A.I. Kholodnykh: Kvantovaya Elektron. **4**, 2225–2233 (1977) [English transl. : Sov. J. Quantum Electron. **7**, 1271–1276 (1977)]

4.468 G.P. Dzhotyan, Yu.E. Dyakov, S.M. Pershin, A.I. Kholodnykh: Kvantovaya Elektron. **4**, 1215–1226 (1977) [English transl.: Sov. J. Quantum Electron. **7**, 685–691 (1977)]

4.469 R. Danelyus, G. Dikchyus, V. Kabelka, A. Piskarskas, A. Stabinis, Ya. Yasevichyute: Litov. Fiz. Sb. **18**, 93–108 (1978) [English transl.: Sov. Phys. – Collect. **18**, No. 1, 62–72 (1978)]

4.470 P.G. Kryukov, Yu.A. Matveets, D.N. Nikogosyan, A.V. Sharkov, E.M. Gordeev, S.D. Fanchenko: Kvantovaya Elektron. **4**, 211–213 (1977) [English transl.: Sov. J. Quantum Electron. **7**, 127–128 (1977)]

4.471 R.B. Andreev, V.D. Volosov, V.N. Krylov: Pisma Zh. Tech. Fiz. **4**, 256–258 (1978) [English transl.: Sov. Tech. Phys. Lett. **4**, 105 (1978)]

4.472 A. Tokmakoff, C.D. Marshall, M.D. Fayer: J. Opt. Soc. Am. **B 10**, 1785–1791 (1993)

4.473 G. Nath, G. Pauli: Appl. Phys. Lett. **22**, 75–76 (1993)

4.474 E.O. Amman, M.K. Oshman, J.D. Foster, J.M. Yarborough: Appl. Phys. Lett. **15**, 131–133 (1969)

400   References

4.475 A. Laubereau, L. Greiter, W. Kaiser: Appl. Phys. Lett. **25**, 87–89 (1974)
4.476 R.L. Herbst, R.N. Fleming, R.L. Byer: Appl. Phys. Lett. **25**, 520–522 (1974)
4.477 A. Seilmeier, K. Spanner, A. Laubereau, W. Kaiser: Opt. Commun. **24**, 237–242 (1978)
4.478 A. Fendt, W. Kranitzky, A. Laubereau, W. Kaiser: Opt. Commun. **28**, 142–146 (1979)
4.479 Z.I. Ivanova, V. Kabelka, S.A. Magnitsky, A. Piskarskas, V. Smilgyavichyus, N.M. Rubinina, V.G. Tunkin: Kvantovaya Elektron. **4**, 2469–2472 (1977) [English transl.: Sov. J. Quantum Electron. **7**, 1414–1416 (1977)]
4.480 R. Laenen, G. Graener, A. Laubereau: Opt. Commun. **77**, 226–230 (1990)
4.481 F. Huisken, A. Kulcke, D. Voelkel, C. Laush, J.M. Lisy: Appl. Phys. Lett. **62**, 805–807 (1993)
4.482 M.J.T. Milton, T.J. McIlveen, D.C. Hanna, P.T. Woods: Opt. Commun. **93**, 186–190 (1992)
4.483 A. Seilmeier, W. Kaiser: Appl. Phys. **23**, 113–119 (1980)
4.484 P.L. Liu: Appl. Opt. **18**, 3543–3545 (1979)
4.485 Y. Tanaka, T. Koshida, S. Shionoya: Opt. Commun. **25**, 273–276 (1978)
4.486 A. Piskarskas, V. Smilgyavichyus, A. Umbrasas, N. Yodishyus: Kvantovaya Elektron. **13**, 1281–1284 (1986) [English transl.: Sov. J. Quantum Electron. **16**, 841–843 (1986)]
4.487 R.W. Wallace: Appl. Phys. Lett. **17**, 497–499 (1970)
4.488 S. Schiller, R.L. Byer: J. Opt. Soc. Am. **B 10**, 1696–1707 (1993)
4.489 W.J. Kozlovsky, C.D. Nabors, R.C. Eckardt, R.L. Byer: Opt. Lett. **14**, 66–68 (1989)
4.490 C.D. Nabors, R.C. Eckardt, W.J. Kozlovsky, R.L. Byer: Opt. Lett. **14**, 1134–1136 (1989)
4.491 D.C. Gerstenberger, R.W. Wallace: J. Opt. Soc. Am. **B 10**, 1681–1683 (1993)
4.492 H. He, Y. Lu, J. Dong, Q. Zhao: Proc. SPIE **1409**, 18–23 (1991)
4.493 S.T. Yang, R.C. Eckardt, R.L. Byer: J. Opt. Soc. Am. **B 10**, 1684–1695 (1993)
4.494 W. Joosen, H.J. Bakker, L.D. Noordam, H.G. Muller, H.B. van Linden van den Heuvell: J. Opt. Soc. Am. **B 8**, 2087–2093 (1991)
4.495 L.J. Bromley, A. Guy, D.C. Hanna: Opt. Commun. **67**, 316–320 (1988)
4.496 Y.X. Fan, R.C. Eckardt, R.L. Byer, C. Chen, A.D. Jiang: IEEE J. **25**, 1196–1199 (1989)
4.497 X.D. Zhu, L. Deng: Appl. Phys. Lett. **61**, 1490–1492 (1992)
4.498 R. Laenen, G. Graener, A. Laubereau: Opt. Lett. **15**, 971–973 (1990)
4.499 R. Laenen, G. Graener, A. Laubereau: J. Opt. Soc. Am. **B 8**, 1085–1088 (1991)
4.500 R. Laenen, K. Wolfrum, A. Seilmeier, A. Laubereau: J. Opt. Soc. Am. **B 10**, 2151–2161 (1993)
4.501 A. Dubietis, G. Jonusauskas, A. Piskarskas: Opt. Commun. **88**, 437–440 (1992)
4.502 S. Burdulis, R. Grigonis, A. Piskarskas, G. Sinkevicius, V. Sirutkaitis, A. Fix, J. Nolting, R. Wallenstein: Opt. Commun. **74**, 398–402 (1990)
4.503 L.K. Cheng, W.R. Bosenberg, C.L. Tang: Appl. Phys. Lett. **53**, 175–177 (1988)
4.504 H. Vanherzeele, C. Chen: Appl. Opt. **27**, 2634–2636 (1988)
4.505 Y.X. Fan, R.C. Eckardt, R.L. Byer, J. Nolting, R. Wallenstein: Appl. Phys. Lett. **53**, 2014–2016 (1988)
4.506 W.R. Bosenberg, W.S. Pelouch, C.L. Tang: Appl. Phys. Lett. **55**, 1952–1954 (1989)
4.507 W.R. Bosenberg, C.L. Tang: Appl. Phys. Lett. **56**, 1819–1821 (1990)
4.508 A. Piskarskas, V. Smilgevicius, A. Umbrasas, A. Fix, R. Wallenstein: Opt. Commun. **77**, 335–338 (1990)
4.509 J.Y. Huang, J.Y. Zhang, Y.R. Shen, C. Chen, B. Wu: Appl. Phys. Lett. **57**, 1961–1963 (1990)
4.510 U. Sukowski, A. Seilmeier: Appl. Phys. **B 50**, 541–545 (1990)
4.511 H. Komine: J. Opt. Soc. Am. **B 10**, 1751–1757 (1993)
4.512 A. Fix, T. Schröder, R. Wallenstein: Laser und Optoelektronik **3**, 106–110 (1991)
4.513 A. Fix, T. Schröder, R. Wallenstein, J.G. Haub, M.J. Johnson, B.J. Orr: J. Opt. Soc. Am. **B10**, 1744–1750 (1993)
4.514 A. Agnesi, G.C. Reali, V. Kubecek, S. Kumazaki, Y. Takagi, K. Yoshihara: J. Opt. Soc. Am. **B10**, 2211–2217 (1993)

4.515   J.Y. Zhang, J.Y. Huang, Y.R. Shen, C. Chen: J. Opt. Soc. Am. **B 10**, 1758–1764 (1993)
4.516   H. Komine: Opt. Lett. **13**, 643–645 (1988)
4.517   G. Robertson, A. Henderson, M.H. Dunn: Opt. Lett. **16**, 1584–1586 (1991)
4.518   G. Robertson, A. Henderson, M.H. Dunn: Appl. Phys. Lett. **62**, 123–125 (1993)
4.519   W.R. Bosenberg, L.K. Cheng, C.L. Tang: Appl. Phys. Lett. **54**, 13–15 (1989)
4.520   F.G. Colville, M. Ebrahimzadeh, W. Sibbett, M.H. Dunn: Appl. Phys. Lett. **64**, 1765–1767 (1994)
4.521   H. Mao, B. Wu, C. Chen, D. Zhang, P. Wang: Appl. Phys. Lett. **62**, 1866–1868 (1993)
4.522   G.P. Banfi, R. Danielius, A. Piskarskas, P. Di Trapani, P. Foggi, R. Righini: Opt. Lett. **18**, 1633–1635 (1993)
4.523   I.M. Bayanov, V.M. Gordienko, M.S. Djidjoev, V.A. Dyakov, S.A. Magnitskii, V.I. Pryalkin, A.P. Tarasevitch: Pros. SPIE, **1800**, 2–17 (1991)
4.524   S.A. Akhmanov, I.M. Bayanov, V.M. Gordienko, V.A. Dyakov, S.A. Magnitskii, V.I. Pryalkin, A.P. Tarasevich: Inst. Phys. Conf. Ser. No 126: Sect. I (IOP, Bristol), p. 67–70 (1992)
4.525   K. Kato: IEEE J. QE-26, 2043–2045 (1990)
4.526   S. Lin, J.Y. Huang, J. Ling, C. Chen, Y.R. Shen: Appl. Phys. Lett. **59**, 2805–2807 (1991)
4.527   H. Zhou, J. Zhang, T. Chen, C. Chen, Y.R. Shen: Appl. Phys. Lett. **62**, 1457–1459 (1993)
4.528   M. Ebrahimzadeh, G.J. Hall, A.I. Ferguson: Opt. Lett. **17**, 652–654 (1992)
4.529   M. Ebrahimzadeh, G.J. Hall, A.I. Ferguson: Opt. Let. **18**, 278–280 (1993)
4.530   G.J. Hall, M. Ebrahimzadeh, A. Robertson, G.P.A. Malcolm, A.I. Ferguson: J. Opt. Soc. Am. **B10**, 2168–2179 (1993)
4.531   M. Ebrahimzadeh, G.J. Hall, A.I. Ferguson: Appl. Phys. Lett. **60**, 1421–1423 (1992)
4.532   G.J. Hall, A.I. Ferguson: Opt. Lett. **18**, 1511–1513 (1993)
4.533   M.J. McCarthy, S.D. Butterworth, D.C. Hanna: Opt. Commun. **102**, 297–303 (1993)
4.534   S.D. Butterworth, M.J. McCarthy, D.C. Hanna: Opt. Lett. **18**, 1429–1431 (1993)
4.535   F.G. Colville, A.J. Henderson, M.J. Padgett, J. Zhang, M.H. Dunn: Opt. Lett. **18**, 205–207 (1993)
4.536   F.G. Colville, M.J. Padgett, A.J. Henderson, J. Zhang, M.H. Dunn: Opt. Lett. **18**, 1065–1067 (1993)
4.537   F.G. Colville, M.J. Padgett, M.H. Dunn: Appl. Phys. Lett. **64**, 1490–1492 (1994)
4.538   F. Hanson, D. Dick: Opt. Lett. **16**, 205–207 (1991)
4.539   Y. Wang, Z. Xu, D. Deng, W. Zheng, B. Wu, C. Chen: Appl. Phys. Lett. **59**, 531–533 (1991)
4.540   J.Y. Zhang, J.Y. Huang, Y.R. Shen, C. Chen, B. Wu: Appl. Phys. Lett. **58**, 213–215 (1991)
4.541   Y. Cui, M.H. Dunn, C.J. Norrie, W. Sibbett, B.D. Sinclair, Y. Tang, J.A.C. Terry: Opt. Lett. **17**, 646–648 (1992)
4.542   Y. Cui, D.E. Withers, C.F. Rae, C. J. Norrie, Y. Tang, B.D. Sinclair, W. Sibbett, M.H. Dunn: Opt. Lett. **18**, 122–124 (1993)
4.543   H-J. Krause, W. Daum: Appl. Phys. Lett. **60**, 2180–2182 (1992)
4.544   F. Huang, L. Huang: Appl. Phys. Lett. **61**, 1769–1771 (1992)
4.545   F. Huang, L. Huang, B.-I. Yin, Y.-N. Hua: Appl. Phys. Lett. **62**, 672–674 (1993)
4.546   G. Robertson, A. Henderson, M.H. Dunn: Appl. Phys. Lett. **60**, 271–273 (1992)
4.547   M. Ebrahimzadeh, G. Robertson, M.H. Dunn: Opt. Lett. **16**, 767–769 (1991)
4.548   Y. Tang, Y. Cui, M.H. Dunn: Opt. Lett. **17**, 192–194 (1992)
4.549   J.T. Lin, J.L. Montgomery: Opt. Commun. **75**, 315–320 (1990)
4.550   K. Kato: IEEE J. **27**, 1137–1140 (1991)
4.551   J. Chung, A.E. Siegman: J. Opt. Soc. Am. **B 10**, 2201–2210 (1993)
4.552   L.R. Marshall, A. Kaz: J. Opt. Soc. Am. **B10**, 1730–1736 (1993)
4.553   W.T. Lotshaw, J.R. Unternahrer, M.J. Kukla, C.I. Miyake, F.D. Braun: J. Opt. Soc. Am. **B10**, 2191–2194 (1993)
4.554   Ch. Grässer, D. Wang, R. Beigang, R. Wallenstein: J. Opt. Soc. Am. **B 10**, 2218–2221 (1993)

4.555 K. Kato, M. Masutani: Opt. Lett. **17**, 178–179 (1992)
4.556 W.S. Pelouch, P.E. Powers, C.L. Tang: Opt. Lett. **17**, 1070–1072 (1992)
4.557 A. Nebel, C. Fallnich, R. Beigang, R. Wallenstein: J. Opt. Soc. Am. **B 10**, 2195–2200 (1993)
4.558 W. Wang, M. Ohtsu: Opt. Lett. **18**, 876–878 (1993)
4.559 Q. Fu, G. Mak, H.M. van Driel: Opt. Lett. **17**, 1006–1008 (1992)
4.560 P.E. Powers, R.J. Ellingson, W.S. Pelouch, C.L. Tang: J. Opt. Soc. Am. **B 10**, 2162–2167 (1993)
4.561 G. Mak, Q. Fu, H.M. van Driel: Appl. Phys. Lett. **60**, 542–544 (1992)
4.562 E.S. Wachman, D.C. Edelstein, C.L. Tang: Opt. Lett. **15**, 136–138 (1990)
4.563 E.S. Wachman, W.S. Pelouch, C.L. Tang: J. Appl. Phys. **70**, 1893–1895 (1991)
4.564 H. Vanherzeele, J.D. Bierlein, F.C. Zumsteg: Appl. Opt. **27**, 3314–3316 (1988)
4.565 S.T. Yang, R.C. Eckardt, R.L. Byer: Opt. Lett. **18**, 971–973 (1993)
4.566 M.J. McCarthy, D.C. Hanna: Opt. Lett. **17**, 402–404 (1992)
4.567 W.R. Bosenberg, D.R. Guyer: Appl. Phys. Lett. **61**, 387–389 (1992)
4.568 D. Lee, N.C. Wong: J. Opt. Soc. Am. **B 10**, 1659–1667 (1993)
4.569 M. Ebrahimzadeh, G.J. Hall, A.I. Ferguson: Opt. Lett. **16**, 1744–1746 (1991)
4.570 M.J. McCarthy, D.C. Hanna: J. Opt. Soc. Am. **B 10**, 2180–2190 (1993)
4.571 L.R. Marshall, A. Kaz, O. Aytur: Opt. Lett. **18**, 817–819 (1993)
4.572 P.E. Powers, S. Ramakrishna, C.L. Tang, L.K. Cheng: Opt. Lett. **18**, 1171–1173 (1993)
4,573 M.G. Jani, J.T. Murray, R.R. Petrin, R.C. Powell, D.N. Loiacono, G.M. Loiacono: Appl. Phys. Lett. **60**, 2327–2329 (1992)
4.574 S.A. Baryshev, V.I. Pryalkin, A.I. Kholodnykh: Pisma Zh. Tech. Fiz. **6**, 964–967 (1980) [English transl.: Sov. Tech. Phys. Lett. **6**, 415–416 (1980)]
4.575 G.I. Onischukov, A.A. Fomichev, A.I. Kholodnykh: Kvantovaya Elektron. **10**, 1525–1526 (1983) [English transl.: Sov. J. Quantum Electron. **13**, 1001–1002 (1983)]
4.576 A. Piskarskas, V. Smilgevicius, A. Umbrasas: Opt. Commun. **73**, 322–324 (1989)
4.577 A. Piskarskas, V. Smil'gyavichyus, A. Umbrasas: Kvantovaya Elektron. **17**, 777–778 (1990) [English transl.: Sov. J. Quantum Electron. **20**, 701–702 (1990)]
4.578 A.I. Kovrigin, P.V. Nikles: Pisma Zh. Eksp. Teor. Fiz. **13**, 440–443 (1971) [English transl.: JETP Lett. **13**, 313–315 (1971)]
4.579 B. Bareika, G. Dikchyus, A. Piskarskas, V. Sirutkaitis: Kvantovaya Elektron. **7**, 2204–2206 (1980) [English transl.: Sov. J. Quantum Electron. **10**, 1277–1279 (1980)]
4.580 W.R. Bosenberg, R.H. Jarman: Opt. Lett. **18**, 1323–1325 (1993)
4.581 V.A. Dyakov, V.I. Pryalkin, A.I. Kholodnykh: Kvantovaya Elektron. **8**, 715–721 (1981) [English transl.: Sov. J. Quantum Electron. **11**, 433–436 (1981)]
4.582 K. Kato: IEEE J. QE-**18**, 451–452 (1982)
4.583 W.R. Donaldson, C.L. Tang: Appl. Phys. Lett. **44**, 25–27 (1984)
4.584 M.J. Rosker, C.L. Tang: J. Opt. Soc. Am. **B 2**, 691–696 (1985)
4.585 M.J. Rosker, K. Cheng, C.L. Tang: IEEE J. QE-**21**, 1600–1606 (1985)
4.586 M. Ebrahimzadeh, M.H. Dunn, F. Akerboom: Opt. Lett. **14**, 560–562 (1989)
4.587 I. Ledoux, J. Zyss, A. Migus, J. Etchepare, G. Grillon, A. Antonetti: Appl. Phys. Lett. **48**, 1564–1566 (1986)
4.588 I. Ledoux, J. Badan, J. Zyss, A. Migus, D. Hulin, J. Etchepare, G. Grillon, A. Antonetti: J. Opt. Soc. Am. **B 4**, 987–997 (1987)
4.589 D. Josse, S.X. Dou, J. Zyss, P. Andreazza, A. Perigaud: Appl. Phys. Lett. **61**, 121–123 (1992)
4.590 S.X. Dou, D. Josse, J. Zyss: J. Opt. Soc. Am. **B 10**, 1708–1715 (1993)
4.591 G. Robertson, M.H. Dunn: Appl. Phys. Lett. **62**, 3405–3407 (1993)
4.592 W. Kranitzky, K. Ding, A. Selmeier, W. Kaiser: Opt. Commun. **34**, 483–487 (1980)
4.593 T. Elsaesser, A. Seilmeier, W. Kaiser: Opt. Commun. **44**, 293–296 (1983)
4.594 B. Bareika, G. Dikchyus, A. Piskarskas, V. Sirutkaitis: In Proc. of 2nd Intl. Symp. on Ultrafast Phenomena in Spectroscopy, Vol. 1, ed. by B. Wilhelmi (Physikalische Gesellschaft der DDR, Jena, GDR 1980) pp. 14–19
4.595 T. Elsaesser, A. Seilmeier, W. Kaiser, P. Koidl, G. Brandt: Appl. Phys. Lett. **44**, 383–385 (1984)

4.596   H.J. Bakker, J.T.M. Kennis, H.J. Kop, A. Lagendijk: Opt. Commun. **86**, 58–64 (1991)

4.597   H-J. Krause, W. Daum: Appl. Phys. **B 56**, 8–13 (1993)

4.598   R.C. Eckardt, Y.X. Fan, R.L. Byer, C.L. Marquardt, M.E. Storm, L. Esterowitz: Appl. Phys. Lett. **49**, 608–610 (1986)

4.599   K.L. Vodopyanov, V.G. Voevodin, A.I. Gribenyukov, L.A. Kulevskii: Izy. Akad. Nauk SSSR, Ser. Fiz. **49**, 569–572 (1985) [English transl.: Bull. Acad. Sci. USSR Phys. Ser. **49** No. 3, 146–149 (1985)]

4.600   K.L. Vodopyanov, V.G. Voevodin, A.I. Gribenyukov, L.A. Kulevskii: Kvantovaya Elektron. **14**, 1815–1819 (1987) [English transl.: Sov. J. Quantum Electron. **17**, 1159–1161 (1987)]

4.601   R.L. Herbst, R.L. Byer: Appl. Phys. Lett. **21**, 189–191 (1972)

4.602   A.A. Davydov, L.A. Kulevskii, A.M. Prokhorov, A.D. Savelev, V.V. Smirnov: Pisma Zh. Eksp. Teor. Fiz. **15**, 725–727 (1972) [English transl.: JETP Lett. **15**, 513–514 (1972)]

4.603   A.A. Davydov, L.A. Kulevskii, A.M. Prokhorov, A.D. Savelev, V.V. Smirnov, A.V. Shirkov: Opt. Commun. **9**, 234–236 (1973)

4.604   J.A. Weiss, L.S. Goldberg: Appl. Phys. Lett. **24**, 389–391 (1974)

4.605   R.G. Wenzel, G.P. Arnold: Appl. Opt. **15**, 1322–1326 (1976)

4.606   E.O. Amman, J.M. Yarborough: Appl. Phys. Lett. **17**, 233–235 (1970)

4.607   D.C. Hanna, B. Luther-Davies, H.N. Rutt, R.C. Smith: Appl. Phys. Lett. **20**, 34–36 (1972)

4.608   D.C. Hanna, B. Luther-Davies, R.C. Smith: Appl. Phys. Lett. **22**, 440–442 (1973)

4.609   Y.X. Fan, R.C. Eckardt, R.L. Byer, R.K. Route, R.S. Feigelson: Appl. Phys. Lett. **45**, 313–315 (1984)

4.610   P.A. Budni, M.G. Knights, E.P. Chicklis, K.L. Schepler: Opt. Lett. **18**, 1068–1070 (1993)

4.611   K.L. Vodopyanov, Yu.A. Andreev, G.C. Bhar: Kvant. Elektr. **20**, 879–881 (1993) [English transl.: Quant. Electron. **23**, 763–765 (1993)]

4.612   K.L. Vodopyanov, L.A. Kulevskii, V.G. Voevodin, A.I. Gribenyukov, K.R. Allakhverdiev, T.A. Kerimov: Opt. Commun. **83**, 322–326 (1991)

4.613   K.L. Vodopyanov: J. Opt. Soc. Am. **B 10**, 1723–1729 (1993)

4.614   J.G. Haub, M.J. Johnson, B.J. Orr: J. Opt. Soc. Am. **B 10**, 1765–1777 (1993)

4.615   Y. Lu, Q. Zhao, Y. Li, H. He, Q. Zou, Z. Lu, Z. Geng: Opt. Eng. **32**, 713–716 (1993)

4.616   I.A. Begishev, A.A. Gulamov, E.A. Erofeev, Sh.R. Kamalov, T. Usmanov, A.D. Khadzhaev: Kvantovaya Elektron. **17**, 971–974 (1990) [English transl.: Sov. J. Quantum Electron. **20**, 889–891 (1990)]

4.617   M.K. Srivastava, R.W. Crow: Opt. Commun. **8**, 82–84 (1973)

4.618   A.A. Muravev, A.N. Rubinov: Pisma Zh. Eksp. Teor. Fiz. **37**, 597–599 (1983) [English transl.: JETP Lett. **37**, 713–716 (1983)]

4.619   B. Bareika, A. Birmontas, G. Dikchyus, A. Piskarskas, V. Sirutkaitis, A. Stabinis: Kvantovaya Elektron. **9**, 2534–2536 (1982) [English transl.: Sov. J. Quantum Electron. **12**, 1654–1656 (1982)]

4.620   K.M. Pokhsraryan: Opt. Commun. **55**, 439–441 (1985)

4.621   A.J. Campillo, R.C. Hyer, S.L. Shapiro: Opt. Lett. **4**, 357–359 (1975)

4.622   P.B. Corkum, P.P. Ho, R.R. Alfano, J.T. Manassah: Opt. Lett. **10**, 624–626 (1985)

4.623   R.N. Gyuzalyan, D.H. Sarkisyan, M.L. Ter-Mikaelyan: Kvantovaya Elektron. **4**, 1138–1140 (1977) [English transl.: Sov. J. Quantum Electron. **7**, 645–647 (1977)]

4.624   D.H. Sarkisyan: Kvantovaya Elektron. **5**, 928–930 (1978) [English transl.: Sov. J. Quantum Electron. **8**, 535–536 (1978)]

4.625   J.O. White, D. Hulin, M. Joffre, A. Migus, A. Antonetti, E. Toussaere, R. Hierle, J. Zyss: Appl. Phys. Lett. **64**, 264–266 (1994)

Appendix. List of Commonly Used Laser Wavelengths (in μm)

Solid-State Lasers

$Cr^{3+}:Al_2O_3$ or Ruby laser				0.6943
Nd:YLF or $Nd^{3+}:LiYF_4$ laser	1.0471,	1.053,	1.313,	1.321
Nd:YAG or $Nd^{3+}:Y_3Al_5O_{12}$ laser	1.0642,	1.3188,	1.338,	1.444
Nd:YAP or $Nd^{3+}:YAlO_3$ laser			1.0796,	1.3414
Cr, Tm, Ho:YAG laser				2.09
$Dy^{2+}:CaF_2$ laser				2.3587
Cr, Er:YSGG laser				2.79
Er:YAG laser				2.937

Gas Lasers

N_2 laser				0.337
Ar laser	0.4545,	0.4579,	0.4658,	0.4727,	0.4765
	0.4880,	0.4965,	0.5017,	0.5145,	0.5287
Kr laser	0.4619,	0.4762,	0.4847,	0.5208,	0.5309
	0.5682,	0.6471,	0.6764,	0.7525,	0.7993
He-Ne laser	0.5434,	0.5945,	0.6046,	0.6118,	0.6328
				1.1523,	3.3913

Excimer Lasers

ArF laser	0.1933
KrF laser	0.2484
XeCl laser	0.308
XeF laser	0.351

Vapour Lasers

He-Cd laser	0.3250,	0.4416
Copper-vapour laser	0.5105,	0.5782

Subject Index

Springer Series in Optical Sciences

Editorial Board: A. L. Schawlow A. E. Siegman T. Tamir

Managing Editor: H. K. V. Lotsch

Springer
and the environment

At Springer we firmly believe that an international science publisher has a special obligation to the environment, and our corporate policies consistently reflect this conviction.

We also expect our business partners – paper mills, printers, packaging manufacturers, etc. – to commit themselves to using materials and production processes that do not harm the environment. The paper in this book is made from low- or no-chlorine pulp and is acid free, in conformance with international standards for paper permanency.

Springer